VOICE AND DATA
COMMUNICATIONS
HANDBOOK

THE McGRAW-HILL SERIES ON COMPUTER COMMUNICATIONS (SELECTED TITLES)

Voice and Data Communications Handbook

Third Edition

Regis J. (Bud) Bates
Donald W. Gregory

McGraw-Hill
New York • San Francisco • Washington, D.C. • Auckland • Bogotá
Caracas • Lisbon • London • Madrid • Mexico City • Milan
Montreal • New Delhi • San Juan • Singapore
Sydney • Tokyo • Toronto

McGraw-Hill

*A Division of The **McGraw·Hill** Companies*

Copyright © 2000 by The McGraw-Hill Companies, Inc. All rights reserved. Printed in the United States of America. Except as permitted under the United States Copyright Act of 1976, no part of this publication may be reproduced or distributed in any form or by any means, or stored in a data base or retrieval system, without the prior written permission of the publisher.

2 3 4 5 6 7 8 9 0   DOC/DOC   0 4 3 2 1 0

ISBN 0-07-212276-5

The sponsoring editor for this book was Steven Elliot and the production supervisor was Clara Stanley. It was set in Vendome by North Market Street Graphics.

Printed and bound by R. R. Donnelley & Sons Company.

McGraw-Hill books are available at special quantity discounts to use as premiums and sales promotions, or for use in corporate training programs. For more information, please write to Director of Special Sales, McGraw-Hill, Two Penn Plaza, New York, NY 10121-2298. Or contact your local bookstore.

This book is printed on recycled, acid-free paper containing a minimum of 50% recycled de-inked fiber.

CONTENTS

Contents

Contents

Contents

Contents

Contents

INTRODUCTION

Welcome to the world of telephony and telecommunications! We are about to embark on a descriptive and narrative overview of the telecommunications industry. This book is designed to help clear the air for you. One of the major problems with this technology and industry is the use of jargon, or "telephonese," which causes confusion and misunderstanding in the industry for users, purchasers, and vendors alike. Even professionals who have been in the industry for years can have difficulty communicating. The reason is simple: Too many acronyms are used, many with multiple meanings. An acronym will mean one thing to a voice telephony person and something completely different to an engineer who has 20 years of experience.

Your rule of thumb, therefore, should be to disallow the use of telecommunications acronyms in any discussions you have with any vendor, carrier, or end user. If the propensity exists for these people to use terms and acronyms, call a time-out. Have them explain all the alphabet soup they are using. You may be surprised to find out that they can't explain the acronyms. This will obviously cause you some concern, but fear not; these folks will ultimately get to the point. Furthermore, since they won't be using all those buzzwords, communications should flow more smoothly.

Now that the stage is set, let's get into a basic discussion of telephony and telecommunications principles. No magic exists here, merely an understanding of what telecommunications is all about: the principles of a telephone, the line connections employed, the forms of communications used, and an understanding of the telephone company networks. We intend to make things as simple as possible as we cover the various techniques and terminology used throughout this book. Be aware, however, that no matter how simple we attempt to make this information and no matter how smoothly we attempt to steer you through the guides outlined, this is a technical subject. Therefore, from time to time, we may start sounding a little "techie." This is not done to impress or confuse you—we just cannot think of a way to make our explanation any more basic without destroying the flow. At any rate, this book is designed to give you a fundamental understanding of the overall concepts used in the telecommunications arena, both voice and data.

THE FORMAT

The format of this book is arranged to walk you right through the evolution of the industry as it has progressed. Therefore, in Chapter 1, we attempt to discuss the evolution of the network as it pertains to the user. The invention of the basic telephone set was a monumental milestone because it marked the birth of the telephony world. From this invention, a network of networks was designed. We approach the invention and the initial deployment of a network throughout the country on the basis of the original Bell System, and we look at the regulatory scene and the various legal issues that arose. As we all know, the monopoly was created to curtail competition associated with the universal access to a telephony network, but also to be a wedge to recover the costs associated with the deployment of the capacities and services. Later in Chapter 1, we discuss the impact of the new Telecommunications Act of 1996 and what it means to you, the end user. This one act has opened the door to competition in the areas of telephony, long distance, cable services, and many other technologies. The next 3 years will be exciting and confusing as the feeding frenzy begins. Users will be inundated with new opportunities to obtain service from a myriad of suppliers. We even hear that a used-car salesperson in the Southwest acquired a license as a competitive telephone carrier. Now, you can order your telephone service, cable services, Internet access, and long distance through this one company. By the way, you can buy a used car, too! Sound confusing? You bet it is. But read on and see what all this means to you in the future.

In Chapters 2 through 5, we touch on the fundamentals of the voice evolution from the telephone company and end-user perspective. In Chapter 2, we look at the basic characteristics of the human voice and how these characteristics formulated the way in which the network was developed. This also implies that certain constraints were put in place to carry an ordinary telephone call. Chapter 3 details how a series of connections was laid out across the country and ultimately across the world. This discussion of network evolution encompasses the distinct dialing plans and how things have changed to accommodate phenomenal growth. Chapter 4 describes the way a voice call is handled through the basic telephone set. We also look at the way this set has changed over the years from a basic nondial telephone to the latest and greatest all-digital display phone with a built-in speakerphone. The discussion also shows how the telephone set converts a sound wave into an electrical wave and prepares the electricity for transmission across the network.

In Chapter 5, we consider the changes that occurred in the network over the past 30 years following the introduction of digital standards. The basic network was designed around an analog transport system, which served us well. However, as frequently happens, we outgrew the services of this analog system. So, it was only natural that appreciation of the capabilities and benefits of digital systems would emerge. We hope that you will understand the differences between digital and analog after reading this chapter.

In Chapters 6 through 8, we look at the players in the industry and the needs they will experience in attempting to serve a user. Chapter 6 deals with the different service providers, their offerings, and some of the financial considerations that make us want to use their services. This new edition also includes a discussion about the new players in the industry, called competitive local exchange carriers (CLECs). However, we go beyond the CLECs and discuss the cable companies that are vying for your voice business. By comparing lines and trunks and explaining the real differences between them, Chapter 7 will clear the air as to why things happen the way they do when we order a pair of wires. We further try to share some insight into why the old days of competition were so difficult and why the newer competitive situation eliminates these roadblocks forever. Chapter 8 is a mathematical discussion of how we and the carriers need to monitor the performance of our networks constantly. Regardless of whose service is being considered, end users and carriers all must try to predict the volumes of traffic that will appear on the network at any single point in time. The ill-conceived past method of trying to get too many calls on too few lines has disappeared. We now expect that at any time we should be able to access this network and get to any other end point in a moment on demand. What happens when all does not go according to plan? What are the risks that we have too few or too many circuits? How do we compensate for traffic loads that are variable, and where can we get the data to conduct our analyses? All of these points should become crystal clear after this discussion.

In Chapters 9 through 11, we compare the capabilities and features of the equipment that we can plug into our telephone network through some form of connecting arrangement. Chapter 9 looks at the larger-market equipment called the PBX, which is used by organizations around the world. This section will explain what the PBX really is and how it is geared to function. From there, we also look at an alternative to ownership of the big guns by exposing you to Centrex, a service offering from the players in the telephone companies that gives you PBX features and capabilities with the cost savings of building your own system. In Chapter 10, we step down to the lower-end equipment and how it works for the smaller organization, the branch office of the larger organizations, or anyone else who has a need for multiple sets but wants to compare costs and conveniences. Next, in Chapter 11, we look at the devices we can add on to our telephone networking equipment. Voice-processing methods such as voice mail, automatic call distribution, and automatic attendants all constitute ways of serving our customers without labor-intensive human resources. However, we caution you about the risks associated with these techniques, because many organizations have already made a mess of implementing these aspects of communications. These organizations have turned their telecommunications systems into the worst public relations representatives to their customers.

We have added a new chapter (Chap. 12) to discuss the changes taking place with computer-to-telephony integration and what that means to the consumer and businessperson alike. Using a combination of communications devices and computers, the industry has changed the way phone calls are handled. Linking the computer database to the telephone system means that information becomes a strategic weapon in building customer loyalty and confidence. See what we have to say about this.

In Chapter 13, we discuss a subject that normally causes even the strong of heart to shudder: the use of an analog telephone line to transmit critical data. How can we make digital data look like a voice call? What are the tools that will enable us to get the information in a usable and understandable form? This section gets quite lengthy and, from time to time, a bit technical; however, you need to understand this concept. The world of today and tomorrow will demand the use of data transmission techniques. We must understand how to keep pace with the demands of the data world across a dial-up telephone voice network. If not, things will grind to a halt. So, take this section slowly and a little at a time; it will all come together at the end.

In Chapter 14, we explore the use of a digital network instead of an analog network. The use of T1 and T3 services is escalating rapidly, bringing users rewards that they could not believe possible. This technique gets us back into the all-digital transport system and makes the data more reliable. In an all-digital world, voice is data just as information is data. We have no choice, so we must learn how the digital network functions and embrace this technology before we get left behind. We attempt to make this discussion simple, but again, there are some parts that get a bit complicated. Read this one in earnest; it is your future.

Chapter 15 is a lot of fun. We enjoyed coming up with the best way to present this to you. This section gets a little comical in the discussion of the standards that we use in just about all aspects of the industry. In the data world, there is a series of services and protocols. These are all based on standards or de facto standards. We compare the standards as they stack up against the granddaddy of them all, the OSI model. We think you'll really enjoy the way we present this very complex model, and how we can try to make all things simple if we take them one step at a time. Other operating standards exist in the data world: most prominent are the SNA world from IBM, the DNA world from DEC, and the TCP/IP world from the makers of the Internet. So, draw your own conclusions and see whether this doesn't make more sense after you get through this chapter.

We next added two new chapters discussing two new and exciting parts of the industry. Chapter 16 deals with the explosion of the Internet around the world and how it works. This particular chapter will be rewarding to read and will help you understand why things happen so slowly when you dial into the World Wide Wait! We also look at the demand rising for Voice over IP protocols and Voice over the Internet. This is not a book on the Internet, but it sure does address many of the current situations and issues at hand. We are rather proud of the way this chapter came out because there is so much going on in this industry that we have to believe that very few people really understand much of it. Our intent is to get you talking and thinking about the services, then do more individualized research on your own.

Chapter 17 takes you beyond the external networks and brings the discussion "in-house" to the Intranet. Many companies are now building internal networks mimicking the Internet. See what this technology can do for you and how the use of browsers can enhance the overall acceptance of technology by novice and experienced users alike.

Chapter 18 deals with the evolution of a packet network using an international standard known as X.25. We are not talking about some sci-fi robot or formula; this is a very well-documented standard for breaking large problems (our data) into smaller problems for reliable delivery across the network. It served us well and will be around for a while, so you will benefit from learning how this can also be taken into account. Our comparison and dialogue in this section is sure to amuse you and also to provide a very simple analogy on how X.25 works for you!

Chapters 19 through 21 all deal with another way of moving data across a communications system, but on a localized basis. In the early 1980s a technique called a local area network (LAN) emerged from the backdrop of our data communications systems. It was designed to make the data communications between and among computers, spread around our buildings, user friendly, and easily accessible. So we consider what a

LAN is and what it is not. Then, we compare the way an Ethernet and a token ring system work. You will find this discussion most informative, but possible a bit tricky. We cover the limitations, strengths, distances, and speeds of each of the topologies used. Also, the way to get the best from your data dollars is included in these sections.

Chapter 22 takes a comparative look at the terms used on our data networks. The baseband versus broadband discussion should really clear this up for you. Too often we hear these words tossed around, but no one understands just what they mean. We will attempt to show you how and why each of the types of cable, and the capacities of the coaxial and the multiplexing schemes, will deliver high-speed communications to the desktop on a single platform.

In Chapter 23 especially, we now have discussion of LANs on steroids—called the Fast Ethernet and switched Ethernet—in an attempt to understand the differences between traditional LANs and the newer versions. Also included is a discussion of the emerging gigabit Ethernet! (That's a billion bits of information per second. *Wow!*) You won't be able to put this book down as you read through this evolutionary progress of raw power to the desktop.

Chapters 24 through 27 look at taking the LAN out of the building and making it more of a public, rather than a localized, service. In Chapter 24, we look at the ways of creating a campus area network (CAN), a metropolitan area network (MAN), or a wide area network (WAN) connectivity solution. Each of these will provide wider area coverage and higher-speed connections. In the SMDS discussion, we look at the MAN solution to get the data across the network quickly and efficiently. In discussion of the fiber-optical connection using a higher-speed token-passing ring, we see the CAN emerging. These are all addressed in some detail, so some added time will be required to understand the concept and the reasons for using these technologies. Allow some extra time when reading these sections.

In Chapter 25, we describe how X.25 has evolved into a faster WAN connection solution. We take some time to describe the industry's fascination with trying to put voice over frame relay so you can understand what this is all about. Let's face it: Sooner or later, you will hear about it, so why not here? Why anyone would want to run voice over a data network is a mystery to us, but we try to remain objective in our discussions.

Chapters 26 and 27 deal with some of the latest and greatest "hot buttons" in the industry today. The first is the beginning of the emergence of ISDN, which does not mean "innovations subscribers don't need," but means that we have several techniques to dial up a lot of voice and data on an all-digital network. The switched services of a digital network will make your current analog modems look like the turtle, and ISDN will represent the hare. This time, however, the hare wins. Yes, it was slow catching up, but it is finally ready for prime time. Services are appearing every day. Still, we do explain what to watch for and what doesn't quite fit yet. ATM is an emerging fast packet system that makes X.25 and frame relay look weak: ATM is packet-switching on steroids. ATM does not refer to the automated teller machines so prolific in the banking industry. But if you read this section and don't scream, "I want it now," you missed something. You'll see!

Following the discussions of these networking concepts, we cannot lose sight that there are still local loop access situations that must be addressed. In Chapters 28 and 29, we discuss the newer competitive way to access the high-speed communications services through xDSL at the local loop from the ILECs and the CLECs. Following the discussion of xDSL, we jump to the new competitor, the CATV company. Offering CATV modems for high-speed Internet access is one of the ways that the cable companies are planning to capture the customer at the door. Using this springboard, we will look at the speeds and capacities of these wired services.

In Chapters 30 and 31, the discussion leads to some of the finer points of networks, such as the SS7 evolution for call setup and teardown. However, this is only one part of what SS7 brings to the networking strategies, because we can use SS7 for all the features and functions that bring new capabilities. Yet, with competition, we also need to look at the use of Local Number Portability (LNP) and the ability to move from carrier to carrier. However, we also discuss some of the other possibilities of using LNP, such as using E-911 to discover where a user is when in the wired or the wireless environment. Problems and pitfalls aside, we can see some of the benefits of using LNP from a wired and wireless integration plan.

After all this discussion of the wired network solutions, we then take a different approach with Chapters 32 through 34. This is a discussion of connectivity without wires. We look at cellular communications and personal communications for today and the future. The telephony and the data capacities of wireless are not equal to those of the fiber world, but there is some movement here. Enjoy learning how the wired and wireless future will share the same trail toward the turn of the century. We also look at wireless radio in the form of microwave (not the ovens) and satellite communications. These approaches are merely for purposes of comparing the connectivity solutions against the old reliable twisted pairs of wire. Then, of course, we added a section on the use of light beams without the fiber, called infrared transmission. That section details how a short-range communications need may be met without the use of wires, and also shows the strengths and weaknesses of this transport system.

The last four chapters are all add-ons about items we felt the average user might have some interest in, but would not be chomping at the bit to get specific details on. Chapter 35 describes how videoconferencing has come a long way in terms of standards, availability, and cost advantages. Then, with an eye to the costs of any communications method, we included Chapter 36, which discusses rudimentary cost justification techniques that will help describe how to sell a system or network concept to management. This covers the basics of financial justifications. Chapter 37 describes the evolution of facsimile, or fax, machines and the trends of today. This is a form of e-mail that most people have gotten very accustomed to. We compare how the systems transmit and receive information, as well as the evolution of the fax cards used in our PCs and notebook computers today. As Chapter 38 unfolds, we discuss the capabilities of our cabling systems, from copper twisted wires to coaxial wires to fiber optics. The speed of fiber brings us closer to the terabit throughput. With all the emphasis on Dense Wave Division Multiplexing (DWDM), we get the higher multiplexed rates and the benefits of different colors of light (so to speak). The capacities and compositions of these wiring systems are outlined to give the reader an understanding of what to look for and what to steer clear of.

We hope that you are intrigued and will read on. This is not a novel; it is not intended to be read from cover to cover. So, allot some time each day and take a chapter or a group of chapters together to gain an appreciation for the overall world of telecommunications. There is no reason this book cannot give you the tools necessary to deal with the novice or pro alike. Take some time to familiarize yourself with the ideas of the book; use the examples and analogies. Enjoy the stories and heed their message.

One closing thought: Many of our readers have sent us messages (e-mail, voice mail, letters) stating that we were technically incorrect with some of our concepts. Upon discussing this with them, we find that we are not incorrect, but we did not provide sufficient technical or engineering-specific detail for their needs or their liking. Wonderful! That is exactly what we were trying to do. Our philosophy with this book is the KISS method (keep it silly and simple). If you are looking for techno-babble as designed by propeller heads, there are many other books on the market. If you want to learn the basics without having a degree in engineering or in tech-speak, you came to the right place. Let's have some fun!

CHAPTER **1**

The History of
Telecommunications

In 1876, an inventor named Alexander Graham Bell was awarded a patent for one of the most significant devices in our lives—the telephone set. Mr. Bell had spent years trying to develop a method of communicating with his wife. Mrs. Bell was deaf, so Alexander was looking for a means of establishing a way to convert sound into some other form of communication so that his wife could understand him. While she was in a hospital for the deaf in Boston, Massachusetts, he was busy at work trying to solve his basic problem of communications.

Having some experience with the workings of the telegraph, by which coded messages could be sent across a cable, Alexander G. Bell decided to mimic this means of communications. Using the basic principle that communications could be converted from sound into electricity, he would be able to speak into a communications device that in turn would convert the sound waves into electrical energy. Literally, what Bell wanted to achieve was the transmission of voice over the telegraph, or by today's standard, voice over data network. This electrical energy could then be used to produce a coded message similar to a telegraph message. This all sounded good, but led to long struggles during development. Many frustrations were experienced by Bell and his assistant, Dr. Watson. No matter how hard they tried, the success of the transmission of voice into an electrical current seemed to elude these two inventors.

Then one day, fate struck. While working in the lab by himself (Watson was there, but had left for a moment), Bell spilled some acid on his worktable. This acid acted as a catalyst to produce a battery effect. Without realizing what had actually occurred at the time, Bell called out for Watson. His call of "Watson, come here, I need you!" activated the experimental device they had set up for the communicator. His voice carried across a wire to a second room, where Watson was working. Hearing the call for help, Watson ran to the aid of his partner.

The two researchers discovered that if a battery is applied across the electrical circuit (the wires) while the user speaks, the sound wave produced by the human voice could be carried across this same pair of wires to a receiver set up to accept this electrical current and convert the electricity back into sound. Hence, on that fateful day, the birth of a new industry occurred. The telephone set was invented!

Bell and Watson were elated. They were the inventors of what we have grown dependent on, the basic voice communications telephone set. This was in light of similar activity taking place elsewhere.

Who Really Invented the Telephone?

There was a race from the outset between Dr. Bell and a German inventor named Dr. Elisha Gray, who had used a similar concept to Bell's but was too late in completing the necessary documentation to do anything about it. Some say that Gray had actually invented the telephone before Bell but was too late in filing at the U.S. patent office. Gray filed suit against Bell for stealing his idea, but lost. All of this led to the birth of the telephony and telecommunications industry, which has proliferated to the point where we can't live without it. We have made the world smaller, and have attached to the telephony network devices such as computers, fax machines, video conferencing equipment, and even interfaces to LANs. All of this has led users to feel that they can't live, or do their jobs, without the use of a telephone.

Since Bell originally invented the telephone, many events have taken place; we'll examine these in closer detail to give you an appreciation of the industry in general. Incidentally, a theory even exists that Thomas A. Edison was the inventor of the telephone. With so many theories about the invention of such a remarkable piece of equipment, it's hard to believe that it proliferated as much as it did.

Evolution of the Telecommunications Industry

Historically, some of the important dates in the evolution of the telecommunications industry are shown in Table 1.1.

The First Telephone Companies Formed

Once the telephone was invented and proliferation started throughout the country, a group of telephone companies began to form independently to allow for interconnectivity. This technology started out with

TABLE 1.1

Summary of Early
Events in Telecom-
munications
Industry

Year	Event
1872	Western Electric was formed; Bell worked for this company while developing the telephone.
1876	Bell was granted a patent for his invention (March 7, 1876).
1877	Bell offered to sell his invention to Western Union Telegraph for $100,000. Western Union rejected the offer.
1877	The Bell Telephone Company was formed.
1877–1878	Western Union formed a subsidiary company called the American Speaking Telephone Company to market their own telephone set (speaker developed by Tom Edison, receiver developed by Elisha Gray), the first competitor to the Bell system.
1878	The New England Telephone Company was formed as another of the first competitors to the Bell Telephone Company.
1878	Bell Telephone Company sued the American Speaking Telephone Company for patent infringement.
1879	Bell Telephone Company and New England Telephone Company merged and formed a new entity called National Bell Telephone Co.
1879	The Bell and Western Union case was settled. Western Union agreed to stay out of the telephone business, and Bell agreed to stay out of the telegraph business in areas where Western Union operated.
1880	American Bell Telephone Company became the new name for the Bell System.
1882	American Bell entered into an agreement with Western Electric. WECO was to be the sole manufacturer of Bell equipment.
1884	The first long distance line was installed between Boston and New York.
1885	AT&T was formed as a subsidiary of American Bell to provide long-distance and telephone service for communities around the country and the world.
1887	The Interstate Commerce Commission was formed to regulate interstate carriers.
1893	Bell's first patent expired, opening the door for competition without patent infringements.
1899	AT&T acquired the assets of American Bell.
1910	AT&T aggressively fought the competition by acquiring controlling interest in Western Union.
1910	The Mann-Elkins Act was added to the Interstate Commerce Act to regulate the activities of the telecommunications industry.

TABLE 1.1
(Continued)

Year	Event
1913	The Department of Justice (DOJ) considered antitrust actions against the Bell System. Woodrow Wilson made this commitment to break up private monopolies.
1913–1914	AT&T agreed to divest its Western Union stock and stop buying up independent telephone companies, in a commitment by AT&T vice president Kingsbury in return for dropping antitrust action.
1918	President Wilson placed the telephone and telegraph systems under control of the Post Office (until 1919).
1921	The Graham-Willis Act established the telephone company as a natural monopoly.
1934	The Federal Communications Commission (FCC) was created to regulate interstate, maritime, and international communications. Congress also established "universal service" as its goal. The FCC investigated Bell System operations and DOJ began to formulate a major antitrust action against Bell. However, World War II postponed action because Bell was considered critical to national defense.
1935	Public Utilities Commissions (PUCs) were formed to regulate intrastate communications and rate setting. Also, the PUCs were instrumental in formulating revenue sharing for calls using more than one carrier.

many saying that the telephone was a frivolous toy and would only be used by the affluent. How wrong they were!

The first competitive telephone company to be formed was the American Speaking Telephone Company. It was funded by the Western Union Telegraph Company and used the pieces developed by both Gray and Edison. Edison had developed a transmitter that improved on the performance of the set, so this transmitter was used. The receiver was an invention of Elisha Gray. The combined transmitter and receiver (Edison/Gray) set was an enhancement over the Bell invention. Edison was also used to assist in the distribution of the wiring to local business and residential users. Edison's knowledge of electricity and the use of power distribution systems in the form of grids were applied to the telephone business. Of course, the Bell Telephone Company was actively challenging the American Speaking Telephone Company's position on the basis that the company's use of telephony was an infringement on Bell's patents.

The New England Telephone Company followed the development of the two previously mentioned companies in building both telephones and network services. This was in direct competition with the Bell Telephone Company. A problem existed with this whole scenario: any user who wished to connect with a local telephone company, of which there were now three, could not communicate with another neighbor or business connected to either the other companies. Interconnectivity did not exist. What this meant is that had I needed to speak to three different businesses that used the services of the three operating companies, I would have required three separate telephone lines and three different telephone sets on my desk. This is ludicrous to imagine, but true.

The Regulatory Scene

The industry was rampant with suits and counter-suits during the early stages of deploying the networks. Bell was active in trying to displace the "fake" companies on the basis of the patent infringements. Many of the cases had a bearing on the future of the industry, in particular with the formation of the telephone companies as monopolies. Some of the key events are listed in Table 1.1.

This was quite a turn of events; AT&T (as we know it) started out as a subsidiary of Bell and ultimately became the dominant force in the market. All of this happened in the first 23 years of the industry.

While all this was happening, many smaller companies started up, either building telephones or networks. By the 1920s, nearly 9000 independent telephone companies were in operation. This number was reduced somewhat because AT&T would not let the independents interconnect to the AT&T network. As a result, many were financially strapped and went bankrupt. AT&T began acquiring these failing operations, causing the number of independents to diminish (about 1400 independent operating companies are still in existence). AT&T had the single largest network in place and began to undercut prices, buying out competitors in an attempt to drive the smaller companies out of business. This left gaps in coverage for customers served by the independents, who could not connect to the major cities where AT&T had the market.

AT&T began to cave in to public pressure. A statement from an AT&T vice president (Kingsbury) outlined new positions to the attorney general. Kingsbury agreed to

- Relinquish its holdings in Western Union
- Stop the practice of acquiring the independents, unless authorized by the Interstate Commerce Commission (ICC)
- Allow the independents to interconnect to the AT&T network

Still another major event in the industry was the Graham-Willis Act, which established the telephone system as a natural monopoly. For years AT&T had dominated the industry, providing all telephone equipment and service through its normal distribution channels, the telephone companies. The network was the sacred cow, with only AT&T products and services allowed for interconnection to the networks. The philosophy was that any other product might harm the network, thereby keeping the monopoly over the products and services to be installed.

"Hush-a-Phone"

In 1947, the FCC permitted the connection of customer-owned and -provided recorders through special protective arrangements. These devices were used to record conversations, but were required to sound a beep when recording took place.

In the late 1940s, other manufactured devices could be attached to the telephone lines so long as a protection device was used. This was an appeasement to the telephone companies, which stated that the connection of a foreign piece of equipment on the network would disrupt the network and decrease its efficiency, as well as jeopardize national security. The carriers, who controlled the network and the revenues achieved through this total control, did not like these devices.

In the 1950s, an innovation known as the "Hush-a-Phone," an acoustically coupled device designed to eliminate noise and increase privacy, was tested before the FCC; the FCC agreed with AT&T's position and moved to prevent "foreign" devices from being connected to the telephone network. An appeals court later overturned this decision. Because the acoustic coupler was a nonelectrical device, it would pose no threat to the integrity of the network. The U.S. Naval Department was one of the users who fought for the use of the Hush-a-Phone. After losing the battle in court, AT&T suggested that if a recording device were to be allowed on the network, then a beep should be sounded on the line periodically (every 15 seconds). This beep would alert the user at the far end that the conversation was being recorded. The Bell system therefore suggested that if a beep were introduced to the line, then an electrical input was required. Bell fought for the right to install the equipment on the line that provided the beep. This was made possible because the Hush-a-Phone decision was made on the basis of a nonelectrical input. The Bell system was then able to charge the Naval department for the equipment that introduced the beep. All was settled.

In 1956, the Department of Justice filed an antitrust suit against the Bell System that had been postponed because of World War II. In general, the suit was aimed at getting AT&T to divest itself of Western Electric. This was finally settled in what was known as the 1956 consent decree. The result was that AT&T could retain ownership of Western Electric if it only produced products for the Bell-operated companies. This decree went on to prevent the Bell System from offering commercial data processing services, limiting Bell to providing telecommunications services under regulation. This still left the Bell System a regulated monopoly with no competition for equipment and services in their operating areas.

The Introduction of Competition

In 1968, however, the first true case was tested by a company called Carter Electronics of Texas. Carter made a device used to interconnect mobile radios to the telephone network via acoustic couplers. The FCC ruling was paramount in that direct electrical connection of devices to the network was allowed so long as a protective coupler arrangement was used. The decision had monumental impact because, prior to the Carterphone decision, manufacturers had a very limited outlet for their products. They typically sold their products to the independent operating telephone companies. Now the floodgates were opened, and these manufacturers had a whole new world of opportunity to market their products. Even the foreign manufacturers sped into the equipment market, offering end-user products that had previously been totally controlled by the Bell System. The European and Japanese marketeers were quick to enter the PBX and terminal equipment business. However, the use of a protective device was still an area of frustration with these equipment vendors. Rent for the Bell protective arrangements was a recurring cost that made the purchase of other equipment less attractive. The manufacturers continued to complain about this. These complaints resulted in the institution in 1975 of an FCC registration program for all products that could be attached to the telephone network. As long as a manufacturer could pass the requirements of the FCC registration (Part 68), then its products could be attached to the network without the use of the protective devices. The stage was set after the Carterphone decision, with many companies competing to build connecting devices. This began what was called the interconnect business.

Still, additional activity took place. In the 1960s, a small microwave carrier company called MCI began constructing a microwave network between Chicago and St. Louis. MCI's initial intent was to provide alternate private line services in a high-volume corridor. MCI took its interconnection request to the courts and won the ability to interconnect its network with the telephone company network. MCI further began to offer switched services, which was immediately reported to the FCC. The legal actions nearly put MCI into bankruptcy, but the little company prevailed and began the "other common carrier" industry—one more nail in the monopolistic coffin for AT&T.

The Divestiture Agreement

In 1982, the most monumental decision to be made since the inception of the monopoly in the telecommunications business was reached. In light of all the computer inquiries and consent decrees that were taking place, the Department of Justice was still after AT&T, Western Electric, and Bell Labs. This was an action that had started in early 1974. It was aimed at the complete breakup of the Bell System. In early 1981, the suit finally made it into the courts, but was surprisingly brought to a halt on January 8, 1982, when an agreement was reached to drop the suit and submit a modification of the 1956 consent decree to the courts.

In order to achieve the goals targeted, AT&T would be relieved of the limitations keeping it from other unregulated markets. However, in return for this relief the courts demanded the breakup of the Bell System. Because the Bell System operated as a monopoly and enjoyed the non-competitive environment, it was to remain under a regulated status. The Modified Final Judgment (MFJ) basically considered the provisioning of local dial tone services as a monopoly and everything else as a competitive operating environment. Therefore, the local dial tone business provided by the local exchange carriers would remain regulated and AT&T could pursue and operate in the competitive long-distance, equipment manufacturing, computer equipment/processing, and sales markets. What was once built as a single entity, owned and operated as an integrated system, was now about to be broken into multiple pieces. The 23 Bell Operating Companies would be divested from the organization of AT&T (which retained its Western Electric Manufacturing, Bell Labs, and Long Lines divisions). Some of the salient points of the MFJ include the following:

- AT&T had to transfer sufficient personnel, assets, and access to technical information to the Bell Operating Companies (BOCs), or whatever new organization was owned by the BOCs, to enable exchange services or access to exchange services to be provided without any ties to AT&T.

- All long-distance services, links, personnel, and other facilities had to be relinquished by the BOCs and turned over to AT&T.

- All existing licensing agreements and contracts between AT&T or its subsidiaries and the BOCs had to be terminated.

- Equal access for all interexchange carriers into the BOCs' switching systems had to be provided. This broke down the special privilege that AT&T enjoyed over its competition. The equal access for all carriers was to be provided within two years, with the only exceptions being electromechanical systems where it was cost prohibitive or companies serving fewer than 10,000 users.

- After the breakup, the BOCs could provide but not manufacture equipment.

- After the divestiture, the BOCs could produce and distribute directories to subscribers.

- Carrying calls between and among offices was defined as local exchange services and interexchange services. The local exchange carriers (LECs) would hand off any interexchange call to an interexchange carrier (IEC). To define the boundaries of who carried the call and who shared in the revenue, a number of local access and transport areas (LATAs) were created. The calls inside a LATA were the responsibility of the LEC, and calls that left the LATA were the responsibility of the IEC.

- Joint ownership or participation in the network between AT&T and the BOCs was prohibited. Although this was a common practice before the breakup, it was not supported after the breakup.

The provisions of the MFJ were strictly aimed at the Bell System. This was a point of confusion for many people, because they thought that the whole network would fall apart—which it did not. The greatest confusion arose when the MFJ was decided in early 1982 with an implementation date of January 1, 1984. Fear, uncertainty, and doubt reigned in the industry.

The business users were prepared for the breakup, but not much happened in the residential market. Although the BOCs and AT&T tried to communicate the message to end users, they were highly unsuccessful. Ten years after the breakup, many residential users still did not know that there was a difference between the suppliers. It is a sad comment on the industry that a communications company was unable to communicate successfully with its customers. The small business and residential users were so accustomed to the "cradle-to-grave" services offered by their BOCs that they did not grasp the significance of this breakup. No longer would one call do it all. Separate companies provided dial tone services and long-distance services.

The divestiture of the Bell system made for advances in the industry that might not otherwise have occurred for some time to come. With the introduction of competition and the ability of other suppliers to enter new markets, the evolution of the network and service offerings escalated dramatically. Digital networks had already been used in the Bell system for years, but with the breakup new digital architectures rolled out to customers at a rapid pace. Further, when divestiture took place, the costs associated with local dial tone increased by as much as 50 to 60 percent on a monthly, recurring basis. However, long-distance services and usage charges decreased by 60 percent or more. This might sound like a break-even analogy, but the ratio of local to long-distance service is a 20:80 rule. Twenty percent of a customer's actual monthly bill is for the fixed recurring charge, whereas 80 percent of the bill is for the variable usage-sensitive costs associated with long distance. From this, you could suggest that we would gladly accept an increase of 50 to 60 percent on 20 percent of our bill, if the remaining 80 percent of our bill is reduced by 60 percent.

To add to the confusion, whenever a customer called a provider for a service that did not fall within that provider's scope, the standard answer given was that the "judge" didn't allow the vendor to provide such services. This reference to Judge Harold Greene of the Federal District Court, who presided over the suit and the MFJ rulings, was a standard escape clause used by LECs and IECs alike. Unfortunately, the MFJ was an agreement reached by the court and AT&T rather than an actual mandate. Thus the burden of setting the rules in the telecommunications industry was placed on the shoulders of the courts.

Tariffs

Another escape clause used on a regular basis is that "the tariffs don't allow this service," which is a clear misrepresentation. The BOCs and the

AT&T organization write tariffs and submit them to their respective judicature, whether it is the FCC for interstate traffic for AT&T or the Public Utilities Commission for the BOCs. An easy way to avoid offering a service or to delay a customer request is to suggest that the tariff doesn't allow something. However, who writes the tariff? If a service or something out of the ordinary is not covered in the tariff, then the logical solution is to write a new one. It is very rare for a new tariff to be written.

What is a tariff? A tariff is a description of a service that offers an appropriate rate of charge for that service and the rules under which the service is to be provided. It is the basic agreement of terms for a service between the customer and the provider that must be submitted to the regulators for approval before any changes in service can be offered. In the United States, there are 50 different regulators that rule on the tariffs, one for each state. This leaves room for plenty of disparity between offerings and rates. Furthermore, the FCC governs the rates and services for the long-distance suppliers, primarily AT&T. This makes it very difficult to understand what the offerings are and what pricing strategies are in force. A user typically does not understand or have the patience to read through a tariff. Therefore we relegated the overseer function to the Public Utilities Commission and the FCC. We can only hope that these organizations are performing a reasonable job of ensuring that the rates and offerings meet the needs of all concerned parties in a fair and equitable manner.

In the rest of the world, rate settings and guidelines differ depending on the country. Many of the telephone companies are under direct control of the local government. The telephone service then is relegated to the post office agencies; these are called Post Telephone and Telegraph organizations, or PTTs. The name stands for the services provided: postal, telephony, and telegraphy. Privatization is now starting to be enacted in several parts of the world; telephone services are being removed from government control and passed on to private organizations. It stands to reason that these organizations will face stiff competition for the services, much the same as has happened in the United States.

The Telecom Act of 1996

In February 1996, the Clinton Administration signed into law the Telecom Act of 1996. This act was the culmination of several years of trying to deregulate and provide a competitive marketplace in the telecommunications arena. This law, when enacted, opened the door to an open communications infrastructure. Essentially, what the administration put into play was the beginnings of the concept of the information superhighway.

The Telecommunications Act of 1996 opens the way for a myriad of new players to compete for the local dial tone service. In the United States, dial tone amounts to a $115 billion a year industry. No wonder all the emerging players want a piece of that action. What this means, however, is that a group of new players will emerge to provide dial tone services while the local telephone companies are unshackled and allowed to penetrate new markets that were unavailable to them in the past.

During the beginning stages of this Telecom Act, elation and overwhelming support for the newly emerging marketplace from all players was the name of the game. However, as one would expect, things don't work as smoothly as they are supposed to when first starting. The Telecom Act of 1996 allows for the long-distance companies (IECs) to enter into new business opportunities. These include dial tone, cable TV services, high-speed Internet access, and two-way video communications capabilities after the infrastructure is in place.

The cable TV companies, on the other hand, are now allowed to enter into telephony and other forms of the communications business. These providers are allowed to offer voice communications, high-speed Internet access, two-way multimedia communications, and cable TV services, all on a single communications platform. As the cable companies look at their infrastructure, they already have a high-speed communications channel running either to or by everybody's door. However, they must recognize that in the past their primary service was the delivery of one-way communications in the form of cable services. In order to provide high-speed Internet access and enhanced capabilities as well as voice communications, these companies were forced to create a two-way communications cable system. This means that they had to either add new cables or provide high-speed fiber in the backbone network to the curb and then coax to the door. Although this sounds fairly straightforward and easy, it does require significant investments on the part of the cable companies. This alternative proved very effective with a hybrid fiber/coax arrangement that, by the end of 1999, the cable TV companies implemented in several cities in the United States and Canada to provide the high-speed access in a bundled pricing mechanism. Moreover, 1999 marked the year when AT&T (the now-streamlined carrier) acquired two of the largest CATV companies, TCI and Media One. Having invested over $100 billion in acquiring these two cable companies gave AT&T access to many consumers' doors.

As the Telecommunications Act continues to be enforced, the telephone companies will break out into new markets such as operating long-distance service for less and providing cable TV services, Internet access, and videoconferencing capabilities on their local infrastructure. What the telcos have to realize, however, is that the local two-wire cable facility (called the *local loop*) is a single pair of copper wires that was not designed to sustain the high-speed communications we're talking about here. Therefore the telcos will continue to update their cable infrastructures. These companies are enamored with the new xDSL technologies that use high-speed digital subscriber links. Using various techniques such as asymmetrical digital subscriber link (ADSL) or very high-speed digital subscriber link (VDSL), the telephone companies can provide high-speed communications to the customer's door. In the ADSL marketplace they envision delivering up to 9 Mbps to a customer's door, whereas outbound from the customer to the network the service will offer plain old telephone service (POTS) and up to 384 Kbps data transmission. Two occurrences have deviated from this scenario to start:

1. Rate-adaptive ADSL (RADSL) was introduced allowing the telco to deliver less than the 9 Mbps downloadable to the door. Instead, they use a figure of 1.544 Mbps downloadable and adaptive rates of 256 to 1.024 Mbps uploadable. If the network is busy, then the consumer will get slower-speed access, but a contracted minimum comes into play. However, if the network is lightly loaded, the consumer will benefit from the higher throughputs.

2. One-meg modems were developed. The ADSL Forum looked at the application and need for the speeds mentioned above. What this forum determined, in the short term, is that the average consumer only really needs 1 Mbps downloadable and approximately 160 Kbps uploadable asymmetrical speeds. Therefore, they developed what was termed ADSL Lite or G.lite specification. Over time, this specification will allow higher-speed access, but for the short term this is sufficient.

In the ADSL marketplace the telcos have been dragging their feet implementing these services. There are many reasons, but the most common is that they do not want to be forced to provide the xDSL service at a discounted rate, like they do the dial tone. Additionally, there is much ado over the standards being implemented in xDSL technology.

In the VDSL marketplace, the telcos envision up to 51 Mbps to the customer's door with a much lower-speed communications channel outbound, or a symmetrical 51 Mbps in each direction. Regardless of the technique used, the telephone companies are in a position to find technologies that will support and sustain these speeds on their local, single twisted pair of wires to the customer's door. This is their challenge. Beyond the high-speed communications, the telephone companies can also enter into manufacturing, long distance, and cable TV service. However, the caveat of the Telecom Act of 1996 is that these companies must first prove that an open competitive environment exists at the local loop. There has been much ado in terms of the Telephone Companies being willing and cooperative in providing this access. One can imagine that for the next few years this will be one of the contention points as the newer players emerge and attempt to get into either facilities-based or a non-facilities-based dial tone provision.

In the facilities-based environment, the carrier will provide its own cables or wireless communications to the customer's door. At that point all of the communications will be carried right out to the wide area network, bypassing the local telephone companies.

In a non-facilities-based environment, the new emerging players will rent or lease facilities from the local telephone company at a discount. The discount ranges anywhere from 17 to 28 percent off of what the local telephone company's tariffs are today. This, however, is an area always open for debate as the telephone companies attempt to select and provide access on a non-facilities basis to their competitors. Clearly, the telephone companies argue, by having to rent the services to competitors at a discounted rate, they are putting themselves in an unfair position. The resellers of dial tone would merely turn around and rent services right back to the consumer at a rate less than what the telephone companies charge. The telephone companies contend that this is a discount that should not be provided, and that if competitors wish to work in this market they should build their own capabilities or rent from the telco at the same rates the telco charges other customers. Embroiled in all of these battles are the other players such as the long-distance and cable TV providers, who are equally distraught because they also have to pay for other access fees to provide services to their consumers. They argue that the telephone companies have been raising the rates for access because of the potential loss of revenue on the basis of the Telecom Act of 1996. An interesting event occurred in 1999 when AT&T acquired the local cable TV companies (TCI and Media One) to get access to the consumers' doors. Shortly after acquiring these giants, AT&T was challenged to offer the access on CATV to competitors at a reduced rate. AT&T immediately balked at that idea and appealed to the FCC because the local utilities commissions were ruling in favor of the competitors. AT&T is screaming about the unfairness of such a move, yet this worked just the opposite when they wanted access through the telco and were asked to pay the access fees.

Throughout all this maneuvering, the consumer may or may not win. Clearly, with competition, prices should fall to a more reasonable base and other discounts might exist. This will work in the major NFL cities.* The pricing mechanisms will probably drop the local cost of dial tone and access to the long-distance network. However, in rural communities where services and facilities have always been limited, there may be no advantage or a slight disadvantage. The consumers in the rural areas will be left in the lurch because few companies will want to serve such regions. In the old days the telephone companies had to serve it as a last resort. Under deregulation, all of this might well change. A couple of the competitive local exchange carriers (CLECs) have actually opened many third- and fourth-tier communities and done very well. Because they are the prime competitor in the town, users who are willing to take a chance will do so with the new provider. This has been a lucrative market for the new CLECs in this area, as opposed to the results in the major metropolitan areas.

As one might expect, several new providers got into the business. As of 1999, over 500 CLECs had jumped into the competitive local dial tone business, yet they have achieved only 3 percent penetration into the overall market. This penetration amounts to approximately $4 to $5 billion annually. Each of the competitors entering this market offers some form of discounts on cable services, dial tone, or long-distance access and services in order to pick up a few market points. Many providers are building out high-speed access on fiber-based networks bypassing the local telco. Newer providers are now offering the one-stop shopping method by offering a bundle of services, including dial tone, long distance, equipment, and internal wiring all on a single phone bill. AT&T, for example, began offering as much as a 26 percent discount overall for the bundled-service packages they now offer, including dial tone, CATV, and long-distance services. These new players use loss leaders to attempt to pick up their market shares. Through this loss leader market, they will make no money. Consequently, the consumer may stand to gain for the short term, until these new emerging competitors realize they are making no money by offering heavily discounted services. At that point, the carriers will

---

*The reference to NFL cities is that any major city across North America with a professional football team has all the services and benefits, whereas the smaller communities that do not have a football team do not get the same level of service and accessibility.

have to raise their rates or offer some other value-added services to gain new revenue. One can imagine what services those might be.

The Canadian Marketplace

In 1994, the Canadian Regulatory Telecommunications Committee (CRTC) endeavored to do something very similar to what happened in the United States. The goal was to deregulate or de-monopolize the local dial tone provisioning service. On May 1, 1997, the CRTC provided its interpretation of how it will deregulate and open the market to dial tone to a competitive environment. One can expect that, one year after the Telecom Act of 1996 in the United States, the CRTC will likely model a good many services and provisions of the law after those accomplished by the FCC. The reason is to provide harmony throughout all of North America. While the CRTC is endeavoring to break up the dial tone monopoly, the cable companies, the long-distance providers, and a rash of emerging players who will be either facilities based or non-facilities based have already filed and petitioned for the right to offer services. Once again, in the Canadian marketplace the influx of new providers and new opportunities will overwhelm the consumers.

In both the U.S. and Canadian marketplaces, all of the dial tone and long-distance access services have traditionally been through the local dial tone loop. Now, with the cable TV companies trying to get into this business, the dial tone could be provided on either a cable or a local loop. However, more and more opportunities exist for a wireless connection to the consumer, whether residential or business. The wireless dial tone providers are constantly springing up around the country as they entertain the thought of the personal communications service (PCS). It will be through a combination of the dial tone, the TV services, the high-speed Internet access, and the multimedia communications capabilities to the local door that all of these activities will come to culmination. No one service alone may warrant all the new emerging players, but as a whole this marketplace is enormous.

Therefore, as the Telecommunications Act is rolled out and the CRTC deregulation is rolled out in the Canadian marketplace, we can expect to see dramatic changes in the way we do business. Through the convergence of wired and wireless communications, dial tone, and long distance, a variety of other services will emerge to create for us a competitive potential. One-stop shopping may well be the way of the future.

Just as the industry started to settle down, with all of the communications providers vying for a portion of each other's business, another change took place. The competitive local exchange carriers (CLECs) emerged providing various new opportunities for organizations that were not already in this business. This includes some of the providers of the competitive access provider (CAP) services of old. Newer players continue to emerge on the basis of either facilities-based or non-facilities-based provision. Resale of dial tone and resale of long distance have all gained a foothold, but a new approach has been added. The power utilities around the United States and Canada that have an infrastructure of fiber in their ground wire (see notes later on the ground wire optics) now have the opportunity to deliver high-speed fiber-optic communications right to a pole line near the customer's entrance. Many of the power utility organizations have applied for licensure to provide dial tone, high-speed wide area networking communications, video, and TV-type services through the infrastructure under their ground wire. One-stop shopping might become a reality from an electric company who can provide cable, TV, Internet access, and your power, all on a single bill. As a commentary to this industry, most people in the United States are apparently enamored of their electric companies. This is a direct contrast to how they feel toward their cable TV and telephone companies. These two providers are always looked on with scorn or distaste. If an electric company does offer dial tone service, one may expect a significant hit rate in terms of customers signing up to use that utility as the provider for communications services also. If the electric companies can provide dial tone and cable TV for less, they will do so across an infrastructure that was used for their own internal process control systems. This means their cost for providing dial tone across this high-speed fiber backbone that was put in place over the years will be marginal.

The Telecom Act of 1996, as well as the events in the Canadian marketplace, is similar to what is happening around the world. In Europe and in many other places, the communications bastion of a monopoly for dial tone services is quickly eroding. Since the European Union started this overall movement in 1987, with completion scheduled by 1998, privatization is taking place in which the governments around the world are unshackling the dial tone providers or divesting them as government-controlled organizations from their telephone companies. This marks a whole new trend that will hit around the year 2000. By the turn of the century, privatization will be standard. The opportunities to reap benefits and to provide the access necessary will abound.

Voice Characteristics

The telephone set, as it was designed, is used to transmit sound to the receiving party. The basic telephone set that was designed by Alexander Bell carried his voice between the set in his room to the one where Watson was working. So it began. If the telephone network and the telephone set were designed to carry voice signals from one point to another, then the characteristics of the voice signal had to be known. A great amount of time and effort went into studying the actual variables associated with sound as the vocal chords create it. From these studies, one can derive the fundamental concepts of a telecommunications network. This implies that the telephone set is designed to carry only voice. In reality, this is correct, because all of the effort was directed toward carrying voice in its truest possible form from the sender to receiver. The word *telecommunications* involves this very basic concept. A definition is in order here because we will be covering the characteristics of the telephone network and its ability to carry the human voice.

Telecommunications is the transmission of information, in the form of voice, data, video, or images, across a distance, over a medium, from a sender to a receiver, in a usable and understandable manner.

The definition is complicated, but it can be simplified by breaking the two pieces apart and stating them differently.

Tele is a Greek root meaning *from afar.* This implies that the information will be carried over some distance, from as little as a few inches to as much as thousands of miles.

Communicate comes from the Latin root meaning *to make common.* The implication here is that if we are to send any form of information across the medium, then it must be understandable or else it is wasted energy.

The Medium

A *medium* is any form of transmission capacity that is used to carry signals. In the telecommunications world, this can be in the form of copper wires, coaxial cables, optical fiber (glass or plastic), or air. There are other forms of media; for example, this book uses paper as the medium to get the information to the reader. But we are dealing with a telephony concept, so we will stick to these types of analogies. Depending on the medium used, the distances over which information is transmitted can vary greatly. However, strides have been made since the late 1870s, when the telephone set was first invented. Now we can transmit any information over thousands of miles as simply as we do over inches. The information takes on the form of electrical energy to recreate the sound waves generated by our voices in a representative form, so that listeners at the distant end will be able to understand the message.

Sound

Sound is nothing more than the banging of air molecules together at a rapid rate. As we generate sound with our vocal chords, we are banging these air waves together to produce intelligible information that can be used and understood by others. This is in the form of air pressure changes. Hold your hand in front of your mouth as you speak. Feel the air pressure hitting your hand? This is the effect of sound as we generate the changes in air pressure. Our vocal chords, therefore, are moving back and forth and banging the molecules of air together.

There is an old question that runs: "If a tree falls in the forest and no one is there to hear it, does it create sound?" The answer must be an obvious yes. Although no one is there to hear it, the air pressure changes caused by the tree falling to the ground—regardless of where it falls—must still be producing sound. If we put a tape recorder in the forest and the tree falls, it is most likely that the sound will be captured by the recorder. Even though no one is there, the recorder still captures the noises created by the air changes.

The telephone converts sound waves to their analog equivalent in the form of electrical pulses that are then carried across the telephone network. We can therefore assume that in order to communicate across these wires we must have some form of sound. The sound will be converted into electricity. The electricity is then sent across the telephone wires.

Sound is the banging together of air molecules at a rapid pace. This is called *compression and rarefaction*. The human voice produces sound at a constantly changing set of frequencies (pitch) and amplitudes (loudness). The human voice changes these variables of frequency and amplitude in cycles per second. The vocal chords compress the air molecules at a rate of between 100 and 5000 times per second. To recreate the sound faithfully, the sound waves (or air pressure changes) are converted from sound into electricity. This is what the telephone set does. For a more detailed explanation of how this happens, refer to Chap. 4.

As the electrical equivalent of sound is created, compare the sound wave to an electrical wave. Electricity is typically generated in an analog form by rotating the electromagnetic energy around a center point. As the energy is on the rise, it increases the amplitude to a peak level in decibels, then begins to fall. Because the wave is concurrent, its electrical field will have both a positive and a negative side. Therefore, as the signal decreases from the peak of the positive energy, it moves back toward the zero line. As with a magnet, there are two poles, positive and negative. Therefore, as the wave gets to the zero line, it will continue to fall to the negative side of the voltage line until it hits some peak at the bottom of the energy field. From there, it will in turn rise back up to the zero line (value) again. This, in effect, constitutes a 360° cycle around the baseline, or one complete rotation. This rotation is called a sinusoidal wave. This sinusoidal wave (Fig. 2.1) is the analogous recreation of the human sound wave in its electrical form. This single wave cycle over a specified period of time, usually one second, is called one hertz, named after the man who discovered this concept.* Hertz is normally abbreviated *Hz*.

The sound wave produced by a human voice will have a constantly changing variable energy in both signal strength (the amplitude) and the number of rotations around the baseline over a period of time (the frequency). The voice creates these waves from 100 to 5000 times per second, as stated. Therefore the voice characteristically creates 5000 cycles of infor-

---

*Heinrich Hertz came up with this concept.

Figure 2.1
The human sound wave produces both frequency and amplitude changes.

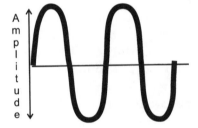

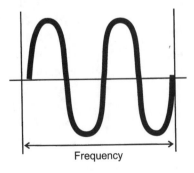

Frequency

mation per second. This is abbreviated into 5 kilohertz. Kilo means thousands and is represented by the *k* in kHz.

The human ear is responsive to variations in frequency and amplitude at rates from 25 to 22,000 Hz. This means that the ear can receive and discern all of the information contained in the human voice. This is important in telephony.

Differences exist in human responsiveness to sound. For example, older humans can discern sounds ranging only up to 7–8 kHz as a result of abuses and deterioration of the eardrums. A person who has fired a weapon (rifle or handgun) without proper ear protection will have damaged the upper and lower frequency responses. A younger person, who has not had the chance to damage the ears, will be able to receive and discern sound waves in the 16- to 18-kHz range. When we see these youngsters carrying those "boomboxes" on their shoulders, however, with the box blaring away and sitting directly next to the ear, we can only imagine how the ear will respond in a matter of time. The youngsters are inadvertently narrowing the range of frequency responses to which their ears will be able to selectively respond.

The telephone company realized that the majority of usable information in a human conversation will fit in a 3-kHz range; therefore, it provides a pair of telephone wires, typically made of copper, that will carry all of the usable information. The telephone networks were built to carry speech, and the most commonly carried signal on the network is the electrical equivalent of speech (voice). The telephone transmitter converts the acoustic signal (sound wave) that is generated in the human speaker's larynx into electrical waves.

Actually, the analog waves can be represented in frequency and analog changes over a broad spectrum (or band) from approximately 30 Hz to about 10 kHz. However, most of the usable and understandable energy falls in the spectrum of 200 to 3500 Hz. If we subtract the differences between the high end and the low end, the spectrum is 3300 Hz, or 3.3 kHz, wide. It is not necessary to recreate all of the speech waveforms precisely to get an acceptable transmission of human speech across the telephone network. This is because the ear is not highly sensitive to very fine distinctions in the frequency changes, and the human brain can make up for any variations in the speech form by interpretation. Of course, if something does not come across the wire clearly enough, the human brain will intervene and cause the mouth to say, "What?" This in turn will cause the speaker at transmitting end to generate the signal over again by repeating the words.

Because the cost of transmitting the signal across a telephone network is directly proportional to the range of frequencies carried, the telephone company uses a bandpass filter on the circuit. Commercially acceptable and usable information is transmitted in what is called a band-limited channel. This means that all of the usable information is allowed to pass onto the circuit, but the extraneous information that does not add significantly to the conversation is filtered off. If the extra energy is put onto the wire and carried from end to end, the costs will go up at an equal rate; this is wasteful.

What Is Bandwidth?

One of the toughest concepts for anyone to understand is *bandwidth*. To the novice, it becomes even more perplexing when bandied about by telephone company personnel, engineers, and others. However, it is the basis of most of what we do in the telecommunications and telephony world.

Think of bandwidth as a water pipe or a garden hose. The greater the size of the pipe, the larger the volume of water that will flow through the pipe. The smaller the pipe, the smaller the flow of water. Now, in telecommunications terms, think of bandwidth as a communications pipe. The bigger the pipe, the more information that will flow through it. The smaller the pipe, the less information that will be carried through the pipe. Assume, for example, that we have a lawn and we need to water it regularly to keep it green and moist. If we have an average-sized lawn, the job can be done with the standard ⅛-inch garden hose, the type that can be bought in any hardware store. When we turn on the spigot attached to our regular water pipes, a sufficient flow of water comes out of the hose. This is a regulated flow with several control mechanisms in place:

- The diameter of the water pipes is perhaps ¾ inches, allowing a certain amount of flow through them to begin with.
- The garden hose is slightly smaller, so that the flow from the larger pipe to the smaller-diameter garden hose is restricted, but pressure allows the flow to be constant.
- The spigot also can be used to turn the water on all the way or to limit its flow to a specific level that is more to our liking.

This is a band-limited pipe that uses several constraints to control the flow of water through the pipe. Enough is allowed through to do the job; any more would probably cause flooding and over-saturation in certain areas. Therefore the limitations meet our needs without being wasteful.

Now let's assume that we want to handle the watering of the Super Dome if its turf were real. This football field is much larger than the average home lawn, so if we use the tools that we have in the home watering scheme, things will be tougher. Imagine trying to water this lawn with the average ⅝-inch garden hose! It would take forever. Not only would it take forever to get from one end to the other, by the time we got to the opposite end, we would have to start the job all over again. The first end would be parched dry, due to the length of time it took to get from one end to the other.

This is ineffective. So to do the job, we will have to purchase a garden hose with a diameter of 6 inches. Now we hook this 6-inch hose to a spigot that also measures 6 inches and is connected to a water pipe at least 6 to 8 inches in diameter. The flow of water through this pipe will be significantly greater than that of the ⅝-inch garden hose. Thus the job can be done in a reasonable amount of time.

Bandwidth is similar to this garden hose analogy, as shown in Fig. 2.2. It is the range of frequencies that can be carried across a given transmission channel. If more information is sent, more bandwidth is necessary. A typical telephone channel (line) is provided by the telephone company that will carry 3 kHz.

Therefore this is a 3-kHz channel; it has 3 kHz of bandwidth. This is fine because, as we previously saw, all the usable information of a voice-grade conversation is contained in this amount of bandwidth. In actuality, the telephone companies break the available electromagnetic spectrum into slices, each about 4 kHz wide. Then these 4-kHz slices (called channels) are limited with bandpass filters, as shown in Fig. 2.3. Consider the spigot analogy, turning the water on faster or slower via the spigot valve. The result is that we receive 3 kHz of the available 4-kHz slices.

Other forms of bandwidth requirements exist. For comparative purposes, we can see the differences in capacities used in various forms of bandwidth allocations as shown in Table 2.1.

Figure 2.2
Bandwidth can be compared to a pipe or hose.

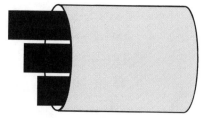

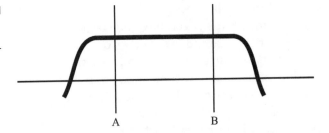

Figure 2.3
Bandwidth of 4 kHz.

A   B

MHz is a new term here. It represents millions of frequency changes per second. You can imagine the amount of information carried in a TV signal, where sound, motion, and voice are all on the same channel. Think of how a video signal on a TV station would look if the channel was restricted to carrying only 3 kHz of information. By the time some moving picture was created on the set, the viewer would have lost interest.

The frequency spectrum of a TV channel as shown in Table 2.1 shows that a community antenna television (CATV) channel is actually allocated 6 MHz of capacity. Yet the amount actually used approximates 4.5 MHz. The difference is the band limitation on the channel, much the same as with the voice channel. In this case, the amount of flow is in the millions of cycles per second (a big pipe is needed here). The band-limited channel uses bandpass filters so that there is a guard band between the TV channels. In other words, as you watch channel 3 on your TV set, the filters are placed on the line to prevent frequencies and information from channels 2 and 4 from overflowing onto channel 3 (Fig. 2.4). These guard bands are placed on every channel, thereby restricting the 6-MHz channel to only 4.5 MHz of usable information.

The same concept holds true for just about all of the channel capacities used in the telecommunications industry. The available bandwidth is a

TABLE 2.1

Summary of Channel Capacities to Carry Different Forms of Information

Service	Bandwidth Allotted
Voice channel	3 kHz
High-fidelity music	15 kHz
CD stereo player	22 kHz
FM radio station	200 kHz
TV channel on CATV	6 MHz
TV channel actually used	4.5 MHz

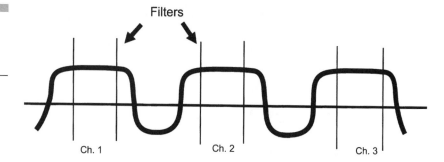

Figure 2.4
The bandpass filter separates the channels.

function of need and cost. You get what you need, but any more will be too expensive.

Voices

The human voice is a consolidation of the waves of electrical energy carried across the given channel capacity. Humans generate a combination of amplitude and frequency changes in a continuing flow. If the changes are held constant, then the conversation becomes monotone, highly unacceptable for the average conversation. Indeed, if everything were held constant, the recipient of the information would be lulled to sleep. We use the variations in our voices to reflect deviations in emotion, accentuation, or articulation and emphasis on certain points. Every voice generates a different pattern of amplitude and frequency changes. This is what gives every individual a unique and recognizable speech tone.

It is interesting to note that the female voice pattern typically generates more changes in amplitude than in frequency. This accounts for the higher pitch in the female voice. Conversely, the male vocal pattern generates more frequency changes than amplitude shifts. This accounts for the low-pitched, more grainy tones of the male. These are averages; every human is different and the norm can be deviated from at any time. Suffice it to say that when someone is highly emotional, there are definite shifts from that person's normal voice patterns. The pitch might go up by varying degrees, showing the results of stress or anger.

The telephone links are designed to handle this widely varying pattern of shifts. However, from time to time the voice pattern might exceed the differences in the allotted bandwidth. Then the frequencies in the voice go above or below the normal range. At this point, the bandpass filters start to remove the excesses. What can result is a pattern of tinny or flattened speech conversations. This can be also heard if someone uses the letters *F* and *S* where the frequency ranges on these two sounds might exceed the bandpass ranges; they get flattened.

Other Services

Because the telephone network was built to carry the analog equivalent of human speech patterns, the other services that we wish to communicate, such as data, facsimile, images, and video must be transmitted within the same constraints as voice calls. The network allows any telephone set to contact any other telephone set across a 3-kHz bandwidth. If you want to communicate anything else, such as data, it must be accommodated on this same 3-kHz bandwidth. This invokes a limitation on the speed of our data communications channel capacities. Although modems can transmit signals more quickly, they must be constrained into the size of the pipe. A video conference transmission will also reflect non-real-time motion because of the channel capacities on a dial-up basis.

If you need to move more information across the channels, you have two choices. One option is to dedicate a high-speed line between the two or more end points we need to communicate with. However, this approach will eliminate the possibility of "any-to-any" connection and will require planning well in advance of communicating with the end points. Leased lines between the points might be underutilized and could be more expensive.

Another option is to have the bandpass filters moved out to (for example) 8 kHz between the end points. Unfortunately, however, there are similar limitations here. We would have to know every location that we would be communicating with, losing the benefits of the any-to-any connection. As well, there is no guarantee that the switching systems will route the call over the same path on a switched dial-up connection every time. This would force us back into a 3-kHz channel along the route selected if it is different every time, eliminating the benefits of the wider channel capacity.

The Telephone Network

A network is a series of interconnections that form a cohesive and ubiquitous connectivity arrangement when tied together. Whew! That sounds ominous, but to make this a little simpler, let's look at the components of what constitutes the telephone network.

Generally, a network is a series of interconnection points. The telephone companies over the years have been developing the connections throughout the United States and the world so that a level of cost-effective service can be provided to their customers. In order to build out this connectivity, the telephone companies install wires to the customer's door, whether the customer is business or residential. The spot where these wires terminate is called the demarcation point (demarc) or the point of least penetration. The position of the demarcation point depends on the legal issues involved. In the early days of the telephone network, the telephone companies owned everything, so they ran the wires to an interface point, then connected their telephone sets to the wires at the customer's end. This was the cradle-to-grave service that allowed the companies to expand the connections throughout their entire operating areas.

New regulations in North America changed this connection to a point at the entrance to the customer's building. From there, the customers hook up their own equipment, items they purchase from a myriad of other sources. In the rest of the world, where full divestiture has not yet taken place, the telephone companies (or PTTs) still own the equipment. Other areas of the world have a hybrid system under which customers might or might not own their equipment. The combinations of this arrangement are almost limitless, depending on the degree of privatization and deregulation. However, the one characteristic that is common in most of the world to date is that the local provider owns the wires from the outside world to the entrance of the customer's building. This is called the *local loop*.

A Topology of Connections Is Used

In the local loop, the topological layout of the wires has traditionally been a single wire pair or multiple pairs of wires strung to the customer's location. This has always been an issue of money. Just how many pairs of wires are needed for the connection of a single line set to the network? The answer (one pair) is obvious. But for other types of service, such as digital circuits and connections, the answer is two pairs. Depending on the customer, the number of wires run to the location has been contingent on the need versus the cost. As a result, the use of a single or dual pair of wires has been the norm. More recently, the local providers have been installing a four-pair (eight-wire) connection to the customer location. This is because the customer (both business and residential) has begun to use voice lines, separate fax lines, and separate data communications hookups. Each of these requires a two-wire interface, so the need for multiple pairs has grown. It is far less expensive to install multiple pairs the first time than to install a single pair of wires every time the customer asks for a new service. So the topology is a dedicated local connection of one or more pairs from the telephone provider to the customer location. This is called a *star configuration*. The telephone company (whether ILEC or CLEC) connection to the customer originates from a centralized point called a central office (CO). In a star configuration, all wires home back to a centralized point, the CO.

Once the hundreds or thousands of wire pairs get to the CO, things change. The provider at that point might be using a different topology. Either a star configuration to a hierarchy of other locations in the network layout or a *ring* can be used. The ring is becoming a far more prevalent method of connection for the local telcos. Although we might also show the ring as a triangle, it is still a functional and logical ring. This star/ring combination constitutes the bulk of the networking topologies today.

At the local telephone company's (whether ILEC, PTT, or CLEC) office, the wires are terminated in what is called a *wire center*. The wire center is nothing more than a very large extension of the customer's hookup. Thousands of hookups come together in this centralized point. From the wire center, a series of spokes is run out to the customer, to other central offices, to higher-level offices in the hierarchy, or to any location desired. The wire center is also called a *frame*; all the wires are connected to the frame.

At this frame, a series of cross-connections is made. These will either be to other wires that go to other locations or to a switching center where the telephone company's central computer (in older offices this can be an electromechanical system) resides. This is called the *switch*. Most of the equipment today is a stored-program common-controlled computer system that just happens to process cross-connections for telephone calls. Remember one fundamental fact: the telephone network was designed to carry analog electrical signals across a pair of wires to recreate a voice conversation at both ends. This network was built to carry voice. Only recently have we been transmitting other forms of communication, such as facsimile, data, and video.

The switch makes routing decisions based on some parameter, such as the digits dialed by the customer. As these decisions are being made very quickly, a cross-connection is made in logic. This means that the switch sets up a logical connection to another set of wires. The connection can be back to the frame, where the wires serve a neighboring pair of wires connected to our next-door neighbor, or to another connection that links another central office. The possibilities are only limited by the physical arrangements in the office itself.

Between and among the offices built by the carriers (the local and the long-distance providers) is a set of connections usually laid out in a ring, but also possibly in a star configuration. These are called the facilities that carry traffic. These are called interoffice trunks. In some COs, these are copper wires; in others, they are fiber-optic wires; and in still others, they may be on radio systems.

Throughout this network, more or fewer connections are installed, depending on the anticipated calling patterns of the user population. Sometimes there are many connections among many offices. At other times, it can be simple and single connections.

Tied all together, then, is a series of local links to the customer locations, through a central office where switching and routing decisions are made, then on out to a myriad of other connections from telephone companies, long-distance suppliers, and other providers. This is the basis of the telephone network.

The Local Loop

Our interface to the telephone company network is the single-line telephone set covered in Chap. 4. It stands to reason that we need to connect this set to the telephone company central office (CO). The pair of twisted wires running from the telephone company's CO is called the local loop. Each subscriber, or customer, is delivered at least one pair of wires per telephone line.

There are exceptions to this rule in rural areas where the telephone company might share multiple users on a single pair of wires. This is called a *party line* and is again a financial decision. If the number of users demanding telephone service exceeds the number of pairs available, a telco might well offer the service on a party or shared set of wires.

The phone company distributes its outside plant, or distribution, to the customer by running large bundles of twisted pairs toward the customer location (Fig. 3.1). This is done using feeders, which are composed of 50 to over 3000 pairs of wires.

The feeders are run to splice points or breakout points called manholes or handholes. At this point, the splicing of two reels of cable will take place, assuming that the cable on a single reel was not sufficient. A lateral

Figure 3.1
The local loop.

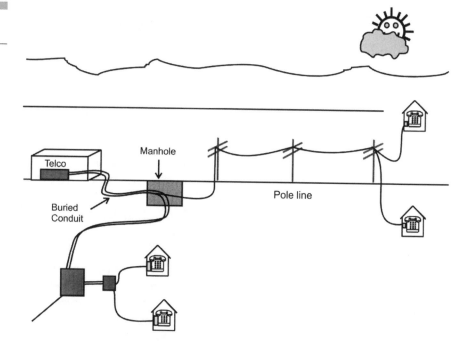

distribution can also take place here. Lateral distribution is the breakout of a number of pairs to run in a different direction. The lateral distribution or branch feeder is then strung to various customer locations. The end of the pair to the final customer location is called the customer pair or station drop.

It is in this outside plant, from the CO to the customer location, that 90 percent of all problems will occur. This is not to imply that the telco is doing a lousy job of delivering service to the customer. These cables are exposed more to cable cuts because of construction (commonly called backhoe fade), flooding at the splice locations, rodent damage, and many other risks.

In the analog dial-up telephone network, each pair of the local loop is designed to carry a single telephone call to service voice conversations. This is a proven technology that works for the most part and continues to get better as the technologies advance. The cables can be delivered via a telephone pole, buried conduit, or direct buried cables. Either way, the service is one that we are familiar with and feel comfortable with. What has just been described is the connection at the local portion of the network. From there the local connectivity must be extended out to other locations in and around a metropolitan area or across the country. The connections to other types of offices are then required.

The Network Hierarchy (Pre-1984)

Prior to 1984, most of the network was owned by AT&T and its local Bell operating telephone companies (BOCs). It evolved through a series of interconnections based on volumes of calls and growth. A layered hierarchy of office connections was designed around a five-level architecture. Each of these layers was designed around the concept of call completion. The offices are connected together with wires of various types, called trunks. These trunks can be twisted pairs of wire, coaxial cables (like the CATV wire), radio (such as microwave), or fiber optics. The trunks vary in their capacities, but generally high-usage trunks are used to connect between offices. Figure 3.2 shows the hierarchy prior to the divestiture of AT&T, with the five levels evident.

The class 5 office is the local exchange or end office. It delivers dial tone to the customer. The end office, also called a branch exchange, is the closest connection to the end user. Think of a tree: all the activity takes place at the ends of the branches, and the customers are the leaves hanging off the branches. Calls between exchanges in a geographical area are connected by direct trunks between two end offices. These are called interoffice trunks. There are over 19,000 end offices in the United States alone that provide basic dial tone services.*

The class 4 office is the toll center. A call going between two end offices that are not connected together will be routed to the class 4 office. The toll center is also used as the connection to the long-distance network for calls where added costs are incurred when a connection is made. This toll center may also be called a tandem office, meaning that calls have to pass through (or tandem through) this location to get somewhere else on the network. A basic arrangement of a tandem switching system is shown in Fig. 3.3. The tandem office usually does not provide dial tone services to the end user. However, this is a variable where a single office might provide various functions. The tandem office can also be just a toll-connecting arrangement that is a pass-through from various class 5 offices to the toll centers. Again, this varies depending on the arrangements made by the telephone providers. The ratio of toll centers that serve local long distance is approximately 9 to 1. Prior to divestiture, there were approximately 940 toll centers.

---

*Prior to the 1984 divestiture, there were almost 66,000 end offices in the United States. The equipment is smaller and can handle more connections now, so the providers have been able to consolidate into fewer locations, giving the opportunity to save money and real estate.

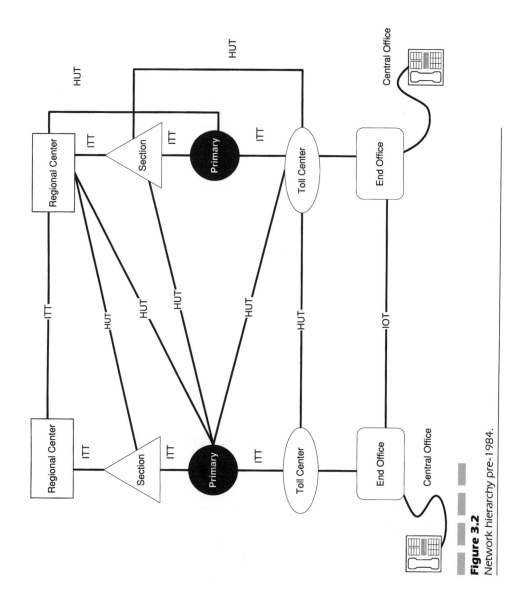

Figure 3.2
Network hierarchy pre-1984.

35

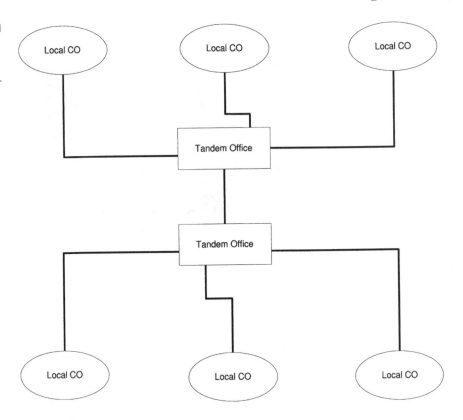

The class 3 office is the primary center. Calls destined within the same state area are passed from the local toll office to the primary center for completion. These locations are served with high-usage trunks that are used strictly for passing calls from one toll center to another. The primary centers never serve dial tones to an end user. The number of primary centers prior to divestiture was approximately 170, spread across the country among the various operating telephone companies (both Bell and independent operating companies).

The class 2 office is the sectional center. A sectional center is typically the main state switching system used for interstate toll connections designated for the processing of long-distance calls from section to section. There were approximately 50+ sectional centers before the divestiture of the Bell System. These offices did not serve any end users, but would serve between primary centers around the country. The class 1 office is the regional center. Ten regional centers existed across the country. The task of each center was the final setup of calls on a region-by-region basis. However, the regional centers constituted one of the most sophisticated computer systems in the world. The regional centers continually updated each other regarding the status of every circuit in the network. These centers were required to reroute traffic around trouble spots (e.g., failed equipment or circuits) and to keep each other informed at all times. As mentioned, this was all prior to the divestiture of the local Bell operating companies by AT&T. The number of regional centers has changed to seven in the AT&T network, and might be consolidated into four mega-centers in the future.

The Network Hierarchy (Post-1984)

After 1984 the network took a dramatic turn, with the separation of the Bell Operating Companies (BOCs) from AT&T. Many users screamed that things would fall apart, with service being affected. None of this came true, however. This doesn't mean that there was not a lot of confusion; there certainly was. However, things just kept humming along for the most part and calls were completed through this series of interconnecting points called the network.

The hierarchy of the network shown in Fig. 3.4 introduced a new set of terms and connections. The BOCs were classified the same way as independent telephone companies. They are all called local exchange carriers (LECs). The seven spin-off companies formed as a result of the divestiture became Regional Bell Holding Companies (RBHCs), which had regulated arms called the Regional Bell Operating Companies (RBOCs). Each RBOC had the Bell Operating Companies in its geographical area. Additionally, the RBHCs also had an unregulated side of the business, where they could enter into new ventures (such as equipment sales, finance, real estate, etc.).

Equal access, or the ability of every interexchange carrier (IEC or IXC) to connect to the Bell operating companies for long-distance service, became a reality. Equal access was designed to allow the same access to other long-distance competitors that AT&T had always enjoyed prior to divestiture. Prior to divestiture, a customer attempting to use an alternative long-distance supplier would have to dial a 7- or 10-digit telephone number to get to this supplier's switch. Then, when this connection was completed, a computer would answer the call and place a tone on the line. From there, the caller would have to enter a 7- to 11-digit authorization code. This code identified the caller by telephone number, caller name and address, and

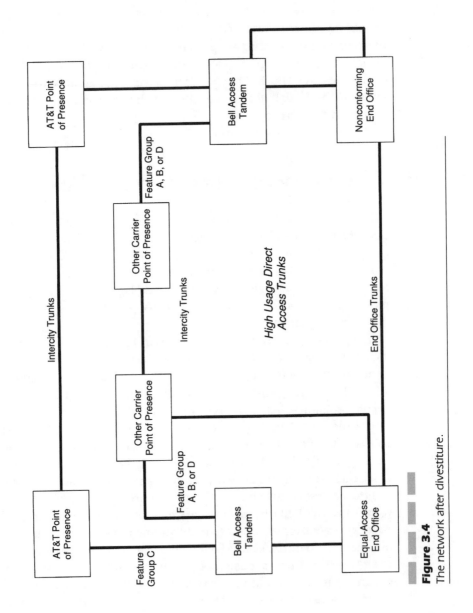

Figure 3.4
The network after divestiture.

billing arrangements. Only after the computer (switch) verified this information would it then send dial tone to the caller's ear. The caller would then have to dial the 10-digit telephone number of the requested party. This could involve very lengthy and frustrating call set-up times—especially when the called number was busy.

Users chose not to opt for these alternative carriers because of the time and the number of digits required and the frustration of attempts to call busy numbers. That is, unless the organization forced the user to dial across the carriers' networks. The reason for all of the digits was simple. The telephone company did not pass on the caller information to the alternate carrier (MCI, Sprint, ITT, etc.) that it passed to AT&T. Thus the choice of many callers was AT&T because it was simpler. Now the caller information is passed on in an equal basis, so the access is equal.

However, some of the independents and BOCs who haven't yet upgraded their offices do not connect to these IECs. These are called nonconforming end offices. The point of presence (POP) is the point where the IEC and the LEC are interconnected. This usually refers to the offices of AT&T, MCI, Sprint, and other carriers. This name is now starting to change to *point of interface* (POI) as opposed to POP. The main reason is that the term *POP* implies that a switching system is at that location, which is not always true. The point of interface is where the two parties, Bell or independent, and the interexchange carrier connect. This can be a closet in a basement or hotel, connected to a dedicated trunk to another part of the country. All of this should be transparent to the end user, however.

The Public-Switched Network

The U.S. public-switched network is the largest and the best in the world. Over the years, the network has penetrated to even the most remote locations around the country. Primary call-carrying capacity in the United States is through the public-switched network. Because this is the environment that AT&T and the Bell Operating Companies built, we still refer to it as the Bell System. However, as we've already seen, significant changes have taken place to change that environment.

The public network allows access to the end office, connects through the long-distance network, and delivers to the end point. This makes the cycle complete. Many companies use the switched network exclusively;

others have created variations depending on need, finances, and size. The goal of the network hierarchies is to complete the call in the least amount of time and over the shortest route possible. The network is dynamic enough, however, to pass the call along longer routes through the hierarchy to complete the call in the first attempt wherever possible.

The North American Numbering Plan

The network numbering plan was designed to allow for the quick and discreet connection to any phone in the country. The North American numbering plan, as it is called, works on a series of 10 numbers as shown in Table 3.1.

The Area Code

Note that there have been some changes in this numbering plan. When it originally was formulated, the telephone numbers were divided into three sets of digits. The first was a three-digit area code or numbering plan assignment (NPA). This started with a digit from 2 to 9 in the first slot of the sequence. In the second slot, the number was set as 0 or 1. In the third slot, it could be any digit from 1 to 0 (with 0 representing 10). The reasons behind this sequence were very clever. For example, the first digit did not use a 0 or 1 because these digits were used for access to operators (0) or operator services such as credit card calling, etc. The 1 was used to send a significant digit to the local switching office indicating that the call was long distance; this enabled the switch to immediately start setting up a toll connection to the toll center or tandem office. Thus the exclusion of 0 and 1 in the first digit of the area code facilitated the quicker call setup. In the second slot the digit was only a 1 or 0. This was used by the

TABLE 3.1	Timing	Area Code	Central Office Code	Station Subscriber Number
The North American Numbering Plan as It Evolved	Original	N 0/1 X	N N X	X X X X
	Pre-1995	N 0/1 X	N X X	X X X X
	Post-1995	N X X	N X X	X X X X

switching office equipment in a screening mode. As soon as the system sampled the second digit and saw a 0 or 1, it knew that this three-digit sequence was an area code. The third slot was any digit; having no special significance, it was just processed normally. Back in the early 1960s, we recognized that we were running out of area codes, given that there were only 160 available ($8 \times 2 \times 10 = 160$). In reality, only 152 were allowed for use by the various states because certain ones were allotted for special services (the N11 area codes, e.g., 211, 311, 411, etc., were always reserved). Very close administration of the area code assignments kept this scheme working until 1995, when the inevitable occurred. From the outset, the use of the entire numbering plan was limited, so it had to grow. In 1995, the area codes were expanded by allowing any digit in the center slot of the three-digit sequence. This expansion created 640 new area codes for the North American Numbering Plan (NANP). The use of these area codes has been so dramatic, there is speculation that even though we got a fourfold increase, these will be depleted by 2010 (or sooner!). Our current rate of consumption of telephone numbers is alarming. When we think back to the beginning days of the networks, customers had a single number that sufficed. Now the average household has three or more numbers (second line for fax, third line for modem, fourth line for teens, separate number for cell phones, etc.). We are facing a worldwide numbering shortage.

The Exchange Code

Following the area code is another three-digit sequence called the exchange code. This is a central office designator that lists the possible number of central office codes that can be used in each area code. The exchange code originally was set up in the sequence NNX, meaning that the first and second numbers used the digits 2 through 9, for the same reasons as in the area code. The digits 1 and 0 were reserved for operator and long-distance access and the 0/1 exclusion in the exchange code prevented this three-digit number from being confused with an area code. In the third number slot of the exchange code, any digit could be used. Clearly the greatest limitation in the exchange codes was that we would run out. With NNX we have 640 possible exchange numbers to use ($8 \times 8 \times 10 = 640$), but these were used very quickly.

In the late 1960s, we began planning the use of an exchange code numbering plan that changed the sequence to NXX, expanding the number of exchange codes in each of the area codes to 800. This added some relief to the numbering plan, but when using the *NXX* sequence the need arose for a forced 1 in advance of the 10-digit telephone number so that the call screening and number interpretation in a switch wouldn't get confused. This met with some resistance, but ultimately customers got used to the idea. The first two locations to use this revised numbering plan were Los Angeles and New York City back around early 1971.

The Subscriber Extension

The last sequence in the numbering plan is the subscriber extension number. This is a four-digit sequence that can use any digit in all of the slots, allotting 10,000 customer telephone numbers in each of the exchange codes. Because the four-digit sequence can be composed of any numbers, the intent is to give every subscriber his or her own unique telephone number. This hasn't changed as yet. However, the possibility that we can still run out of numbers always exists. Therefore, a couple of ideas have been bandied around: add two, three, or four more digits to the end of the telephone number or add one or two more digits to the area code or exchange code.

In either method, users will be asked to dial more digits, an idea that is never popular but that might become a necessity. However, this is far more complicated than just adding a few more digits here and there. The whole world will be impacted by any such decision, and the length of time required to implement such a global change will be extraordinary.

Currently in North America, several new area and exchange codes have been installed where only one existed. This forces users in the area to dial 11 digits (1 plus the 10 digit telephone number) for local, as well as long-distance, calls. This plan has met quite a bit of resistance, because people do not like to dial that many digits, and they have trouble remembering that many numbers. However, regardless of their feelings on the subject, there are not a lot of choices.

Private Networks

Many companies, depending on their size and need, create or build their own networks. These networks are usually justified on the basis of cost, availability of lines and facilities, and special need. Often these networks employ a mix of technologies, such as private microwave, satellite communications, fiber optics, and infrared transmission. The initial and ongoing costs of private ownership can be quite high (Fig. 3.5). Because the costs for a private network must be borne by the end user, often the telecommunications department managers are under pressure to cross-subsidize the costs in other ways. Many companies with private networks have been subjected to criticisms from end users that the costs are higher than those of the public-switched network. Individual sites or departments, therefore, begin reducing their own use of the private facilities. This increases the burden on remaining users, who then foot the bill for lightly used networks.

Still others who have built their own networks have stated that they are amassing huge savings. When placed in a position of defending their networks, they can produce statistics of the savings reaped on a company-wide basis. If, however, they have a problem with underutilization, they can sell off some of their excess capacity.

This gives these organizations a little breathing room. Collections, maintenance, and administration costs tend to increase when reselling takes place. There is no single best answer here; each organization must look at its needs, availability, costs, and services before deciding which network to use.

Hybrid Networks

Some companies have to wrestle with the decision of whether to use a private or a public-switched network. This decision is not an easy one because often the numbers do not play out well. The return on the investment just will not be there.

As a result, these organizations use a mix of services based on both private and public networks (Fig. 3.6). The high-end usage services (heavy volumes between two or more major company locations) will be connected via private facilities; the lower-volume locations will utilize the switched network. This usually works out better financially for the organization, because the costs can be fully justified on a location-by-location basis. Pressure to install private line facilities comes from the integration of voice, data, video, graphics, and facsimile transmissions. Only by combining these services across common circuitry will many organizations realize a true savings.

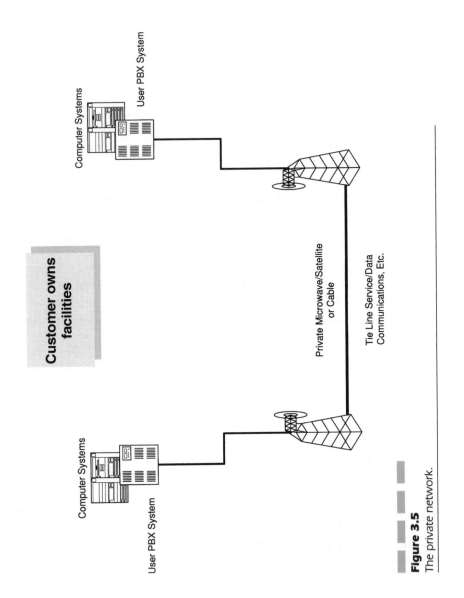

Figure 3.5
The private network.

Computer Systems

User PBX System

Customer owns facilities

Private Microwave/Satellite or Cable

Tie Line Service/Data Communications, Etc.

Computer Systems

User PBX System

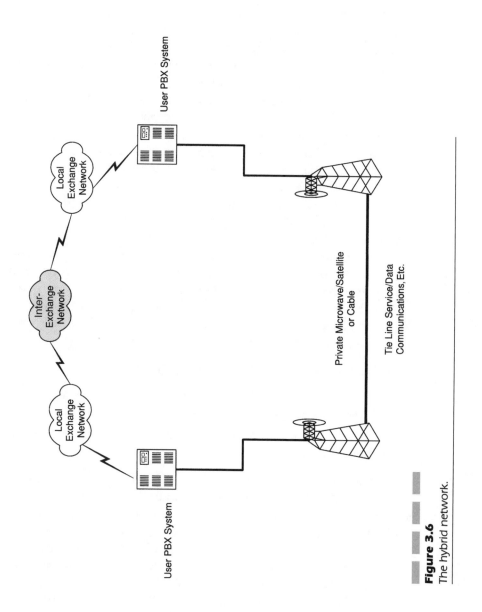

Figure 3.6
The hybrid network.

Local Access and Transport Areas (LATAs)

Local access and transport area is a term introduced with the divestiture agreement in 1984. One of the problems facing the court system was dealing with the long-distance versus local calling areas. It was a revenue-sharing concern more than anything else. AT&T forced the issue by demanding that some form of revenue sharing be put in place so that a single LEC would not have the option of handling all calls and cutting out the IECs.

To solve this problem, an agreement was reached that stated the LECs would still maintain a monopoly on local dial tone and local calling. They would be restricted from carrying long-distance traffic, which would fall under the domain of the IECs. However, many states cover very large geographical areas, and calling from one end to another would be considered long distance. The country was broken down into 195 separate bounded areas for local calling (some local tolls are allowed) based on the density of population in each area. This again was financially motivated. The interconnection between two LATAs would be done by the IECs. A whole new set of acronyms emerged as a result of the divestiture agreement; LATA is only one of them. To complicate things even more, there are four types of calling under the LATA concept:

- Intrastate—Intra-LATA belongs to the LECs
- Intrastate—Inter-LATA belongs to the IECs
- Interstate—Inter-LATA belongs to the IECs
- Interstate—Intra-LATA can be either/or, but was originally given to the LEC

The result of this is a very confused public, and sometimes very confusing tariffs. For example, a call from Philadelphia to Los Angeles can be carried on the IEC network for as little as $0.10 per minute, yet a call from Philadelphia to Wilmington can cost as much as $0.40 per minute. The difference here is that the 3000-mile call is regulated on the long-distance basis and FCC jurisdiction. However, the 29-mile call between Philadelphia and Delaware is regulated by BOC tariff and the PUCs. These anomalies will change and are doing so now, but this is the confusion that can crop up when various players are trying to protect their interests.

After the Telecommunications Act of 1996 in the United States, this LATA boundary began to erode quickly. The local telephone companies and competitive local exchange carriers (CLECs) are all extending their reach in the local areas. At the same time, the IECs are offering services intrastate and intra-LATA. Many of the IECs are registered as CLECs now, offering a variety of services that were heretofore unavailable from the single source.

Wiring Connections: Hooking Things Up

The telco uses a variety of connections to bring the service to the customer locations. The typical connection is the two-wire service that we keep talking about. This two-wire interface to the network is terminated in a demarcation point, as required by law. The DEMARC is the point of least penetration into the customer's premises, typically within 12 inches of where the telco cable comes up into the building. Normally, telco terminates in a block; this can be the standard modular block for a single-line telephone. If the customer has multiple lines, telco will terminate in a 66 block, or an RJ-21X. These are fancy names for their termination points. The typical modular connector uses an RJ-11C for telephones connected to a two-pair interface (not to be confused with the two wires) or an RJ-45X as a four-pair interface for both voice and data. Another version of connector for digital service is an eight-conductor (four-pair) called the RJ-48X.

When a telco brings in a digital circuit, it will terminate the four-wire circuit into a newer RJ-68 or a smart jack. There is no major mystique in any of these connectors. The number is strictly a uniform service code so that the telco can keep it all straight. However, when ordering a circuit, the telco will ask you how you want it terminated. The rule of thumb in a multiline environment is to use the RJ-21X (which is a 66 block with an amphenol connector on it). Sounds complex, doesn't it? A single line will terminate in an RJ-11C or RJ-12.

Types of Communications

The next area is the directional nature of your communications channel. Three basic forms of communications channels can be selected. These are as follows.

One-Way (Simplex)

This is a service that is one way and only one way. You can use it to either transmit or to receive. This is not a common channel for telephony (voice) because the occasions when we speak and everyone else only listens are very rare. Feedback, one of the capabilities we prize in our communications, would be eliminated in a one-way conversation. Broadcast television is an example of simplex communications.

Two-Way Alternating (Half Duplex)

This is the normal channel we use in a conversation. We speak to a listener, then we listen while someone else speaks. The telephone conversations we engage in are normally half duplex. Although the line is capable of handling a transmission in each direction, the human brain can't deal well with simultaneous transmit and receive.

Two-Way Simultaneous (Duplex or Full Duplex)

This is used in data communications, where a device can be sending to a computer and receiving from the computer at the same time. The direction of the information can be at differing speeds, such as 1200 bits/s toward the computer from the keyboard (humans cannot type much faster than that) and 9600 bits/s from the computer. This can also be done on a conversational path, when two people try to overtalk each other.

Equipment

Equipment in the telephony and telecommunications business is highly varied and complex. The mix of goods and services is as large as the human imagination. However, the standard types are the ones that constitute the end points on the network. The computerization of our equipment over the years has led to significant variations, but to improvements. The devices that hook up to the network are covered in various other chapters, but a summary of some of these connections and their functions in the network gives the following list:

- The single line set
- The multiline set (called a key set)
- The PBX (private branch exchange)
- The modem (data communications device)
- The multiplexer (allows more users on a single line)
- ACD (automatic call distributor)
- VMS (voice mail system)
- AA (automated attendant)
- Radio systems
- Cellular telephones
- Facsimile machines
- Centrex (central office exchange)

This is a sampling of the types of equipment and services you will encounter in dealing with vendors in this business.

The Telephone Set

Most people take the telephone set for granted. How can this device that we have known and used since we were old enough to stand up be worth any form of mention? Why should we even care? This attitude is not uncommon. Since the early days of the telephone network installation, users have accepted the device as the norm. There was no real concern. We had the device, and all we had to do was pick it up and use it. It always worked!

For the most part this is true. However, from time to time telephones do break, and when they do we feel lost. We do not have the foggiest idea of what went wrong or what to do to fix it. Yet the mystique is not just a carryover from the good old days when the telephone set was sturdy, dependable, and reliable. It seems that we were always told, prior to divestiture, that we were not allowed even to take this set apart. Because we did not own it, we were not allowed to tinker with it. The poor user who might have taken the set apart and not known how to reassemble the set would rue the day that he or she had to call the local telephone company and admit to committing the worst of all possible telephone crimes. Users would conjure up a vision of being put away for life, or, worse yet, losing our privileges to use the telephone network. Telco would turn off our dial tone, take the set away, and leave us without a means of communicating. How awful!

This scenario might sound far-fetched to many novices in the industry, but surely the old-timers around can remember the stories well. The authors (not to imply that they are old-timers; that's a matter of mind) have heard of many similar cases. One story goes like this:

> Back several years, a farmer in St. Louis, MO wanted to know just what was inside the telephone set. So one night he decided to check it out for himself. Family members, realizing his plans, recognized that this was a major problem. They knew from the local telco that you were not allowed to open the cover of the set—to do so would violate all warranty and utility commission rules. They couldn't bear the thought of having no telephone in their household. They begged and pleaded with the farmer to let it go, not to fall prey to the mystique of the set.
>
> Unfortunately, the farmer would not be daunted by fear, uncertainty, and doubt. The decision was made. As a means of acquiescing to his family, he agreed to do his investigating after dark, when no one could see from outdoors. That night the farmer covered all of the windows of the farmhouse with blankets and proceeded to

quench his thirst for knowledge about the telephone set. As he opened the set, he found a nest of wires running from here to there. There was no magic to be found in that. Not to be dissuaded, he quickly screwed the set back together, then proceeded to unscrew the mouthpiece from the handset. Alas, when he untwisted the mouthpiece, it came off the handset. When the set came apart, a number of small black particles fell out of the mouthpiece and rolled around on the floor. Before the farmer was able to catch these little black pellets, they rolled down between the slots on the floorboards—back then the floorboards were installed with spacers in them, leaving large cracks between the boards. Well, as one might imagine, these pieces of whatever-they-were fell through the floorboards and were lost forever. Now the farmer was in a spot. What could they have been? Why were they inside the telephone set and why were they loose? Regardless of the answer, another thought loomed up in the farmer's mind. The dreaded thought of having to call the local telco and ask them to send someone out to fix the set wrought fear in the deepest crevices of the farmer's mind. How would he explain that he was merely inquisitive? What would they do? Would he lose his telephone privileges? Would they take away his dial tone and telephone set? For how long?

Being a quick study, the farmer recognized that the pellets, whatever they were, resembled something he had seen before.... Aha! He remembered that this stuff looked just like gunpowder. The problem was solved. He would merely repack the mouthpiece of his set with the granules of gunpowder and be done with it. So he did.... We heard that his next phone call was a real blast!

All kidding aside, this scenario is probably unfounded. But certain issues were brought up here to make us stop and think. When the original telephone network was installed under the monopoly and controlled by AT&T and its Bell Operating Companies (the independents were not as severe in their treatment), we were left with the fear that if we tinkered with the telephone set and broke it, we would lose our phone privileges. Can you imagine that? Of course, since the breakup of the Bell System, things have changed considerably. The set now belongs to the end user. We can do anything we want with it or to it. If we break it, we buy a new one. Simple?

Despite this silly illustration, there must be some significance to just what this set is all about.

The Function of the Telephone Set

Chapter 2 covered the characteristics of voice. We stated that the voice can create sound wave changes at a rate ranging from 100 to 5000 times per second, and that amplitudes and frequencies are constantly changing. From this understanding of how the human voice works, we then must understand that the network was built to carry voice. How then does the voice get put onto a telephone company network and carried to the other end of the line? The sound waves must travel down the one pair of wires that the telephone company has provided. The challenge is to take sound and convert it into something that the telephone network deals with: electricity. Yes, electricity is what runs across copper wires. The telephone company delivers a pair of copper wires to our door and then attaches a telephone set to those wires. So, the function of the telephone set is to convert sound into electricity. It will therefore act as a change agent (Fig. 4.1).

Conversely, when a voice is sent through the wires as electricity, the process must be reversed at the receiving end. Therefore the telephone set must also function to convert this electricity back into sound. The change process must work in two ways. The "from" and "to" functions are what the set must accomplish. Figure 4.2 is a representation of the reverse process.

But wait, there's more. On the top of the telephone set, there are two little buttons that are spring-loaded. They must have a function too. The switchhook, as these buttons are called, is a mechanism that originates and finalizes calls. When you press the springy buttons on the top of the set, the dial tone goes away. If you lift the handset off the base of the telephone, you hear the dial tone (this is called *going off hook*). So, these buttons "request" and "relinquish" the dial tone from the telephone company.

By now, you will notice that the telephone is a very complex piece of equipment. The invention of Alexander G. Bell has come a long way from

Figure 4.1
The telephone set converts sound into electricity.

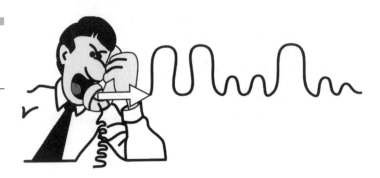

Figure 4.2

The set also converts electricity back into sound.

its founding. But functionally it still does the same thing. The first telephone sets had no dial pads; all calls were handled by a telephone company operator. In order to make a call, you would have to "flash" the operator and request the called party. The next evolution was a crank phone: you lifted the handset and turned a mechanical crank, which sent a ringing voltage down to the telephone company operator (Fig. 4.3). The operator still had to place the call, but the crank was a means of signaling the operator when you wanted service. Other variations appeared as time went on. This was inefficient, so the manual call placement was eventually replaced with the rotary telephone set. Later came the Touch-Tone

Figure 4.3

The crank phone allowed the user to generate a signal to the operator by turning the crank.

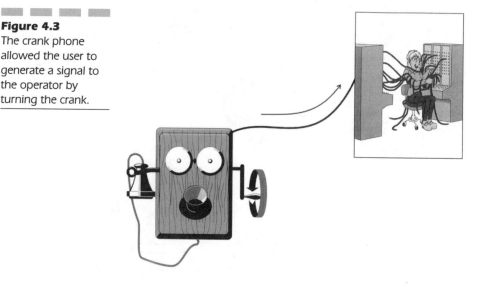

set, which was designed to improve efficiency and speed the process. Still later, the digital telset created a whole new wave of interfaces to the network. Figure 4.4 shows a group of variations on the telephone set. The set's evolution has added features and functions, but the primary function is still the same after these 120+ years. So, let's back up a little and look at the pieces of a telephone set, at least generically. The telephone set consists of multiple pieces, each of which serves at least one function. Without one or more of the pieces, communication might not take place. They are all intertwined to work together so that we have constant service at any time.

The Pieces

The telephone set is made up of the following major components.

The Base

All of the components are either housed inside or attached to the base (Fig. 4.5). It is typically a plastic or space-age healed plastic case that will withstand many impacts, such as being dropped from a table, desk, etc.

Figure 4.4
A potpourri of telephones has evolved.

Figure 4.5
The base of the set houses a far more complex system.

This case protects the inner workings and other components from exposure to the most dangerous part of a telephone network: humans. We tend to spill things on the set. We also drop it or knock it over. When we are frustrated on the phone, we slam it down or beat it on the desk, taking out our anger and irritation with other humans on this device that is always supposed to work. Without the casing to protect the telephone, we would never have service. This casing also protects the inside of the phone from dust and other elements that are floating around in the air.

The Handset

The handset (Fig. 4.6) houses two separate components, the transmitter and the receiver. The handset is ergonomically designed to fit the distance

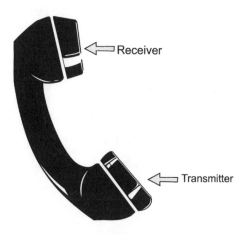

Figure 4.6
The handset houses the transmitter and receiver.

Receiver

Transmitter

from the mouth to the ear. The transmitter is located at the bottom portion of the handset, so that it is positioned in front of the mouth when we hold the handset to our face. The receiver is located at the top of the handset so that it rests against our outer ear when we hold the handset to our face. The first handsets consisted only of an ear mechanism; the transmitter was a stationary mount on the front of the base (Fig. 4.7).

Now to the inner pieces of the handset. Figure 4.8 shows the handset with the transmitter and receiver indicated. Note that the wires also connect the pieces. There are two wires running from the earpiece and two wires running from the mouthpiece. Remember that the telephone is going to convert sound into electricity. The wires carry the electricity created at the mouthpiece to the telephone set. The second set of wires carries electricity from the telephone set to the earpiece. So, there is a four-wire circuit in the handset: two wires for the transmitter and two wires for the receiver. Each side of the function has its own two-wire electrical circuit at this point. Modern-day functions of the following parts are still the same. Although newer solid-state components have replaced the pieces, they perform the same functions as the pieces described in the following paragraphs.

The Transmitter

As mentioned in the discussion about the farmer wanting to see the inside of the telephone set, the transmitter houses some very sophisticated components. The first portion of the transmitter houses a diaphragm just under the outer casing of the mouthpiece. The diaphragm is a very sensitive membrane that will vibrate with air pressure changes. When we speak into the mouthpiece of the telephone set, we are really creating air pressure

Figure 4.7
Older models had a stationary mouthpiece but a movable earpiece.

Earpiece
(movable)

Mouthpiece
(stationary)

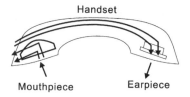

Figure 4.8
The transmitter and receiver are connected with two wires each inside the handset.

changes with our vocal cords. The sound pressure we create causes air pressure changes that move at the same frequency as the changes we produce by vibrating our vocal chords. The pressure changes cause the membrane on the diaphragm to move back and forth at the same frequency (Fig. 4.9). Behind the diaphragm are loosely packed carbon particles (the black pellets the farmer dropped). When the diaphragm moves back and forth, it causes the carbon particles to move back and forth. This produces an electrical charge that occurs at the same frequency and amplitude as the sound pressure changes created by the voice. The result is an analog sine wave that is directly (mostly) proportional to the frequency and amplitude of the voice, or the electrical equivalent of sound. This electrical energy is carried away from the transmitter across the wires behind the mouthpiece.

The Receiver

The receiver is shown in Fig. 4.10. When voice or other telecommunication in the form of electrical energy is being transmitted to us from the far end (the other party on the phone call), the network carries this electricity to the telephone set. The wires behind the receiver are run up to an electromagnet that is mounted inside the receiver portion of the handset. The electromagnet is attached to a diaphragm similar to that in the mouthpiece. The electricity comes through the wires to the magnet, and the magnet will vibrate at the same frequency as the received energy. As the magnet vibrates, it causes the membrane on the diaphragm to move back and forth. This back-and-forth motion causes air pressure changes

Figure 4.9
The transmitter had a diaphragm and loosely packed carbon particles that create the electricity.

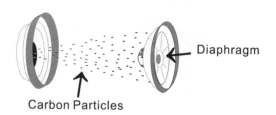

Figure 4.10
The receiver uses an
electromagnet to
change the electricity
back into air pressure.

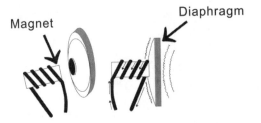

Figure 4.10
The receiver uses an
electromagnet to
change the electricity
back into air pressure.

(vibrations) that are at the same frequency as the transmitted energy. As the air pressure changes occur, they cause the air waves to bounce off the inner eardrum, which is what produces sound to our ears. See Fig. 4.11 for the eardrum comparison.

The transmitter and receiver complete the conversion of sound to electricity and electricity back to sound. We often refer to the handset functions as E&M, or ear-and-mouth communications process. The E represents the direction of flow to the ear or to the receiver, and the M represents the direction of energy from the mouth or the transmitter. In actuality, E&M means something different, but that is for a later discussion, possibly the sequel to this book.

The Connector or the Handset Cord

Coming from the handset and proceeding to the base of the telset is a coiled wire. It doesn't have to be coiled; that just makes it easier to manage. Inside the curly wire are four wires: two from the transmitter and the receiver, respectively. The modern sets use a connector that is called an RJ-11C. This is the four-wire connector that has a little spring-loaded clip on it to plug and unplug the connector into the female receptacle. The cord is double-ended with male RJ-11C connectors. This is also a matter

Figure 4.11
The eardrum receives
air pressure and
converts it into
sound.

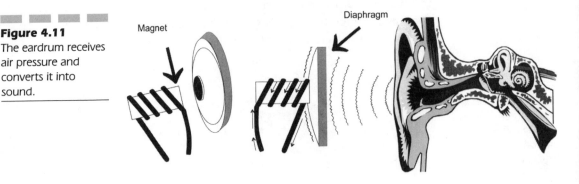

of convenience: because the cords and the connectors are the most likely pieces to wear out, it makes sense to use a plug-and-play module. Older versions of this cord used a screw-down version (called a spade lug), but this required that the telephone set be opened up to get at the screw mounts. Remember, this was an area that we were not allowed into.

A funny thing happens along this connector cord, however. When it gets to the base of the telephone set, it gets plugged into the female version of the RJ-11. But inside this jack, a hybrid arrangement takes place. One of the transmitter and one of the receiver wires are connected to each other (Fig. 4.12). The other two wires that continue into the set are covered in the next section. The important thing here is to understand why the transmitter wire gets connected to the receiver wire. This creates a loopback arrangement of sorts. The transmitted sound is looped right back to the ear. We can hear everything we say in the conversation because the energy is transmitted back to the earpiece. You are probably asking yourself why this is done. The process produces what is called *side tone*. This side tone is a means of letting us know that an electrical connection across the network still exists. When you speak into a telephone during a conversation and you hear yourself, you recognize that there is still a connection to the distant end. You can also regulate your volume when you hear yourself speak. When you speak into the set and you hear nothing or the sound is dead, you recognize that there is no one out there—in other words, that you have been cut off. This is part of your basic training in telephone usage. Although no one ever explained this to you, it was learned through experience.

Figure 4.12
As the four wires enter the set, they go into a hybrid arrangement.

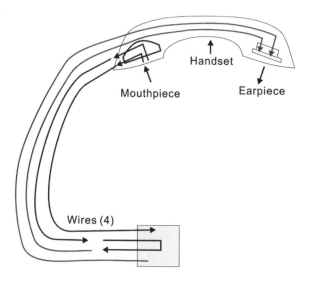

Handset

Mouthpiece

Earpiece

Wires (4)

The Inside of the Telephone

Once the wires are run from the handset to the base of the telephone, two wires go into the set. Actually, these wires meet up with a whole network of cross-connects (connections from place to place) inside the set. But we are concerned with the two wires. As shown in Fig. 4.13, one of the wires, the transmitter wire, is connected inside the set to the dial pad (whether rotary or tone dial). This is used as an addressing mechanism. The telephone network uses a decimal numbering system in the form of a sequence of digits that we dial to the called party. The dial pad is used to send these digits out to the telephone company for delivery to the called party. This wire connected to the dial pad is also cross-connected to the switchhook (the spring-loaded buttons). The same wire is used to connect to another jack (RJ-11, RJ-12, or other), which is connected in turn to the outside world.

The second wire from the handset cord is connected through the network of wires to a ringer. The ringer can be in the form of a bell or a *turkey warbler* (electronic ringer), but the function is what is important here. The ringer is a receiver wire function. It is used to alert you that someone wishes to speak to you. If you did not have a ringer, you'd have to pick up the handset every now and then and listen to see whether someone was on the line. This would be inconvenient for you, not to mention for the caller, who would never know whether you would pick up the handset. Callers would be in limbo forever. So the ringer is a crucial component of the set and the process.

Figure 4.13
The transmitter wire is connected to the dial pad and the switchhook. The receiver wire connects to the ringer.

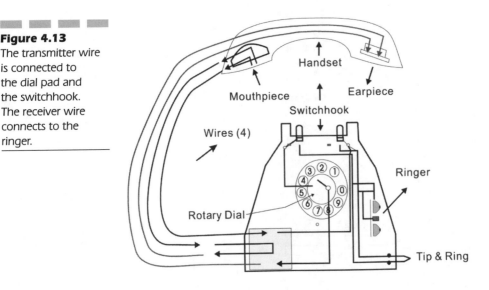

The Switchhook

Already mentioned and equally important is the switchhook function. This is the portion of the set that actually gets the dial tone from the telephone company or, when we are through with the call, gives the dial tone back. In reality, what happens is that an electrical wiring circuit is physically connected from the telephone company office to our telephone set. This circuit gives us access to the services offered by the telephone company, mainly the dial tone and the ability to make and receive calls. Under normal conditions, we do not have a dial tone at our set. This is a service that we must request. To request the service, we lift the handset off the switchhook (called "going off hook"). When we do this, the little buttons pop up. At the same time, they "make" the circuit by adding a resistance across the wires. The telephone company equipment constantly monitors all of its connections to end users. When the central office computer (now they are computers) sees a change in the resistance across the wires, it recognizes this as a request for dial tone. If dial tone is available, it will be provided immediately. In the time it takes us to go off hook and bring the handset up to our ear, the dial tone is there. Actually, it gets there much faster than that if everything is functioning properly. As we use the line and make calls, we will use the other components in the telephone set, but the central office computer will monitor the line (wires) for the time when we are finished with our calls. When we disconnect, we place the handset back into the cradle. This pushes the spring-loaded buttons back down, breaking the circuit. The central office computer sees this change on the wires and recognizes that we are done. Therefore the central office computer takes the dial tone away from us and provides it to other users who require it. Then the process starts all over again. The phone company uses a 10:1 ratio of users to dial tone. For every 10 users, the company knows that statistically only 1 user requires service.

The Dial Pad

It really does not matter whether you have a newer set or older one. The dial pad uses the set to address the network. When we have our dial tone, we then start a sequence of dialed digits indicating to the central office with whom we wish to speak. As mentioned, the first telephone sets did not have a dial pad. Everything was manually performed through the telephone company operators. If you needed to speak to someone, you got the operator by going off hook. The off-hook process created the "make" on the circuit that lit a light or created a visual indication to the telephone company operator. The operator would answer and converse with you. At this point, you would pass on the name or number of the requested party and the operator would provide the connection. It was quite simple, but as things got busier, the human involvement became an issue. You could be on the line waiting for an operator forever if things were busy at the central office.

The rotary telephone set was the first attempt to automate the process (Fig. 4.14). Initially it was well received; however, over time it became a nuisance. The rotary set is an electromechanical device. As you dialed the number, you would rotate the dial in a clockwise direction. When the dial was released, the counterclockwise return to the normal position created a series of electrical interrupts or pulses across the line to the telephone company office. The CO would interpret these pulses and make the necessary routing decisions. The main problem that led to frustrations was that statistically four attempts were necessary to make a call and speak to the intended party. Either the network was busy, the called party was busy, the phone at the receiving end went unanswered, or some other event got in the way. Thus users felt frustration with this set, especially because it took so long to dial a number.

The pulses are generated at 10 per second (10 PPS), which also created long call setup times because of the slowness of the dial pad. Additionally, because it was electromechanical, the system was prone to failures. The rotary set might stick or misdial the digits, or otherwise thwart the user. Users began to complain, and the manufacturers had to respond. The primary manufacturer at the time was Western Electric, a division of AT&T.

Why even bother to mention a set that is so old? At least 35% of U.S. telephones still use pulse dialing. There are many reasons for this, but some of the major ones are:

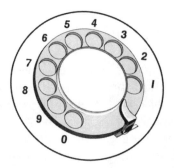

Figure 4.14
The rotary dial pad.

- *Cost.* The cost of Touch-Tone dialing is typically $1.00 more per month. Many people who make calls only infrequently don't want to pay for a service they seldom use.

- *Technology.* Many independent LECs have old technology and have not yet planned to upgrade their equipment. Therefore the service is only rotary.

- *Gimmes.* A lot of folks who now own their own phones have subscribed to magazines and received free phones as a reward for signing up. These phones might have push buttons on the set. However, during dialing, the conversion from tone to pulse occurs inside the telephone set. The distinctive click-click in the ear identifies such phones as rotary sets.

The Touch-Tone phone was the second edition of the dial pad. This changed the look and feel of the dialing process. Instead of using the rotary dial to electromechanically produce the digits, this phone created by AT&T's Western Electric Division employed a tone dialing sequence. Using a group of tones that fall into the spectrum of voice frequency, the tone dialer has a set of buttons that send out frequency-based tones instead of electrical impulses. The Touch-Tone telephone set was created to speed up dialing and call processing, reduce user frustration, increase productivity, and allow for additional services through the dial pad. The touch tones are actually a dual-tone multifrequency (DTMF) system. Each of the numbers represented on the dial pad has both an X and Y matrix or cross-point. This was a decision that Western Electric made to create a number of tone generators that could be used at the set and tone receivers to be used at the central office. Western Electric did not want to have too many of these tone generators/receivers, because they are expensive. Therefore, under a dual-tone arrangement, they could get away with using only half of the generators needed to represent the same number of digits. See Fig. 4.15 for the tone dial pad and note the frequencies as they appear on the chart. When you press a number, two separate tones are generated and sent to the CO simultaneously. Each number on the dial pad has its own distinct set of dual tones, so the system recognizes each as a discrete number. The reasoning behind the dual tones is that the human voice could not generate a harmonic shift in frequency quickly enough to recreate these tones. Therefore the telecommunications systems can recognize the tone shifts as dialed digits instead of conversational path information (voice). For the most part this works; however, from time to time humans can generate these quick harmonics—it is rare, but it can be done. In particular, a female voice generating more amplitude shifts will generate a quick shift in vocal pattern and create the digit 8, whereas the male voice will create the digit 7. This of course is rare, but it does happen and is called "talk off". This is important when we discuss later evolutions, such as voice messaging systems, where the digit 7 or the digit 8 means something to a computer system when a voice is being used. An example would be the case where the male voice pattern generates the digit 7 to a voice messaging system and the digit 7 means *delete*. We must be aware that if we leave a message for someone on a voice messaging system and a reply never arrives, it may well have been that we deleted the message before sending it. This "talk off" arrangement can be controlled through the use of voice bandpass filters that can be acquired from the voice messaging system supplier.

Tone dialing is faster than rotary because the tones are generated at 23 pps, much quicker than the rotary set.

As the tone dial was introduced, a 12-button pad was created, adding two extra numbers to the telephone. The older rotary set only had 10 numbers (decimal). The two new buttons allowed for new features and capabilities to be introduced in the telephone world, using the pound (#) and star (∗) keys as delimiters when dialing into computers, etc. The extra keys allow for feature activation in Centrex and PBX-type systems (for example: to put a call on hold dial #4; to retrieve the call dial ∗4). These extras also squared off the dial pad to make it easier to form.

Tone dialing also helps to get the call through the network more quickly because the tones are sent at a much faster rate than that of rotary pulsing. This improves network performance because the network processes the call setup and teardown faster, creates happier users because they can dial their numbers much more quickly rather than with the tedious rotary dial sequence, and promotes overall satisfaction.

Figure 4.15
The Touch-Tone
phone.

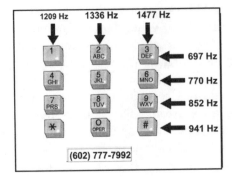

One example of improvements is with the call setup time. Setup for a rotary call was 43 seconds, whereas with tone dialing the call can be set up in approximately 6 to 10 seconds. Other improvements, such as improved signaling systems, have reduced this setup time further, but the tones were the first big thrust in gaining time back from the network. Is this important? Sure it is! Think about the hundreds of millions of calls placed in any given month. If you had a system that took 43 seconds of call setup (the older system not only took that long, but tied up a line across the network for the same setup period) and you could reduce it to 10 seconds, you have reclaimed 33 seconds of line utilization (which was not billable time) per call for other users. This is big bucks no matter how you look at it. There is a paradox in this situation, however.

Think about this for just a second: you as a user of the telephone network have a rotary dial set. You want to place a call from Boston to San Francisco (approximately 3000 miles). You place the call and tie up 3000 miles of wire for as much as 43 seconds. This is nonbillable by the carrier. The clock only starts when the call gets answered. So, now the carriers offer you the ability to speed up the process by using a tone dialer. You are obviously going to be happy with this arrangement because you can dial so much faster. The carrier stands to save 33 seconds (remember back then the cost per minute was $0.65) and get the call billing going that much faster. For this savings of approximately $0.30 to $0.35 per call that you make (which all rolls back into the carriers' profits), you have the privilege of paying $0.75 to $1.00 per month more on your phone bill for the convenience of tone dialing. Isn't this a wonderful industry?

The Ringer

Already discussed, the ringer serves the purpose of alerting the called party that someone wishes to converse. The ringer has no mystique other than it must be "told" via an electrical current to ring. The current from the central equipment out at the telephone company office is sent to a ringing generator that will produce the current across the line. A -48-V direct current from a battery in the office produces an 80-V alternating current at 20 Hz output across the telephone line to the customer location. This is as simple as can be. Occasionally, differences appear in the way these things are laid out or the voltages that are applied to the line, but we are using norms in our examples.

Tip and Ring

Tip and ring is a phrase that telco installers or repairers use all of the time. They love to talk about tip and ring. This is a description of the two wires that are connected to the telephone set from the outside world. You will recall that the four wires that left the handset went through a hybrid arrangement as they came into the telephone set. At this entrance point, only two wires were carried into the set to connect to the signaling systems inside the phone. The two wires, one from the transmitter and one from the receiver, now exit the set and proceed toward the telephone company connection. They call the transmitter wire the "tip" and the receiver wire the "ring." Tip and ring is better described as the transmitter and receiver wires, but telephone company personnel held the words over from a bygone era. Tip and ring comes from the use of the old operator cord boards. When the operator tested the line, the tip of an RCA jack was used. If a call was already in motion on a particular line, the operator would hear static on the line when "tipping." If the line was not in use, the operator would not get any static from the tipping; the operator would then insert the jack on a cord board into the designated extension port on the board. This insertion would then cause a ringing voltage to be generated across the wires and your phone to ring. This system has all but disappeared here in the United States, but in several other parts of the world cord boards still exist. That's the main reason for the mention of where this all came from.

Newer Sets

The telephone is a remarkable piece of equipment that evolved over time. Conceptually nothing has really changed, but the pieces continue to advance. Today's digital telephone sets provide far more than just a plain old analog telephone call. The older set functionally achieved interface to the network provided by the various carriers. This interface, with all of its electrical and mechanical components, got us to where we are today. However, progress continues. Users wanted more features and functions from the devices placed on the desk or wall in offices and residences. Consequently, the migration to solid-state, microprocessor-controlled telephone sets became the wave of the future. The newer sets are digital. That means that they perform the same functions with less voltage, very discrete events are preprogrammed into them, and other characteristics of digital transmission will add to the quality of conversation. Digital transmission is covered in later chapters (Chap. 5 discusses the differences between analog and digital, and Chap. 14 discusses the T1 digital transport system).

The telephone set was a clear target for digital technology. The newer sets are varied by the number of manufacturers. Anyone can now manufacture a telephone set, or system for that matter, so long as they abide by certain registration techniques and parameters. A newer set is shown in Fig. 4.16. On this set, several added features that enhance the call placement process are incorporated into the telephone set rather than being served by an external device. For example, some of the features that are now becoming the norm are as follows.

Speed Dialing

A user can preprogram from 1 to 100 numbers into the memory of the set. When you use a feature button or a preprogrammed button on the set, the telephone will dial a much longer string of digits (usually up to 33 digits) and connect you with the requested party.

Call Hold

A hold function will place the handset in the mute mode: both the transmitter and the receiver function are disengaged for a period of time. When you use this hold feature, a visual indicator will alert you that you have done something. In the case of hold, a flashing light will let you know that the call is still there but on what is called *hard hold*. This serves

Figure 4.16
Newer digital sets are feature-rich but still perform the same functions.

a mechanical function by keeping the link active but disengaging the set until the line button is pressed again to reconnect the parties.

Call Transfer

Assuming this feature is available at the central office or PBX equipment within your area, a call can be transferred to another line. This function is available in many central offices today. To activate it, you must press the hookswitch down and draw what is called recall dial tone. Then you must dial the third-party number (the number transferred to) and wait for the line to ring. It's a good idea to wait for the new called party to answer and announce that the call is being transferred, then hang up. However, many people don't provide this courtesy. A problem exists with activating this feature; it requires that you press the hookswitch down for a short period (usually 100 to 200 milliseconds) to draw the recall dial tone. Most people fall into one of two categories: "fat fingers," who hold the hook switch down for 1 to 2 seconds and cut off the caller; or "skinny fingers," who press and release the buttons too fast and never get the recall dial tone. Consequently, a preprogrammed feature button on this newer set will perform these functions for the exact amount of time and prevent the problems associated with using transfer.

Conference Call

This is similar to the transfer feature, but in this case, rather than transferring the call to a third party, the third party called will be added into the call. The conference feature allows the three parties to converse without the need to worry about cutting each other off.

Redial Last Number

Sets with this feature have their own internal memory, so they can automatically dial the last number called—even if it is not a speed number. This is great when you encounter a busy tone on the network. The problem with last-number redial comes into play when the call requires a string of digits longer than the normal 11 digits for a phone call. Pauses are required in some cases, but this feature will send the 11 digits plus the credit card number and any other digits dialed in the last sequence. This can cause some confusion, but it is a time saver. Redial can also be a pre-programmed feature where some sets, on encountering a busy tone, will redial up to 99 times in a preset time period (over the next 60 to 90 minutes, for example). Not all sets will do this. The degree of sophistication is directly proportional to the dollars spent for the set.

Built-in Speakerphone

Many sets have the ability to turn on a speakerphone for hands-free two-way communications. This can be simplex, which means that one must be careful not to overtalk the other end. If overtalk occurs, then the conversation becomes choppy because the set is bouncing between the transmitter and receiver, cutting off words or syllables. The other option some sets have is a duplex speakerphone. This means that simultaneous talk and receiving can occur without cutting in and out. This duplex circuitry has separate paths for the transmitter and receiver.

Hands-Free Dialing

Similar to speakerphone service, but not the same, hands-free dialing allows just that. Without picking up the handset to the face, a user can go off hook electronically by pressing the line button or an "on" button, then dialing. This frees the users' hands for other tasks while the connection is being made. However, once the called party answers the phone, the caller must pick up the handset and converse. The speaker is only a receiving speaker not a two-way. Many people get confused over this feature and cause a lot of grief or hangups when they attempt to use it improperly.

Displays

Where features are available in the central office, a display set can be used to show the incoming caller identification. This is not ubiquitous or even legal everywhere, but it is coming. Where the incoming caller display is not available or used, the display can still be used to display the number that the caller dials. Further, when the call is dialed, a timer can be used to keep the chronology of the length of call and so on. This can be used for client billing and chargeback or for reminders of what time it is in case you have a meeting.

Analog versus Digital Transmission

Introduction

In the data and voice communications field, one of the most common points of confusion is between the words *analog* and *digital.* Most people recognize that the term *digital* refers to the expression of information in terms of 1s and 0s. But few can easily make the mental connection between this fact and the real-world requirements of moving voice signals expressed in a 1s-and-0s format. In dealing with the term *analog,* everyone merely refers to voice communications. This chapter will help you to make and understand the connection between these two terms. Not that you will become an electrical engineer, but you will be able to converse with the best of them! If there is a single element that must be understood, it is that analog and digital formats are means used to move information across any medium. Some media typically deal with analog, others deal with digital, and the rest deal with both. Thus this understanding will help you to understand the reasons why the different transmission systems are used.

Analog Transmission Systems

As mentioned earlier, "the network" was originally developed solely to provide voice communications services. The communications circuits AT&T built through the Bell Systems and their own communications capabilities used strictly analog technology. Yes, it changed into a digital world, but it was built around analog communications for voice. But what does this mean?

In an analog communications system, the initial signal (in this case, the spoken word) is directly translated into an electrical signal. Chapter 4 describes how the telephone set accomplishes this task in more detail. The characteristics of an analog signal deal with two constantly changing variables: the amplitude and the frequency of the signal. The strength (amplitude) of the electrical signal varies with the loudness of the voice, and the frequency of the electrical signal varies with the pitch or tone of the voice. Both variables (amplitude and frequency) change proportionately with the original sound waves.

If the signal were to be monitored with an oscilloscope, the displayed pattern would visibly and recognizably change as the sound changed. This is shown in Fig. 5.1, where the signal characteristics can be seen on an oscilloscope. For a more graphic example, think of the visual displays in some discos, where the music is fed into a light display that varies as the music

Figure 5.1
Using an oscillo-
scope, you can see
the characteristics of
an analog signal.

changes. In an analog communications system, the electrical signal, not light, changes. In Fig. 5.2, an analogy to this concept using a musical beat is shown. The lights of the disc jockey's equipment will change and flash with the beat of the music. In this case, the lights are changing on the basis of the electrical characteristics of the signal. What is actually happening? In telecommunications terms, you could say that the actual sound is pro-

Figure 5.2
Comparing the
analog wave to how
the beat of the music
changes lights in a
DJ's equipment.

duced by banging air waves together. This banging of air waves is really the movement of the air molecules. Technically this is called compression and rarefaction. But who wants to get technical?

The human voice is an interesting phenomenon. As we speak, we generate sound. We actually bang the air waves together quite a few times in a relatively short period of time. This was discussed in Chap. 3, but, to stick with this thought, we create an analogous sinusoidal wave. This wave is then converted into its electrical equivalent. Quite simply, if you were to take the sounds created by the voice and modify them on the basis of the pitch and the strength of the sound, you could produce this sinusoidal wave (Fig. 5.3). The wave is what would happen electrically if we were to use a magnetic field to change the sound into electricity. Working around a baseline of zero electrical voltage, we then would have a 360° rotation of electrical current. This 360° wave around the zero line is called a hertz, named after the electrical engineer who documented this concept. The wave starts at the zero line and rises as the amount of energy increases. This will ultimately peak at some point and then begin its descent. The electrical energy will then proceed from the height on a downward slope to the zero line. From there it will continue on its slope below the zero line to a point below the zero line (this is called the negative side of the wave). Continuing to a peak voltage on this side of the line, the energy will hit its peak value, then begin an upward slope back to the zero line. The wave has made a complete 360° cycle. One complete cycle constitutes one hertz (1 Hz), used in a reference against one second in time. A 1-Hz signal produces one complete revolution in one second, whereas a 100-Hz signal completes 100 revolutions in one second.

Figure 5.3
The sinusoidal wave rotates 360° of electrical current around a magnetic force. This is one wave or revolution, called a hertz.

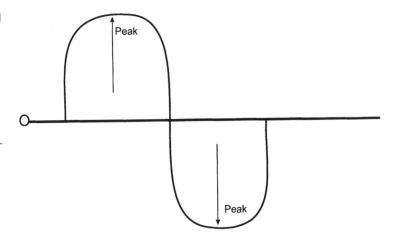

Peak

Peak

The human vocal chords can bang the air waves together between 100 and 5000 times per second. Simply stated, then, the voice is producing up to 5000 cycles of information per second. When this is converted into its electrical equivalent, there are up to 5000 hertz. We do not like to state such big numbers, so we abbreviate this to 5 kilo (meaning *thousand*) hertz—5 kHz, for short. From this electrical equivalent, we now have an analogy or look-alike to our sound waves. This is a constantly changing variable of electrical energy. Both the amplitude and the frequency will change from 100 to 5000 times per second. This is something that the telephone companies learned to deal with from the beginning of our communications industry. Over time, however, they found that the human voice will generate from 300 to 3300 cycle changes per second or 3 kHz of electrical cycles as a norm (Fig. 5.4). As with any communications channel capacity, the telephone company did not want or need to give users any more than they needed to carry on a voice conversation. Over the years, as the network expanded, the telephone companies limited the bandwidth of a telephone call to a 3-kHz channel capacity. This was a money issue. Given a limited amount of bandwidth, how can we allow a human conversation to take place across the network and produce reasonable representations of the original voice? And how can we do it cheaply? The range of frequencies generated goes from 300 to 3300 Hz (Fig. 5.5). The telephone company therefore limits our use of the channel to just that amount. In the radio frequency (RF) and electromagnetic spectrum, the telephone companies divided all of their capacities into 4-kHz slices. On each of these 4-kHz slices (wires or radio channels), they installed frequency bandpass filters at 300 and at 3300 Hz. Anything that falls in the middle of this allocation of RF spectrum will be allowed to pass. Anything that falls outside of this range will be filtered out (thrown away). This is called a band-limited channel. Because the voice can go as high as 5 kHz, there will be cases where we hit the higher range of frequencies with our conversation (such as words with the *S* and the *F* sounds) and it will be flattened out through the filters. This might sound unreasonable, but what it really produces is a little fuzziness on the line. Neither the human ear nor

Figure 5.4
The human voice produces usable information in a range of 300 to 3300 Hz.

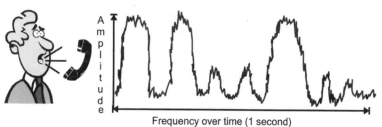

Frequency over time (1 second)

Figure 5.5
The telephone companies band-limited their 4-kHz channels to a usable 3 kHz with bandpass filters.

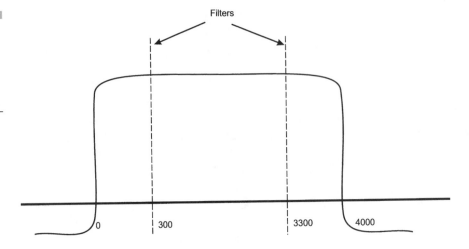

the telephone equipment is sensitive enough to recognize any significant problems. Once the electrical equivalent of the sound is created (through the telephone set or another device), the electricity is sent down the wires.

When a call is proceeding down the wires, resistance in the wires to the electrical signal immediately begin to diminish its amplitude. The signal gets weaker and weaker. This weakening of the energy will eventually lead to the total absorption of the electrical energy, or the loss of the signal beyond recognition. This is called *attenuation* of the electrical signal. The signal can only travel so far before it runs out of strength and disappears. An analogy to this is a relay race runner who is trained to run around a quarter-mile track as fast as he or she can. By the time the runner gets to the end of the quarter mile run, all strength and energy is expended. The runner is out of steam (Fig. 5.6). If the runner did not have someone to pass the baton off to, but had to go around the track again, the second lap would take forever. That is, if the runner made it around the track at all before passing out and falling flat on the ground.

In order to keep the signal moving along the wires, amplifiers are used to boost the signal strength (Fig. 5.7). These amplifiers are usually spaced about 15,000 to 18,000 feet apart. Typically, we only need one amplifier between the customer location and the telephone company central office—two, at most. The central office is usually located close to the customer, within 5 to 7 miles on average. Only in remote locations will the central office be farther away. This may not be the case where the telephone companies have changed their architecture and closed the CO in an area. The consolidation of COs across the United States specifically has led to greater distances from the CO.

Figure 5.6
The runner around the track is "out of energy," so he must hand the baton to a new runner.

A second activity takes place on these wires concurrent with the transmission. Noise (introduced line loss, frayed wires, lightning, electrical inductance, heat, etc.) is ever present in the form of white noise or hiss. It also begins to cause a deterioration of the signal. Noise is always on the line, but its increased presence from cabling problems causes the degradation of the signal. Unfortunately, as the signal proceeds down the line the noise and the signal begin to intertwine. Amplifiers cannot discern the noise from the actual conversation. Thus the amplifier not only boosts the signal, it also boosts the noise. This creates a stronger but noisier signal, which leads to some extremely noisy circuits (Fig. 5.8). The results of amplification are cumulative over distance. The more amplifiers that must be used, the worse the overall signal gets. Many of the longer-distance circuits were constantly being amplified. This again is not to imply that the telco doesn't want to do any better. It is a function of the equipment, the electrical characteristics, and the finances combined.

It is interesting to note here that some of the latest advances in digital signal processing (DSP) could now be used to clean up an amplified voice transmission somewhat if analog technology were still in wide use. But a better approach to producing clean signals was developed many years ago.

Figure 5.7
Amplifiers are used to boost the signal strength along the wires.

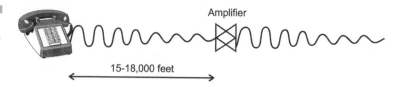

Amplifier

15-18,000 feet

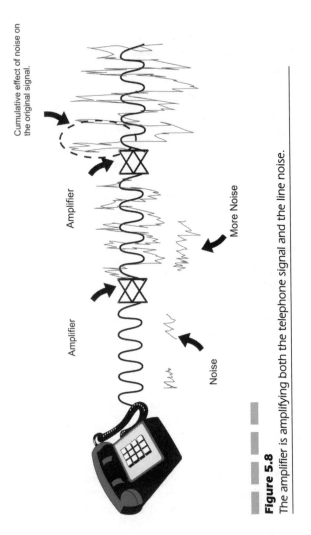

Cumulative effect of noise on the original signal.

Amplifier

Amplifier

More Noise

Noise

Figure 5.8

The amplifier is amplifying both the telephone signal and the line noise.

Analog-to-Digital Conversion

Because the analog version of the network was a problem for the network suppliers and the telcos alike, both turned to a digital form of communicating. In order to convert the voice conversation from analog to digital, a device called an analog-to-digital (A/D) converter employs a sampling technique.

Sampling refers to the process of measuring (see how we are creeping up on quantifying a signal?) representative portions of a signal over time. We make the assumption that chronologically adjacent portions will differ only slightly. If the samples are taken frequently enough, and played back faithfully at the other end, the ear will not be able to differentiate the playback from the original. A (nondigital) sampling technique is used in movies and other video applications. When a movie is made, there is no truly continuous record of the images; instead, a series of still images, sampling the reality at 30 samples (or "frames") per second, is recorded and later presented to the viewer. Normally, the viewer cannot distinguish between playback of the samples and the real thing.

As mentioned earlier, the bandwidth of the audio signal we wish to transmit is 3000 Hz (3300 minus 300). On the basis of the Nyquist theorem (which states that sampling should occur at a rate at least twice the maximum frequency of the line), the minimum sampling rate would be 6600 Hz (2 × 3300 Hz). In fact, a somewhat higher rate of 8000 Hz (samples per second) is used. This addresses the higher range of frequencies in a conversation, such as those *S* and *F* sounds that were filtered out in the analog world (Fig. 5.9).

Figure 5.9
During the analog-to-digital conversion, a sampling rate of 8000 Hz is used.

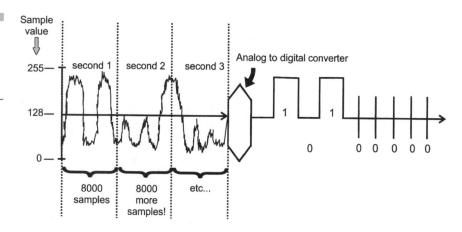

Each sample measures the amplitude level of the voice signal at a particular point in time. One sample comprises eight bits, where a bit represents a 1 or a 0. An eight-bit character or byte can represent any decimal number from 0 to 255 (00000000 is zero, 00000001 is one, 00000010 is two, 00000011 is three, 0000100 is four, 0000101 is five, and so on up to 11111111 which equals 255). Therefore there is a total of 256 possible levels, sufficient to recreate the analog signal faithfully at the receiving end. More samples would produce a higher-quality replication, but the ear is not sensitive enough to discern the differences.

Eight thousand samples per second, with each sample requiring eight bits, generates a digital stream of data at a rate of 64,000 bits per second. We know this as the digital signal 0 (DS0), the digitized equivalent of one voice channel. The bits are each in the form of a square wave, in contrast to the familiar sinusoidal wave that is typically seen on an oscilloscope.

The square wave travels down the same pair of wires we are accustomed to, but in order to handle the digital signal, the amplifiers are removed. Additional equipment (such as loading coils) is also removed. In short, the entire circuit is reengineered from end to end. In place of the removed equipment, a digital regenerator (or regenerative repeater) is used. The strength of the digital signal is based on a 3-volt pulse of a very short duration; therefore, the repeaters must be placed closer together, typically every mile (Fig. 5.10). This requires more equipment, which is an expense to the telcos and network suppliers alike. But when the samples are played back at the far end, customers receive the transmitted signal in a form indistinguishable from that sent from the original A/D converter at the sending location.

Figure 5.10
Repeaters are standard at approximately each mile of line between the telephone company and the end user.

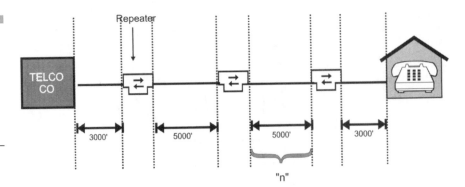

Digital Signaling

As mentioned earlier, digital signals consist of 1s and 0s. But even this statement can be a bit confusing; how does one insert a "1" or a "0" onto a wire? Unlike the varying levels used in analog transmission, signaling on a digital circuit is very simple; two different voltage levels are sent to represent 1s (i.e., ±3 V), while no voltage represents a 0. This is called a unipolar signal. The receiving equipment measures the voltage level once each bit time.

A bit time is the length of time that it takes to transmit one bit. For example, at 9600 bits per second, one bit time equals 1/9600 of a second (about 104 microseconds). The measurement is taken in the middle of the bit time, when the likelihood of getting a precise level is maximized (Fig. 5.11).

There are actually two types of digital signaling. The previous description presents the type of signals used for local (that is, non-telco-based) digital signaling. A slightly more sophisticated scheme, called either alternate mark inversion (AMI) or bipolar signaling, is used for wide-area digital transmissions. Chapter 14 includes a detailed discussion of AMI.

Because voice is inherently analog (i.e., it varies continuously, rather than occurring in a discrete, numerically related pattern), an analog-to-digital (A/D) conversion must be performed if you want to transmit voice digitally. Later paragraphs cover how to change an analog signal into a digital format. But first we must answer the important question, "Why bother to convert to digital?" The answer is quality.

Analog signal paths must use amplifiers if the paths cover any significant distance. Because amplifiers retain and even strengthen noise in a conversation, the more amplifiers used in a transmission path, the noisier the background. So, when you use an analog circuit to call your grandfather in the old country to wish him a happy birthday, you will be most

Figure 5.11
The measurement is taken in the middle of the bit time, when the likelihood of getting a precise level is maximized.

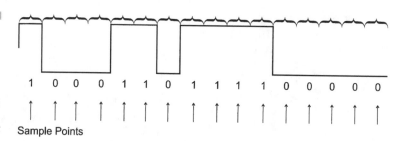

1 0 0 0 1 1 0 1 1 1 1 0 0 0 0 0

Sample Points

fortunate if either of you can understand the other over the background noise. To compensate for this problem of a noisy overseas (or long-distance) call, we use an age-old philosophy: yell!

By yelling into the phone to overcome the noise, we do little to help the situation. When we raise our voices to new heights, we increase the amplitude. This makes the signal very strong (stronger than what the telephone companies expect). The telephone companies have, however, installed equipment on the line that will hold the height somewhat to a level. These *pads*, as they are called, sense that the amplitude is much higher than normal, so they introduce loss to compensate and bring the amplitude down. This just serves to compound the problem, because the loss in decibels (dB) will steal away the amplitude but do nothing for the noise. So we have gained nothing (Fig. 5.12).

The digital equivalent of an amplifier is a repeater. We commonly refer to this as a "schizophrenic cookie monster." Why? Well, let us explain.

As the digital pulse is placed onto the wires, only the 1s (+3 V) need to be of any concern. But as the pulse is placed onto the wire, it immediately begins to propagate down the wires. The wire will immediately begin to act on this pulse by diminishing its strength. The result will be a constantly decaying pulse of electricity (Fig. 5.13). Because this decaying process begins immediately, the likelihood of the pulse falling into obscurity is very high. So this "cookie monster" is placed on the link at approximately every 5000 to 6000 feet. As the pulse declines in strength, it must be kept at sufficient strength to be recognized as a pulse and not noise. This distance is used to keep the signal above what is called the pulse detection threshold level. If the strength of the pulse falls below this detection threshold, the pulse will be mistaken for noise and ignored.

When the pulse arrives at the repeater, the repeater will act like the cookie monster. It will literally absorb the signal—eat it up and make it disappear. The cookie monster has done its job. However, the other side of this cookie monster is schizophrenic. Therefore, it will basically say, "there

Figure 5.12
When we yell into the set, the amplitude is increased, but equipment at the central office "pads" the signal to a norm. Therefore nothing is gained by yelling.

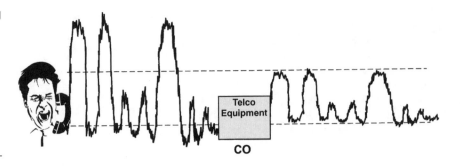

Figure 5.13
The pulses of electricity constantly decay as they travel down the line.

was a pulse there, so I'll be a sport and put a brand-new pulse on the line in the original's place." Thus, a brand new 3-V pulse will be introduced. The signal begins to propagate immediately down the next leg of the circuit. And the process starts all over again.

But whereas an amplifier simply strengthens the input signal, a repeater detects the 1s and 0s of the original signal and retransmits them out the other side as good as new. In particular, any noise introduced into the signal between the source and the repeater is completely ignored as long as it stays below a certain tolerance level. Amplitude decreases will still occur between repeaters, and noise will be introduced, but so long as a repeater can recognize the original pulses (1s), those are the only parts of the received signal that will be used to build the new output signal. Thus, if a communications path is built with digital technology, it is entirely possible to deliver to the ultimate receiver a signal that is identical to that sent from the first digital transmitter in the path (Fig. 5.14). Because the pulse is short in duration and a discrete value of energy, rather than a constantly varying voltage level, more repeaters are required on the link. Telcos must use three to four times as much equipment on the same distance of wire with digital communications at the local loop.

Note the last part of the previous statement "...from the first digital transmitter in the path." Because voice is analog, there will be a conversion at some point. From the voice source to that first conversion point, noise can be introduced that will remain in the signal. Obviously, the closer to the voice source that the A/D conversion takes place, the better.

If you have a conversation with someone in another state or another country, and that conversation is as clear as it would be if the person to

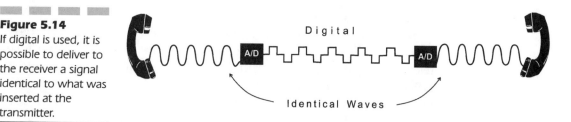

Figure 5.14
If digital is used, it is possible to deliver to the receiver a signal identical to what was inserted at the transmitter.

whom you are speaking were right in front of you, you can be sure that most if not all of the technology involved is digital rather than analog. Such quality results in happier customers, cleaner transmission, improved data communications, and happier carriers. This is why all the hype exists in the industry about the benefits of digital transmission.

There is another benefit to using digital rather than analog transmission, although it directly benefits the service providers more than their customers: digital technology facilitates combining and manipulation of signals, providing a carrier more flexibility in how it moves the signal from source to destination.

Digital Data in an Analog World

We have been focusing on voice transmission, a communications technology that starts and ends with an analog signal—a voice. But what about data communications? Because it starts out as digital, doesn't that simplify things? The answer is no, not so you'd notice.

Until the last few decades, digital service was not available to users. Because of this, an entire technology came into being that provided methods for carrying digital signals across the analog voice network. The critical component in this technology is the modem, a contraction of the terms modulator and demodulator. In a nutshell, a modem's function is to provide digital-to-analog conversion (and of course the reverse) for a digital device that needs to communicate with a remote digital device across the analog telephone network. Figure 5.15 shows this multiple conversion process.

Because in fact telephone companies' backbones are now mostly digital (with the local loop signals being digitized at the CO), a communications path using modems includes not one, or even two, but rather a minimum of four digital-to-analog or analog-to-digital conversions—and that's only one way!

Modems and other data communications equipment are addressed in more detail in Chap. 13.

Most voice and data transmission systems now use digital technology, with a few key exceptions. An exception that has particular impact is the telephone local loop. The local loop (sometimes called the "last mile" or other less printable terms) has been held back because of the vast installed base of analog telephones and amplifiers. However, even this last bastion of analog transmission is under assault.

Figure 5.15
Four analog-to-digital
or digital-to-analog
conversions!

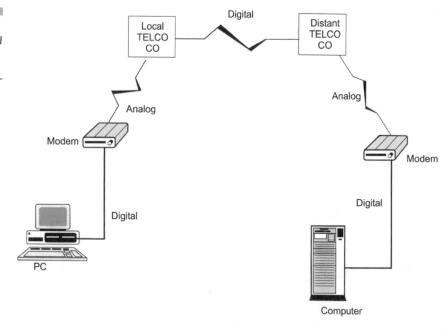

A digital technology and telco service offering called integrated services for digital networks (ISDN) is slowly spreading. ISDN can provide digital voice transmission right up to and including the telephone. Chapter 26 covers ISDN in greater detail.

Digital transmission is the way the world is going. While in the past the cost of digital devices exceeded that of analog equivalents, the price curve for the analog devices has long since flattened—whereas the cost of digital components has steadily dropped, and continues to do so. Until the local loop is fully digitized, we will operate in a world of mixed analog and digital components. But except for that local loop, expect virtually every other communications medium to employ digital technology.

CHAPTER

Carriers

Incumbent Local Exchange Carriers

The incumbent local exchange carrier (ILEC) is the telephone company serving your area and providing you with dial tone services. The term *local exchange carriers* (LEC) came into existence after the breakoff of the Bell System from the AT&T network. The modified final judgment decree specified that the local telephone company, or dial tone provider, would be kept separate after divestiture. Prior to 1984, the carriers were called the Bell Operating Companies (BOCs), and they still are referred to this way. The names of the telephone companies attached to the older Bell System have changed a little, but some have tried to keep their identity as much as possible. Examples of this identity are names still used in the industry today, such as Bell Atlantic, and so on. However, the telephone companies were fighting to keep the Bell logo at the same time the parent organization (AT&T) was also trying to keep certain identities. The local carriers won out in this battle and managed to keep the logo and the Bell name. Surprisingly enough, even though the divestiture of the AT&T and Bell Systems occurred in 1984, many people still do not separate the two entities. For the public, this is the toughest part of the whole scenario. People still think of the system as one big entity, rather than organizations that were separated as a result of the court decisions.

In early 1999, 15 years after the breakup occurred, there are still users who do not recognize that the change ever occurred, although fewer than ever. This, of course, leads to a lot of confusion, because the problems that occur on a daily basis are always blamed on the telephone companies, regardless of where the problem actually resides. Additionally, the independent telephone companies, of which there are hundreds (1400+), who were always referred to as independent, are also incumbent local exchange carriers. These organizations, although not part of the Bell System, still provide the basic dial tone for many communities. Some of the larger of these independents include GTE Telephone (including their acquisition, CONTEL Telephone Systems), Commonwealth Telephone (now part of two different companies, RCN cable company and Frontier Communications), Standard Telephone Company, Centel, and so on. The numbers and the areas served by the various independents differ, but the function is the same. These companies all provide local dial tone service as their primary charter. Now they are all considered ILECs because the term fits better with the service they provide.

The ILECs provide more than just the dial tone. Since the divestiture, when they were required by law to separate the customer premises equipment part of their service offering, the ILECs have a clearly defined demarcation point where the telephone company service stops and the customer takes over. This demarcation point, also referred to as the DEMARC, is the point where the rate and tariff process starts and stops. Figure 6.1 shows the termination of an ILEC circuit in a demarcation point. This will fall apart over time as the ILECs and CLECs all get into every aspect of communications services in the future. All the providers are trying to be the one-stop shopping provider of the future. In this particular reference the demarcation point is a network interface unit (NIU), which is a smart version of the RJ-11 jack. To give an idea of how this all evolved, remember that the telephone companies provided "everything from soup to nuts" in the early stages of the industry. They provided:

- The dial tone
- The local loop of wire connecting to their equipment and the customer location
- The telephone set to interface to the network
- The interface to the local and long-distance portions of the networks and the access to the long-distance network
- The billing and collection functions for all services
- The installation of all related pieces and components to allow the end user to access the network
- The maintenance and repair functions

Figure 6.1
The LEC terminates the two wires at the NIU (DEMARC), where the LEC stops and the customer takes over. In this case, the LEC may provide a hard-wired lug connector or an RJ-11 jack. The added RJ-11 jack on top of the NIU is for testing.

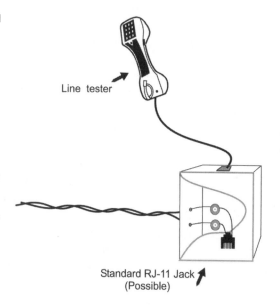

Line tester

Standard RJ-11 Jack
(Possible)

Because this was a single interface, the customer's only responsibility was to make a single call that did it all. The use of the network led to additional billings, so the only actual responsibility of the customer or user was to pay the bill. This all changed in the 1984 divestiture of the network and telephone company interfaces. Consequently, the customer now must decide whom to talk to regarding services and equipment needs. However, with the Telecommunications Deregulation and privatization occurring throughout North America (and other parts of the world), the way we used to do it is becoming the way we will do it again in the future. What goes around, comes around. Soon the ILECs will be able to compete and offer all the services (and more) they did prior to 1984!

Some of the primary services that the incumbent local exchange providers or ILECs provide follow. This is a representative sampling of their services, not an all-inclusive or exhaustive list.

Local Dial Tone (Single-Line and Party Line) Service

This is the primary bread-and-butter service and the main service the telephone companies are currently chartered to provide. The local dial tone can be in the form of a single-line service, with one registered user per pair of telephone wires, or a party line service where multiple registered users share a single pair of wires. The use of party line service was always a financial consideration, particularly in the rural areas of the country. Many telephone suppliers did not have a financial justification for installing multitudes of wires outside of the major downtown areas. Therefore, as the rural users requested dial tone, the cost of running the wires out to their locations was exorbitant. Consequently, the telephone companies allowed for connection to multiple customers (two to four) on a single cable pair, with a distinctive telephone number and ringing tone for each user. This should not be confused with a multiplexing scheme. Only one of the users could use the dial tone line at a time. Others on the same physical pair of wires had to wait for the line to be available before they could either make or receive a call. If a local community was on party line service and user A (Fig. 6.2) was on the phone, the wires were busy. When user B wanted to make a call, the process required B to pick up the handset and listen to ensure that the line was free and dial tone was heard. Because A was already on the line, B would be able to hear the entire conversation on the wires, but would not be able to make

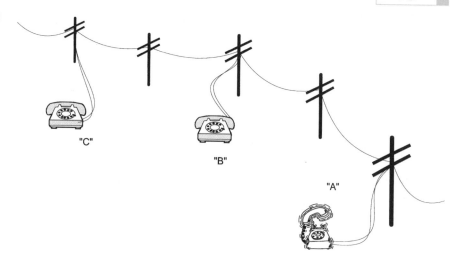

Figure 6.2
With party line service, multiple users share the same pair of wires. This arrangement was used in many rural areas when the LEC did not have enough pairs to serve all subscribers.

"C"

"B"

"A"

a call until A got off the line. Additionally, if C was expecting a call, as long as A was on the line an incoming call could not make it to C because the line was busy. This required the cooperation of the users sharing the service and a form of mutual respect. Because any one of these users could listen to the conversation, there was a good neighbor policy that said that the user, on hearing a conversation on the line, would hang up and not eavesdrop on the other party. Also, a form of cooperation or scheduling was informally used so that when B wanted to make the call, he or she would tell A, who was already on the line. Therefore, A would hurry the call to its conclusion so that B could have the use of the line.

This all worked well in rural America, where neighbors were more dependent on each other and were far more cooperative in their communities. However, in the cities where this was also an offering at a reduced rate, the system really could become a major problem. Suppose that A had a teenager who liked to talk on the line for hours. B and C would continually be blocked out from the line. Or, as they would continue to pick up the phone and hear a teen on the line, would ask the teen to get off the line so that they could make a call. This could lead to significant hard feelings among the neighbors and even to confrontations. Thus the use of party line service was discontinued as much as possible. The rural parts of the country still have this service because, given the limited facilities in these areas, no other option exists. This is a variable depending on the local provider services and the geographical location.

Centrex Service

Centrex (central office exchange service) is a technique the ILECs offer as their flagship service for business users. The nature of Centrex is the subject of several discussions in the industry. When the Bell System was originally united with the AT&T organization, the rental of telephone service included the private branch exchanges (PBX). Because this service was rented, or leased over longer periods of time, the Bell System always had the edge over competition. After all, the supplier of the telephone dial tone also provided the interface equipment at the customer's location. Therefore the customer needed only to use the service and pay the bill. However, as competition was gaining ground over the existing installed equipment base provided by the Bell system, the telephone companies realized that their own limited product offerings were contributing somewhat to the move toward competitors' products. Many customers were locked into long-term equipment agreements and the AT&T systems were not advancing as fast as the competitors' products. This was the result of trying to keep pace with the market and at the same time protect the embedded base of equipment without mass replacements that would be expensive and difficult to cope with. While this was happening, the central offices were moving ahead of the PBX marketplace in that they were replacing the old electromechanical systems with the newer electronic switching systems (ESSs). Rather than replace both ends—the customer hardware and the central office equipment—the logical selection of a central-office-based telephone system was offered. The customer therefore could rent a partition of the central office, both hardware and software, to act as the PBX surrogate. Further, the Centrex offering allowed the customer to get away from the hardware system roller coaster. Hardware changes can be very expensive, particularly with the constant upgrades that were being introduced by the vendors.

The use of Centrex service was therefore a stepping-stone for the Bell operating companies and the independent telephone operating companies, but once the breakup of the system occurred, the Centrex service became their flagship product to serve the large business customer. As it evolved, the offering became far more viable to the smaller customer with as few as two or three lines. The Centrex offering provides the features and functions of a PBX without the heavy investment requirements.

For some time after the divestiture of the Bell System, Centrex was the primary way the ILECs pushed services. However, as more business customers bought hardware (PBX and Key Systems), the use of Centrex started to wane. Initially, Centrex was a North American service only. A major drawback was that the ILECs originally offered Centrex to the larger customers only, usually with 100 lines or more. After the divestiture and the decrease in interest, this changed and became available to any business, large or small. Companies with only one business line could now rent Centrex service. As the renewed flexibility was introduced, the ILECs saw new growth occur. Now over 34 countries offer Centrex services, and the growth rate is averaging 15 percent per year.

Business Service (Direct Inward Dial and Direct Outward Dial Lines and Trunks)

Beyond the basic dial tone services for business and residence customers, the ILECs offer the rental of inward and outward dial services for the business user. The use of direct inward dial (DID) services allows the customer or outside caller to dial a user's telephone directly via an extension number. The call is passed into the telephone system (PBX) from the central office switching system and is redirected to the called party's extension telephone. Just as the caller can dial directly into the called party's extension on the telephone system, an added capability, called direct outward dial (DOD), exists on the telephone systems to allow the extension user to dial directly outside the system. This provides the users an access code to dial (number 9, for example) that tells the telephone system to find a line or trunk to the outside world and return dial tone. On receipt of this second dial tone, the station user then dials a 7- or 10-digit telephone number. No operator or manual assistance is required via this access method. Hence the term direct outward dialing. DID and DOD services are covered in greater detail in Chap. 8.

Residential Service

Of course, the ILECs provide both business and residential services; the delivery of residential dial tone is a primary objective. Residential services include the single-line and party line services. However, the ILECs are always looking for newer sources of revenue. Therefore they offer various features and functions in the residential package. The first and foremost choice the residential user might have is the standard flat-rate residential service. In some cases, the ILECs also offer a measured usage service at a reduced rate. Flat-rate service can cost between $10 and $28 per month, depending on local regulatory rates. To help reduce the high monthly costs for the nontypical user (a user who does not make very many calls per month), the ILECs offer the residential user the ability to rent a reduced-rate dial tone line for $8.50 to $10.00 per month, with a usage charge added for every call that is made (a message unit might be $0.05 to $0.07 per call) in excess of some predetermined number of calls. The third option is to use a further reduction of the monthly charge to a rate of $4.50 to $6.00 per month, with a usage charge for every call. Regardless of the selection made, the customer typically tries to match to specific individual needs. Although this is attractive, there are some cases where the telephone company will offer the reduced-rate service as opposed to the flat rate, but with the stipulation that once a measured rate is selected the customer cannot go back to the flat rate. This would be a critical service decision if the caller has a variable usage pattern, because the usage might be higher than the flat-rate service. In Table 6.1, the cost for service is compared at various usage levels.

In this table it is assumed that the cost of the monthly flat-rate service is $15.00; measured service with the first 50 calls free is $10.00; and measured service with no free messages is $6.50. Each call above the allowance is set at $0.07. The rates for these services vary from state to state because the tariffs are all different. However, the rate per message is a new variable because the cost per call is now becoming distance and time sensitive. A call, for example, might cost $0.07 for the first five minutes, then an added $0.07 for each additional three minutes. Other variables exist, and the combinations are innumerable.

Local Calling Services

Direct distance dialing (DDD) is simply the ability to place a call at any distance within the ILEC's area of service. This includes the basic residen-

TABLE 6.1	Description	Flat Rate @ $15.00/mo.	Measured Rate @ $0.07/call	
Comparison of the Various Residential Service Offerings with Measured versus Flat-Rate Service			50 calls	No calls
	10 calls per month	$15.00	$10.00	$ 6.50
	50 calls per month	$15.00	$10.00	$10.00
	100 calls per month	$15.00	$13.50	$17.00
	200 calls per month	$15.00	$20.50	$20.50

tial and business offerings at a monthly rate for the line and the usage. This is not unlike a regular business line except for the fact that the business user can access the dial tone and dial wherever he or she wishes without the assistance of an operator. The acronym DDD characterizes the changes that have occurred over time. In the late 1960s, an operator was required to make some calls on a long-distance or international basis. Evolution of the network beyond this point was a must because the manual intervention was expensive and labor intensive. In the earlier days of the telephone networks, this was not a major problem, but as the usage picked up and dependency on the telecommunications networks increased, the carriers realized that this access was a must.

Pay Phone Service

Clearly evident anywhere in the ILEC's service area will be the existence of pay phones (coin telephones). The LEC offers the access to the local and long-distance networks via either a public telephone service, a semipublic telephone service, or a private telephone service. Each of these is handled differently on the basis of the access method chosen. The public pay phone is an obvious landmark on the streets of any city or town. These are the phone booths that are positioned in full view for use and access by anyone who wishes to make a call. The pay phone or coin-operated phone is designed to allow access to the network for the masses. If a user does not have a phone in the home or office, then the public phone is available for outgoing calls to the network on a pay-as-you-go basis. There is no monthly rental charge for this service, because it is shared by the general public. However, the cost per call or per minute will be more expensive because operator or automated assistance is typically necessary to complete a call from a coin-operated phone. This is the most expensive type of call available on a dial-up basis. Along with this dial-up service, the caller can use a credit card to complete a call through the carrier of choice or can make an 800 toll-free call without the need to deposit a coin. The dial tone accessing the ILEC's network is present, so a call can be initiated without the use of a coin.

The semipublic telephone service allows an organization or business to rent the set and line through the ILEC and place it within their offices or lobby, thereby denying access to the general public. The rental will normally be based on a flat rate per month, less a portion of the income that is generated by the pay phone. Normally there is a minimum amount of revenue guaranteed to the ILEC by the renter of the pay phone. If, however, the minimum is not met, the renter must pay the difference. The variables of the pricing arrangement are a complex aspect of the local tariffs.

A new twist to this part of the billing is the possibility that all customers using the pay phone may be initiating toll-free (800, 888, and 877) calls. This obviously doesn't generate any revenue for the pay phone owner, so now the cost of the connection to the toll-free number is being passed back to the holder of the toll-free number!

The final type of pay phone service is the use of a private coin telephone service. In this scenario, the pay phone can be rented from the ILEC or purchased from any third party by the customer. The customer also rents the access line to the ILEC's network for a fee (flat rates are still available in some areas) then selects the long-distance carrier to carry any calls from this phone. The customer can allow selected users access to this phone, which is normally located well within the confines of the customer's building, perhaps in the cafeteria or lobby. Therefore anyone off the street cannot just walk into the building to use the phone, because it is not in a public place. From this phone, the customer can select the method and the price to be charged to the user on a cost-per-call or a cost-per-minute basis. The ILEC carries the local call and keeps track of the minutes of usage for billing and collection purposes. Under this arrangement, the ILEC might get a percentage of the revenue for the administration and handling of the service. Another option is that the ILEC merely bill the customer for the basic service at the tariffed rate, with all revenues going to the owner of the pay phone. This has become a competitive tool for customers since the divestiture of the telephone system.

Private Lines and TIE Lines

These are leased services rented from the ILEC on a flat-rate basis. A *private line* indicates that the service is dedicated to the user rather than accessible to others. This is a contrast to the public-switched telephone network, where many users compete for and share the lines and trunks from the ILEC. The private line is typically a flat-rate service connecting two locations of an organization. In the case of a TIE line (terminal interface equipment), the ILEC will install a private line connection between two customer locations located in the same local access and transport area (LATA). The customer can use the TIE line to link two telephone systems (PBXs) together. The connection can be via a two-wire circuit with E&M signaling (which actually uses four wires) or a four-wire circuit with E&M signaling (which uses six or eight wires). The variable nature is caused by the customer equipment being used. Clearly, the four-wire E&M circuit will provide a higher-quality service, because twice the bandwidth is available. Using six or eight wires allows two wires for transmitting, two wires for receiving, and the remaining wires for signaling outside the bandwidth of the circuit. After drawing internal dial tone from the PBX internal extension, a user then dials some selected dial code (i.e., 7 for TIE lines). Once the dial code is entered, the PBX will draw a second dial tone from the remote PBX on the other end of the tie line. From this second dial tone, the end user then dials a three- or four-digit extension number to reach the called party at the other end. This access method allows for the flat-rate connection between two company locations. There is no switching or routing of the call across the network; it always takes the same path on the dedicated circuit. The ILEC only has to provide the dedicated pair of wires from location A through the wire center at the local central office and then right out the back end to another central office or directly to a cable pair to the second location. This is shown in Fig. 6.3 for the connection between the two customer locations in different central offices. In Fig. 6.4, the connection is the same but is served by a single central office. In both cases the central office only provides the service through the wire center rather than routing the connection through the switching center. The end user gains the access to the remote site without having to dial an outside line (the ninth level is typical in a PBX) and then a seven-digit direct inward dial extension number. This reduces the number of digits the user dials and eliminates the additional usage-sensitive costs of message units for local calls. Depending on the number of interoffice calls made between the two sites, the savings realized can be significant. A flat-rate service might prove advantageous over usage-sensitive rates. The ILEC charges for the service on a mileage basis for the dedicated pair of wires between the two locations.

Foreign Exchange Service

Foreign exchange service (FX) is a leased-line service that allows a customer to draw dial tone from a remote central office in the ILEC's service area. The primary use of the foreign exchange is to provide a local telephone number for customers to call. The local seven-digit telephone number is connected to a leased line that goes to a central office in a remote location and is then passed along to the subscriber's location. This will eliminate a long-distance call from a customer to a supplier, allowing the appearance of a local presence. The user pays the ILEC for a local dial

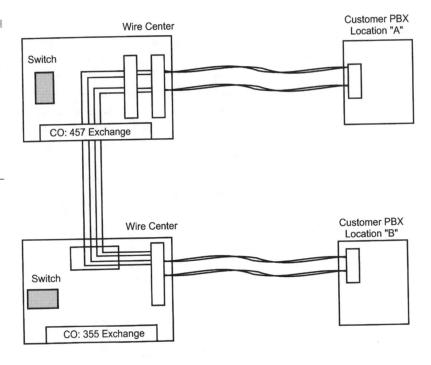

Figure 6.3
A tie line linking two customer locations in different COs in the LEC area. The wire centers are connected with physical connections rather than access switched services.

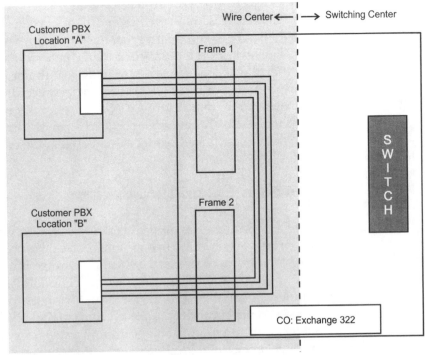

Figure 6.4
Two locations linked together by tie line in the same CO.

tone line at the foreign central office and the mileage charges for the dedicated private line to the user's location. An example of this service is a customer located in Wilmington, Delaware calling a supplier in Philadelphia, Pennsylvania. Normally, this is a long-distance intra-LATA call between the two cities served by the ILEC. The distance between these two cities is roughly 29 miles. At the current tariff rates, this call costs $0.40 per minute. Because the customer calls the supplier on a regular basis to place an order, it is advantageous from the supplier's perspective to allow the customer to call free of charge. Therefore the supplier rents a foreign exchange service from the ILEC. Normally the customer calls (215) 555-1111, but using the FX service the customer now calls a local seven-digit telephone number (302) 777-1234. By calling this local number, the customer does not incur any local long-distance charges and will not hesitate to call the supplier on a regular basis. From the supplier's perspective, the leased line from Wilmington to Philadelphia will cost in the neighborhood of $200.00 per month. Given the cost for the toll call at $0.40 per minute, the break-even point is 500 minutes. Because the customer makes calls that can last up to 10 minutes each, then 50 calls justify the circuit. Taking this to the next logical step, the supplier who is paying the cost of the monthly circuit charge can also use the FX when calling the customer back. The service can be set up on a one-way in, one-way out, or two-way basis. Therefore the return calls will help to justify the expense between the two sites and reduce the overall cost per minute for the usage fees. Although there are other ways to handle this same service, than with FX, this is an example only. Keep in mind that the service in this context is offered by the LEC and is a local toll replacement. If this capability is required between locations in other parts of the country, the ILEC will only provide the last mile on each end, while the interexchange carriers (IEC) provide the interstate, inter-LATA portion of the circuit. Table 6.2 shows a summary of the one- or two-way traffic, comparing it with local calling costs. This is calculated for a standard call of 10 minutes in duration. Other variations of the pricing arrangements can be calculated on the basis of the ILEC's tariffs in the specific areas around the country. Each variation carries different pricing configurations and significantly different cost justifications.

Table 6.2 reflects a total number of 10-minute calls made between the two locations and only two parties. If the number of customers increases, the total number of calls and minutes used on a single line will be limited. There is only one line between the two central offices and the supplier location. Therefore only one conversation can be accommodated at a time. The numbers of calls and minutes will be limited, because the total

TABLE 6.2

Summary of
Possible Savings
with FX Service

Number of Calls per Month at 10 Minutes per Call	Tariffed Toll Calls	FX Service Fixed Rate	Difference
50	$ 200.00	$200.00	$ 0
75	$ 300.00	$200.00	($ 100.00)
100	$ 400.00	$200.00	($ 200.00)
150	$ 600.00	$200.00	($ 400.00)
200	$ 800.00	$200.00	($ 600.00)
300	$1,200.00	$200.00	($1,000.00)
400	$1,600.00	$200.00	($1,400.00)

usage on a single circuit will not be equal to eight hours per day. This is because of the downtime when one caller hangs up and the other initiates a call. The circuit will only provide a certain percentage of availability to both ends. Typically, a single circuit will only result in 3 to 3-$\frac{1}{2}$ hours of total talk time availability. Therefore, if the number of calls exceeds those reflected, another FX line might be required. One can see, however, that the pricing differences and savings opportunities are substantial. The use of FX lines is a mainstay for the ILEC in providing cost-effective solutions for businesses.

WATS (intrastate, intraLATA)

Wide area telecommunications services (WATS) is a special tariff that the Bell System created back in the 1970s. The original purpose of a WATS line was to allow large customers to access the local and long-distance network at a reduced rate. In effect, it is a volume discount offering that AT&T (the Bell System) created for the very large user or for the smaller business having large volumes of calls to specific geographical areas of the country. The way the WATS tariff originally worked varied depending on the volumes of calls that the customer had. This included the following types of service:

BANDS The country, exclusive of the state the customer is located in, was originally broken down into concentric circles or areas of coverage depending on the user location. In Fig. 6.5, the original banding is shown for the east coast, using Massachusetts as a reference point. In each case,

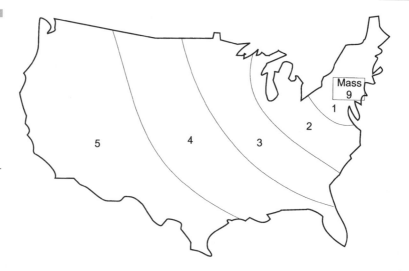

Figure 6.5
Original WATS bands consisted of five contiguous bands emanating from the originating point. Band 1 lines cannot be used to call areas beyond that contiguous band.

the band selected by the customer allowed for calling into states by area code contained in the band. In the example of the east coast, the band 1 line selected allowed calls into the eight New England states. Any call made into the area codes covered by this line was carried without any restrictions. However, if a caller attempted to place a call into a state that was not part of this band, a reorder (fast busy) tone sounded, meaning that service to that area was denied.

If the need to call a greater area of the country existed, then the next higher band was offered. The band 2 line would allow expanded calling coverage to the next concentric circle of states. Included in the band 2 service area is the area covered in band 1. Thus the coverage remained inclusive as the concentric circles got larger. Using a band 5 WATS line allowed calls anywhere in the contiguous United States. The costs for the lines varied by banding, with band 1 bearing the least cost and band 5 bearing the highest cost. Companies using these services had to select the zones that they planned to call into. The burden was on the users to configure the lines on the basis of the communities of interest in their calling patterns.

Obviously, the selection of a mix of bands 1 through 5 was the preferred choice. A company did not want users to arbitrarily call a band 1 state (or area code) on a band 5 line because of the difference in cost for the service. The carrier, in this case AT&T, created this configuration to appease the large customers without offering the same discounted rates to the smaller user. The FCC and the local public utilities commissions were very sensitive to the differentiation of pricing among groups of users.

Therefore, AT&T developed a technique of offering WATS without the call detail that was associated with the standard long-distance direct distance dialing service. The customer was responsible for capturing the details of the calls for allocation purposes through extended call detail recording capabilities on the PBX or Centrex service, and the customer paid a flat rate for the service whether the line was used or not. Thus a difference in the service justified the difference in prices.

FULL-TIME WATS   A full-time WATS service offering allowed the customer to pay a flat rate per line regardless of the number of calls or hours used. Actually, the cost was not truly a flat rate; up to 240 hours of usage per month (8 hours per day at 30 days per month) was allowed. When 240 hours of usage was reached, a premium was applied, but at an extremely low rate per additional hour. The flat-rate costs were dependent on the band selected. In Table 6.3, the original pricing is shown for the rates from the New England area. These prices differed depending on the location of the user.

MEASURED-RATE WATS   Measured-rate service became the option for either the smaller user who had certain volumes or the larger user who had a mix of flat and measured-rate lines. The measured rate allowed a customer to use the first 10 hours at a fee. Each additional hour of usage bore an hourly rate at a cost that was higher, but still less than direct dial rates. Many customers used the measured-rate service as an entry level into WATS service. On reaching a sustained volume at some break-even point, the customer converted to flat-rate services. This was more complex to manage and monitor on a regular basis. However, the typical break-even point ranged around 80 hours of usage. Once again, the onus rested with the customer to monitor and reconfigure the mix of services on the basis of sustained use.

TABLE 6.3	WATS Band	Flat Rate Monthly Cost
Original WATS Flat-Rate Pricing for 240 Hours Usage per Month	1	$ 900
	2	$1,100
	3	$1,265
	4	$1,440
	5	$1,675

ILEC INTRASTATE WATS   Where AT&T offered the interstate calling services, the local operating telephone companies offered the in-state service. Because the ILEC is constrained to offer services in its LATA only, distances covered by WATS line capability must be limited. The ILECs offer long-distance intra-LATA WATS at a reduced rate over the toll call rates. As with the IEC version of WATS, when this was first introduced, the intrastate service had to be separated from the interstate service. The reason for this separation is obvious, because the FCC controls the rates for the interstate inter-LATA offerings whereas the local PUCs control the LATA and state rates. To offer some volume discount to their large business customers, the ILECs had to come up with a discount plan through a tariff filing that would be approved by the local PUC. WATS on an intrastate basis was the answer. The use of WATS on an intra-LATA intrastate basis is founded on the basic line charge per month (typically $37.50 per month) plus usage charges based on hours of usage. The use of the line increases and therefore the cost per hour decreases. Many organizations have been able to reduce their costs per minute on WATS to a reasonable $0.10 to $0.15 per minute, as opposed to the direct dial rates of $0.20 to $0.25 per minute or more. Clearly, the benefits of volume discounts are the savings that can be achieved.

ONE PLUS WATS   Once the costs for WATS service became a commodity, the carriers had to come up with a new offering to appease the smaller business. To do this, the One Plus WATS was offered. Using an access line to the long-distance network within the ILEC's domain, the user can place long-distance calls on a single access line to any point in the LATA. Therefore separate lines are not required for the customer to use a discounted plan. As a matter of fact, the ILEC can provide the same service to route local calls alongside toll calls. Consequently, the business user can use the service even though the telephone system might not accommodate added lines for special purpose use. Discounts can be gained through a billing arrangement rather than through special circuits. This makes the service readily available to the masses and saves the customer a significant amount of time and money over the current mode of operation.

800/888/877 Service

Introduced over 25 years ago, the toll-free arrangement became the service to vie for. While businesses were reaping the benefits of reduced outbound calls through the WATS tariffs, the other side of the business dealt with the incoming traffic. As users recognized, the cost of telecommunications expenses was escalating dramatically. Customers located far from the supplier had some hesitation about calling long distance to place orders. Further, the supplier's salespeople out on the road were forced to carry pockets full of change or a telephone credit card. The credit card was kept as a perk for some; only senior executives were issued a credit card to place calls back to the office from the road. Consequently, many users of the telephone network set up a work-around procedure to overcome the cost of long-distance calling back to the office. A customer or an employee of the organization would dial an operator-assisted call, collect person-to-person. This was the most expensive cost to make a call. However, the caller would also ask for him- or herself or some fictitious individual. This would be used as an indicator that the called party answering the incoming collect call should refuse the call by saying that the individual was not there. The operator would then suggest that when the asked-for party returned, he or she call "the caller" back at the number. The called organization would then hang up and grab an out WATS line and return the call immediately. This saved the user a good amount of money on the telephone bill because the call was sent out at a reduced rate. Further, the incoming collect call based on a person-to-person calling rate would not be billed to the recipient of the call because the call was literally not accepted. To overcome this use of the network for the incoming fictitious numbers, the carriers had to come up with a solution. They could not bill these calls, even though completing them tied up the network and human resources. This arrangement was also used by residential customers who had children in schools or who were traveling: the student could call home asking for himself—the signal that he or she had arrived all right.

Thus the 800 or toll-free calling service was introduced. This is nothing more than a reverse-billing WATS service where the callers can enjoy the benefit of calling at any time without having to worry about the cost of the call. The owner of the 800 WATS service had to worry about the volume of calls and the origin of the calling party. The ILECs offer in WATS, as part of the potpourri of services for the intrastate intra-LATA areas, to allow the reverse charges of the calls. This is a tariffed item similar to the outward service where a monthly access fee is charged for the line plus usage. When the 800 service was first introduced, the same flat- and measured-rate services were available. As the commodity business of WATS caught on, the same evolution took hold whereby a customer could use the same access line for incoming regular calls and for incoming 800 calls. The limitation is that only one call per line can be active at a time. If the customer calling a supplier uses the line for an 800 service call, then others trying to dial direct into the called party will encounter a busy tone. With the possibility of having an 800 number, the supplier can accomplish the receipt of a call without inconveniencing customers through an expensive call to the location. This allows the customer to call toll free to the supplier regardless of the distance, feeling no remorse whatsoever about the long-distance call. 800 area code numbers are rapidly being depleted, so the carriers now offer 888 and 877 as additional toll-free numbering schemes.

Directory Services

The ILEC also offers a directory assistance for a fee. The typical arrangement offered by the ILEC is to look up called parties' numbers using operator assistance. Once the number is obtained, a voice response unit takes over, playing the number out to the interested party. The typical fee for this type of directory assistance is from $0.35 to $0.50 per call. Some companies offer an initial three free operator assists for directory assistance. Newer versions of this service are emerging in which the operator still keys in the name of the requested number and then the voice response unit offers to complete the call on delivery of the message. The fee for this connection and automatic dialing is from $0.35 to $0.85, depending on the location. These are newer methods of offering assistance and at the same time generating new revenue streams.

Although it might sound excessive or impractical to have a calling party agree to this arrangement, one need only think about trying to get a telephone number from a phone booth while on the road. The pen always runs out of ink at the time the number is being announced, or the rain starts to pour while you are trying to write on a piece of paper. Either scenario is both frustrating and normal. Trying to remember a number in your head compounds the problem all the more; therefore this added service is not as far-fetched as it might sound.

Off-Premises Extensions

Another form of leased-line service offered by the ILEC is the use of off-premises extensions (OPX). Let's assume that an organization is in a large office building. This user has plenty of growth from a telephone system (a la PBX) perspective, but not as much square footage to sustain growth of human resources. If, in fact, the organization runs out of space, it becomes necessary to move people off site for either short-term periods while reconstruction efforts can be accommodated or longer-term periods that could be months or years. In either case, moving people off site becomes a problem in terms of managing the telecommunications flow. A new phone system might be required as a purchase at the off-site location. Along with the new phone system, new lines will be brought into the building for the branch or department that is relocated. This means that customers might have to remember two different numbers when calling the organization. This can be inconvenient for the customer, and any such inconvenience could lead to an unhappy customer. By all means telecommunications should be friendly, not inconvenient.

Above and beyond the problems caused for customers by this arrangement, one also must look at what it might do to the internal communications flow of the organization. When dialing between offices in the same primary building, one merely dials a four- or five-digit telephone extension number. However, when calling this remote group after getting internal dial tone, the user must dial 9, followed by a seven-digit telephone number associated with the newer building. At this point, the call might just go to a receptionist who screens all incoming calls, because direct access to an individual is probably impractical. The receptionist will then either "buzz" the called party on an intercom or page through the building for the called party to pick up line 1 (or 2, 3, etc.). Frustrations with the delay inherent in this process might build. The process can also fail through constant cutoff conditions or through extended hold times while waiting for the intended party to come to the phone. Therefore morale can be affected within the organization. Another issue is that the remote group will have to make long-distance and local calls from this new system, which might be more expensive. This gives these people the feeling of being orphaned children as far as the organization is concerned.

To overcome this problem, the use of off-premises extensions might fit the need better than the separate system as described above. This means that every user being relocated will maintain his or her extension number from the primary telephone system, but that the ILEC will run (or install) private lines between the two locations. Every extension will have a two-wire interface to the telecommunications system and a single-line set at a remote building. The ILEC merely provides the wires from the initial location to a demarcation point at the remote location (Figure 6.6).

Figure 6.6

Using off-premises extension, the LEC provides a two-wire facility (leased line) between the two locations through the wire center. Dial tone at location B comes from the PBX location A.

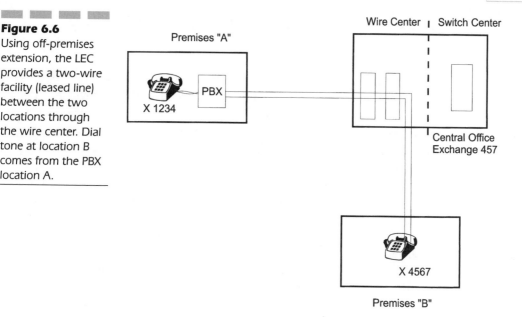

The ILEC bills this on a per-station mileage base per month, plus a one-time installation fee. Mileage is handled differently by each provider, so the pricing is deliberately avoided in this discussion. The customer then plugs into the outside lines from the ILEC with a piece of equipment (i.e., a single-line set). The ILEC runs the wires through the wire center at the central office(s) involved, as with other private line services. The dial tone at the end unit is drawn from the customer's switch (PBX), not from the local central office. All features and functions available to the primary location, local and long-distance services, and extension users are accessible from this OPX. The only difference is that the buildings are separated and the single-line set might not give these remote users capabilities identical to those available in the office. This is an analog telephone set for the most part, because digital display sets and feature buttons are not available on a single-line analog desk set. However, the users are still treated as part of the organization and it is not apparent to the callers that this separation is in effect. That is the primary goal of the OPX, and it does work. Some suggest that the OPX is a better solution, while others suggest that separate systems work better. This is a matter of personal preference and should be analyzed in the actual circumstance.

Access to Interexchange Carriers (Equal Access)

The ILECs have also provided the access to the long-distance suppliers on an equal-access basis since 1984. Prior to 1984, the ILEC provided access to the interexchange carriers (IEC) based on the interconnection arrangements that they had set up. Typically, this was through the AT&T network because that was the way things were done back then. But when equal access was mandated, the ILECs had to allow connections to any carrier of the customer's choosing. This is done according to user selection, not as a fee-based service. Any long-distance carrier marketing in a specific area is granted access to the customers on an equal basis, as was provided to AT&T. The IECs will be charged for the access fees that they build into their rate structure and pass along to the user. The access is based on the type of connection available and the distances separating the carrier's point of presence (POP) and the ILEC's central offices involved. The customer is given a choice of primary (or presubscribed) interexchange carrier (PIC) for each line installed. The ILEC then sets a database on how to switch or hand off calls destined to the IEC from each customer. The connections are shown in Fig. 6.7. The ILEC might hand the call up to the

Figure 6.7
The LEC hands calls off to the IEC depending on the customer's selection of which IEC will carry the call. Various forms of media can be used to interconnect the LEC and IEC offices.

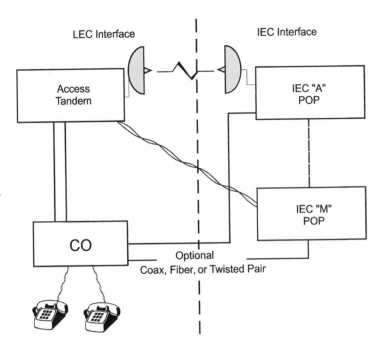

access tandem switch (class 4 office) or optionally have connections directly to the IEC in an equal access end office (class 5 office).

Competitive Local Exchange Carriers (CLECs)

Because of the deregulation and the Telecom Act of 1996 as discussed in Chap. 1, the competitive local exchange carriers (CLECs) can offer services very similar to those of the ILEC. As mentioned in Chap. 1, the CLEC may well choose to offer both a facilities-based or a non-facilities-based service. The competitive LECs will be offering dial tone as their primary vehicle to attract customers. This may include newly emerging entrepreneurial organizations, the IECs, electric companies, local telephone equivalents, other utilities (gas, water, etc.), or cable TV companies. Regardless of who the player actually is, these providers will emerge and appear throughout the country. As they build their infrastructure or choose to use the telephone company facilities, they will begin to offer dial tone and a whole range of services as discussed above. One must then decide who to use for the local access and dial tone services. Will the choice be a pricing arrangement? Or will the telephone companies have frustrated their end users sufficiently to warrant consideration of a new provider? In either case, the new providers will be in a position to provide services in a limited fashion to a localized audience.

These new audiences will be looking for one-stop shopping—the ability to use a single source for dial tone, long distance, cable services, equipment, internal wiring, desktop devices, LAN components, and any other new technology that comes along. It will be these CLECs that will attack the last bastion and the last mile. Many of these CLECs are now offering Centrex service in their operating areas. With the Centrex they are offering local dial tone, long distance, equipment, and wiring all on a single bill. The equipment offerings are based on monthly or annualized rental agreements. This represents the first step toward the one-stop shopping mentioned throughout this book. The telcos obviously are fighting fiercely to prevent such a competitive situation. Their posture on this situation is that the competitors will be *cream skimming.* What this means is that the competitors will offer services to the large organizations where they can sell hundreds of access lines, as opposed to trying to provide rural communications services to the masses. With the Telecom deregulation, these CLECs may well come in and rent the facilities from the local telephone companies, then turn around and resell them to the end user. The kick here is that the reseller will be able to sell the services at a reduced rate. The telephone company that builds, provides, and installs the service may charge, for example, $20 per month for local dial tone. By law, the company must sell this to its competitors for a discount of somewhere between 17 and 28 percent. Using an average of 25 percent, then, the telephone company must sell dial tone to the competitor for $15. Now the competitor can turn around and sell it to the end user for $18 per month. The telephone company still installs and maintains all of the service, but now has lost $5.00 in revenue. The competitor, on the other hand, has bought the service for $15 and sold it for $18. The $3.00 profit is gravy! The end user has bought dial tone for $2.00 cheaper per month or approximately 10 percent off. Although 10 percent may not sound like a lot, it has been determined that a lot of users will jump ship for a mere 10 percent. This is especially true when the customers find out that the same provider still owns and operates the network. It is strictly a billing arrangement.

As the CLECs begin to spread out throughout the country, many users will be exposed to four, five, or six providers within a community offering services. As mentioned in Chap. 1, the opportunity for the end user to barter or bargain may well exist. As the new players are vying for market share and market points, they'll begin to offer some of these services at or below cost. At a cost or below-cost arrangement, the end user stands to gain significantly. One caveat to this, however, is that it can't go on forever. In a matter of a couple of years these providers will recognize that in order to profit and prosper, they must raise their rates. Therefore they will be raising their rates or adding new value-added services for a premium, much like telcos now do, and the end users will likely accept it.

These new value-added services may include the ability to access the Internet at high speed or the ability to have two-way simultaneous conferencing at the local loop, or, best yet, video dial tone. It will only take a few years to see how these new challenges and opportunities will play out. In the meantime, customers will be jumping from provider to provider in an attempt to reduce their monthly cost. This was evident after the Divestiture Agreement of 1984 when the long-distance marketplace was deregulated. Users would sign up for one carrier for a period of say, three or four months; then, after all the rewards and benefits of using that provider had been used, customers would change to a new vendor. Bargaining, discounts, and giveaways were all present throughout the past 15 years. This still goes on today; however, it has tailed off somewhat.

Interexchange (IEC/IXC) Carriers

The interexchange carrier (IEC or IXC) is the long-distance carrier. In general terms, this carrier connects to the ILEC or CLEC with circuits from its point of presence (POP) to the central office at the local telephone company. Most of the major players in this arena are IECs such as Sprint, MCI/Worldcom, AT&T, Cable & Wireless, and so on.

The IECs can carry either long-distance switched and/or private line service. The major players provide both. Some of the IECs, however, primarily provide private line service, depending on their charter. Of course, MCI and British Telecom (BT) were going to merge, but Worldcom jumped into the fray and acquired MCI, keeping them in the number two slot for now! This leaves British Telecom without a partner, so watch for the AT&T/BT partnering. There are some benefits to providing switched services, which at this point is a commodity service.

Typical services provided by the IECs include interstate and inter-LATA capabilities as follows.

Switched Long Distance (DDD)

With direct-distance dialing (DDD), the IECs provide for the end user to pick up a phone and dial anywhere in the continental United States. So long as the end user has selected the IEC as the primary interexchange carrier of choice and the end user is served by an office that has equal access, the user merely has to dial 1 plus the 7- or 10-digit telephone number desired. The carrier will then use the capability of the switched public network to route the call to the appropriate end point. Many of the carriers have the ability to let the user access their network through the one plus network access or through a leased line into the carrier's point of presence (POP). The POP can be either the access point into a switching system that will route the call to the appropriate state, or a closet in which the carrier has terminated equipment on a leased high-capacity circuit.

To better explain this concept, let's assume that a long-distance provider wishes to set up an operation in downtown Philadelphia. However, the closest points of access into the carrier's long-distance facilities (the microwave radio systems or the fiber-optic cabling system owned or leased by this particular carrier) are either in Washington, DC or New York City. To begin service in the Philadelphia area, the carrier could rent space, buy a switching system, and run cables or radio systems out into the area (Fig. 6.8). A second option is to rent space and put some equipment (such as a multiplexer or a rack of dedicated wires) in a closet or a hotel room. As the carrier begins to add customers, the LEC connections from the customer location to the IEC location will be on a two- or four-wire local loop. Rather than terminating into a switching system, the ILEC's or CLEC's* wires will be terminated onto a frame or backboard in the rental space defined above. The IEC will then take the wires from the LEC demarcation point and cross-connect them to a different pair of wires running from Philadelphia to Washington or New York (Fig. 6.9), where the carrier has a closet approach. At this point the IEC will bring the customer's connection into the switching system located within the boundaries of the network connections.

Although this is not a major problem, the IECs using this system are carrying all the traffic from a single point to maximize their usage of the circuits emanating from their offices. However, if things go wrong, the testing and troubleshooting can be prolonged because the connections are run across multiple extra miles of wires to get to the switch, adding to

*For the sake of simplification, references to both CLEC and ILEC will be used as a generic LEC throughout the rest of the book.

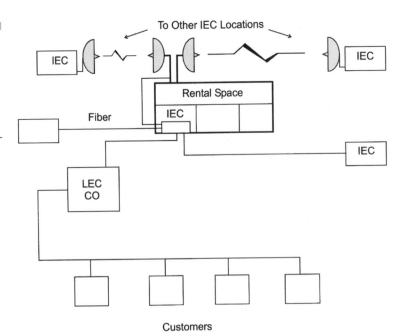

Figure 6.8
The IEC rents space, buys a switching system, and connects to POPs located in its area. This is a true switching center.

■■■ ■■■ ■■■ ■■■

Figure 6.9

To open a new area, IEC A may rent closet space only to provide a physical interface to the LEC wires. From there, the IEC can lease lines from a competitor or install its own lines to the city where a switch is located.

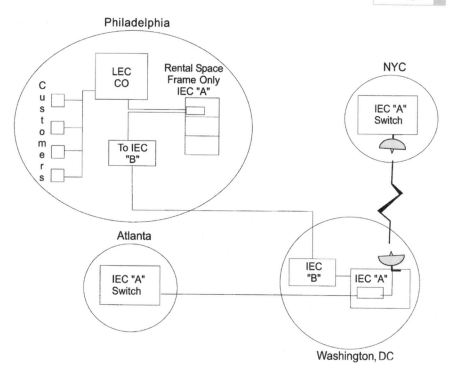

prolonged delays and possible complications with the integrity and quality of the circuit. Many of the IECs start out using this closet approach until they increase the volumes in the location; then they build a switch in the area when the volume has sufficiently justified the expense. Others keep the arrangement set up this way to maximize the total throughput at their primary locations and never plan any additional sites, yet they still market in the areas where they do not really have a switch. They do advertise that they have a POP in the area, which gets to the crux of the definition. This point of presence is different than others, however. One should at least understand just what the differences are and ask the carrier how it handles the interconnection.

Credit Card Service (Calling Card Service)

All of the IECs (and the LECs for that matter) offer the ability to take advantage of their network services using a credit card or calling card. This is a matter of convenience for the customer. When traveling, people need to use hotel/motel phones, pay phones, and generally any other service that allows access to the long-distance networks. Rather than have a customer run around the country with a pocket full of quarters ($0.25 pieces), which would be very difficult and would discourage the use of the network, the credit card allows the caller to access the network at will and use the service unhindered. This service carries a surcharge with it. For every call placed on the carrier's network, an initial surcharge of between $0.50 and $1.00 per call is tacked onto the first minute of the call. Given that the average length of a call on the network is five minutes, this adds a penalty of $0.20 per minute on the call. If a caller does average more than five minutes, the cost of the call becomes close to the direct distance dial rate. The benefits and discounts of using a single carrier help to offset this surcharge.

When the calls initially had to be handled by an operator-assisted arrangement, the surcharge was introduced to offset the cost of the personnel needed to process these credit card calls. However, this is a misnomer today because the automated process is now used where a customer dials 0 plus the 10-digit telephone number and waits for a tone (or bong). After receiving the tone, the customer then continues by dialing the credit card number to be charged. From there, the call gets processed across the network and completed. For billing purposes, this is straightforward. But, no operator was ever involved, since the automation took care of the entire call. Why then do the carriers continue to charge the operator-assisted rates? To add insult to injury, if the need arises for an operator to complete the call for whatever reason (not on a Touch-Tone phone; the dial pad is shunted out after the 0 and the 10 digits are dialed), then the carrier bills a premium for the operator to assist in the completion of the call. This sounds like a double whammy in terms of the billing mechanism, and technically it is. The carriers would have the user believe that the more expensive service of using a human in the call completion has been reduced under the automated process. One never knows. As a management function, it is very difficult to know, because the calls being placed are all being made by others on the road. The end user is primarily interested in ease of call completion, and so does not worry about the cost issue. Moreover, these end users will not be as intent on reporting problems that occur on the line. Many of the carriers will offer a credit for calls that are poor quality or for cutoff situations where the end user has to dial the called party a second time and incur the credit card setup cost for a second time. However, the users either are unaware that this is an option or are uninterested because this is just another inconvenient step. One could possibly save a significant amount of money by educating the user and by following up on the billing at the end of each month.

WATS Service

Just as with the LECs, the IECs all have some form of WATS service as a volume-discounted offering. The more of this service a customer uses, the less expensive the cost per minute will be. To use the WATS services of many of the carriers these days, customers have several options (Table 6.4).

In each of the cases shown in the table, the options are many, indicating that the carriers are concerned with offering some form of discounted service to users, depending on the volume of calls and the distances that will be called. In every case, the user still benefits over the regular long-distance (DDD) rates, but the variables are really becoming commodity items again. When looking at the options, customers can easily become confused; yet, with all of the changes going on in this industry, it cannot be assumed that the carrier's representative will be able or willing to point out added options or discounts. The best way to take advantage of this type of service is to consider no more than an annual contract period or a month-to-month contract with the carrier. This way, the user will not be locked into a pricing structure that prohibits on-the-fly changes and adjustments as usage changes. The carriers will, however, dis-

TABLE 6.4

A Summary of the WATS Variations and Pricing Arrangements

WATS Service	Initial Cost	Variable Costs
Dedicated lines	$37.50	Based on usage by bands. The usage follows a rate structure: First 10 hours Next 25 hours All over 40 hours
Dial One WATS		Cost per minute based on distance and time of day.
Reach Out America	$10.00	Cost per minute based on distance and time of day. Minimal discounts applied.
Virtual WATS (SDN/VPN)	Access fee into VPN or SDN	Special billing based on distance, time of day, and volume used.
Special Tariff 12 WATS service	Access fees at the T1 or T3 rates	Special pricing based on long-term contracts of three to five years. Greatest discounts apply for WATS.
Pro WATS	$5.00	Plus cost per minute on distance and time of day.
Multilocation WATS	$2500	Flat rate to consolidate bills plus largest discounts possible on WATS.

courage this option by stating that the customer stands to benefit from longer-term contracts based on reductions offered over and above the standard offerings. Keep this straight and in perspective. The value of the end user is critical to the IEC as a result of the fact that equal access has been administered and implemented because there is little or no price sensitivity on the cost per minute. The differences between carriers can be as little as $0.002 (stated as two mills of a cent), but if the carrier sincerely wants the business the differences can be far more significant. Let's assume that an IEC can deliver a cost per minute of $0.20 to a customer for long-distance calls. What is the carrier's actual cost, as opposed to the rate charged? Interestingly, the carriers (at least the major players) can deliver a call at their cost for $0.03 or less per minute. Now, in order to generate profits, they might mark this up to approximately $0.06 per minute (representing a 100% gross profit).

So, how can they sell service to the end user at $0.20? Well, first they have to pay access fees to the LECs to gain the access on the LEC's wires to the customer's premises. The typical access fee charged on a call-by-call basis comes to $0.07 per minute per end. Looking at the total, the cost per minute is shown in Table 6.5. This table covers a call from San Jose, California to Philadelphia, Pennsylvania.

Obviously, the numbers reflected in this table are for the casual user of a WATS service, not the larger users who have thousands to hundreds of thousands of minutes of usage. However, to overcome these costs and to add some degree of new attractiveness to the end user community, the IECs are now offering bundled services on T1 access lines. This works out to mean that the IEC or the customer rents the T1 line as a dedicated link. The cost of the leased line is fixed, doing away with the usage sensitivity of the calls. At a cost of between $350 and $400 per month, the cost per minute is reduced. In Table 6.6 the costs of the calls are shown. In this scenario, costs are compared for 24 individual WATS lines versus a T1.

TABLE 6.5

The Total Cost Picture Charged at $0.20 per Minute to the End User

Description	Fees
Local access charges at the LEC end in San Jose	$0.07
Interexchange carrier cost	$0.03
Interexchange carrier profits	$0.03
LEC access fees in Philadelphia	$0.07
Total	$0.20

TABLE 6.6

Comparing
Dedicated WATS
(Analog) Lines to a
T1 Service

Description	Using Dedicated WATS Lines	Using a T1 in Lieu of WATS Lines
Line rental	$24 \cdot \$37.50 = \$900$	$400
Cost per minute on access	$\$0.07 \cdot 24 = \$16.80$	0
Cost per minute for WATS	$\$0.15 \cdot 24 = \$3.60$	$3.60
Cost of 100 hours usage	$[60 \cdot 100 \cdot (3.60 + 16.80)] + 900$	$[60 \cdot 100 \cdot 3.60] + 400$
Total	$123,300	$22,000

800/888/877 and 900 Service Offerings

As mentioned earlier, the 800/888/877 service is a derivative of WATS line service. When initially introduced in the AT&T network 25 years ago, the 800 service was called In-WATS. This was a mechanism to replace the inward collect calling services. As the name changed, the service was made more flexible. A call coming into the 800/888/877 number is billed at a usage-sensitive rate based on the band from which the call originates. Further, as a call is routed and delivered into an 800/888/877 service, the customer can specify that the network deliver the call to different locations or to specific agents in a call group on the basis of the call's origin. This inbound service can be called 800/888/877 service, 800/888/877 Megacom service, or other names. Further, small business services (such as Readyline or 800/888/877 Starter line) are available. In each of these cases, the service is priced according to the needs of the customer. For example, the 800/888/877 service will be used by a medium-size customer who selects dedicated lines for the inbound service. The 800/888/877 Megacom service will be chosen by larger customers who use a T1 line to bundle 24 In-WATS lines together. The Readyline service would be used by the small branch office that does not have line space on its existing telephone system and so routes the calls into an existing incoming line that is shared with a regular incoming dial tone line. Finally, the Starter line would be used in a very small office or a home office where the customer uses the smallest volumes routed to a specific telephone number shared with a business or residential line. The prices vary with the choice selected, but the flexibility is what is being highlighted here. Table 6.7 shows the different uses as orders of magnitude pricing, rather than specific pricing arrangements.

900 services are pay-per-use offerings that have drawn mixed emotions and reviews. When first introduced, the 900 offerings were the domain of

TABLE 6.7

A Summary of
Various in WATS
Offerings

Service Offering	Average Pricing
In-WATS 800 service lines	$37.50 per line per month plus usage at $0.15+ per minute
800 Megacom service	$400 plus usage at $0.10–0.12 per minute
Readyline service	$20 per month plus $0.20 per minute
Starter line service	$6 per month plus $0.26 per minute

pay-per-call providers, such as the lottery services, dial-a-porn, dial-a-joke, and so on. They could be accessed by all callers willing to pay for the service charge associated with the call. The owner of the 900 service offering establishes an account with the IEC. From there the owner then establishes a rate for the service ($2.00 per call, $25.00 per call, $0.50 per minute, or any other derivative of this pricing scheme). The IEC assigns a 900 area code number that the owner then advertises. The next step is that the IEC establishes a rate to be charged to the owner. Now the customers start to call the 900 number, based on the inducement to call, regardless of the offering. As the caller dials the number, the call is delivered to the advertised location of the owner. This is then billed to the caller on the basis of the advertised rate. The IEC bills the call, collects the revenue, holds out a piece of the revenue based on the agreed-upon price (10% or more), and, after collecting the revenue, sends a check for the difference to the customer (owner). In Fig. 6.10 the flow of this operation is shown. This service has created many millionaires in the industry.

However, to override these premium services, many organizations disallow the access to 900 area code numbers. Some schemes use the features of the telephone company to block this area code from access by individual users; others use the features of internal telephone systems to deny access to the area codes. However, new uses of the 900 service are being introduced. For example, Novell, a manufacturer of network operating systems software for LANs, now has a 900 service for customer support and maintenance. This, of course, means that the telecommunications managers in business must allow some calls but deny others. Confusion will reign in this environment.

International Access

The IECs provide gateway access into international direct distance dialing (IDDD) through their service offerings. The call is carried across the IEC's network, then delivered to the international carrier at a midpoint. In this

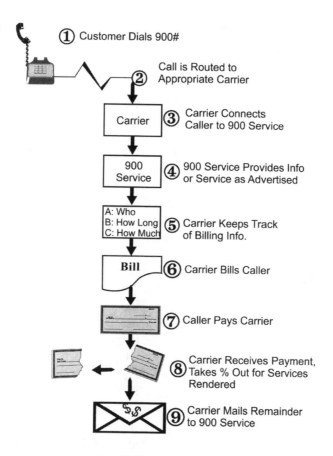

Figure 6.10
The flow of a 900 call process.

① Customer Dials 900#

② Call is Routed to Appropriate Carrier

Carrier ③ Carrier Connects Caller to 900 Service

900 Service ④ 900 Service Provides Info or Service as Advertised

A: Who
B: How Long
C: How Much ⑤ Carrier Keeps Track of Billing Info.

Bill ⑥ Carrier Bills Caller

⑦ Caller Pays Carrier

⑧ Carrier Receives Payment, Takes % Out for Services Rendered

⑨ Carrier Mails Remainder to 900 Service

arrangement, the IEC and the international carrier share the revenue. IDDD is a service that will allow a user to pick up a phone anywhere and dial an international country code sequence, followed by a city code and then the variable-length telephone number of the called party. The IEC does the billing on the basis of cost per minute, then collects the revenue for the international call and disburses it according to a revenue-sharing arrangement with the international carrier. This makes it simple for the user of the service; because all the information is on one bill, the call can be logged and appropriately charged without having to pull together different pieces from various bills.

Foreign Exchange (FX) Service

From an IEC perspective, the foreign exchange service that spans across LATA or state boundaries falls into the IEC's domain. A foreign exchange line from Dallas to New York (or any other two points under the IEC domain) is provided as a leased line with dial tone coming from the foreign central office. The IEC can bill this at a monthly rate for the mileage points between the two cities, and will also be responsible for the end-to-end connection if the customer so desires. Many IECs and LECs will allow for the long-haul portion of the circuit to be the responsibility of the IEC and for the billing for the local dial tone service in the remote city to become a billing responsibility of the LEC. This is less desirable because the customer must then assemble various portions of bills to gather all of the facts and costs of an FX line. The service will be billed primarily, at a flat rate; however, originating and terminating minutes might bear a cost from the LEC at the end of the circuit. One should verify all costs associated with the line as a means of cost justification. The use of a foreign exchange line between the two points mentioned might have applications other than the obvious benefit to a customer. Initially, the calls made by a customer to a supplier's location can be made on a foreign exchange. This allows the customer to place a local call that is then connected to a long-distance line 1,000 miles away, avoiding the toll call for the customer. However, another application of the FX is for, say, a Dallas-based company that is tentative about opening an office in New York. Rather than renting an office, buying or renting furniture, hiring a staff, and trying to break into a new market, this company can take a special group of agents to deal with future New York customers. Customers can call a local seven-digit telephone number—implied with this is the local presence of the supplier. As some companies feel more comfortable dealing with organizations that are local, the supplier can present the image of local presence using the FX. Therefore, for a minor investment, the supplier can then begin marketing in the New York area. If the volume of calls is sufficient to warrant a staff in the New York area, then the Dallas-based organization can staff an office. However, if the volume and acceptance level do not warrant a full-time staff, the supplier can still service the customer base in New York without the investment in real estate and office furnishings. This is an effective use of telecommunications as a strategic corporate resource.

Off-Premises Extensions (OPXs)

When a customer of an IEC has many locations around the country, or around many other LATAs, the IEC can provide private line service in the form of an off-premises extension. This has already been discussed in the LEC environment, so the IEC's involvement only includes the extension of this private line across boundaries that relegate the service offering to the IEC domain. Once again, the private line draws dial tone services from inside the customer-owned and -operated PBX or Centrex. The IEC merely provides a quality circuit to any point outside the originating LATA at a flat rate. Figure 6.11 is a representation of the use of an OPX in two different states. The cost is dependent on the mileage associated with the circuit. There is no magic in this offering; the IECs provide many connections with this service.

Operator/Directory Assistance

Just as with the LECs, IECs also provide operator services to assist with call completion. A fee is charged for the intervention of the operator. More and more services are being displaced from the venue of human intervention to the use of interactive voice response and recognition services (IVR). In the systems of more than one carrier, operator assistance and directory assistance are initially obtained through contact with a human but then revert immediately to IVR. As more of these services are provided, IVR will provide the number, offer to complete the call automatically for an added fee, and, if a busy tone is encountered, allow the caller to leave a voice message on the network and redial on a regular basis until the message gets through. This is also a fee-added service. Other systems are now using voice response so that callers on the IEC network can use voice-activated dialing to the 10 most frequently called numbers in a preestablished database. The Foncard2 offering allows a user to preestablish calls to "home," "office," "lawyer," etc. When the caller simply speaks the word into the IVR, the call is automatically dialed. This is a convenience, but not a showstopper.

Figure 6.11
IECs can provide OPX between two different states on a mileage (month-to-month) basis. This is a standard offering from the IECs.

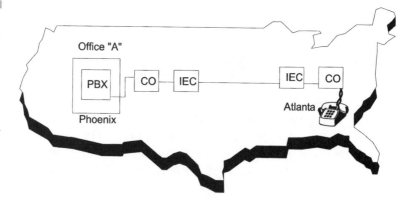

Newer operator and directory services emerged that will allow a customer to dial a special code (00) and get access to anywhere in the country. For a fee, the carriers are offering the ability to work with the caller to help find a number, even if the customer has limited information about the desired company. The operative words here are "for a fee." Nowadays, you can get almost anything you want from your carriers if there is a revenue stream involved.

Remote Call Forwarding (RCF)

Remote call forwarding, as the name implies, is the ability to forward calls dialed into a number from the original location to a remote location across the IEC network. For a fee, the customer rents a dial tone line from the LEC, then establishes an automatic call forwarding arrangement to a remote site in another city outside the LATA or in another state. The customer then pays for the long-distance call from the original site to the forwarded site (typically about $0.25 per minute) on the basis of distance and time of day. The situation is similar to that with the FX in that the caller is dialing a local seven-digit telephone number but the call is routed to a remote site in another city (Fig. 6.12). This is particularly useful in lieu of a private line service, where the private line can be an expensive proposition. If the calling volume is low, the use of call forwarding on a call-by-call basis might be less expensive. Thus the option is straightforward as regards the outlay of money each month. If the volume stays low, then the RCF remains in place. However, if the volume of calls increases and the price increases proportionately, then a private line FX or other service can be used. Another option is the use of the 800 service, but this automatically implies that the called party is far away or long distance—even if this isn't the case.

Value-Added Carriers

Other carriers, called value-added carriers or value-added network (VAN) suppliers, emerged into the network business. Many of these initially were data communications suppliers. They provided services such as dial-up data communications, packet switching, and other miscellaneous services. Initially, most were based on circuits leased from AT&T. The

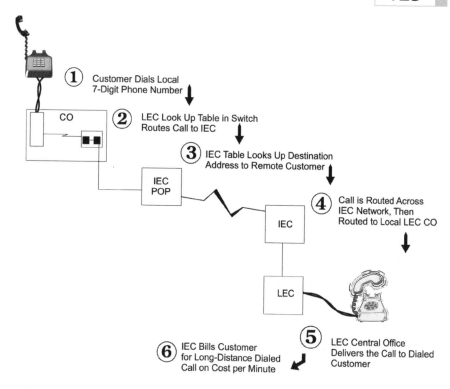

Figure 6.12
The remote call forwarding process.

1. Customer Dials Local 7-Digit Phone Number
2. LEC Look Up Table in Switch Routes Call to IEC
3. IEC Table Looks Up Destination Address to Remote Customer
4. Call is Routed Across IEC Network, Then Routed to Local LEC CO
5. LEC Central Office Delivers the Call to Dialed Customer
6. IEC Bills Customer for Long-Distance Dialed Call on Cost per Minute

primary VANs were Telenet (acquired by Sprint) and Tymnet (acquired by British Telecom). As long as these suppliers provided additional services (or value added) that could be reasonably priced, a niche existed for them.

However, many of these terms have slipped into the background because every carrier is offering some form of IEC and VAN service as part of its portfolio of products today. Even the LECs are building out these services on an intra-LATA basis. So the distinction is becoming very fuzzy. The terms are still used, so be aware.

Newer offerings, such as service bureaus that offer voice messaging services, store and forward message switching, voice to text/text to voice, voice recognition services, and other variations are also considered VANs.

Alternate Operator Services

Alternate operator services (AOSs) began when divestiture and deregulation became a reality. Pay phone services, which were normally part of operator-assisted calling, became a whole new battlefield. Anyone who wished to become an AOS was free to do so. You could buy a pay phone, rent a local dial tone line from the LEC, select the IEC who would carry the call, and be in business. The IEC would bill you for the long-distance calls made on your line. However, the caller would be billing the call on a credit card, so you could charge any rate you wished. For example, a long-distance call at $0.30 per minute could be billed to the credit card at $2.00 per minute. Although this sounds unrealistic, this scenario was a reality. Users were unaware of the rates they were being charged because there was no human operator involved. They found out about the rate when the bill appeared at the end of the month. The AOS providers got deeper into the long-distance business by making arrangements with hotels and motels to carry their long-distance traffic from guests and then offering commissions to the hotels/motels. The rates were even higher by now; the hotel was charging a surcharge for every credit call made in a guest's room ($0.50 to $0.75 per call) and getting a commission at the same time.

Although the customer was perplexed, options were limited. Many of the AOSs and the hotels blocked the customer from the 10XXX access. You were stuck with the carrier and the rates. Your only option was to find a different phone that allowed access to the IECs directly or through the 10XXX route. The FCC has finally stepped in and demanded that the AOS carriers allow the caller a choice, and so the AOS carriers have begun to adjust their rates and practices. Watch your phone bills and compare rates; opportunities exist to enjoy large savings by educating your users and frequent travelers.

Aggregators

A still newer phenomenon in the industry is the *aggregator*. Aggregation is a spin-off of AT&T's tariffs for large companies. AT&T offered larger discounts to organizations with multiple locations. The smaller locations of these big firms were billed for their long-distance and WATS usage as single entities. In an effort to keep the customer and be competitive against the competition, AT&T began offering a multilocation WATS tariff. All charges were consolidated and billed at the greater volumes, regardless of the number of locations and the volume of calls per site. There is a fixed fee for this service and certain stipulations of volume. But to the large company, these conditions were easy to meet. What then of small- to medium-size companies? They had no benefit, even if multiple locations existed. The volumes, guarantees, and fixed fees were so severe that the smaller companies had no way to use this service.

Enter the aggregators. These started out as consultants and entrepreneurs who saw an opportunity. If they could pay the fixed fee, then recruit small and midsize companies to use aggregation rates, the smaller organization could save some money and the aggregator could make a profit. The carrier typically charges the aggregator an annual fee of $30,000. The aggregator in turn recruits hundreds of smaller businesses to sign up. The aggregator then calls the IEC and tells it to bill the customers usage for all sites at the reduced rate.

Some aggregators have even convinced their new customers that they should sign long-term agreements (one, three, or five years) with them, and possibly even split the savings (50/50).

Competitive Access Providers

The local loop has been protected as a monopoly since early in this century. The LEC was the only carrier allowed to deliver dial tone. Regardless of the cost, service, or quality of the lines, users had no options except to bypass the LEC with their own equipment. This might include microwave, copper, or satellite communications. However, bypass only served to get access to the IEC or to another customer premise. The telephone company still provided dial tone and local-switched services. However, a group of entrepreneurs began to lay fiber optics in the ground, connecting multitenant buildings in major cities. The vendors began to offer all-digital fiber-optic bypass solutions to the customers. Bell's turf was being invaded by competitors offering reduced rates, volume discounts, and quick delivery times for the service. These vendors are beginning to offer some attractive options to the end user. They might also offer some relief to the LECs, even though they are perceived as the enemy today.

The LECs might gain more from the presence of these carriers/vendors than they believe possible. Many of these CAPs are also CLECs, whereas others are strictly CAPs providing access and transport. The choice is up to the provider.

Resellers

Resellers have been around for a long time, although they were considered OCCs at one time. The reseller usually owns systems, or switches, that are designed to route calls at the least possible cost. Resellers rent or lease their lines from a variety of carriers (IECs) in various configurations. For example, they will look at volumes of calls to specific cities and rent FX, tie lines, WATS services, etc. They then resell the long-distance service to the end user, who can be business or residential. The volume of calls carried gives the reseller the ability to offer switched services to its customers at rates lower than those of the other IECs. This is like a combination of aggregation, IEC, other common carrier (OCC), and VAN all in one.

The resellers' networks are becoming quite sophisticated. Using the least expensive services from a variety of carriers, they have built very elaborate dial plans, alternate routes, and so on. They do bill the customer and provide a mix of custom billing arrangements including reports, locations, account codes, and the like. They make money and the customer saves at the same time. The larger resellers are also connected to the LECs via the equal access capability, thereby making it easy for the customer to use the service.

At one time, many of the resellers were considered fly-by-night operations. Today, the shadier companies have been shaken out, and the reseller is now a respected part of the network.

Newer opportunities include services, such as switchless resellers. A reseller of long-distance services can negotiate volume pricing deals with the big 4 suppliers. Rather than owning equipment, the switchless reseller merely resells the long-distance services and has the actual carrier handle the call processing. Call detail can be passed on to the reseller so the reseller can bill the user.

Lines versus Trunks

Introduction

In this chapter we present a number of service offerings available from various carriers. Note that, with few exceptions, the offerings in this chapter are based on a set of identical technologies. They differ only in exactly where and how the connections are made, how they are billed, and what information is transmitted over the circuit in addition to the voice signal itself. Underlying all of these offerings are the concepts of *line* and *trunk*. We will use the term *circuit* to refer to a general point-to-point voice-grade physical connection. We will also see that there might be no physical difference between a line and a trunk because a two- or four-wire facility might be used for either. The real difference is the functional use of the wires that is applied to this linkage. The easiest way to define *line* and *trunk* follows.

The term *line* unfortunately refers to more than one type of circuit. In most cases, it includes a connection configured to support a normal voice calling load generated by one individual (typically about 10 minutes per hour for a business user). But in the case of a PBX (see Chap. 9), the term *line* usually corresponds to one connection from the PBX to a desktop. In the case of Centrex, a line is normally one physical connection from the customer site to the CO. With a key system, a line corresponds to one telephone number—but it might also be referred to as a trunk (Fig. 7.1).

Figure 7.1

Places where lines are defined.

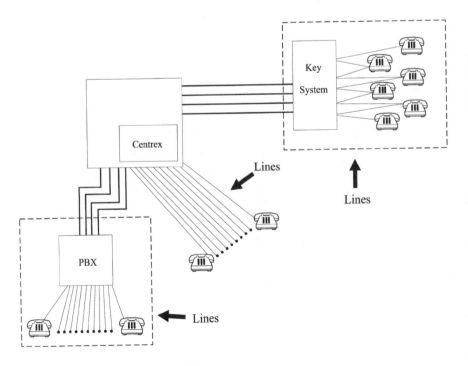

The term *trunk* normally refers to a circuit configured to support the calling loads generated by a group of users—possibly numbering many thousands. Thus a general-use circuit from a PBX to a CO would usually be described—and billed—as a trunk (but see the discussion of DID lines). Connections between COs or offices higher in the network hierarchy would also be referred to as trunks. But note that these trunks are (or at least can be) physically identical to lines. Why then the different terminology? This is shown in Fig. 7.2 with reference to a trunk.

The ability of any given switching system, such as a CO or a PBX, to establish connections is limited. For example, although a PBX might be able to support 200 connections or ports, it might only actually provide 80 paths at any one time. In such a case, if 80 people were to connect to 80 other people (some of them possibly off site), that would account for 160 of the ports; if any of the remaining 40 telephones or ports attempted to access service, they would fail. That is, a user could pick up the telephone and not receive a dial tone. Some systems are configured so that no such failures can happen. In the previous example, if only 160 physical connections were made to the PBX, then it could provide simultaneous service to all of them. Such a configuration is described as *blocking* (Fig. 7.3). Figure 7.4 represents a nonblocking environment.

Normally, a PBX's connections to the CO are configured so that a much higher utilization than 10 minutes per hour is achieved on those ports; a primary benefit of a PBX is the ability to buy fewer telco connections than one has telephones. The CO must be configured so that it can provide connection services to such trunks at this higher utilization

Figure 7.2

Trunks are used between intelligent switching systems.

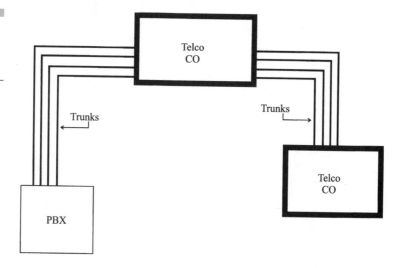

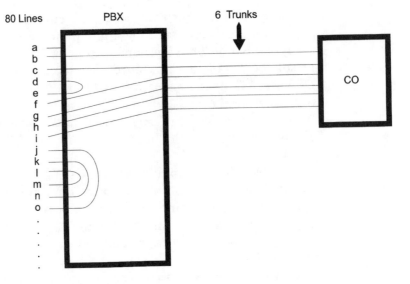

Figure 7.3
Blocking in the PBX arena.

rate, thus using more of the CO's overall switching and connection capacity (COs are not normally configured as nonblocking switches). So, the telco will naturally bill a PBX trunk at a higher rate than a single business line—even though the PBX trunk might be physically identical to that single line.

A final comparison of lines versus trunks would be as follows.

A *line* is an end point from a central switching service, such as a CO or a private automated branch exchange (PABX). The line is represented as

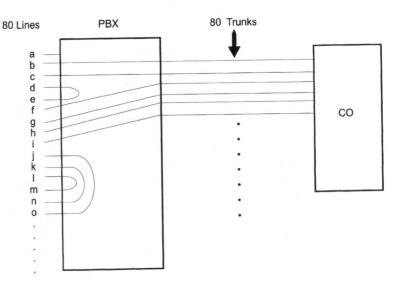

Figure 7.4
Nonblocking represents a 1-to-1 ratio.

the end point on the pair of wires regardless of where the intelligence resides. A line carries one single conversation at a time on the physical channel capacity. It is a billable location for the telephone companies.

A *trunk*, on the other hand, connects two intelligent switching systems. The trunk might be a single circuit carrying a single call at a time, or it might be a bundled service that is multiplexed and carries multiple simultaneous conversations. The difference is that a trunk will be used for switching and routing decisions from the switching offices (CO or PABX). The trunk is continually rather than occasionally used. It is a billable address that can have additional subaddressing capabilities behind it. In the telephone company world, it is the connection between and among other offices in the hierarchy discussed in Chap. 3. In the private user (customer) world, it might be a single connection to the intelligent PABX from the CO. These distinctions offer some variations in billing and utilization.

With this in mind, here are some common configurations.

DID

DID refers to direct inward dialing. From a caller's point of view, this service is in place if the caller can dial a 10-digit number from the outside and reach a specific individual without operator (live or automated) intervention. Thus Centrex normally inherently supports this capability without any additional configuration—everyone already has their own telephone number. A true key system (where telephone numbers are normally shared) can only allow DID if any given telephone number has only a single appearance.

But DID is usually referred to in the context of a PBX. It is a specific PBX feature that must be enabled and configured, with elements set up both within the PBX and also with the telco. Consider as an example a new site intended to support 1100 employees, each with his or her own telephone connected to a PBX.

The first step in arranging DID is to reserve the telephone numbers for all those employees. Let's say the main company telephone number is 555-1234. The telecommunications manager will request a block of DID numbers from the telco, probably about 2000. The telco might say, "Your DID numbers are 555-2200 through 555-4199." Notice that while there is a good chance the block will have the same exchange as the main number, it probably will not include (and we would not want it to include) the main number. The company will pay for these numbers on a monthly basis, but they will not cost anywhere near as much as actual telephone lines. So far, the only thing arranged is the reservation of the block of numbers themselves. These numbers will not be given out by the telco to anyone else. The telecommunications manager will assign each employee one of the numbers in the DID block.

Next, the telecommunications manager must determine how many trunks (or DID lines) in the trunk group will be required to support the calls from outside to the company's employees. These are inbound only, and are in addition to the normal in/out or inbound trunks that serve the main operator, so they must be engineered to a very low level of blocking indeed (see Chap. 8 for more information on traffic engineering). With DID, the telco passes on to the customer PBX the responsibility of handling answer supervision (e.g., busy signals). The DID link is shown in Fig. 7.5.

So, if an external customer calls "Jane" at extension 2313, the customer will dial 555-2313. The telco CO will seize the next available trunk in the DID group (if no trunk is available, the caller will receive a busy signal) and signal along it that there is a call for extension 2313. At that point, if extension 2313 is busy, the PBX must deal with it; the CO is merely passing along the signals. Possible PBX actions include forwarding to a message center, generating a busy signal, or forwarding the call to a specified alternate extension.

DID is most often used to reduce or eliminate the manpower required for a central answering position. The more calls customers can place directly, the fewer must be answered by the company operator. On the other hand, some companies prefer to have all incoming calls answered by someone trained in the way the company wants its telephones to be answered (e.g., "Thank you for calling Kay's deli! How can I help you," vs. "Hello?"). It would generally be a mistake for a telecommunications manager to make decisions regarding whether DID is to be implemented without consulting company management.

Figure 7.5
DID trunks link the PBX and CO together.

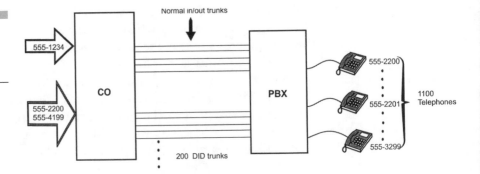

DOD

DOD refers to direct outward dialing. If an employee can dial and reach an outside number without internal operator intervention, then the company has implemented DOD. In the past, when less sophisticated telephone systems were available, it was not uncommon for a company to route all of its outbound calls through an internal operator. The operator's responsibility was both to screen calls ("no, you may not call Australia from that telephone") and to route the calls over the appropriate facilities (e.g., the right WATS line—see below). With the advent of modern PBXs and Centrex, such limitations can be programmed on a telephone-by-telephone or even user-by-user basis, eliminating the need to involve an operator in outbound calls. DOD is a term not often used these days because few companies consider not providing it.

FX

FX, not to be confused with FAX, refers to a foreign exchange circuit. In this case, *foreign* refers to a CO other than the local CO, not to a location outside the country.

Consider the case of an airline that wishes to locate all of its reservations clerks in Atlanta. It cannot expect all of its customers to pay long-distance charges to make reservations. What are its alternatives? One possibility is a group of 800 circuits. Indeed, it will probably have a large number of those, but 800 trunks cover large areas (and are priced accordingly). What about service for customers calling from large, high-density metropolitan centers, such as Chicago? Perhaps a more focused service might be more cost effective. The FX grouping is shown in Fig. 7.6.

Think of an FX line (or trunk) as two-thirds of a dedicated point-to-point (or tie) connection. It starts at the customer's location, connects to the local CO, and extends from there to another foreign CO anywhere in the country. There is a fixed monthly charge for all that mileage, but there are no usage-sensitive charges for these miles. At the foreign CO, it is open. It has a telephone number associated with that foreign CO. Calls made to that number ring at the customer's location. Calls made from the customer's location over the FX line emanate from the foreign CO, incurring only local charges for the call from the foreign CO to the called location.

FX lines are often used by companies to provide local numbers that customers can call in cities where the companies do not in fact have offices. In the airline's case, it could arrange a group of FX lines from its

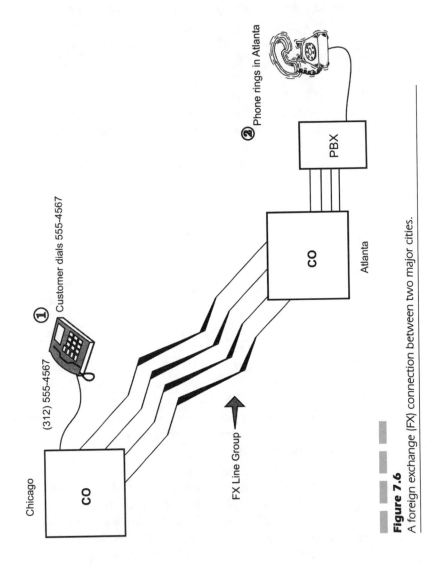

Figure 7.6
A foreign exchange (FX) connection between two major cities.

Atlanta offices to a Chicago CO. All of the lines could share one Chicago local telephone number. People from anywhere could call the number, but normally only Chicagoans would, because it would appear only in their telephone book—and it would be a local call only for them. If the airline wished to allow it, service representatives could also place calls from Atlanta to Chicago over the FX lines. The calls would be billed as though they were placed from within Chicago. Perhaps calls notifying customers of changed flight information might be placed this way.

OPX

OPX refers to off-premises extension. An OPX line permits a telephone not at a company's location to function to all intents and purposes as though it is located at the company's location. This capability becomes particularly interesting with the recent increase in telecommuting. Suppose an employee plans to work at home. One of the problems to overcome in such a case is the isolation such a worker might experience. Providing the employee a telephone that looks like an internal line at the company might help to reduce the problem. Others calling the line within the company will dial an internal extension, which will ring at the employee's home; if the employee wishes to make a long-distance call, he or she usually just dials 9 and the rest of number just as at a desk at the company's location. An OPX link is shown in Fig. 7.7.

Figure 7.7
A typical off-premises extension (OPX).

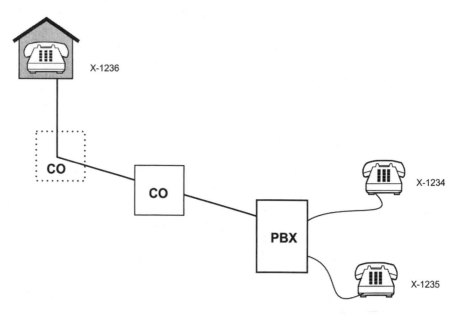

X-1236

CO

CO

PBX

X-1234

X-1235

As with an FX line, an OPX connects from the company's location to the local CO, then continues via whatever intervening COs are necessary until it terminates directly on a telephone at another location. A key difference from an FX, however, is that on the PBX an OPX is connected and configured as a telephone rather than a trunk. This results in a limitation on the type of service provided: normally, only an analog telephone can be used at the end of an OPX because the digital signaling between a PBX and its proprietary telephones will probably not successfully make it through the various analog and digital circuits that make up the OPX. This limitation is not normally a show-stopper; rather, it just imposes on the telecommunications manager the need to configure the PBX to support a certain number of analog telephones as well as the digital telephones that might be used in house.

TIE Lines

In Chap. 3 we discussed the general categories of dial-up and private leased or dedicated point-to-point circuits. In the voice communications industry, a *TIE line* (also sometimes called a *TIE trunk*) refers to a private point-to-point circuit used to connect two voice facilities. For example, a dedicated link between customer PBXs at two different locations would be referred to as a TIE line as shown in Fig. 7.8. Other examples of TIE lines might include a link between a PBX and a Centrex system, as shown in Fig. 7.9, or one between two Centrex systems. In all of these cases, it would be equally correct to refer to the circuits as private or leased lines. On the other hand, if one of the connected systems is not a voice system, the term *TIE line* would not normally be used. *TIE* stands for terminal interface equipment (Fig. 7.10).

WATS

WATS is an abbreviation for wide area telephone service. WATS lines come in two flavors: in-WATS and out-WATS. Another name for in-WATS is 800 service. When most people refer to a WATS line, they mean an out-WATS facility. Both services are merely billing arrangements for reduced billing of long-distance calls based on a fixed monthly fee and discounts for larger calling volumes. 800 service also has the characteristic of reversing the charges to the called party.

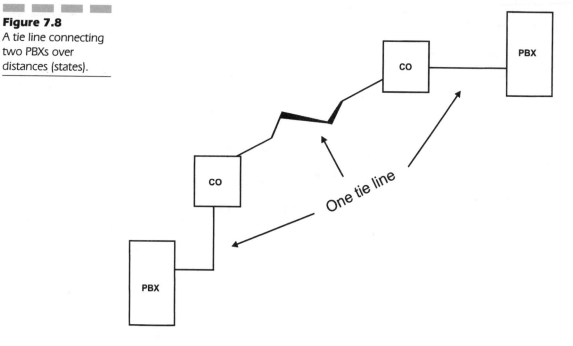

Figure 7.8
A tie line connecting
two PBXs over
distances (states).

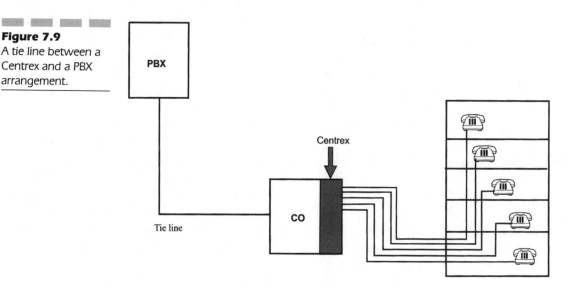

Figure 7.9
A tie line between a
Centrex and a PBX
arrangement.

Figure 7.10
A tie line connecting
two PBXs locally.

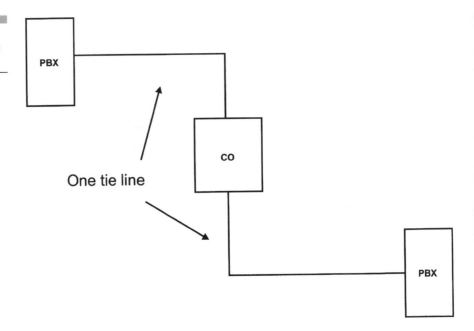

One tie line

Historically, WATS lines have been separate facilities (physically identical to local PBX trunks or private lines). Their geographic coverage was also banded; thus a user might have had a WATS line that only reached adjacent states (band 1), or all of the lower 48 states (band 5), or some intermediate variation. For out-WATS, either the PBX had to be smart enough to recognize the dialed area and choose the correct outgoing facility or users had to dial special codes to select the right WATS line. Any given band included all closer bands (but, of course, billed the calls at the higher rate for the wider band), and this caused a certain amount of difficulty in either configuration or training. Because different physical facilities went to different regions, this also resulted in a traffic engineering nightmare. This complication was all the more unreasonable because WATS calls (both in and out) are handled identically to all non-WATS calls; WATS is really only a bulk billing arrangement for calls that would otherwise be considered direct distance dialing (DDD or toll calls).

WATS service has never been free, although some of the older tariffs did specify certain (rather large) volumes above which all calls were free. Those tariffs are long gone; all calls are now charged on a per-minute basis. The only variable is the per-minute charge, which does decrease as the calling volume increases.

One significant improvement is that WATS-type volume discount billing can now be set up on existing trunks; no longer are separate facilities into the local CO required for such an arrangement.

Private Line

Any circuit leased from a carrier from a point on one customer's premises to another point on a customer's premises (even the same premises) can be described as a private line or circuit. Of course, if an organization builds its own facilities (e.g., a microwave link across a metropolitan area), these facilities would also be described as private circuits. In either case, the alternative is normally a dial-up link.

Many factors go into the decision as to whether to set up a private facility (of either type). Some reasons why a company might set up a private link include:

- Private analog circuits can be tuned for higher performance (both in terms of speed and reliability) than dial-up facilities can.
- Many types of digital facilities are only available on a private basis.
- Management and troubleshooting of private facilities can be more tightly controlled than in a dial-up environment.
- High volumes of calls or data would generate higher charges on the public dial network than on a non-usage-sensitive private network.

Reasons to go with the public dial network include:

- Volumes of calls or data too low to justify a leased link.
- Unwillingness or inability to coordinate and manage a private network (do you really want to be your own telephone company?).
- A large number of small locations that would be uneconomical to connect with private links.

With few exceptions, there is not a "right" decision on this issue; rather, any conclusion is based on a set of trade-offs involving economic, management, and performance considerations. What might make the most sense today might be obviously uneconomical tomorrow.

A classic example of this kind of change is the decision as to whether to build a private voice communications network of TIE-line-connected PBXs (called a tandem network—this is unrelated to Tandem computers). Many large companies built such networks for sound economic and functional reasons between the early 1970s and mid-1980s. But the emergence of virtual networks, such as AT&T's Software Defined Network (SDN), made many of these private networks uneconomical by comparison. They continued to function, but many companies have retired them because, in most cases, the virtual networks cost far less while delivering most of the characteristics that justified the construction of the original private networks.

Comments on Line
and Trunk Networking

Some of us might remember the early days of competition in the long-distance arena. Remember the way we had to connect to the alternate long-distance suppliers, like Sprint and MCI. The sequence was covered in Chap. 3, but it is worth restating here to show the reasons why things occurred the way they did. This scenario will clear up how the networks all came together after divestiture and how things have improved with the use of trunks instead of lines. The sequence of events leading finally to a call was as follows.

A customer might sign up with a long-distance supplier other than AT&T prior to 1984. This new customer of the competitors was offered several discounts over the long-distance tariffed rates from AT&T. So that the customer could use the service, the long-distance supplier would issue an 800 number to call its network, or a special number. This special number was a seven-digit telephone number that could be a regular local number in the area, a 950-XXXX number, or a foreign exchange telephone number from a major metropolitan area. The choices were based on the density of the carrier's services in the customer metropolitan or geographical area.

The customer would issue this telephone number to all internal users. Along with the 7- or 10-digit telephone number for entrance into the carrier's network (we'll use MCI from this point on for simplicity), another 10-digit number, called an authorization number, was issued. This might be a unique number for every individual in the organization or a global number used by the entire organization.

The caller (end user) now wants to make a long distance call from his or her office. So the sequence begins like this:

- Pick up the phone and get a dial tone, then dial 9 for an outside line.
- Dial 1-800-Cal-1MCI (numerically, 1-800-225-5624), for example.
- Wait for a connection. As the call proceeds, a ring tone is heard; then the MCI system answers and provides a "computer tone," which sounds much like a steady, high-pitched tone.
- On getting the computer tone, dial 1234567890 or whatever 10-digit authorization number is assigned to the organization. Wait for computer to confirm this number.
- After the computer acknowledges and verifies that the 10-digit authorization code is valid, it returns a dial tone to you.
- Now dial the 10-digit telephone number of the party you wish to speak with: (602) 555-0121.
- Wait for the call to proceed and ring. Hope and pray that the call is answered and the line is not busy and is clear enough to hold a conversation on. Otherwise, start all over at square one.

Users would obviously become very frustrated with this procedure. This is especially true if the called parties were busy or if the user (for instance, a telemarketing group) needed to make multiple calls. The need to dial 32 or 33 digits just to get a call through was frustrating—especially since the users did not follow the company guidelines and dialed AT&T directly, they only had to dial 12 digits. This was significant, particularly when there were thousands of calls being made per month. The accumulated waste of time might have cost the organizations more in productivity losses than MCI's service saved them.

So, why did MCI require all of these digits? The answer is simple: they had no choice. When competition first began, AT&T was the owner of the Bell System. To preclude the competitive threat, AT&T controlled how the network was set up. MCI had to rent telephone lines from the local Bell telephone company. At the central office, these lines were connected from the CO to the MCI computer. The call was a completed call the minute the computer answered the incoming request. MCI did not get any of the information that is passed along from CO to CO or from CO to long-distance supplier, because they were on the wrong side of the switching system. They were on the line side, not the trunk side. AT&T was on the trunk side of the switch, so all of the caller ID information was passed along from switch to switch; thus no extra digits were required.

When divestiture took place and the Bell System was broken apart from the AT&T network, then equal access was allowed. Prior to that,

AT&T controlled the network and made sure that equal access would not be a reality, or they priced the equal-access connectivity so high that no vendor could afford it.

Now that all things are equal in the eyes of the MFJ, the carriers (such as MCI, Sprint, LDDS Worldcom, etc.) can all be connected to the trunk side of the system. Now caller ID information, called automatic identification of outward dialed (AIOD) or automatic number identification (ANI) and many other names, is passed on to any carrier that is connected to the local or toll switches on the trunk side of the network. Calls are passed from intelligent switch to intelligent switch, routed through the network to an end point before a termination takes place. The world is a better place for this. MCI and its peers are now all able to offer the same limited dial sequence that AT&T has always enjoyed. Further, now they get even better access to the systems and are offered services called feature groups, allowing for flat-rate billing, call screening, and multiplexed services on high-speed trunks. This makes them as attractive as any of the long-distance services that they were competing with in the past.

Traffic Engineering

The art of conducting true traffic engineering studies has all but died. In the early days of telecommunications systems, the telco planning and capacity engineers spent days and weeks designing their switching and trunking systems to provide optimal performance. Since the inception of computer technology, programs have been written that can perform all of the calculations and the various iterations necessary to fine-tune a network. What used to take the telco engineers weeks now can be accomplished in minutes of processing time. Beyond the telco engineers, telecommunications managers and designers were equally concerned, and they attempted to model their networks for the best accessibility and greatest utilization of their lines and trunks, at the most reasonable price. From the discussion in earlier chapters, we can see that options for usage-sensitive services do exist. Using the best mix of WATS lines, for example, could reap exceptional savings in the past. In all actuality, this is not an engineering case, but a mathematical computation of all probable events that can take place in a given time frame. There is no engineering taking place, but rather a modeling of the traffic needs of the organization and of the access lines necessary to support those traffic needs with a given level of service. It is the calculation of a random number of arrival events taking place over a specified period of time. In the telecommunications arena, this is the number of possible first attempts to get an outside line, or possible times for an incoming line to ring into an organization on the first try. Because there are a finite number of time events, but an infinite number of possible attempts that can be made in a time period, we attempt to define the best possible level of service within a specified period of time. For the installation of lines and trunks, the goal is to serve the maximum number of callers at a single point in time (usually a 1-hour period). Because the arrivals of calls are random, the design must equate to an hour's worth of possible arrivals, that number being 3600. This is obtained by multiplying the number of 1-second increments in a minute times the number of minutes in an hour (60 s × 60 min). From this possible random 1-second arrival event, the concept of serving the worst-case scenario in a 1-hour period (the busy hour) is calculated to arrive at the number of lines or trunks needed to serve the heaviest load in a 1-hour period. For the rest of the day, there will be too many facilities, but for the busy hour, there will be an adequate number to deliver the level of service that has been decided upon.

An example of this would be similar to the work of road system traffic designers. These people are charged with the responsibility of creating road systems that will allow the maximum number of cars and trucks to traverse the roads in a city, town, or state. Because a good deal of money will be invested in building the road system, the engineers are requested to allow for the greatest number of vehicles to travel the road system within a reasonable cost estimate. From there, the engineers create a model for the traffic based on the hourly distribution of the vehicles on the road. They conduct counts at random times, clicking off the number of vehicles passing an area within a specified period of time. Then they allow for a certain amount of new traffic or growth as traffic patterns change. Armed with this information, they design a road system with the appropriate number of lanes to support the steady flow of traffic at a constant rate of speed (whatever is allowed by law in the area). Knowing pieces of the daily and hourly usage patterns, the engineers then design their roads to maximize the total throughput from end-to-end on the road that will be built.

Reality starts to set in when we perceive the actual daily conditions on our current roads. In the early morning and late afternoon rush hours, traffic gets snarled and grinds to a halt for extended periods of time. The engineers have accomplished their mission: they have maximized the traffic on a road system—on every square inch of road, there is a car. Unfortunately, during rush hour, one critical ingredient is forgotten—these cars were supposed to move across the road system at a constant rate of speed and then exit within a certain time frame. However, as we all know, they are all parked on the highway system—no one is moving, so congestion exists. Taking this one step beyond the rush hours, though, the engineers have designed a road system that allows users to get on the road and move at a steady rate of speed from an entrance point to an exit point. The level of service during nonrush hours is much better. Actually, the system is overdesigned for these off-hours, when there is a somewhat inefficient use of the roads. But, you cannot add and delete lanes of highway for selected portions of the day; they are fixed in concrete or asphalt forever.

This is exactly the same scenario that a telephone traffic study must follow. During an average hour of the day, all calls should be processed into or out of the organization with limited delays. The numbers of incoming and outgoing lines are crucial to ensure that the communications process works efficiently. However, the number of access lines is going to be fixed all day long because the copper or other medium brought into the building is not something that can be taken away or added on an hourly basis. For certain periods of the day, more calls might be required. This is called the *busy hour*. The telco or telecommunications analyst typically tries to design the number of lines to support the worst hour of the day. Yet for the rest of the day, too many lines are available. The more lines that are sitting around at a fixed cost, the more expensive the organization's costs are. This is based on a mathematical probability, and the dynamics of the calling patterns of the users or customers must be looked at. Consider, though, the risks of a poorly designed network:

- If the customer attempts to dial into an organization to place an order and all the lines are busy, then that customer might hang up and place the order with a competitor.
- If the outgoing traffic within an organization is severely blocked due to busy conditions, then the users might not be able to get mission-critical information as it is needed.

When busy conditions arise, users might find alternative ways of getting their jobs done. They might rent or lease their own lines, bypass the cost-effective facilities provided, or place calls via credit card or third-party arrangements through an operator-assisted call. In any case, all of these options will be far more costly to the organization.

Therefore, the services designer must consider the levels of service to be provided. This requires a thorough understanding of the traffic to be handled. From this perspective, a designer will have to consider the information that is available for study. This plays out in the paragraphs that follow. Information gathering is essential.

Where Can the Information Be Obtained?

Analyze the traffic that is coming into and going out of the organization. Access to this information might be a difficult portion of the desired study. To obtain the information there are several points that can be used:

- The telephone bills for all incoming 800/888-service calls are good indicators of the total volume. But, this bill will only show the completed calls, not the ones that encounter a busy tone when trying to call into the organization.
- The telephone bills for all outgoing calls from an organization can be useful. These might be difficult statistics to obtain, depending on how the billing systems work. For example, if departments each receive their own bills and process them for payment, the information might be difficult to get. Or, as another situation arises, the billing cycle might be different for different types of lines and trunks. In many cases, the main billing for direct-dialed calls is received on the

first of the month, whereas the long-distance portions of the bill might be received on the tenth of each month. This requires an inventory of all the line and trunk billings that are processed each month. Working with either the providers or the internal accounting department might well get this all started in the right direction.

The call processing time must be considered when analyzing any requirement for access lines. Often, the novice telecommunications analyst ignores the amount of setup time for calls and, therefore, can miss a sizable chunk of time to be factored into a calculation. This figure can be obtained by dialing several calls and determining the average amount of time necessary until the call is completed to its destination point. Remember that the telco and long-distance carriers will bill on the duration of a call from the time it is answered until it is terminated, with some rounding (up to 6 seconds) possible. Therefore, the added time that a circuit is held up for call setup is important in calculating the total usage time, because the line or trunk is inaccessible to others during this time.

Many telephone systems, call distribution systems, and so on can provide management and statistical reports. These might not be generated on a regular basis, but they can be obtained when necessary.

Telco and long-distance suppliers alike can gather statistics through what is called a *busy study and peg count*. The busy study can usually be conducted for a week and can reflect all call attempts to the incoming line group of an organization, whether completed or not. Further, this study can indicate the hourly distribution of the incoming calls so that some hourly sensitivities are afforded in the design study.

The wrong mix of telephone lines will be frustrating for customers and users alike. Yet, the overkill situation might be too expensive. The charter of the telecommunications department within an organization has always been "to provide the proper mix of goods and services to accomplish the calling needs of the organization, increase productivity, and reduce the bottom-line expenses to the organization." All too often, we have made the mistake of trying to get to a bottom-line figure, forgetting the other portion of the mission statement.

Accomplishing the Mission

Armed with the calling information, the next step is to look at the role being accomplished. What will the design effort accomplish? You can create a design to improve customer service, to reduce costs, or to improve user connectivity. Another option is just to attempt to configure a new setup. Regardless of the choice, the basic elements are the same. Upon successfully garnering a mission, several steps can be taken to execute the design. This is the traffic engineering function at its best. The steps that should be taken are highlighted in Table 8.1. These are the basic steps that one would follow for just about any design of a configuration. Each of these is covered in greater detail later.

TABLE 8.1	Basic Steps	Action
Steps in Accomplishing the Traffic Engineering Function	1. Determine the traffic demand that will have to be satisfied.	Gather the data on incoming calls, peg counts, busy studies, and other sources of information.
	2. Convert the information from a monthly basis to a distribution table based on daily and hourly statistics.	From the monthly information, plot the data into an hourly or daily distribution so that the design can be used.
	3. Determine the appropriate traffic engineering tools to be used. The tools will include Poisson, Erlang, or extended Erlang.	Choosing a tool that will assist in creating a rudimentary design, given the historical data that can be assimilated into a design tool.
	4. Apply the information into the modeling tool.	Using the tools and tables, apply the data to the tool to obtain the desired format.
	5. Select the grade of service to be delivered.	By establishing the grade of service, the percentage of calls blocked vs. service within a specified period can be predicted.
	6. Extract the information from the tools and the tables, depending on the method chosen.	Actually, once the data is submitted and the grade of service is established, extracting the information is fairly simple.
	7. Apply the information back to a monthly distribution.	From the hourly distribution, the monthly traffic that will be handled, as opposed to the blocked and overflow traffic, can be assessed.
	8. Try various iterations of grade of service.	Since the data is known, the iterations will allow for usage sensitivities and variable conditions that can be accommodated.
	9. Select the right configuration based on the available information.	Using the information gathered in the iterations, select the design that best fits the need. From this design, costs can be estimated and equipment needs can be established.

Using the Information

From these steps, the information can now be set up for the calculation process. This will entail all of the steps to calculate the actual number of circuits needed to serve a need. There is no mystique here, just a formal mathematical computation of the data against a given set of rules.

Gather the Data

As already stated, the ability to gather the data will be contingent upon our ability to assimilate the various bills, reports, and studies that are available from the local telco and from the in-house hardware systems. From these reports, the amount of traffic to be served will be easier to determine. Differences will exist between the outgoing and the incoming traffic. In all cases, it might be preferable to give a better grade of service to incoming customer calls than to outgoing administrative calls. This all depends on the individual organization.

Assuming that all of the data is gathered, let's play out an example of an outgoing calling pattern for an organization. This example is based on a typical bill where the average usage is 500 hours per month. Note that this is the aggregated amount of calling in the 1-month period. The data are then accumulated and estimated averages are placed on the total number of calls. For the sake of this example, the average length of a call is estimated at 4 minutes. The total hours are then divided by 4 minutes per call to come up with the total number of calls.

$$\frac{\text{Total hours} * 60 \text{ minutes per hour}}{4 \text{ minutes per call}} = \frac{500 * 60}{4} = \frac{30000}{4}$$

$$= 7500 \text{ calls per month} \qquad [8.1]$$

A certain amount of extra time must be added to the total calls per month. The 500-hours figure reflects what the telco will bill for calling time, not for processing time. As a matter of simplifying this, the average time allotted per call for setup and tear-down, which consumes time on the circuit but is not billed, will be 30 seconds per call. The 30-second figure is derived from a tone-dialed call from its beginning until the call is answered and billing begins. This is broken down into the components shown in Table 8.2.

TABLE 8.2

Added Time to Call Processing

Action	Time Allotted
1. Dialing an 11-digit telephone number, from a Touch-Tone phone	8 s
2. Delay in telephone system (PBX) network processing	6 s
3. Delay in network processing and regenerating digits	7 s
4. Ringing, assuming three rings (1 s ring, 2 s pause)	9 s
Total time allowed	30 s

Thus, the added call processing time must be accumulated with the hours of usage for the month. Therefore, these two numbers are added together. If this step is omitted, the engineer might design a system with a certain blockage level (grade of service) and not be able to deliver that level of service. Look at the amount of time added to the hours of usage.

Total hours usage + (total calls * added processing time)

$$= \text{Total time}$$

$$= 500 \text{ hours} + (7500 * 30 \text{ seconds}/60 \text{ seconds})$$

$$= 500 \text{ hours} + (62.5 \text{ hours}) = \text{Total time of } 562.5 \text{ hours} \qquad [8.2]$$

Note that if the extra 62.5 hours were not added into the total number, the design could be off by a factor of 12 percent in the overall traffic. This is a common mistake that people make when designing networks.

Convert the Data to Daily or Hourly Usage

Although the telephone billing records are possibly a source of the information, the calling patterns on a daily or hourly basis can become quite tedious. However, if a design is to be effective, this information must be converted into a usable and understandable format. The indication of traffic occurrences will assist in determining just how much traffic arrives within a given arrival time. The best way to achieve a good result is to look at hourly distribution. From the telephone bills, the total hourly distribution will likely vary from day to day and hour to hour. Consequently, scanning the bills into a computer database, or receiving the information in some electronic format (i.e., on diskette, tape, CD, etc.), will allow a telecommunications manager or designer to read the information into a PC or computer system. From this read, a sorting of dates and hours can be achieved. This will give the best results. However, when the information is a combination of paper records and other forms of data, the choice is to draw some averages of the calling patterns.

Typically, a pattern of distribution will exist that can be used in assimilating the information into a calculable form. The call distribution for many organizations is shown in Table 8.3. This table reflects the hourly distribution averages for a typical example; it is not to be construed as applicable to any specific organization. Further, this distribution will be based on the services offered by the organization. Call centers might well experience an average hourly distribution, whereas a manufacturer might see a dual-mode distribution curve (Fig. 8.1). The example used in Table 8.3 is geared toward a manufacturing environment, with no traffic being carried during the lunch hour. This is usually a fair assumption.

The distribution would then be spread across a dual-mode distribution curve. This attempts to define the number of circuits required to satisfy the distribution of calls in the worst hour, as already stated.

These assumptions are made on the basis of various inputs and can be modified. Given that the information is now placed in an hourly percentage, the next step is to convert the data into hours of traffic in the hourly distribution. From there, the process can begin to take place.

TABLE 8.3

Hourly Distribution of Calls for a Manufacturing Organization

Hourly Distribution	Percent of Calls per Hour
08:00—09:00	8
09:00—10:00	13
10:00—11:00	17
11:00—12:00	12
1:00—2:00	12
2:00—3:00	16
3:00—4:00	15
4:00—5:00	7
Total	100

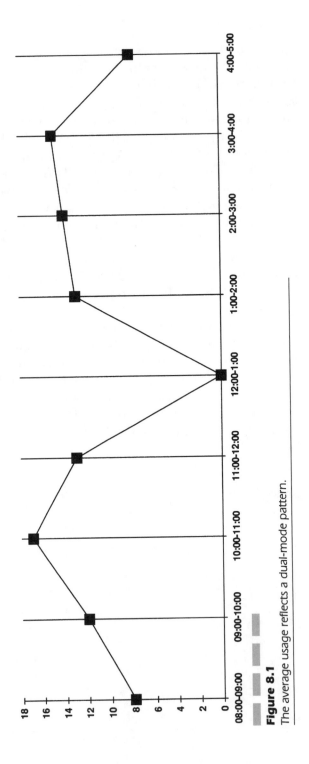

Figure 8.1

The average usage reflects a dual-mode pattern.

154

Choose the Appropriate Tool

Several tools exist to model the data. The primary ones include Poisson, Erlang B, and extended Erlang B distributions. The best way to describe the tools is the way the iterations take place and why they do so. In Poisson distribution, a mathematical distribution of the loads placed on the number of circuits is not calculated. Rather, the numerical values assigned with the overall distribution as set in the tables are merely placed into a grade of service level and a number of circuits is selected. The load is assumed to be even over the number of circuits chosen. This is okay for some calculations, but if additional distribution is required, then the other tools will be used. For a comparative look at these tools, first look at how the Poisson distribution will work. Remember that the monthly traffic is 500 hours. From a daily perspective, some added criteria are needed. The added parameters are shown in the following list:

1. The number of business days in the month is 22.
2. The average day is 8 hours long.
3. The hourly distribution is shown in Table 8.3.
4. The average length of a call is 4 minutes.
5. The call processing time is 30 seconds.

Poisson Distribution

From all of these numbers, a set of parameters can be established. One added factor is that the Poisson distribution model uses a slightly different calculation. Remember in the earlier section of this chapter that the statement was to consider the random arrivals of calls (incoming or outgoing) based on a 1-second arrival rate. Therefore, the number of possible arrival rates in an hourly distribution is 3600 (60 min × 60 s). The Poisson distribution takes into account all 3600 possible arrival rates and calculates these on the number or circuits necessary to support this traffic load. The telco engineers use a term called *Centum* (or hundred) *call seconds* (CCS). Rather than divide and multiply the hourly traffic by 3600, the theory is that if we divide the 3600 by 100, we arrive at a traffic load of 36 CCS per hour of traffic. Therefore, to use the Poisson distribution, all traffic is converted to CCS prior to submitting it into the model. Once the traffic is converted into CCS, the model will determine the number of CCS that can be carried within a specific design.

Therefore, if you use these assumptions, the traffic load can be converted and then inserted. The resultant traffic load is therefore summarized in the following list. The traffic is based on a worst-case scenario in the busy hour. A grade of service level will be assigned to determine the number of circuits necessary.

1. Total traffic hours carried   500

2. Added traffic for call setup   62.5

3. Total hours divided by   25.57 daily hours
   22 business days (562.5/22)

4. Daily hours multiplied by busy   4.3469 busy hour traffic
   hour percentage (25.57 * 0.17)

5. Busy hour traffic multiplied   156.48 CCS
   by 36 CCS (4.3469 * 36)

6. Grade of service desired   P.05

Determine the Grade of Service Desired

As already mentioned, the grade of service (GOS) desired is the worst-case statement for the busiest hour of the day. The GOS is used to determine the number or percentage of calls that will experience a busy tone on the first attempt during the busy hour. A grade of service of P.05 (probability of 5 percent) means that 5 of 100 callers might encounter a busy tone on the first attempt. The service actually experienced might be much better than the P.05, but the worst case is 5 of 100. Using this information and the average busy hour load of 156.48 CCS that must be carried, the Poisson distribution model can be applied. A sample Poisson distribution table is shown in Table 8.4. This table is read in the following way:

Read across to the grade of service desired (P.05). At the column headed P.05, read down until the calculated load value is found. In many cases, the actual value of the load will not be even; therefore, round the number in the table up to the next value on the table. From the load number, read left until the column labeled *number of circuits,* where the numeral is the number of circuits that will be required to carry the traffic. In this case, the number of circuits required to carry this traffic is 9.

Using the information in the highlighted area, reading across from left to right, then down, and back across from right to left, the answer is 9. This means that the number of circuits necessary to support the busy hour traffic is shown. However, during nonbusy hours, fewer circuits are

TABLE 8.4

An Excerpt of
a Poisson
Distribution Table

Number of Circuits Needed	Grade of Service					
	P.01	P.02	P.03	P.05	P.07	P.10
1	0.4	0.7	1.1	1.8	2.6	3.8
2	5.4	7.7	9.6	12.7	15.4	19.2
3	15.7	20.3	23.8	29.4	33.9	39.7
4	29.6	36.5	41.5	49.1	55.2	62.8
5	46.0	55.0	61.4	70.9	78.3	87.6
6	64.3	75.2	82.8	94	103	113
7	83.9	96.6	105	118	128	140
8	105.0	119	129	143	154	168
9	126.0	142	153	169	181	196
10	149	166	178	195	208	224
11	172	191	204	222	236	253
12	195	216	230	249	264	282
13	220	241	256	277	292	311
14	244	267	283	305	321	341
15	269	293	310	333	350	371
16	295	320	337	361	379	401
17	320	347	365	390	409	431

used, which should mean the delivery of a far better grade of service. Rather than go into the details of the actual grades of service for each of the other hours, these are shown in Table 8.5. These grades of service are extrapolated from the information in the percent of calls in each of the hours and the number of CCS loads for each of these loads.

To conduct some sensitivities on this design, the number of CCSs can be kept constant and the grade of service can be either increased or decreased, with the results reflecting increased or decreased circuit requirements necessary to support this new service level. You can play this out in various sensitivities, which is why the excerpt from the Poisson distribution table includes P.01 to 0.05, 0.07, and 0.10. Bear in mind that the sensitivities still include the differences of the expected hourly load. Worst case, this is the level of service. You might well expect that better

TABLE 8.5

Sensitivity of the
Same Data with
Differing Grades of
Service Reflects
Additional or Less
Trunks Needed

Grade of Service	Number of Trunks @ 156CCS	Difference
P.01	11	+2
P.02	10	+1
P.03	9–10	0, +1
P.05	9	0
P.07	8–9	0, –1
P.10	8	–1

grades can be achieved, because this is not a perfect science and the rounding effects of the calculations can become a factor. Further, in arriving at the hourly percentages of the monthly calls, some estimates were used. If the organization has periods of fluctuation during certain months of the year, or days of the month, then these calculations could be off significantly. But the data supporting the way the calculations were performed did not reflect these monthly deviations. A true monthly detailed distribution would be more precise, but this would take much longer to analyze the actual call details.

Erlang Distribution

Mathematical computations of the traffic offered to a group of circuits can take different forms that use different assumptions. One such variation is the use of Erlang B techniques as opposed to the Poisson distribution. Erlang B was developed in the early 1900s by K. Erlang of the Copenhagen Telephone Company. The assumption used is that if a caller encounters a denial or busy tone when attempting to place a call, the caller will be routed to a more expensive circuit (such as the long-distance network called DDD) or will give up trying to make the call. This last would be devastating for a call center operation, if callers who experience a busy tone hang up and call the competition. However, the assumption is that some alternative facility would be available. Erlang B is more effective in calculating call completion and attempts on an outward dialed basis, where callers can be routed through an automatic route selection process to a more expensive link. The network would not accommodate a queuing arrangement in the assumptions modeled with Erlang B computations.

Other models were created using Erlang assumptions; these introduced various options in the design of a circuit group. A summary of these variations is shown in Table 8.6, which compares the techniques and the basis for using each variation. The Erlang table comparisons assume that a queuing technique can be introduced and accepted.

For the purposes of describing a queuing technique, you can imagine a busy branch office of the local bank. During periods of heavy load, the bank might have two or three tellers available to service its customers. However, on paydays for many major corporations, the lunch hour becomes a very congested period for the banks. Even though three tellers are there to handle the customers, there will be more customers at any one time than there are tellers. Thus, the bank installs the old cattle-car line arrangement. Using a series of weaves or redirection cords, the bank attempts to get the customers to line up one behind the other. This allows the bank employees to serve customers on a first-come, first-served basis. Because the arrival of the customers is serial, as they form (or queue) in the line, they should be serviced in the order in which they arrived. Of course, the length of time it will take to get service depends on the arrival rate and on what position the customer holds in the queue. During busy conditions, the bank manager, seeing the queue building, might decide to assign more employees to servicing the queue. This might require that other employees with different functions be pulled into customer processing. Supervisors and managers of other departments might also be assigned until the queue is handled. All of these arrangements work on

TABLE 8.6

Variations of the Erlang Modeling Techniques

Model	Handling	Choice Considerations
Erlang B	No queuing	Use a route advancement technique automatically routing the traffic to the next available route, even if it is more expensive to service the call.
Extended Erlang B	No queuing	If all circuits are busy, the caller must hang up and try again later. No advancement of routes is used.
Erlang C	Queuing allowed	Calls are queued by a computer or an operator and wait until the next available circuit becomes available.
Equivalent Queue Extended Erlang B	Queuing allowed	Calls are queued for a specific time; if a circuit is still not available, the system may then route the call to a more expensive route or give the caller a choice to try again later.

the basis of creating additional resources for the handling of the traffic. However, if the customers were to arrive during this rush hour, and the bank had taken a position that only one customer at a time would be allowed in the bank, the picture would change dramatically. Instead of receiving service, the customers would be told to go to a different branch office for service. This would infuriate the customers and could prove to be very costly. It would only take one occasion like this, and the customers would find a new bank to handle their accounts. So goes the Erlang technique—you can be queued for the same group of facilities, or you can be sent out across the network to a more expensive solution.

Each of the Erlang theories is a better gauge of the blockage and grade of service than is the Poisson distribution theory listed earlier. Thus, they have been used in the most recent modeling techniques, since computers have been used in the modeling of traffic systems. An Erlang is equivalent to 1 hour's traffic in a 1-hour period; or, stated against the Poisson model, an Erlang is 36 CCS.

Using the same data as was applied to the Poisson distribution model, a computation can be made with the Erlang method. Therefore, the data to be used in the Erlang B formula is 500 hours of usage, with the average length of a call set at 4.5 minutes (this is the 4-minute average plus the 30-second overhead). The application of the formula actually works out as shown in Eq. [8.3]. This formula is used to calculate the number of circuits required to deliver a specified grade of service.

$$E(C, T) = \frac{\dfrac{T^c}{C!}}{\displaystyle\sum_{i=o}^{c} \dfrac{T^i}{i!}} \qquad [8.3]$$

where T = traffic attempts
       C = circuits available

To make this a little simpler, the following shortcut can be applied using tables that have already been created, then applying a simpler format to the data and deriving the number of circuits from Table 8.7. To create the scenario, look to the following formula:

$$\text{Erlangs} = \frac{N \times A}{3600} \qquad [8.4]$$

where N = number of calls handled in the busy hour
       A = average length of calls, seconds

TABLE 8.7

Erlang B Tables

Number of Circuits	Grade of Service				
	P.01	P.02	P.03	P.05	P.10
1					
2					
3					
4	0.87	1.09		1.52	22.05
5	1.36	1.66		2.23	2.87
6	1.91	2.28		2.96	3.76
7	2.50	2.94		3.74	4.66
8	3.13	3.62		4.54	5.60
9	3.78	4.34		5.37	6.54
10	4.46	5.08		6.22	7.50
11	5.16	5.84		7.08	8.49
12	5.88	6.62		7.95	9.47
13	6.61	7.40		8.83	10.46
14	7.35	8.21		9.72	11.47
15	8.11	9.02		10.64	12.48
16	8.87	9.83		11.54	13.51
17	9.65	10.66		12.46	14.52
18	10.44	11.49		13.37	15.55
19	11.23	12.33		14.31	16.58

As stated, the Erlang result from the formula is then applied across a table in the same way as the Poisson distribution model to arrive at the appropriate number of circuits needed. Use Table 8.7.

You can see from these tables that some sensitivities exist, and that the difference between the number of circuits needed and the various blockages that could be experienced will produce differing results. An organization that is in the mode of trying various iterations can get between P.01 and P.05 with the possibility of minimal additions or deletions to the calculations. However, this is not just about the number of circuits needed. The obvious deduction that can be drawn revolves around the difference in "customer service" and the ensuing numbers of devices required to deliver that grade of service. What also goes with this is the possibility of staffing a call center operation. The number and cost of circuits are variables that are not necessarily that important anymore. Because the cost of renting additional circuits has all but become a commodity pricing arrangement, the number of circuits is irrelevant, other than with respect to the degree of blockage allowed. The cost of a WATS line, or an 800/888 service, these days is based on the monthly worst-case charge of $37.50. Thereafter, the usage cost is based on an average of hours used per circuit. However, as shown in Chaps. 6 and 7, these costs are really based on the averages derived from the total number of hours used divided by the number of circuits, then applied to certain rate steps (0 to 15 hours, 15 to 40 hours, and over 40 hours). With this in mind, assuming that each circuit will carry at least 40 hours of traffic, each added hour is a marginal increment, costwise. But the degree of frustration delivered to the caller will be the actual deciding factor. Now, applying this degree of frustration and the number of other ancillary devices, is where the calculation of traffic demand will be used.

For example, if, after using the tables and drawing the conclusion that a 5-percent (P.05) blockage level will be delivered to the first-attempt callers, the telecommunications manager decides to see what the impact of a 1-percent blockage will require, the calculation will take on new meanings. This can be extended to the following requirements:

- Ports required on the automatic call distribution system
- Agents required to service the calls
- Terminal devices required for the agent answering positions
- Floor space and desk space required for the agent to conduct business
- Benefits associated with hiring the agents
- Payroll costs associated with each agent
- Personal computers or mainframe terminal devices required for customer lookup and inquiry
- Wiring required

The costs associated with each of these devices, and the human resource costs, can amount to a significant number. The marginal cost to add one agent could be as much as $100,000 per annum (benefits and payroll taken into account) plus the capital costs associated with the equipment purchase and maintenance of the devices. This size number is not to be taken lightly, so a true analysis will be required to prove the benefits. The issues are not always related to costs, but a cost-to-benefit ratio could be applied that will either justify or prove that a better grade of service is warranted.

If the decision to improve levels from a P.05 to a P.01 grade of service is considered, then the additional costs can be compared to the expected increases in sales, revenues, or service benefits. For general discussion, many organizations use a 3:1 cost-benefit ratio before selecting a new system or a change to an existing one. If an organization meets or exceeds the 3:1 ratio, the decision is usually easily justified. The greater the ratio, the easier the sell to management. A representative cost-benefit ratio equation is:

$$Cost\ benefit\ ratio = \frac{\text{Benefit}}{\text{Total cost}}$$

$$= \frac{450,000}{100,000} \qquad\qquad [8.5]$$

$$= 4.5$$

However, if the cost-benefit is below the 3:1 ratio, then one will have to struggle to gain management's acceptance, or have some alternate plans ready in case the funding cannot be accommodated. The example used in Fig. 8.1 is for a new design that will cost a total of $100,000. However, sales is expecting to create $450,000 of new revenue from the installation of this system. Bear in mind that this is on the basis of cost to benefit. If one were to create a profitability index on this installation, then the picture could change dramatically. What is the true benefit? What other choices would the organization have to implement a new program to generate $450,000 of new sales? If the differing projects pointed to the cost of implementing some other technique (such as a marketing campaign, promotions, etc.), then the cost comparisons could be used. For example, if sales would have to spend $600,000 in marketing campaigns and promotions, then the difference would be between the two choices ($100,000 versus $600,000 yields a 6:1 benefit-cost ratio). This allows a business decision to be made based on the facts, rather than strictly on some guesstimate or on emotional facts.

This is what traffic engineering is really all about. When this can be fully justified, then a business decision can be supported and defended before the senior management people who are charged with the custodianship of the organization's spending and the maximization of shareholder wealth. Therefore, the selling of communications lines and equipment is not for the benefit of the telecommunications department, but really is in support of the organization. Consequently, this is in support of the call center operation, whose personnel should have an input into the budgeting process, followed by the benefits expected from the installation of the new configuration or equipment. The use of traffic engineering is a means to get to an end, not the end in itself. Once this is clearly understood, the telecommunications department can proceed from there.

Equipment: Private Branch Exchanges

Private Branch Exchange (PBX)

A private branch exchange (PBX) is the typical telephone system for large organizations. In this environment, an organization that is served by a central office dial tone from the local exchange company might need the capacity of high-volume calling and handling services. Clearly, a single-line telephone set with a dial-tone line for each user will work. But, it will only just work! It will not satisfy the needs of the organization.

In addition, it will be expensive. Assume that a dial-tone line costs $20 per month. If the organization has a multitude of users, the cost per month will be significant. Table 9.1 highlights some of the typical costs associated with basic dial-tone service for various numbers of employees. These numbers are only representative, but they should get our point across. The table reflects the basic monthly cost and the annualized cost of renting a dial-tone line from the local carrier.

You can clearly see from these numbers that the use of a basic dial-tone service can get quite expensive. As a matter of fact, many organizations now say that telecommunications is the number-two expense item in their corporate expense registers, second only to personnel costs. This is both good and bad. It is good that organizations are depending on telecommunications more, as opposed to more expensive alternatives (such as travel, personnel, and other sales and marketing costs). Pound for pound, telecommunications still produce a greater return on every dollar spent.

But back to the point. The costs can be staggering to a financial or senior managerial person in an organization. But the dial-tone line costs listed in Table 9.1 give the user only dial-tone access. This is a full-time dedicated access line for two-way service for every single user. If you add just a single-line telephone set for each of these users, then there are some capital costs associated with the ownership of these lines. Table 9.2 shows the costs of a single-line set for every user, at a base price of $60 per single-

TABLE 9.1	**Number of Users**	**Monthly Cost @ $20.00**	**Annualized Cost**
Summary of Dial-Tone Line Costs for the Organization	100	$ 2,000	$ 24,000
	500	$ 10,000	$ 120,000
	1,000	$ 20,000	$ 240,000
	2,500	$ 50,000	$ 600,000
	10,000	$200,000	$2,400,000

TABLE 9.2

Summary of Single-Line Set Equipment Costs

Number of Users	Cost of Equipment
100	$ 6,000
500	$ 30,000
1,000	$ 60,000
2,500	$150,000
10,000	$600,000

line telephone set. These are, again, basic assumptions on the purchase of these sets; one could do better.

Again, you can see that the equipment costs can mount quickly. But what is wrong with this picture? Well for starters, the single-line set limits what the user can do with the basic dial-tone service. Also, the single-line set does not allow for intercommunication between the users within the organization, unless they tie up their dial-tone lines as follows:

- Grab the dial tone by going off hook.
- When dial tone is received, dial the digits (seven or ten*) of the desired internal party.
- When the ring is generated and the party answers, conduct a conversation.

But this completely ties up two outside lines for the two parties to converse. If a customer tries to call either of these two parties, the customer will get a busy tone. That is, unless the call hunts to some other number. If the call does hunt, then a third outside line is occupied while a message is taken at the rollover line. Customers can be denied access, and can get frustrated. All of this while the two parties could be talking to each other in the next office (Fig. 9.1). Note that however long the wires are that run back to the central office where the dial tone is provided, the call uses twice that to get the two conversationalists together. Clearly, this is not an optimized use of telecommunications services.

It should be obvious from the preceding discussion that larger organizations require the larger capacity and capability of a private branch exchange (PBX). These systems have names that come in many flavors,

---

*Many parts of North America now must dial 10 digits for local calls because of the addition of new area codes in their operating area. This overlay of area codes may cause some concern for many of the PBX owners because they have to program the new area code without the use of 1+ dialing, whereas other area codes require the 1+.

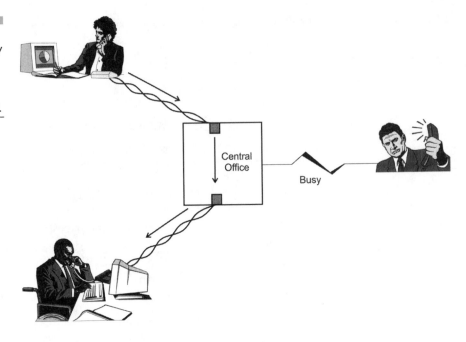

Figure 9.1
The single line is busy
when A talks to B;
therefore, when the
customer calls in, a
busy tone is received.

such as Private Automated Branch Exchange (PABX), Computerized Branch Exchange (CBX), Digital Branch Exchange (DBX), Integrated Branch Exchange (IBX), and Nippon Electric Automated Exchange (NEAX).

These names basically mean the same thing. They are just different vendors' acronyms used to differentiate their specific products. The generic term *PBX* is a private (customer owned and operated) branch exchange (like a central office, it switches and routes calls internally or externally and provides a dial tone to the internal users). The PBX marketplace is inundated with acronyms and features. However, they all do similar things: they primarily process voice calls for the organization. These devices are computer systems that just happen to do voice. Now they also do other things, such as provide data communications and data access.

On average, the all-digital PBX will cost approximately $750 to $1000 per station. A *station* is the end-user device, and the figure includes the cost of all the associated hardware to support that telephone set. Included in this generic price is the card inside the computer that provides the dial tone and the logic, a portion of the common equipment that serves many users, and the telephone set, the wiring, and the installation.

The components of the PBX are shown in Fig. 9.2.

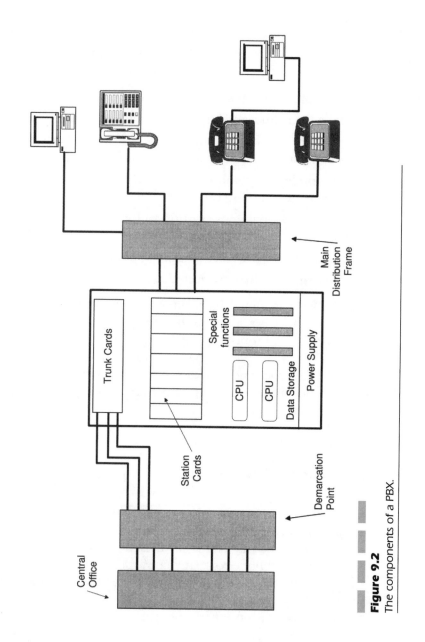

Figure 9.2

The components of a PBX.

- The central processor unit (CPU) is the computer inside the system—the brains.
- The memory—any computer needs some amount of memory.
- The stations, or telephone sets, are also called lines.
- The trunks are the telco CO trunks that terminate into a PBX.
- The network switches calls inside the system.
- The cabinets house all the components.
- The information transfer, or bus carries the information to and from the computer.
- The console or switchboard allows the operator to control the flow of incoming calls, and so on.
- The common logic, power cards, and so on facilitate the system's operation.
- The battery back-up insures against power failures.
- The wiring infrastructure connects it all.

The PBX is a stored-program, common-controlled device. As a telephone system, it is a resource-sharing system that provides the ability to access a local dial tone and outside trunks to the end user. This stored-program controlled system today is an all-digital architecture. In older versions, the PBX could be an analog system, but newer systems are all digital. It would not make sense to produce an older technology for a modern-day telephone system.

Analog Systems

The analog system used analog components to handle the call setup and tear-down for the entire system. A voice call is introduced into the system in much the same way that a business or residential user's input is introduced to the telephone company network. As the user generates a call, the telephone handset is picked up from the cradle. At this point, an input/output (I/O) request signal is sent to the main architecture of the PBX, which is usually a computer. Once the signal is sent to the common control, the system then returns a dial tone. The user then dials the digits for the party desired. This dialing sequence is done in-band on the wires in the talk path of the caller. The digits, either rotary (pulse) or tone (DTMF), are sent down the wires to the telephone system.

From there, the telephone system kicks in and generates a request through the architecture to a trunk card. The trunk card serves as the interface to the central office (CO) to request an outside dial tone. The PBX, upon receiving dial tone at the trunk card interface, then regenerates the pulses or the tones across the line to the central office. The CO processes these digits in the same manner that it processes individual line requests from a residential user. From the telephone company's perspective, this is the easiest way to process the information.

Digital PBX

All newer systems are basically digital. As a computer architecture, the system processes the information in its digital format. A digital coder/decoder (codec) in the telephone set converts the analog voice conversation into a digital format. The digital signals are then carried down the wires to the PBX heart (the CPU) for processing. If a call must go outside to the world, the PBX has to determine the best route to process the call onto. In the case where the call will be traversing the telephone company's central office links on an analog circuit, the PBX must format the information for the outside link. In this case, a digital-to-analog conversion will take place. Even if the call is to traverse a digital link to the world, the PBX might have to go through a digital-to-digital conversion. This is because the digital signal at the PBX interface is a unipolar signal, whereas the signal to the telephone company is a bipolar signal (Fig. 9.3).

The list of vendors selling and supporting PBX systems is quite lengthy. The manufacturers offer them to the customer directly or through a distributor. The options are many. The two largest suppliers of systems in the United States are Lucent Technologies and Northern Telecomm Inc. (NORTEL). This ranking is based on number of systems sold, rather than a qualification of "best," although you might establish that the quantity

Figure 9.3

The digital signal from the PBX is converted to analog to the telco.

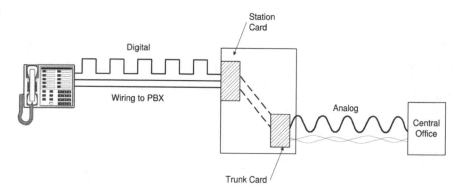

TABLE 9.3

Summary of the
Big Players in the
U.S. PBX Market

Top Players in the U.S. Marketplace
Northern Telecomm (NORTEL)
Lucent Technologies
Rolm/Siemens
NEC
GTE
Intecom
Fujitsu
Hitachi
Mitel

sold is a reflection of some qualitative measure. Table 9.3 shows the top players in the United States, based on sales volumes. It is interesting to note that the top 2 command better than 50 percent of the U.S. market.

The PBX market has recently been plagued by soft sales. This is a function of the recession, the rightsizing and downsizing of corporate America, and the overall unsettled market from a technological standpoint. End users are uncertain of what to buy and when on the market curve they should buy. Therefore, the vendors have had to resort to major markdowns, and they often throw in several other goodies. The buyer's market prevails in the PBX industry. As a result, significant discounts can be obtained if you work with the vendor and understand the product being offered. Many vendors will also compete severely with their distributors. Remember, this is a buyer's market. In Table 9.4 is a summary of how the costs would look for the acquisition of a digital PBX, the basic telephone system for an organization. This table reflects three important pieces of the billing arrangements. It would not be unethical to see how the ven-

TABLE 9.4

Summary of Costs
for a 1000-Line
Digital PBX

Item	Price
Cost of hardware, software, training, all telephone sets, and interfaces with installation of the hardware	$ 350,000
Cost of wiring and installation for the building infrastructure	$ 350,000
Markup and profit	$ 300,000
Total	$1,000,000

dors price out their systems against this model. In Table 9.4 we use an average price per port of $1000. The costs associated with a 1000-user system would, therefore, be as they appear in Table 9.4.

Another item of note is the third line item, that being profit. We always want our vendors to survive for another day, no two ways about that. However, we do not want to pay a 30 percent total markup on a system for profit. In actuality, the margin is 37 percent, and we will see why later. This is unheard of. So, the discounts that might be passed along from the vendor might well be from the profit picture. Suppose that the vendor offers a discount of 20 percent off the top of the price. The total price is $1,000,000 and the discount is 20 percent, so you can expect to pay $800,000. That should make you feel pretty good, to get a $200,000 discount off the top of your system. But, wait! What if the vendor came back and said that the total discount is only $70,000? Where did we go wrong? Well, the issue is where the numbers are being calculated. The vendor discounted the 20 percent from the top of the system cost ($350,000 × 0.2 = $70,000). Now, you are paying around $930,000 total for the system, installed. That is not exactly what you thought you were getting a discount on! The vendor will explain that the cost of the wiring cannot be discounted because they use a subcontractor and have to pay this third party for the installation. True, but the vendor also marks up the cost of the wiring and installation. That $350,000 fee to install and wire the system is probably only a $280,000 to $300,000 charge from the subcontractor. So, the manufacturer or distributor is getting a piece of the pie for the installation too!

Yes, this is true. Regardless of how we slice and dice the numbers, this is still a very lucrative sale for the vendor. With a $50,000 to $70,000 markup on the wiring (which is conservative), a $300,000 profit margin, and the remaining cost of the system ($280,000), you can imagine just how much the vendor is making on this system, can't you? Well now look at the margins based on this new evidence (Table 9.5).

TABLE 9.5

Summary of New Profit Margins with a 20% Discount on the System Cost

Item	Original Cost	New Cost	Profit	Percent Margin
PBX system	$ 350,000	$280,000		
Wiring and installation	$ 350,000	$350,000	$ 70,000.00*	20
Margin and profit	$ 300,000	$300,000	$300,000.00	30
Total	$1,000,000	$930,000	$370,000.00	37

*Excludes the wiring contractor's profit margin for the installation and wiring

Can you see anything wrong with this picture? Even though the vendor has given a 20 percent discount to you, and you feel so special for negotiating such a difficult deal for the vendor, and a great one for the organization, the overall margin of profit that the vendor has achieved is still 37 percent. This still leaves a lot of room for negotiation before the deal is done. If you consider that there is still room to cut the cost in the profit margin, the profits on the subcontracted piece of the wiring, and the overall system cost, then the dealing has only begun. In many cases, the ability to subcontract the wiring (for example) might produce more productive and competitive results. In this case, many organizations will act as the general contractor for the overall telephone system and then contract for the wiring separately from the telephones. An example of the wiring costs might look like the numbers shown in Table 9.6, where a separate contract is issued for the installation of a four-pair cable installed at 1000 user locations, the horizontal wiring between the telephone closets and the main distribution frame, and any ancillary cabling needed to implement the system.

Keep in mind that these figures are generic, and will require separate bids from various installation companies. If, however, you now consider this figure, and recognize that the wiring contractor has already built in the necessary profit margins to make money on the installation, then the PBX price now has a different perspective. The margin for the hardware, installation, and warrantee on the PBX is now subject to serious negotiation (Table 9.7).

TABLE 9.6

Summary of Wiring Costs

Cost per User Location	Extended Price
Cost of wiring a 1000-user system @$250–280	$250,000–280,000
Cost for PBX manufacturer @$350	$350,000
Difference	$70,000–100,000

TABLE 9.7

Summary of New Cost Structure and Margins

Item	Cost	Percent Margin
PBX	$280,000	
Markup	$300,000	
Subtotal	$580,000	115
Wiring (as a separate contract and separate margins)	$280,000	15
Total	$860,000	

TABLE 9.8

Final Configuration of System Pricing

Item	Original Pricing	Revised Pricing	Difference
PBX	$ 350,000	$280,000	($ 70,000)
PBX markup (@30% of contract)	$ 300,000	$ 84,000	($216,000)
Wiring	$ 350,000	$280,000	($ 70,000)
Totals	$1,000,000	$644,000	($356,000)
Percentage			35.6%

As you can now imagine, the cost for the telephone system is $280,000 with a profit margin of $300,000 (over 100 percent markup). No vendor will ever approach this structure; these are comparative pricing scenarios. However, if you consider that a 30 percent markup is what the vendor is entitled to, the following summary gives us a whole new structure to deal from. The intent is not to jeopardize the stability and profitability of the supplier, but to maximize the comfort between the two parties. This case will obviously consume a lot of time and effort. But, the overall results are significant (Table 9.8).

Clearly the price has changed significantly! The system is now being considered at approximately $644 per user instead of $1000. This accounts for a $356,000 discount overall. This is the way you can look at using the system pricing, rather than just accepting standard pricing. The pricing can vary quite a bit from the original proposal.

Central Office Centrex

An alternative to the purchase of a telephone system is called Centrex. Centrex is a service offering from the local exchange carrier (LEC) and now the new players, such as CATV companies and CLECs. As a matter of fact, Centrex is one of the flagship offerings from the new players, such as the CATV and CLEC companies, in competition with the telephone companies. They, in turn, will bundle their services at a price that is very competitive with the incumbent telephone company. In many cases these CLECs and other new players will have service offerings that add features that are not typically available from the LEC. This will put the pressure on the LEC to compete, or to lose a significant portion of its market share. The consumer will stand to benefit from this one. Centrex stands for *central exchange,* or PBX services provided from the central office (exchange). The LECs, either the Bell Telephone Companies or the independent telephone companies, all have a service offering. The Centrex service is a partition inside the CO, which provides telephone service on a private basis to businesses (Figure 9.4).

Centrex is usually rented on a line-by-line basis, month to month. Some companies have long-term agreements to hold down costs, but charging a monthly rate is more prevalent. Costs per line can range from $20 to $25 a month, depending on the company and the serving telco. The user must still buy the station equipment (telephone sets). However, with very large organizations, a long-term contract can be negotiated with the provider. This will result in costs that are in the range of $10 to $12 per month for a Centrex line. The overall result is a monthly savings of $10 to $15 per month per line. With a 1000-line system, this can be a significant savings. An example is the 1000-line system that we were using in the PBX comparison. Here, a system can cost approximately $12,000 per month on a long-term agreement. The $12,000 is for the dial tone and features; anything else is an extra.

The components of a Centrex are as follows:

- Central office, which is the serving office
- CPU or computer
- Station cards
- Switching network
- Line cards
- Memory
- Common logic and power cards
- Bus to carry information to and from the CPU

Centrex Service

The primary suppliers, as already stated, are the LECs. They provide the service as part of the central office function. The most commonly used systems in North American COs are Northern Telecomm (Nortel) and Lucent Technologies, although NEC, Ericsson, and others might be used.

The list of Centrex providers includes:

- *Bell Operating Companies.* The seven regional Bell Operating Companies and their 23 operating telephone companies are the largest sellers of Centrex.

- *Independent Operating Companies.* These are such companies as Contel GTE, Centel, and Commonwealth.

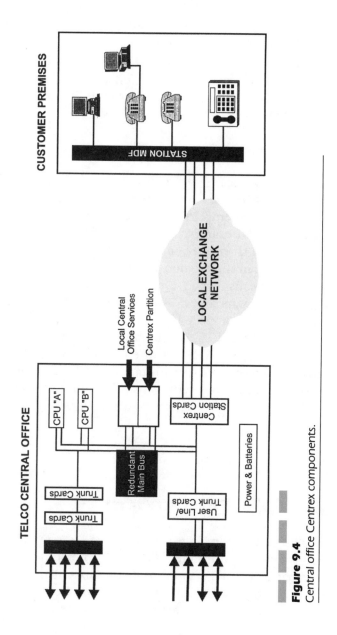

Figure 9.4

Central office Centrex components.

- Resellers of Centrex service. Many arrangements have been made with resellers around the country)
- *Bell authorized agents.* These are authorized agents of the LEC, but are not employees of the company.
- *Consultants.* These are special resellers who have a vested interest in their customers' behalf, and a vested interest in the commissions they might receive from the LEC for the additions to the Centrex:
  - CATV companies
  - CLECs
  - Electric companies

Centrex service is typically found in North America. However, it is now available in the United Kingdom (and 34 other countries) due to the appearance and emphasis of the competitive providers of telephony services there.* An example of that is NYNEX's appearance as a cable TV operator and then offering a competitive telephony using Centrex as its basis. It is interesting to note that Centrex services in the United Kingdom have grown considerably, even though they were only introduced within the past few years. The possible reason for this is the fact that many people always feel compelled to move away from their local telephone company provider or PTT. Where British Telecom was offering service as the local provider, NYNEX was the new kid on the block and was a more attractive alternative. Notwithstanding, NYNEX was very aggressive in its pricing in competing with British Telecom. As a result, many British Telecom customers moved over to the Centrex service being offered by NYNEX.

Peripheral Devices

The list of peripheral devices for the key systems, PBX, and Centrex markets is virtually unlimited. The devices range from items as simple as an external bell to very sophisticated management systems. The pieces are too numerous to list here (and change too frequently!), but there is still a lot of negotiating room for any c°omponent you might need.

---

* NYNEX merged with Bell Atlantic and the company name was set to Bell Atlantic.

Some of the devices that might appear in the picture are as follows:

- Automatic call distribution
- Voice mail
- Automated attendant
- Call detail recording
- Modem pools
- Multiplexers
- Head sets
- Display sets (telephones)
- Paging systems
- Least cost call routing
- Network management systems
- Design tools
- Answering machines

10

Key Telephone Systems

Another form of equipment is used in a smaller environment. Where the user office is small or a branch office is involved, the equipment used is called a key telephone system. This is another form of resource-sharing device, used to reduce the number of outside telephone lines and provide access to many end users.

The equipment selection depends on the size of the organization and the need for communications connectivity. Many small branch offices, small businesses, home businesses, and other such organizations require only a limited amount of telephone sets and can get by with a key telephone system. This type of system usually consists of a predetermined number of telephone sets to outside lines. Note the reference to lines in this example; the telephone company sees the system as an end user connection, much the same as a single-line telephone set. Therefore the customer gets by with the use of lines instead of trunks. Although they functionally do the same things, lines are typically less expensive monthly than are trunks. The nomenclature for a key system is normally outlined in the number of lines and the number of sets; for example, 1648 equates to 16 outside telephone lines, and 48 telephone sets comprise the system. These are maximums on the systems; smaller amounts can be used where appropriate.

A key telephone system is composed of the following pieces:

- The *key service unit* (KSU) is the heart of the telephone system.

- *Line cards* are the interfaces to the telephone company lines. They effectively provide the off-hook and on-hook signals to the central office in lieu of the individual telephone sets.

- *Station cards* are the cards that control the intelligence and interface to the end user's key system. The station card interfaces with the end user and the line cards for the access control of the system.

- *Intercom cards* are not always required. Some systems use an intercommunication card for internal connectivity, whereas others have this built into their functionality on the backplane of the system.

- *Telephone sets* are variable-user interfaces, whether they are single-line or digital multiline sets. The intelligence that a user is allowed to access within the system is controlled through the telephone set.

- *Power supplies* and *logic cards*.

- The *wiring infrastructure* that is used to connect the sets to the KSU central processor unit (CPU), the brains of the system.

These pieces are shown in a graphic representation in Fig. 10.1. The graphic is a representation of several different pieces that might well be

Figure 10.1
The key telephone
system components.

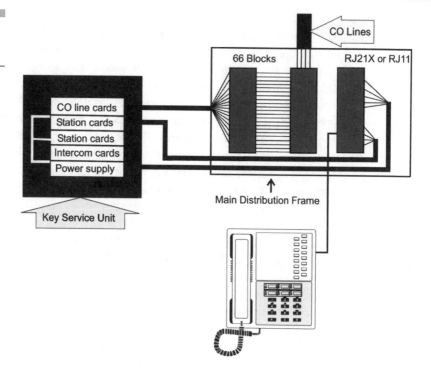

included. For example, we see that the telephone company has brought
in central office lines and terminated them in either an RJ-21X or an
RJ-11. From there, the graphic shows a 66 block, sitting side-by-side with
the RJ-21X, where a straight cross-connect arrangement is taking place.
On the left side of that 66 block a fanned-out 25-pair cable is punched
down on the 66 block. We can refer to that 66 block as the vendor block,
whereas the RJ-21X would be the telco block. From the vendor block
with the fanned-out 25-pair cable, the cable is run into the key service
unit, or the intelligence of the key telephone system. Here we might use
an amphenol connector, a 25-pair prewired connector, for termination
in the key service unit. This merely simplifies the process of providing
the connectivity. Looking at the other pieces here, the system therefore
has CO line cards as opposed to the trunk cards that would appear on a
PBX. Station cards are then installed, giving the intelligence to the individual telsets that are installed within the system. Intercom cards allow
for intercommunication to prevent the problem that was mentioned
earlier in the PBX section. One does not want to tie up all of the outside
lines for two parties to be able to talk to each other internally. Therefore
the intercommunications device, or intercom, provides for the station-

to-station talk path. Many key systems have an unlimited number of intercoms, whereas others have a very specific number of intercom paths available. This depends on the manufacturer. From the key service unit, we see two lines drawn over to another 66 block that would be run in a separate telephone closet. These would be what we call *horizontal runs*. The horizontal run connects from the key service unit to the telephone closet, where the 66 block is a termination point. The purpose of this horizontal run is to bundle cable pairs together using 25, 50, or 100 pairs of wire to that telephone closet. On termination on the 66 block, a separate station drop is then run to the individual desktop device. At this particular point the station drop is used on the basis of a 2-, 3-, or 4-pair wiring system. Each key system does require a different number of wires to the telephone set. For the key telephone systems available on the market today, a single-pair cable system can be used. However, it would be imprudent for the individual telecommunications person to install only one pair of wires to every desktop. The reason is probably obvious; however, it would leave the possibility of being cable limited. Once a telephone set was installed on that station drop using a single pair of wires, the inevitable would happen. The very next day, users would complain that they needed additional wires for fax or modem connectivity. This places a bind on the telecommunications manager who has not installed enough wires to meet such needs. Therefore a recommendation at this point is that, regardless of the number of pairs a contractor requests, always install a four-pair cabling system. The four-pair cabling system will allow flexibility for additional wires for fax, modem, or any other type of external communications connection. Moreover, the authors typically recommend the use of 2 four-pair cables to every desktop. This will allow enough future flexibility so that the expensive cost of labor to rewire the building can be avoided. Each of the four-pair cables should be terminated in what is known as an RJ-45 jack. The RJ-45 jack, an eight-pin interface, is the standard plug for the future. All four-cable pairs should be terminated into the eight-pin interface. All too often telecommunications personnel use jacks with less capacity than the RJ-45 and do not terminate all the wires. This means that extra wires are in the wall, but are not useable without a lot of labor to reinforce it. Most key telephone systems today are computer-stored, programmed-controlled systems, similar to the PBX discussed in Chap. 9. As a matter of fact, it is becoming far more difficult to differentiate the PBX and the key system. Key systems can be as large as 200+ ports, whereas PBXs now come with as few as 50+ ports. Functionally, these two types of systems provide the same services and features. The differences are so subtle that it is becoming almost a moot point to call these systems by different names. These systems are used extensively throughout the industry, so they are fairly perfected in terms of connections being simple, are rich in features, and have significant penetration in the business community.

The key systems are used by larger organizations too. In departments where groups of people require connectivity to each other, or where a consolidation of features is required behind a Centrex or PBX, the key system offers some solutions. Although many organizations try to emulate the key system in a PBX, with multiline sets, it is often more prudent to install the key service unit. Many dual systems provide the access to features of the PBX but the clustering of workers of a key system.

Why Key Systems?

It is not uncommon to hear business users question the need for a key system. The obvious choice is to go to the telephone company and rent a single line for each user in the organization, branch office, sales office, or home-based business. However, the same arguments that were discussed in the opening of Chap. 9 hold equally true for the key system justification. It would be frivolous to rent a single line for every user in a sales or service office. The intent of the sales office is to provide a location where the sales force can schedule work and visit from time to time for meetings and report writing. However, the sales force might be road based, with primary emphasis on visiting the customer and meeting the customer's needs. According to this analysis, there should not be a one-to-one relationship of employees to outside lines. This, of course, assumes that the sales force is out calling on customers. If the charter is to telemarket, the entire scenario changes dramatically. When the office force does drop in at varying intervals, it is unlikely that everyone will be around at once. Therefore, a 3-to-1 ratio can usually be achieved, with three telephone sets being used for each outside line. This can save a significant amount of money in a small office, and more in the larger office environment. Table 10.1 shows a comparison of the use of outside lines on a 3-to-1 ratio. This does not include the costs for the key system; these will be shown in a later comparison. Suffice it to say that the comparison here is based on the monthly recurring telephone company costs.

TABLE 10.1

Comparison of Line
Rentals with and
Without a Key
System

Number of Users	Without Key System	With Key System	Difference
10 @$20	$200	$ 80	$120 monthly
20 @$20	$400	$140	$260 monthly
30 @$20	$600	$200	$400 monthly

The obvious reason an organization would choose a key system is the monthly recurring savings of $120 to $400, as shown in the table. However, the cost of the key system is not shown. The primary savings shown in this table are the cost of renting a telephone line for each user as opposed to sharing the number of lines with the resulting savings. Now we can add the cost of the capital equipment—the key telephone system. A typical key system with installation, hardware, software, and warrantee will run an average of $500 per user. A comparative analysis of the costs associated with the acquisition of the system is shown in Table 10.2

You can see from this table that the costs have been applied in a linear fashion. There might be better discounts involved as the system gets larger. Further, the system costs are associated with a basic telephone system. Other features that will increase cost include:

- Voice mail
- Station message detail recording
- Automated attendant
- Least-cost routing
- Data and voice simultaneously
- Automatic call distribution

These will all raise the cost of a telephone system. The wiring is typically sold to a user on the basis of the vendor's approach to installation. Many of these key systems, for example, use a two-wire (one-pair) connection to the desktop; others might use four wires (two pairs) and so on. In

TABLE 10.2

Summary of Key
Telephone System
Costs for Various
Numbers of Users

Number of Users	Key Equipment Costs
10	$ 5000
20	$10,000
30	$15,000

general, a good rule of thumb is to install a four-pair connection to every desk, whether you believe you need it or not. This will increase the cost of the wiring, but it is not that bad a deal. The cost of the wire is approximately $0.03–0.05 per foot for the added pairs. All things considered, the little add-on for the wires makes for an insignificant increase to the overall system. However, over the life of the system, the extra pairs of wires will pay for themselves in no time. If a user needs a modem or fax machine, for example, then the extra wires will be necessary. To have an installer come back and pull a whole new pair of wires will cost from $250 to $350, most of the expense being labor. This can also be disruptive in the workplace. Table 10.3 compares the equipment costs and the monthly rental costs for the telephone lines to see just what the impact will be overall. Taking the comparative numbers from the original tables (Tables 10.1 and 10.2), we see that the cost structure is as follows.

Although this table might look like it is a break-even to use or not use a telephone system in lieu of the monthly rental of a line per user, there are some basic assumptions being made here. The equipment cost for the nonkey system is for the purchase of basic analog single-line sets. No multiline sets are involved in this scenario.

The monthly equipment costs are associated with a three-year depreciation cost on a straight-line basis. In reality, the finance department would likely write the capital off over five years. The finance department would also write the equipment off at an accelerated rate based on the IRS and financial policies accepted.

The one-year differences will have to be spread over the number of years the system is in place. After the payback period is achieved, the numbers change significantly.

Table 10.4 covers the same system costs over a 5-year depreciation cycle, with the system life being a 10-year period. In this case, after the fifth year, the equipment costs disappear and only the monthly recurring costs

TABLE 10.3 Summary of Costs for Key System versus Nonkey System	**Without Key System**		**Total 1-year Cost**	**With Key System**		**Total 1-year Cost**
Number of Users	**Line Cost**	**Equipment Cost**	**Depreciation @ 3 Years**	**Line Cost**	**Equipment Cost**	**Depreciation @ 3 Years**
10	$200	$ 600	$2600	$ 80	$ 5000	$2660
20	$400	$1200	$5200	$140	$10,000	$4980
30	$600	$1800	$7800	$200	$15,000	$7400

TABLE 10.4

Five- and Ten-Year
Comparative Costs

Users	Non Key Costs	Line Costs Monthly	Total 5 Years	Total 10 Years
10	$ 600	$200	$12,600	$24,600
20	$1200	$400	$25,200	$49,200
30	$1800	$600	$37,800	$73,800

Users	Key Costs	Line Costs	Total 5 Years	Total 10 Years	Difference	
10	$ 5000	$ 80	$ 9800	$14,600	($2800)	($10,000)
20	$10,000	$140	$18,400	$26,800	($6800)	($22,400)
30	$15,000	$200	$27,000	$39,000	($10,800)	($34,800)

remain. These are simple calculations—any increases in the line costs are not included and no factoring for maintenance is built in. The assumption here is to look at the overall differences between having a system or not. The same assumptions from Table 10.3 are used: the number of users, the cost of equipment at $60 per single-line set (written off, in this case, for over 5 years), the line costs held constant over the 5- and 10-year life cycle, etc.

Using this table results in a much different picture of the financial impact of using the key telephone system. On the basis of an example of 30 users, the difference in raw costs over the 5-year period is $10,800; over 10 years, the key equipment will save $34,800. These are significant savings over the life of the equipment when you consider that the entire organization might only be 30 people. The equipment will obviously pay for itself over the 5- and 10-year period. Remember also that the key system includes the features and functions of a telephone system and will include multiline sets. The nonkey system comparison includes only single-line sets and no added features. Each of these will change the equation and show a better payout for the key telephone system. Having drawn this to a conclusion, you can now see why an organization might well consider the installation of a resource-sharing arrangement such as a key telephone system. However, there is more to this than just a financial calculation.

As covered in Chap. 9, there are other advantages to the systems approach. One, for example, is the use of the intercommunications channel, called the intercom. The intercom frees up the outside lines for incoming customer calls when two workers in the same office need to communicate. Using the intercom, party A calls party B on the internal communications paths in the system. Some systems have a fixed number of intercom paths, whereas others have an unlimited number. This is a function of the manufacturer equipment. When using the alternate system (the nonkey telephones), the choices are far more restrictive. If user A wants to talk to user B, then the two outside telephone lines associated with these two individuals are busy. The customer calls coming into the building for either party will be blocked until they terminate their conversation. The customer might become frustrated and take his or her business elsewhere if this is a constant occurrence.

Moreover, it is far easier to add features and services in the key telephone system than in the single-line telephone set. Users will ultimately like the features that can be achieved in the key system, such as:

- Call transfer to another party
- Call hold by pressing a feature button
- Three-way or conference calling
- Consultation hold, where a call can be placed on hold while the user confers with another user in the system
- Speed dialing on an individual or system basis
- Speakerphones
- Hands-free dialing

Some of these features are displayed in Figure 10.2, where the buttons are allocated for the use of the specific features. No one telephone set or system offers everything the user wants the way he or she wants it. However, strides have been taken in the past to accommodate what users have asked for. Newer key telephone systems will use a soft key approach, where a visual display panel such as a light-emitting diode (LED) or LCD (liquid crystal display) is used on the telephone set. As features and functions are activated, instead of fixed keys on the telephone set the display will have a series of buttons whose functions change as you move through the menus on the system. Although these are a little more expensive, they may add flexibility over time. In viewing the actual key telephone system as shown in Figure 10.2, the feature keys are shown. Note the differences as they're shown here on the drawing. When a user activates a feature, a new set of menu options appears for each one of the soft keys. This makes for a very flexible arrangement, but we must be aware that the end user's ability to grasp and use these features will make the sole decision. Merely throwing features and functions at an end user does not assure ease of use or connectivity establishment. Users may feel intimidated by soft keys with LCD-driven menus and therefore not use the capabilities of the system. We must

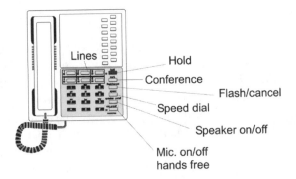

Figure 10.2
Features on sets for key system.

Lines

Hold
Conference
Flash/cancel
Speed dial
Speaker on/off
Mic. on/off
hands free

be concerned with this risk. Therefore it is highly recommended that, before going to a state-of-the-art telephone system, the telecommunications manager select a few users to be pilots (guinea pigs) for testing this system. This will check ease of use and the users' ability to absorb the information and use the features. One should not select the super or power users, but take a mix of power users and novices alike to guarantee a good cross-representation of internal users. This will help ensure a win/win situation when installing a key telephone system. Without it, one may well install a system that is not fully utilized and therefore is expensive due to the limited features being used. We do not want to waste corporate monies by buying features and functions that people won't be able to use merely for the sake of having glitzy equipment.

Vendor Interfaces

The telephone interfaces in Fig. 10.3 show differing ways that a key system can be connected to the outside lines. Often the telephone company will bring in individual two-wire analog circuits (lines) and terminate these in an RJ-11C jack. This is the case for smaller systems of less than 10 lines. If, however, the system is larger and more connections are required, the telephone company might well be asked to terminate the lines in an RJ-21X. This is a call that can be made by the vendor who is requesting the service for its customers. Frankly, there is no major difference between the connections other than some not-so-obvious benefits.

For example, using the RJ-21X consolidates the RJ-11s into a single block that will support 25 two-wire connections. This is a little cleaner than having 25 individual RJ-11s installed on a wall. The convenience of testing and troubleshooting at one spot on the wall has some benefit.

Figure 10.3
Telco interfaces to a
key system vary.

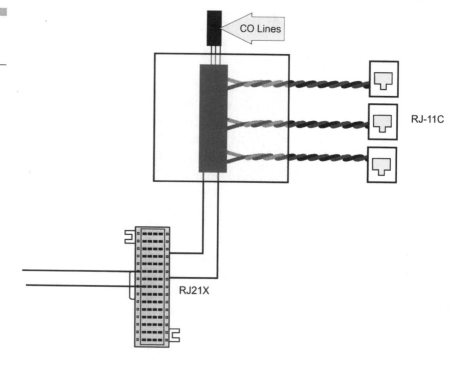

Figure 10.3
Telco interfaces to a
key system vary.

However, the individual RJ-11s can be used in lieu of the larger block. Although these might take up more wall space, if the telephone system ever fails, a user can go into the telephone equipment room and unplug one of the circuit connections to the key system, plug in a single-line telephone set (2500 set), and make calls to obtain repairs from the telephone company or equipment vendors. Furthermore, if critical calls must be made, many users can take advantage of this special access method into the lines.

These systems have been around for a long time, so there is no mystique in dealing with this connectivity. Vendors have pretty well gotten these systems down to a science. Users also have them under control. There are no major management issues in dealing with a key system, except when the system fails. This can usually be rectified quickly by the vendor and telephone company when problems arise. The interface is a two-wire analog connection similar to the residential connection. The main problems a user will have will concern the power. We always recommend that a backup battery be provided with the installation so that a critical component of the system's performance can be preserved with the supplemental power. The backup battery adds some minor costs, possibly a few hundred dollars at most, but buys a lot of comfort and protection for the end user.

Key Players in the Key System Marketplace

The market for these systems is quite large. In many cases, manufacturers will sell directly to the end user, or they might have a distribution agreement with several installation and maintenance companies in the area. Variability in this market is extensive. We have seen situations where we had the manufacturer and two or three distributors in an area competing for our business. The competition is fierce when this happens. In some cases, the distributors offered the system less expensively than the manufacturer's representative. This says a lot about the market, because it is almost as soft as the PBX market. In the following list, we look at some of the key players in the key telephone market. This ranking is on the basis of the number of systems sold or the number of lines delivered. This is in no way designed as a rating of good, better, best. It is strictly a numerical value of total systems. However, there will be no surprises in this picture. The two top players are familiar names and have the majority of the market.

- Nortel
- Lucent Technologies
- NEC
- Inter-Tel
- Telrad
- Siemens
- Tie Systems
- ComDial
- Hitachi
- Iwatsu

The top three players combined control approximately 70 percent of the total U.S. marketplace. This again is not a reflection of the quality of their systems, merely of the percentage of systems sold. Many of the other systems are equally good or fit a specific niche market, and are equally rich in features, competitively priced, and well supported.

When dealing with the purchase of the key telephone system, you will be better equipped if all of the known benefits and features are discussed. The growth capacity of the system before a major upgrade is required is a beneficial piece of information. Never assume that the best deal is placed on the table on the first proposal. As a matter of fact, we prefer to do the following when considering a key system, a PBX, or any other piece of equipment:

- Conduct a needs assessment. Determine what the needs of the organization are—to meet the core business functions of the office or plant, or other objectives.

- Determine what the buying motivations are. Is the system being considered to allow growth, replace obsolete or broken equipment, or just deliver a needed feature?

- Conduct user surveys to see whether any hidden needs exist.

- Document the needs and the buying motivation.

- Send out a request for information (RFI) to gather information about the players in your area, the systems they offer, the average number of systems they have installed, and the typical sizes.

- Prequalify the bidders that you will use on the basis of the responses to the RFI.

- Determine future needs as they pertain to system growth.

- Develop a request for proposal (RFP) based on the consolidation of all the information gathered.

- Obtain budgetary approval in advance for some value of the system intended for purchase. Allow some discretionary funding for added features or growth that you didn't anticipate. Never buy a system that only satisfies today's needs; make sure that plenty of room to grow is available.

- Release the RFP and formulate a target due date for responses.

- Analyze the information and begin to select a short list of possible suppliers that you will feel comfortable with. This includes checking references of other customers using a system similar to the one you are considering.

- Negotiate the contract. Be firm. Make sure all costs are shown and documented.

- Allow plenty of time for the whole process.

These steps are generic enough that the process can be used straightforwardly for any system or service being considered. You can never do too much homework to evaluate a system or a supplier. However, you can get caught into a "paralysis by analysis" syndrome, where the process takes too long. This whole process for a key system can be completed in a matter of weeks before getting to final negotiation. Nowhere is it written that such a study should take months to years. Technological advances and the product life cycles are such that you can only decide on the technology on the basis of the information that is available today. Things will change at a very fast rate.

Voice Processing

Introduction

Voice processing encompasses a variety of technologies that can be categorized into two major areas: directing calls and message manipulation. As we will see, each application in this chapter focuses on one of these two major areas, yet also includes elements of the other area. Moreover, as time goes by and the vendors of these products enhance their offerings, the offerings become more and more alike. So bear in mind that these descriptions are of the essential functions present in each application; in fact, when you go out to buy, you might have a difficult time distinguishing one from another.

The technologies presented in this chapter include the following:

- Automated attendant
- Automated call distribution system
- Voice mail/messaging systems
- Interactive voice response

Other technologies that are still evolving fall into the voice processing category. These will include, for example, voice-to-text conversion and text-to-voice conversion.

Each of these technologies was designed to automate the processing of information, with one major limitation in mind: human beings. The major pitfall in the expansion of our telephony and telecommunications networks has been the inability of humans to keep pace with faster and more reliable communications. Organizations attempting to provide improved services with fewer people have implemented technologies to take up the slack. Many of these technologies have enhanced the price/performance ratios but have subsequently caused user frustrations. No technology is good if the implementation is for the wrong reasons, or if the acceptance isn't there.

Before getting into the details, however, let us discuss two methods that allow human beings to interact with some of these technologies.

Control Alternatives: Touch-Tone or Voice Recognition

Most voice processing technologies depend on Touch-Tone (AT&T) input from callers, if any input is collected at all. As described in Chap. 4, dual-tone multifrequency (or DTMF, the technical term for Touch-Tone signals) tones are specifically designed to be carried well by the analog telephone network. (Pulses, on the other hand, are generally not transmitted beyond a telephone's closest central office; it is impractical to use them for controlling voice processing.) Human beings can also press DTMF telephone buttons far more quickly than they can dial numbers on a rotary telephone.

An obvious alternative to any kind of finger control is voice recognition. After all, people generally prefer speech to pressing keys. And, in fact, some systems now do implement limited voice recognition. But voice recognition technology is not quite ready for prime time in the area of general fielding of calls from the public. The problem is vocabulary. Voice recognition systems generally can either recognize many words spoken by one user (or a limited set of users) or just a few words spoken by many users. Both approaches are being built into production voice processing systems. For example, the voice transcription systems now used by some medical examiners are an ideal application: a very small user population and a precisely definable vocabulary. Another example: some telephone companies are replacing some operator functions with voice recognition. Again, a relatively small vocabulary is required. Have you heard this on the telephone lately?

- "If you are making a long-distance call, please say 'operator' for operator assistance, 'credit card' for calling card service, or 'collect' for collect calls now."

- "The number you called is not answering; please call back later."

- "The number you called is busy; to leave a message…"

These are forms of interactive limited vocabulary. They work well. If you mumble and do not say the word correctly, you will hear the response, "We did not understand your request; please say it again clearly."

Increasingly sophisticated algorithms and more powerful computers are driving up the meaning of *many* and expanding the definition of *just a few* in the first sentence of the previous paragraph. As time goes by, voice recognition will be seen on more and more systems. But to minimize the vocabulary that must be recognized, the inputs will initially be limited to little more than what callers could have entered on their keypads; that is, numbers, letters, and a small number of special characters. But today, most systems accept control from Touch-Tone only.

Automated Attendant

An attendant, in telephone lingo, is an operator for a private telephone system. At your organization, if you have Centrex or a PBX and dial 0, you will reach your attendant if you have one. But if you have an automated attendant, there might no longer be a human in that role. Or if there is an operator there, the operator might be the last resort.

The use of automated attendants to supplement normal operators was intended to assist customers in cutting right through a telephone system if they knew who they wanted to speak to. Furthermore, if customers didn't know who they wished to speak to but knew the department, they could expedite their connections without human intervention. How often have you called an organization only to be greeted by a human with "Good afternoon, ABC Company. Please hold."? The operator had more calls to handle than time to do so. Thus you were put on interminable hold.

Automated attendants were designed to take the rote or redundant answering function away from the operator, so that quality time could be spent with customers who legitimately needed assistance. Some examples of this might include items shown in Table 11.1.

However, too many organizations implemented the system incorrectly. Instead of giving their customers a choice (dial 555-1234 for the automated attendant, 555-1000 for the operator), these users installed the system without warning. Customers who called on Friday afternoon and spoke to a human (as they were accustomed to doing over the years) called in on Monday morning to be greeted by a machine. This shock to the customers' egos and psyches created frustration and a reluctance to deal with this company. Thus, a system designed to improve customer service and relations was instead destroying both. Good installations, coupled with an understanding and concern for the caller, would have prevented the problems.

Automated attendant systems fall into the "directing calls" category identified previously. Automated attendants usually perform some or all of the following functions:

TABLE 11.1

Applications for
Automated
Attendant

Organization	Application
TV station	Newsworthy items such as weather conditions, parade information, repetitive information regarding TV news and community interest
Radio station	Contest information, record requests
Mortgage companies	Current rates on mortgages, information on applications
Colleges/universities	Student activities, registration, class schedules and fill rates
Movie theaters	Schedules, costs, specials

- ▦ Answering the telephone (usually only calls coming in from the outside).

- ▦ Announcing options to the caller. These might include numbers of departments that can be selected, hours of business, and other information useful to callers.

- ▦ Accepting Touch-Tone input from the caller to direct the call either to other information provided by the automated attendant or to a human attendant.

- ▦ Forwarding the call to a human being if no Touch-Tone input is received.

Organizations typically use automated attendants to reduce payroll costs for live attendants. Some have eliminated live attendants entirely; others reduce staffing for the function but allow callers to escape the Touch-Tone system and reach a human attendant. A frequent application is as a night-answer attendant. In this mode, live attendants answer the telephone during business hours; after hours, callers can select an internal extension if they know one.

Another possible use is to provide the daytime functional equivalent of direct inward dialing (DID) without implementing this with the telephone company. With this approach, callers must dial an additional four digits—but no additional trunks, DID or otherwise, need be acquired. The use of DID might not be practical if the key system or PBX is not equipped with these cards, or if an older system is not upgradable. This is a supplement to using the DID trunks, which are more expensive. Depending on the calling population, this can be a very effective supplement to a live operator. Callers who know the extension they want simply enter it; others either select the live attendant from the prompts or time out and speak to the attendant.

Note that an automated attendant is not itself a PBX; it fits logically and architecturally into the same role as a live human attendant. Figure 11.1 illustrates an automated attendant configuration. Although from caller's points of view the automated attendant is in front of the PBX, in practice it might be integrated into or reside behind the PBX as shown in the figure. When the automated attendant is active, the PBX initially routes all calls that would normally go to the human attendant (usually all calls, unless DID is implemented or the automated attendant is present to handle overflow only) instead. The automated attendant attempts to determine the caller's real destination. It then uses the normal PBX call transfer capabilities to route the call to the correct extension, or to the operator if appropriate. The conversation with the automated attendant might go something like this:

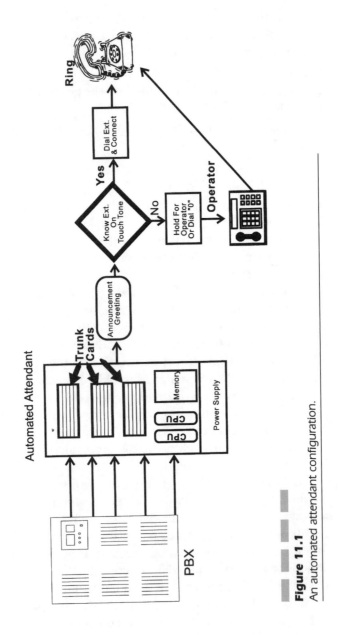

Figure 11.1

An automated attendant configurration.

AA: Thank you for calling XYZ Company! If you know the extension of your sales representative, please enter it on your Touch-Tone telephone at any time. For a list of department extensions, please press the pound sign. If you would rather speak to an operator, please press 0, or just stay on the line and one will be with you momentarily.

Caller: (Presses pound)

AA: For sales, press 31. For accounting, press 32. For installation scheduling, press 33. For customer support, press 34. To hear this list again, press the pound sign. To go back to the first greeting message, press 39. If you would rather speak to an operator, please press 0, or just stay on the line and one will be with you momentarily.

Caller: (Presses 32)

AA: Your call is being transferred.

Note that in this example, no internal extensions would begin with the digit 3. The sample dialogue illustrates some important design considerations:

- Unless it is impossible or you are certain that it is not necessary, always give callers an out to a human operator. If you do not, you risk angering two groups of callers: those whose telephones do not generate Touch-Tones (and there are still many of these) and those who refuse to interact with any kind of telephone answering machines.

- Be friendly—it costs you nothing and will please some callers while offending few or none.

- Keep it short. The longer any one message (or the entire set through which callers must proceed), the higher the percentage of callers that will hang up (also known as balking or reneging). We have encountered one system where the menuing function took us through 12 layers. (We only called that number once!)

- Keep it simple. Some calling populations are very technical (many of those calling technical support for a local area network vendor, for example). But more are not. A lowest-common-denominator approach, simple but not insulting, is usually justified.

- Give people information on how to repeat messages, and never leave them in a position where they will be stuck unless they remember the prompts.

- Do not use telecommunications jargon. You understand it; your callers do not. For example, do not refer to the attendant as such; he or she is the *operator.*

■ When possible, use obvious codes. For example, try to use 0 for the operator. Also, try only to assign a single use to a single code. In the example, the pound sign (#) always provides the department list, no matter where in the dialogue it is pressed.

■ When providing a list of alternatives, provide the more specific choices first, then the more general ones. Not unreasonably, people will often select the first choice that sounds as though it might fit. "Customer support" can cover a lot of ground. If the department list had started with customer support, many callers who might better have been routed to one of the other departments would select customer support instead.

■ If your system permits it, allow callers to enter choices at any time; do not make them wait until a message finishes. People who listen to the end will not notice either way; but those who move more quickly and already know the choices from frequent use will appreciate the flexibility of the dialogue.

A Few Important Points Not Directly Illustrated in the Dialogue

When a machine answers a call, callers often immediately drop into a fault-finding mode; many will be extremely easy to irritate. (Some will already be irritated, a price of using this technology. As evidence that this technology can be a two-edged sword, there is at least one mail-order hardware vendor that includes in its advertising the fact that a human always answers the phone—"no annoying machines!") So consider designing your dialogues to be as inoffensive as possible, even to the point of blandness.

If your system is to be operational both during the day and after business hours, use different dialogues for different conditions. One of the more frustrating things you can do to callers is to inform them that they can escape to an operator, and then prevent them from doing so because the operator(s) went home. It is truly amazing how lousy some recorded greetings sound, even those of billion-dollar companies, and what a bad impression such recordings make on callers. Use a professional studio, announcer, and recording equipment to produce your messages, or your announcements also can fall into the "lousy" category. Remember that this is the doorway to your organization. First impressions are long-lasting.

Dialogues must be designed and programmed (written out) much like computer programs. A poor flow of either words or choices will also reflect badly on the called organization. Although it is important to list choices in an order that will cause the correct choice to be made, saving the most frequent choice for last is one of the surest ways to irritate people. If a large proportion of callers will correctly want a particular choice (e.g., customer service), consider providing a different telephone number for those people to call in the first place.

Try to incorporate a bypass through the prompts if the system allows it. As callers get used to the system, they will not want to listen to every offering. Rather, they'll try to dial the sequence immediately, expecting to get the party they always call.

If you make a material change in the choices, and your system has been in use for some time, announce early in the first menu that the choices have changed. Otherwise, your experienced users will blithely key their way down incorrect paths.

By itself, an automated attendant normally does not take messages or, without direction from callers, route calls other than to a default live answering position. But, as mentioned above, the lines between voice processing systems are somewhat blurred. Do not be surprised if you find a voice mail system or automatic call distributor that encompasses all of the above functions.

Automatic Call Distributor

Automatic call distributors (ACDs) are the grandparents of voice processing systems. Although automated attendants are a fairly recent development (they have been around for a decade, more or less), ACDs have been around for several decades. When you call an airline, if you do not immediately reach a human being, you will reach an ACD. If you respond to a television advertisement that states, "operators are standing by," those operators will be reached via an ACD. ACDs are used extensively for high volumes of incoming calls. This is a load-balancing system to let the agents on the receiving end of the calls spread the workload evenly among the agents. Distributors used in the early days only used a *hunt system*. The hunt would always start at the first agent and proceed down the line until an available agent was found. Unfortunately, if you were in the lead group of numbers, you would be swamped all day long, whereas an agent at the end of the group might only get an occasional call. This was an uneven distribution and caused a lot of stress and frustration in the workforce. The ACD rectifies this by leveling the load by number of calls, amount of time on the line, or other parameters set by the administrator. Typical users of ACDs include telemarketing organizations, the airline industry, travel-related industries (auto rental, travel agencies, hotels), and the like. The purpose of the ACD is also to create a smooth and orderly flow of incoming calls into an organization—for example: by customer, by region, or by product line. Primary responsibilities are:

- Prioritizing the calls
- Routing calls to the appropriate agent
- Delivering any announcements, queue
- Directing busy calls to announcements
- Handling messages for call backs

Because the machine is doing so much in terms of routing the calls to appropriate agents, or overflow to other groups of agents, companies can have far more incoming lines than they have people to answer them. This might or might not have some impact on the customer service function, because frustrations can arise. Think of how you feel when you get to an airline and are told: "All agents are busy. Your call will be handled in the order of receipt; please hold on." The ACD system can be configured in a stand-alone mode (see Fig. 11.2) where all calls bypass the organization's PBX and go directly into the ACD for processing. This is similar to a PBX in that it has its own trunk cards, station (or agent) cards, CPU, printer, memory, and power supply. The system can be simplex or redundant, depending on the finances of the organization. These systems can be as simple or as sophisticated as the company is willing to buy. The ACD can also be a part of the company's PBX system, either integrated into or as an adjunct to the PBX manufacturer's product line. The integrated system uses the same architecture, CPU, memory, power supplies, and so on as the PBX. The agent's terminal is merely an extension off the PBX. There is some merit in having a single product for additional coverage, space considerations, and so on. Again, this depends on the customer's willingness to integrate into a sole source. However, the PBX manufacturers are in the business of making PBXs and the ACD is a spin-off of this product line. Therefore the system might not be as powerful or as fully featured as a stand-alone system designed strictly for the ACD function.

Another variation of the ACD function in PBXs is uniform call distribution (UCD), a basic system with less functionality than the normal ACD. Its function is to uniformly distribute calls to a group of agents, but without the high-end services of a full-fledged ACD.

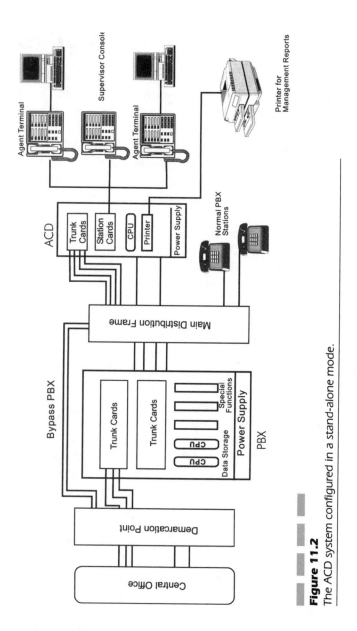

Figure 11.2

The ACD system configured in a stand-alone mode.

ACDs can also be used with Centrex service (Fig. 11.3). The LECs will offer a UCD function, as the ACD capability of a central office is somewhat limited, or, as shown in the figure, the customer can run Centrex lines, 800/888 lines, 900 lines, or the like directly through the Centrex into a customer premises system. The variations do not limit your ability to serve your customers.

Keep in mind the customer service side of the equation. Too few agents can create ill will toward your organization; too many will waste money. Thus a traffic engineering calculation is required to come up with the right mix of agents and lines for the maximization of service (see Chap. 8). Like automated attendants, ACDs fall into the "directing calls" category. But although automated attendants typically are used to replace one to six live attendants and interact with callers, ACDs typically provide access to dozens or hundreds of live agents and do not give callers any choices (although the ACD might make choices itself). An ACD routes calls primarily on the basis of characteristics of the call other than input from the caller. These characteristics might include the line on which the call arrives, the time of day at the answering location, what agents (not "attendants" in this case) are busy, and so on. A typical call to an ACD can go something like this:

ACD: Thank you for calling SuperSoft customer support! All of our support people are serving other callers at the moment, but please don't hang up! Calls are answered in the order they arrive, and the average wait time at the moment is five and one half minutes. (music on hold for 30 seconds)

ACD: SuperSoft also operates a support bulletin board at area code 212-555-7654, with modem speeds up to 14,400 bps at settings No parity, 8 data bits, 1 stop bit. (music on hold for 30 seconds)

ACD: The probable wait time is now about four minutes. (more music on hold)

ACD: Your call is being transferred.

You get the idea. The above monologue illustrates some key characteristics and capabilities of some ACDs.

Unless integrated with other systems such as voice response units, ACDs are not prepared to accept input from callers.

Because the main purpose of an ACD is to smooth calling loads to make the most efficient use of agents' time, a key purpose of an ACD script is to keep callers from hanging up. One way to do this is to provide useful information to the callers—for example, information on how soon

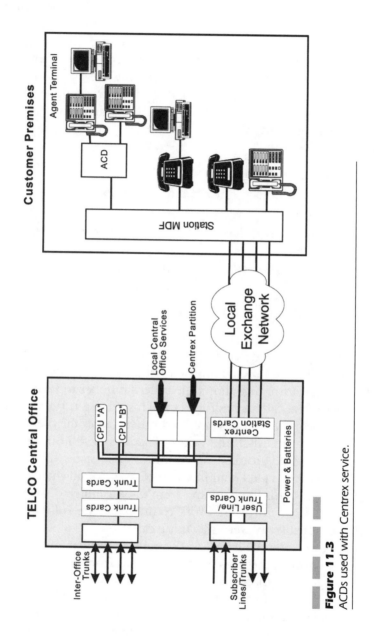

Figure 11.3

ACDs used with Centrex service.

they might speak to an agent (a recently developed capability of some of the more advanced ACDs on the market) or other information related to the company's services.

The preparation of an ACD's response sequences is called *scripting*. Scripting is every bit as critical to a successful implementation as the dialog in a television commercial. Usually the process begins with the construction of a flow diagram (Fig. 11.4). An automated attendant passes on a call and "forgets" about it. An ACD might also function as a PBX (or with one), and might track, retain, and report on numerous details about calls, including:

- The originating numbers
- How long each caller had to wait before an agent was assigned (*queue time*)
- How many callers give up (*balk* or *abandon*) before being assigned to an agent
- The agent to whom each call was passed
- How long each agent took to answer each call
- How long the calls took to handle

As can be seen from the above list, an ACD is a management tool as well as a sophisticated call director.

ACDs are now being used with the automatic number identification (ANI) feature that allows the caller's number to be displayed to the agent when the call arrives (Fig. 11.5). This service is typically offered with 800 Megacom (AT&T) services and other names for the bundled 800 services from Sprint, MCI, and others. The jury is still out on the caller identification issue around the country. Some states have allowed it, whereas others have ruled it an invasion of privacy and therefore disallowed it. On business calls there doesn't seem to be as much dispute regarding this service. When future network technologies are fully deployed, ANI might be available on every incoming call.

Using the ANI feature, the caller's number can be routed to specific messages on the basis of who is calling, from where, and so on. Priority can be given to special clients using this service. Another slant on the use of ANI with an ACD is to create a computer to PBX/ACD interface, an example of computer-integrated telephony (CIT). Figure 11.6 shows a graphic on this integration. With the customer's number delivered (inband) with the call, the system can route the caller ID information to a computer via an X.25 packet switching link. The computer reads the incoming telephone number, looks up a database of numbers, searches and retrieves a

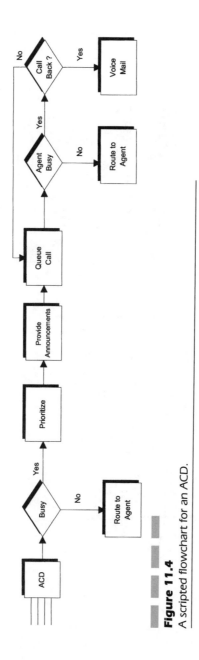

Figure 11.4

A scripted flowchart for an ACD.

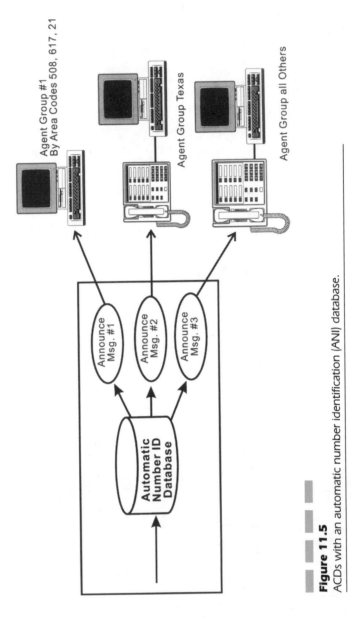

Figure 11.5

ACDs with an automatic number identification (ANI) database.

Figure 11.6
A computer-to-PBX/ACD interface.

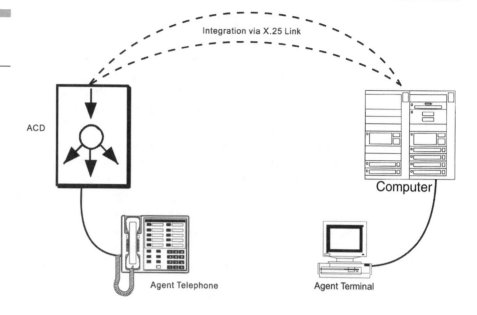

customer database file, and delivers the call and a screen of customer information to a computer terminal simultaneously. This allows the agent to get right to the customer's inquiry, saving from 20 to 30 seconds per call by eliminating the exchange of caller information, the agent looking up customer information, and so on. Many organizations have felt that the efficiency gained by the use of the computer-to-PBX interface (CPI) have paid for themselves in customer service, improved agent performance, and better morale.

ACDs are not appropriate for all businesses. Typical requirements that can justify acquisition of an ACD include:

- Large volumes of incoming calls
- Callers needing to speak to anyone in a particular group (e.g., order takers, customer support personnel, travel agents)
- Need to fairly assign incoming calls among a group of agents

Although large communications environments will often acquire a dedicated piece of hardware built and programmed for use as an ACD, several PBX manufacturers have come out with add-on software packages that allow their PBXs to function as low- or mid-range ACDs while performing their normal PBX functions.

Voice Mail

With fewer people in the office, organizations began to seek new technology to enhance their operations. A company in Dallas offered a new system called *voice mail*. Voice mail was originally designed to reduce the amount of busy tones in the office. Users were absent from their offices more frequently, and the old method of a secretary taking the name and number of the caller was not meeting the needs of the organization. Frequently, the called party diligently returned the call, only to find that the calling party was away from their desk. This game is called *telephone tag*. However, some of these messages were information only; they did not require an interactive conversation. Thus voice mail could provide a means of leaving one-way information without the telephone tag. Further, the messages could be delivered at any time, at the convenience of the recipient. Figure 11.7 illustrates some of the ways voice mail can enhance productivity:

- It can be used as a remote dictation machine, although this is not its primary mission.
- It can be used to deliver a message to a recipient who is unavailable (or on the telephone) when the caller makes his or her initial call attempt.
- It can be used to remind oneself of important tasks; the reminders can be triggered at a specific time, as can other messages.
- It can be used to deliver identical messages to a target group with a single action.
- It can be used to dial out and deliver a message remotely to someone not even on the system!

Users have been somewhat reluctant to accept voice mail. From an internal perspective, the system can be effective. However, from an external standpoint it can be a point of irritation. Users sometimes begin to hide behind the system, screening their calls before answering. Coupled with this problem is the response mode. If a caller has taken the time to leave a detailed message and expects a return, the called party owes them a return call. Human nature states that the called party will not return the call if he or she is busy or if the call is of an unpleasant nature. Thus the frustrations begin.

Voice mail does have benefits. It can gain additional hours of availability for your users, especially when dealing with callers from other time zones. Users can be satisfied that although they are not in the office, they can still receive information. Thus the technology does have its

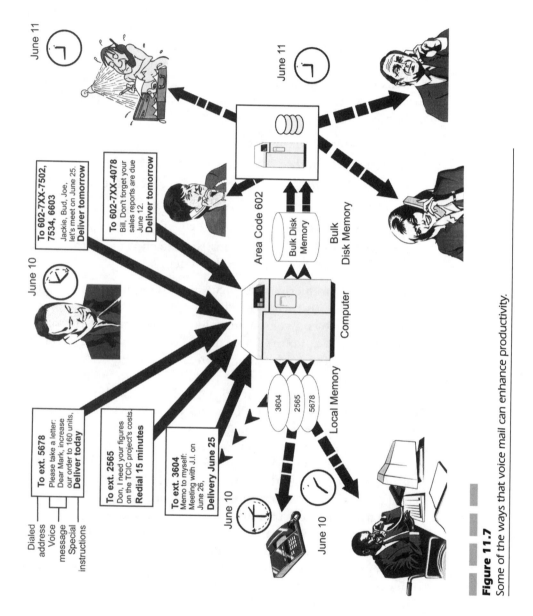

Figure 11.7

Some of the ways that voice mail can enhance productivity.

positive side. As long as the users are properly trained in the use of the system and the administration provides a follow-up on the appropriate use of the system, it can work and save an organization both time and money.

Voice mail, like automated attendants, is based on separate equipment. But voice mail falls into the message manipulation category. Voice mail systems were originally designed to do for voice what electronic mail or messaging does for typed communications. That is, using Touch-Tone pads, users can:

- Dictate messages to the system
- Specify to whom those messages will be delivered
- Listen to, forward, or delete messages directed to them
- Build and use distribution lists for messages
- Modify various characteristics of their "mailboxes"
- Provide a verbal buckslip (routing slip) attached to another voice mail, to be forwarded to another on the system for action and comment

To be most productive, voice mail systems require a fairly high degree of integration with their host PBX or Centrex. Perhaps for this reason, most of the high-end voice mail systems on the market are sold by PBX manufacturers (e.g., Rolm's Phonemail, AT&T's Audix). In particular, it is important that the voice mail system be able to indicate to users when they have new messages. For users with indicator lights on their telephones, the voice mail system must be able to tell the PBX to light the light. For other users, the PBX should provide a "stutter" (intermittent) dial tone to indicate new messages when the handset is raised.

Figure 11.8 illustrates a typical voice mail installation. Like OPXs, voice mail systems are normally connected on the line rather than the trunk side of a PBX (although they will also usually have some of their own external trunks to allow users direct access from off site).

As with other telecommunications configurations, the number of trunks serving a voice mail system is a critical success factor. More than with most systems, the usage pattern on voice mail systems is not a smooth one. Instead, users tend to check their voice mailboxes at similar times of the day: on arrival at work, when returning from lunch, and shortly before departure at the end of the day. The first of these, because it is often the same for the entire user population, often generates a peak demand for circuits far above the expectations of the system designer. Consider the reaction of your user population if 30 percent of them try to reach the system at the same time...and only 10 or 15 percent succeed.

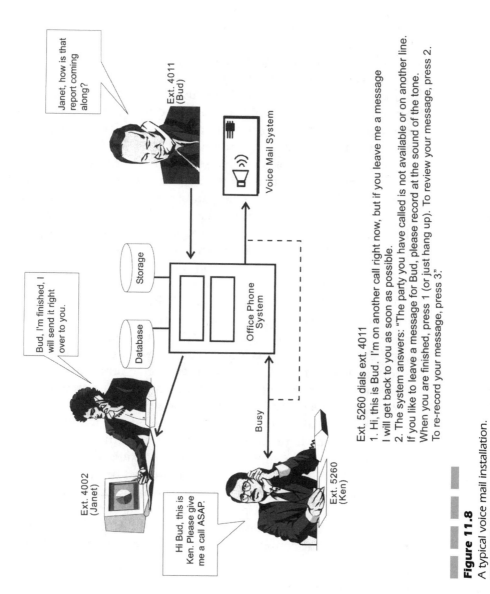

Figure 11.8
A typical voice mail installation.

Low-end PC-based voice mail systems (typically limited to one or two physical telephone lines) are now available. An additional circuit card, along with appropriate software, is required for the voice processing; however, such systems can be implemented for little more than $100 (on top of the cost of the PC, of course).

Large systems are configured similarly to a PBX or key system. To avoid the access problem described above, a traffic study is conducted on the inbound and outbound traffic. Then, using the traffic engineering tables from the Poisson distribution, the number of lines and ports can be determined. Typical configurations are in increments of four trunks/lines. Systems come in configurations of 4, 8, 16, 32, or 64 lines, and so on.

An additional key factor is message retention capacity. High-end systems are normally configured by the number of users they will support. This drives assumptions as to the number of hours of messages that must be stored. Because these systems digitize voice and compress it on hard disks, large capacities are used for the storage of voice messages. These drives are usually described in hours of storage. Examples would be 8, 16, 32, or 64 hours of storage, and so on, although capacities can also be in hundreds of hours.

Because all of the voice messages are digitized (although usually at less than the 64,000 bits per second of "toll-quality" voice—the messages are compressed, sometimes down to as little as 9600 bps), a key consideration is the amount of disk space. So whoever is configuring the system must estimate an average message length, then decide how many messages the average user will retain (or will be allowed to retain) in his or her mailbox. Knowing the bits per second at which messages are stored on a particular system will then allow a calculation of required disk resources. For example, the figures in Table 11.2 represent a sample of the calculations used.

Doesn't sound like much, does it? But consider that at the standard 64,000 bps for toll quality, the total would instead be 720 megabytes. Compression helps! This calculation was much more important a few years ago, before the cost of disk storage dropped to its current levels. (A 1-gigabyte disk can easily be acquired for less than $500.) But understanding it can assist you in selecting a configuration for a voice mail system, as well

---

TABLE 11.2

The Typical Calculation for the Hours of Storage Required

200 users retaining 5 messages each, averaging 90 seconds, digitized at 9600 bits per second requires $200 \times 5 \times 90 \times 9600 = 864,000,000$ bits, divided by 8, divided by 1,000,000, gives 108 megabytes.

as illustrate the impact of some of the choices you can make when configuring the system. Note also that when a vendor touts the number of hours its system stores, and justifies the cost on the basis of that storage, if you can discover the digitization rate you can calculate how much disk storage you are actually getting for your money. Those could be very expensive disks.

Vendors will, of course, be happy to recommend a configuration to suit your requirements. But understanding that one of the key considerations is disk storage capacity should arm you in your discussions.

There are a few pitfalls to watch for in the implementation of voice mail systems. Some organizations have implemented voice mail systems as their primary answering positions. Like automated attendant systems, voice mail systems have the potential to irritate (or infuriate) callers who either do not ever want to talk with a machine or just take exception to the structure of menus on your system. Organizations have lost customers this way, so be careful! What is even worse than the structured messaging service is winding up in the voice mail loop. You have probably already been exposed to this one:

Caller dials called party number and gets "Hi, this is Janet. I'm not available to take your message right now, but if you leave a name and number I'll get back to you as soon as I can. If this is an emergency, please dial 0 and ask the operator to connect you to Bud, who is covering for me."

Caller dials 0 and asks for Bud.

Caller gets "Hi, this is Bud, I'm not available to take your call right now, but if you'll leave a name and number, I'll get back to you as soon as I can. If this is an emergency, please dial 0 and ask the operator to connect you with Don, who is covering for me."

Caller dials 0 and asks for Don.

Caller gets "Hi, this is Don. I'm not available to take your call. If this is an emergency, dial 0 and ask the operator to connect you with Janet, who is covering for me."

Caller hangs up!

Some organizations initially target these systems at selected individuals. Such implementations almost always are abysmal failures, because unless everyone has the facility almost no one will use it. Few have the patience to ask themselves, "Does this particular employee have voice mail?" before every message. The problem is similar to what might happen if the office mail room only delivered to selected individuals; everyone would deliver his or her own mail to ensure it got there. So if you want to put in voice mail, give it to everyone.

If you do give it to everyone, watch out! With a good kickoff, good training, and a good implementation, voice mail might take off to the point where your original traffic estimates err on the low side. So be sure to get a system that is not near its maximum configuration, as you will likely need to add disk capacity or trunks to keep users from receiving busy signals.

Interactive Voice Response

Interactive voice response (IVR) refers to technology supporting the interaction of users with the system (similar to voice mail or automated attendant), but also implies that information will be provided to the caller by the system itself.

The line between IVR and the previously described voice processing technologies is a fuzzy one; after all, an automated attendant often can provide telephone numbers to a caller. But typically, an IVR system is used to provide callers with information unrelated to the system or the technology itself. The data is normally either accessed directly in another system or loaded into the IVR in advance from that other system.

For example, many mortgage companies have implemented IVR to allow clients access to their own mortgage records. A typical dialogue follows:

IVR: Thank you for calling Millie's Mortgage Machine! For current mortgage rates, press 1. To access information about your own account with us, press 2. To repeat these choices, press 9. To speak to an account representative, press 3 or just stay on the line.

Caller: (Presses 2)

IVR: Please enter your account number, followed by a pound sign.

Caller: (Enters account number and a pound)

IVR: Thank you. Please enter your personal identification number, followed by a pound sign.

Caller: (Enters PIN and pound)

IVR: Thank you. Your account is current through October 31st. As of November 1, your remaining balance is $83,432.21. Your monthly payment is $997.92; at 9 percent, you have 132 payments remaining.

To have a balance payout statement mailed to you, press 1. To hear your account information again, press 2. To return to the previous menu, press pound sign. To hang up, press 8. To repeat these choices, press 9.

Many of the points of voice processing systems also apply to IVR systems. The above dialogue illustrates some additional key characteristics and capabilities of IVR systems.

Many systems require the entry of a PIN. Any system capable of providing confidential information to callers should have at least this level of security. Normally, PINs are a minimum of four digits. Some service providers allow users to set their own; other organizations assign PINs to authorized users. Either way, the system's organization must administer the PINs, a requirement sometimes initially overlooked when planning a new IVR system.

Whereas an automated attendant or an ACD is primarily dedicated to providing communications between a caller and another human being, an IVR system's main reason for existence is often to directly provide the caller with automated information. In fact, many organizations install IVR units to reduce the number of agents required on the phones. Even if the initial setup cost of the system is tens of thousands of dollars, the savings in salaries can easily result in a payback within one year. IVR systems always incorporate computers. Therefore, in addition to simple retrieval of data, they can also perform operations or calculations on that data. In the preceding example, most of the numbers were probably calculated during the call rather than retrieved from storage. In another popular application, computers at banks can trigger electronic funds transfers (in effect, pay bills with paperless automated checks) under control of a caller's telephone call. The computer, probably interacting with another computer, adjusts the caller's balance as part of each such transaction. You can bet that both account numbers and PINs are required for such calls! The integration of computers and telephones is sometimes referred to as computer-integrated telephony (CIT) or computer-to-telephony integration (CTI). Figure 11.9 illustrates a possible architecture for such an application. In the figure, the IVR units access the computer database as needed to satisfy caller inquiries. The IVR is separate from the computer, a configuration that might appear less and less frequently as computer manufacturers begin to offer voice interfaces and processing capabilities on their own machines.

IVRs can also be used in combination with ACDs. Figure 11.10 shows a representation of this arrangement. As an example, an IVR can be used in

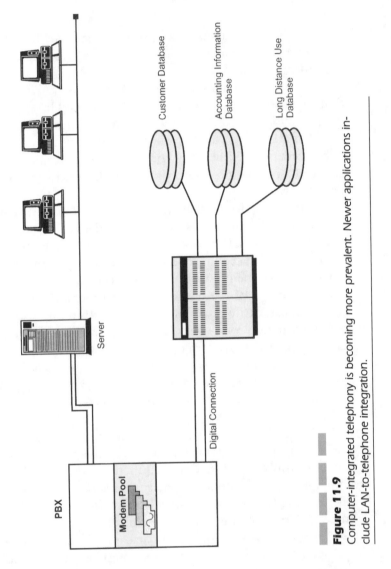

Figure 11.9

Computer-integrated telephony is becoming more prevalent. Newer applications include LAN-to-telephone integration.

220

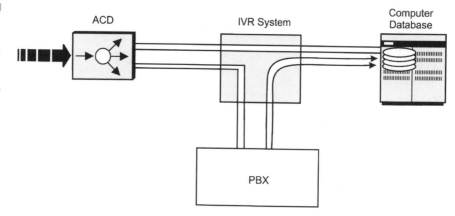

Figure 11.10
Access to computer systems through ACD and IVR integration is now a possibility.

a customer support application to optionally provide callers access to a recorded database of common questions and answers. If the caller elects to inquire against the database (using his or her Touch-Tone telephone to specify the questions to be played), an agent might not be required at all. If the caller selects *Speak to a customer support person* from the automated IVR menu, then the call can be directed to an ACD for queuing and ultimate connection to an agent.

Some ACDs now have built-in IVR capabilities as an option. We expect an expansion of such combinations as ACD providers seek out new ways to be competitive.

12

Computer-to-Telephony Integration (CTI)

In the previous chapter we mentioned the topic of computer-to-PBX interfaces in a cursory manner. In this chapter we will look at the computer-to-telephony interface (CTI). Over the years, much time and effort has been spent in trying to develop interfaces that would tie the computer and the telephone network together. As technology has changed, many new interfaces and applications can be created through the integration of these two techniques. As a result, a whole new industry has emerged. The ability to link computer systems and voice systems together offers some new possibilities on how we approach the office. An example of this was mentioned in an earlier chapter. When one thinks of the automatic call distribution systems with the ability to link the automatic number identification (ANI) and a database to provide screen-popping capabilities, it becomes an exciting opportunity. In Fig. 12.1, a representation of the computer-to-telephony integration capability using a screen-popping service is shown. In this particular case, as a call comes into the building it is initially delivered to an ACD. At this point the ACD captures the automatic number identification (ANI) of the calling party. When the called party's number is packaged into a small data packet, it is then sent to the computer as a structured query language (SQL) inquiry into a computer database. Assuming the database already exists for the established client, the SQL will open up the client file and respond to the query. In this particular case the telephone number of the calling party is being used to create the query from the database. Once the database entry is recognized on the basis of telephone number, a computer screen will "pop" to the agent's desk. This occurs at the same time the telephone call is being delivered to the agent's desk. Now the agent taking a telephone call does not have to ask the caller for all of the associated and applicable information, such as:

■ Name

■ Address

■ Telephone number

■ City and state

■ Account number

■ Any other pertinent information

Of course, if the customer's name and address is not in the database, it can immediately be entered. As the agent answers the call, the absence of customer information attached to the telephone number (ANI) is a clue that this is a new caller. The agent can then ask the appropriate questions and build the entry into the database. By reducing the time spent gathering this information because it is already available in the database, a savings of

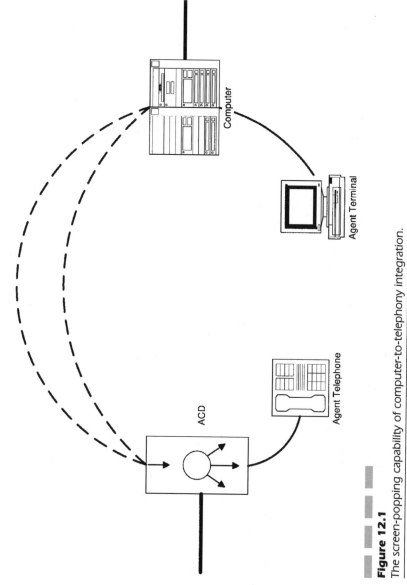

Figure 12.1

The screen-popping capability of computer-to-telephony integration.

20—30 seconds of call processing time can be achieved, as stated in Chap. 11. However, the 20—30 seconds is not the critical part. Instead, the ability to satisfy the customer and to provide better service levels is what is important. By saving the 20—30 seconds of call processing time, an organization will save significant amounts of money—absolutely! When one thinks of other alternatives that might exist here, it can also initiate a couple of thought-provoking ideas. For example, assume that an agent has received a call from an established customer and a screen-pop of information. In this regard, the established customer has saved all the time by not having to give his or her name, telephone number, and so on. The agent gets right to the task at hand—identifying and verifying the customer. Now that the agent has the information, historical files on such things as buying or usage patterns can be developed so that the agent can suggest products and services to the customer on the basis of past buying experiences. Moreover, if there is a problem with a particular customer, such as delinquent payment, the agent can readily obtain this information while talking to the customer.

Let's extend that thought a little bit further. Assume that the agent does encounter a delinquent customer who hasn't paid his or her bill for three months. While talking to this customer and trying to sell through suggestion, or taking an order which is what the customer called for, the agent sees on the screen that the customer is delinquent and that collection action must be taken. Rather than placing him- or herself in the position of being a collections manager, the agent can immediately suggest to the customer that the call must be transferred to the accounts receivable manager. When the agent switchhooks or transfers the call, not only does the call go to the accounts receivable department, but the screen from the database follows. Now, when the accounts receivable department receives the call, the manager there will have a full screen of information about why that call has been sent to him or her. Answering the call, the accounts receivable manager can then immediately and proactively get to the job at hand, collecting the money. Instead of picking up the phone and asking for the same information (i.e., name, address, telephone number, and so on), which would totally frustrate the customer, the accounts receivable manager can immediately introduce him/herself and suggest that payment terms can be worked out. The dialogue may be along the lines of: "We really appreciate your business, and we would like to continue to ship you our products. So if you will Federal Express us a check for the past-due amount, we will immediately ship (after receiving the check) all of the goods and services you have just ordered." This is proactive as opposed to negative collections. Now the accounts receivable manager, after obtaining agreement from the customer to pay, can transfer the caller immediately back to the agent. The agent's record and the screen have been kept by the system. When the call goes back to the agent, the screen plus all of the documentation or notes that the accounts receivable manager has entered return as well. Once again, the agent sees who is on the phone and what has transpired. This prevents the replication of information gathering and keeps the entire conversation in a proactive mode. It is this positive handling of calls, particularly those of a negative nature, that helps to boost customer service and influence the continued buying relationship.

The above scenario assumes that certain things have taken place. It must be assumed that the caller is calling from a prespecified location and that the database already exists. If in fact the database does not exist, the initial data gathering can be accomplished at the receipt of the first call. Thereafter, any time the customer calls from the same telephone number (which is displayed at the agent's console) the database will be hooked and brought with the screen-pop. The first time is the most labor intensive. Each successive contact between the customer and the organization selling the products and services will be much more simple. Using this database and screen-pop arrangement, companies have successfully achieved several things, including:

- Improved customer service
- Improved customer satisfaction
- A reduced number of incoming lines
- Reduction in the number of agents necessary to handle the same volume of calls

Through the combined savings of 20–30 seconds for each successive attempt, the average hold time (talk time) can be reduced significantly. In Chap. 8, where we looked at traffic engineering, the discussion of how many agents are required to handle a particular volume of traffic is covered. When the average hold time is reduced, however, the amount of agents and the grade of service changes dramatically. The number of agents declines, while the probability analysis for serving the customer quickly increases. This is a win/win situation.

Other Applications

Although the screen-pop capability of an ACD is one of the primary uses today, other applications exist that can take advantage of the computer-to-telephony integration (CTI). The newer integrated PBX will become a voice server on the local area network (LAN). In Fig. 12.2, the integrated

PBX is shown as a voice server residing between two separate LANs. Now that the various communications vendors have produced an open applications interface, the ability to use a PBX for several components exists. One of the primary applications would be to install a third-party bridge or router card inside the PBX. The PBX now serves as either the bridging or routing function between the two LANs, even if they are disparate in nature (token ring and Ethernet). Using the PBX also as voice server allows an extension of PBX access to the local area network services such as printing or file serving applications. A remote user who is too far away from the LAN can be serviced through the PBX wiring. All of the PBX manufacturers today are devising schemes to implement high-speed multimedia communications to the desktop. This way a computer user, as shown in Fig. 12.2, can be remotely attached over the same cabling system as the telephone system. Inside the architecture of the PBX will be a high-speed communications infrastructure such as ATM, SONET, or FDDI (each of these are covered in later chapters). Each manufacturer has chosen a different route in providing this capability. With a high-speed communications backbone residing inside the PBX, the remote PC is attached into the PBX and onto the high-speed backplane. Then the router card or bridge card acts as a bridging and routing function between the high-speed infrastructure (like FDDI) connecting to the installed LANs. The PBX manufacturers have devised means of running 10 Mbps or 100 Mbps to the desktop. When dealing with the high-speed communications to the desktop, the user now has access to all of the voice features, LAN features, and multimedia applications that may be running inside or peripheral to the PBX. The cost of this PBX architecture is considerably greater than that for a normal LAN and a PBX environment. But as a combined investment, the differences are minor. By virtue of the high-speed multimedia applications to the desktop, the user also has access to the PBX-type features. In this case, the user can use the PBX as a communications server. A modem pool can be attached to the PBX and bridged onto a user's connection on demand. This is shown in Fig. 12.3. All of the features and functions of the modem pool now become available to the users on the PBX as well as the LAN. As a true communications server, the PBX can also bring other services such as trunk queuing, automatic route selection, least-cost routing, and station message detail recording to the data communications environment. These added features could be available on a LAN, but they are easier to implement on a PBX, where the features and functions have been available for decades. Now, with the convergence of communications and computers, the administrators within organizations have many choices.

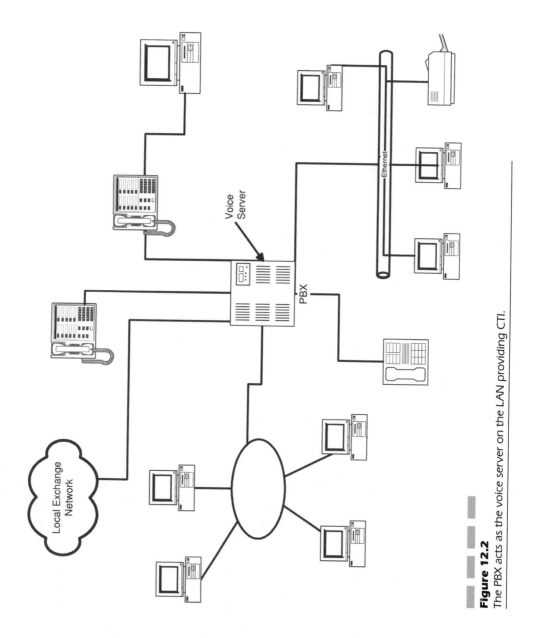

Figure 12.2

The PBX acts as the voice server on the LAN providing CTI.

229

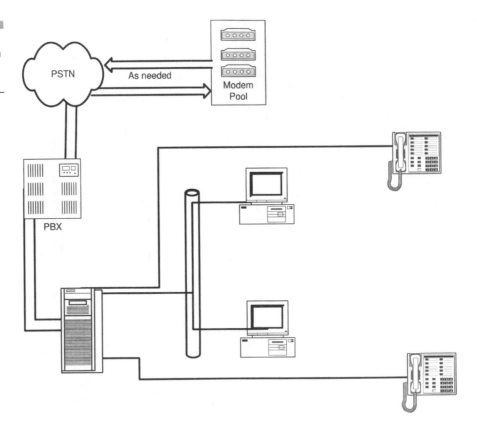

Still other features can be included with this integration of CTI. Voice
messaging, as integral or peripheral to the PBX, can now become available
as a feature or a function hooked to the LAN. As an outside caller dials
into an organization's PBX, the ability to capture the automatic number
identification (ANI) and hook to the database server on the LAN allows
for the delivery of the caller ID to the desktop. This makes it possible for
a user who has a screen (PC or other) on his or her desk to be able to see
the voice messages in the queue. In reality, what the user sees is not the
message but information regarding the message. This helps to prioritize,
or select from a queue, calls that are necessary or important. Moreover, the
end user can print out a list of voice messages, a customer database of
people who have called, ANI information to be included in a database not
yet built, or many other useful and productive types of data.

Beyond the voice messaging capability, the ability to hook to an automated attendant can also be integrated into this CTI application. When an
outside caller attempts to reach a particular extension or department, a
database can be built regarding who the callers are and what extensions are

typically used most frequently. This also extends the ANI right to the desktop of the called party so that the database can also link to some form of a screen-pop. By delivering the ANI to a database server on the LAN, the user can immediately run a structured query on the database to pull up the customer file. A simpler way would be to have the database automatically deliver all the information about the calling party to the LAN terminal device. This doesn't work in all applications, but it does offer some unique capabilities and opportunities that never before existed.

Each of the services that have been discussed, as well as many other opportunities that have not yet been thought about or produced, can all be tightly coupled in the combination of computer and telephony integration. CTI offers the ability to link multiple databases, multiple communications channel capabilities, and the LAN to PBX server functions. The PBX becomes the voice server, whereas the LAN server becomes the PBX's database. Corporate directories, departmental lists, product lines, and other types of usable information can all be made available to the desktop device when a called or calling party needs access. Although the cost is incremental, overall the increases in productivity and satisfaction from both employees and customers may well be worth the effort. This is one of those decisions that a user must make.

In the same manner, you take higher risks by putting all of your eggs in one basket—having a single-server environment. However, the distributed computing architectures that are emerging minimize the risks. Distributed computing architecture can also add to the depth of the PBX by using servers on a LAN that can be voice servers; the functions and the real estate required for a PBX architecture can now be combined and reduced. No longer will a very large room with special air conditioning be required. The servers can just be deployed wherever they are needed to serve a group of users. We can see the excitement and the benefits of the computer-to-telephony integration when we look beyond the basic functionality of each individual box. Hence the decision making process becomes a little bit more complicated than the purchase of a PBX or a server when thinking about acquiring a combined infrastructure to support the communications needs of the organization. But isn't that what communications excitement is all about?

Why All the Hype?

A lot has already been said about what is going on within the industry and organizations today. Business has reached an intensity and a pace that is extraordinary. Competition and changes in technology are driving the pace exponentially. All of these factors combined have driven organizations to seek and find improved means of dealing with their customers and to answer the needs for increasing productivity, decreasing costs, and enhancing the competitive edge over the existing market segment. Keeping pace with the industry and maintaining the competitive edge has become a crucial element in the survivability of an organization. Therefore the demands on organizations have escalated dramatically to basically provide for:

- The ability to deal with corporate decision making processes at a very rapid pace
- Readiness to deal with issues regardless of the time of day, day of week, or availability of Executives
- The continued restructuring and re-engineering of the organization, requiring fewer people to perform more functions
- The changing workforce, in which talent and skill sets are no longer as available as they were, making technology a necessary solution
- Preparing the administrative assistants who have supplanted the administrative and support staff of old (secretarial support), to deal with issues when the senior staff is not available
- Ensuring the availability of the information that is needed to achieve these goals

Therefore, in order to effectively use the resources that still remain within the organization, employees must have information readily available at their fingertips (a few keystrokes away). Using technology of both computers and communications provides the edge to maintain the link between the employee and the customer. It is not necessarily a given that these technologies will always be the solutions, but the right implementation can have significant benefits if done correctly. There are several examples in this book that show how organizations have tied their computers and communications together through various network technologies to facilitate the timeliness and usability of information. Using a computer-to-telephony integration system, employees have the ability to share information on the spot by using updated information on a call-by-call basis. Each of these tools and capabilities is an absolute must if a business is to remain competitive.

Many authors in the industry have written about the re-engineering of the organization, the rightsizing, downsizing, and capsizing of the organization, or just a restructuring in a target market where companies have focused more clearly on serving the niche needs of their customer. Examples of this are clearly demonstrated in both the financial community (credit card organizations, stock brokerage firms, the mortgage banking industry) and the airline industry. Customers can no longer be looked on as a nuisance; rather, they must be seen as the most valuable resource and asset that an organization has (excluding its own internal personnel). Providing the necessary treatment and care of the customer becomes one of the primary goals of the organization. It should have been the goal all along, but the focus on driving costs down and increasing productivity began to erode the customer and organizational relationships. Now that the computer-and-telephony integration has begun its movement within the industry, it is much more feasible to restructure and reaffirm the relationship between the two parties. Customers expect to be treated as individuals who are special for the size and volume of the business that they do. Competition in any segment of industry is fierce! Therefore, if the necessary care and treatment is not provided, customer loyalty shifts dramatically. Changing market demands due to the rightsizing, downsizing, and capsizing in external organizations places more demand on your organization to meet the customer's needs. In planning our technological innovations, we must therefore look to how we can better satisfy the customer's needs by:

1. Giving our customers unlimited access to our employees to place orders, check on status, or make inquiries.

2. Assuming that the employee/agent will have that information readily available and not put the customer on interminable hold.

3. Anticipating the demand for products on the basis of sales projections and customer demands so that "out of stock" or back-ordering can be curtailed.

4. Having technical support available to the agent so that a question can be easily answered either by the technical or the administrative/sales person.

5. Considering that the customer's success in business and/or product lines can be matched to your ability to meet the customer's demands for products and services.

6. Anticipating customers' needs and satisfying their demands, so they will not be exposed by the diminishing inventories and just-in-time manufacturing processes of today.

If the above conditions can be satisfactorily met, the customer will feel better about placing orders or checking on the status and availability of products with your organization. This equates to better sales. Telecommunications and computer integration can therefore increase productivity and also assist in increasing sales. Using the right technology in the right mix at the right time can enhance a relationship that is ever so fragile in today's competitive marketplace.

Linking Computers and Communications

Today, more than ever, organizations are using distributed computing and what is known as *client/server architecture*. This architecture of using a three-computer technology is the result of many years of evolution and revolution within the computing platforms. The LAN emerged in the early 1980s as a means of moving away from control by a single entity known as the MIS department (a more detailed discussion of the LAN is contained in Chap. 19). As this evolution moved away from a controlled single computing platform, users were empowered to manipulate their own data, handle their own information, and share that information with others on a selective or exclusive basis. Because of this movement, a lot of pressure was placed on the communications side of the business to provide interconnection using a cabling infrastructure to the various desktop computers and departments that would interrelate. Management saw an opportunity to produce ad hoc reporting, allow flexible data manipulation, and provide more timely information to managers and departments alike.

As LANs emerged, they were primarily tied to a desktop and an individual server within a department. Later in its evolution, the client/server architecture started to emerge more strongly by allowing multiple departments within a single organization to fulfill their data communications and data access needs by accessing each other's databases or sharing the files that they had created on their customers. It is this sharing that created additional appliance affiliations within the different groups. The personal computer (PC), as it attached to the network, basically allowed users to store their applications and files on a single device. By linking these computing systems together with a communications medium, the user can best seek the information wherever it resides within the network. If a new application is needed, either a server can be adapted to fit that need or a new server can be added. From a communications perspective, devices emerged that included such things as electronic mail within departments, fax servers for transmitting information directly from a desktop to a customer, and voice recognition and response systems that enabled customers to call in and literally talk to the server as opposed to having to talk to a human. Furthermore, the linkage of all of these systems through a communications medium became the glue that bound everything together.

The PC of old, although structured as strictly a desktop device, has changed dramatically. Now, new features include:

- Extensive processing power
- Sophisticated operating systems
- Integrated voice boards or sound cards
- Integrated video conferencing capabilities
- The client/server software that allows all of the pieces to be pulled together

With the use of CTI, an organization's needs can easily be met using the buildout of the desktop devices connected to the LAN. As already mentioned, the client/server architecture has the ability to bring together the computers that reside on the LAN and telephony services that can be built into either the PCs or these servers that reside on the LAN. This all becomes possible by bonding these two technologies that began in the mid-1980s and now have matured in the mid-1990s. The architecture within an organization has dramatically changed! In place of a single provider of information, such as an administrative assistant or a particular sales group, there now exists an integrated homogenous work group that can share information, transfer calls among members, or add other departments on calls as needed. This has already been shown in our examples.

While this information process was taking place, organizations began to connect to the outside world and to their customers through the use of their telephony services. In the early 1990s a new explosion took place within the industry. The emergence of the World Wide Web (WWW) and the use of the Internet as a commercial resource added to computer telephony development. Attaching an organization to the outside world through the Internet allowed customers to use their telecommunications services to enter the organization and check inventories, catalogs, and other services that would otherwise not normally be available to them. The ability to introduce fax-back servers also allowed customers to obtain catalog information or technical brochures immediately instead of having to wait for an agent to mail products or brochures through the postal service. This increased the availability of information by linking the communications infrastructure to the computing architecture. Moreover, as organizations started to use electronic document interchange, (sometimes called EDI) they created the ability to place orders, perform functions as dynamic work groups, and facilitate the ordering and payment process. These services enhanced the true organizational capability. Now, the mid-1990s has marked the innovation and growth of this computer and telephony integration through the use of *intranets*. The intranet is discussed in Chap. 17. For now, think of it as the internal Internet to an organization. Using the intranet and communications infrastructures within an organization's client/server architectures, internal organizations can access the same information that might be available to a customer. Furthermore, internal organizations can share files, technical notes, and client notes regarding their customers as well as other valuable information that would not be readily available to the masses. Specific files or bulletin boards can be set up within the organization to achieve that end result. So, it is through the integration of our communications systems and the capabilities of the far more powerful and processor-intense desktop device that the pieces are coming together quickly. There can be no distinction anymore between computing and communications, because the two really draw on each other's resources to facilitate the organization's day-to-day mission—to serve the customer and maximize shareholder wealth. Serving the customer increases sales. Increased sales and, hopefully, decreased costs, raise profits. This maximizes the shareholder's wealth, which is the charter for all business organizations.

The Technology Advancement

Through the integration of computer and telephony-type services, the old philosophy of telephony being a necessary evil has passed. Advances in computing technology have enhanced the telephony world. The age of mass pools of telephone operators is gone. Newer technologies allow customers to call and proceed through an organization without ever communicating with an operator or a secretary. The customer can get directly to the person or information that is desired.

In the key system marketplace, the old labor-intensive electromechanical system is now a computer-driven telephone system. However, these computer-driven systems are now rich in features and laden with capabilities undreamt of in the past. The integration of voice messaging, ACDs, and automated attendants in key systems is now commonplace. Key systems are also emerging as voice servers within the organization.

The PBX, serving hundreds if not thousands of employees within an organization, allows an internetwork of services and capabilities to spread throughout the organization. What was once just a stand-alone telephone system is now the high-end computer that just happens to handle voice. Today's technology allows for full digital transmission systems or services and permits linkage to all devices that were exclusively used by the elite. PBXs now integrate tightly to our computing systems. The PBX also acts as the high-end server on a digital trunk capability to the outside world. By linking to a high-speed digital communications trunk from the outside world (called a T1 and referred to in Chap. 14), users can now access ANI. Using inbound 800/888 services and a common channel signaling arrangement enables delivery of caller ID directly to the called party. Moreover, enhancements in software such as call forwarding, overflow arrangements when agents or other individuals are busy, and other services can all be chartered into a single PBX architecture. Another service called directory number information service (DNIS) can be used on the 800/888 service to direct the call to a specific group, such as technical support or marketing.

The Final Bond

As the client/server architecture and the communications systems all were developing throughout the 1980s and into the 1990s, it was the software vendors who became the aggressors. Through the use of operating systems that could work on a LAN-based server, the software vendors were aggressive in finding the integration tools necessary for the organization. Microsoft Corporation developed a de facto protocol called telephony application program interface (TAPI) in a Windows environment to bring the telephony services right to the desktop LAN-attached device. A TAPI interface on a Windows platform allows users to access information through what is called a graphical user interface (GUI, pronounced *gooey*). Throughout this whole architecture, a server application can handle the distribution of calls to members within work groups or departments. This includes such services as screening calls, rerouting calls to new agent groups if the primary agent group is busy, or routing calls to the voice messaging and voice response systems as necessary. Similarly, Novell and AT&T developed what is known as the telephony services application programmers interface (TSAPI) as a means of providing the computer to telephony integration capabilities from a local area network. TSAPI basically works with Novell's NetWare telephony services. The basis for using this particular product is to gain what is known as *third-party call control*. Third-party call control uses the CTI application on behalf of any clients in a work group or department. The application is running a shared environment, typically on a server, so there is no direct contact or connection between the user's PC and the telephone interface. What happens, then, is that a logical connection is produced where the PC applications talks to the server, which in turn controls the telephone switch. The server then is the controller and sends the order to the PBX to make calls or connections on behalf of the end user. The shared-server environment can handle individual as well as dynamic work group applications, such as directories, individual personal information managers (PIMs), and other work group functions that would occur within a larger organization. Therefore the server provides all of the linkage for all calls being handled within a dynamic work group. This is a more powerful arrangement as regards call control. A central server can handle the distribution of calls to any member in a work group and provide such services as call screening, call answering groups, backing up agents, or routing calls to a supervisory position in the event of overflow. Many of the CTI vendors have seen this as the benefits and strengths of the CTI applications.

These are the primary applications that have been used within the CTI environment. Developers and manufacturers alike therefore have seen the application working like a telephone set. This is instrumental in designing the capabilities of a call answering/call processing environment, but more can be done through the use of the computer-based technology that is available on the market today. The applications that run in server or computing platforms can use call monitoring features within the PBX and collect information of any type. Using the call monitoring features, a CTI can watch every keystroke that an agent group enters. This can include such things as dialed digits, when agents "busy out," answering a call, and how long the call is off hook. Additionally, by monitoring the trunk groups within the PBX, the CTI application can see all the incoming calls and collect the data associated with each call. This uses the service described earlier, called ANI, and DNIS to watch where the call was directed and when it was answered.

The selectivity available in a CIT application therefore allows the supervisor to monitor the activity of each agent within a work group and get a clear picture of what transpires during the course of a day. This helps in determining workload effort, staffing requirements to meet a specific demand, or any seasonal adjustments that must be made on the basis of time of year. About 65–70 management reports can be generated on an ad hoc basis. Supervisory personnel can monitor the work flow, the productivity of each agent in a group, abandoned calls that were not answered soon enough, or any other anomalies that might occur within a given day. Through these useful tools and reporting structures, the supervisors know whether they have enough, too few, or too many agents on board at any one time. This can aid in work flow scheduling as well as in determining the productivity of the individual agents or of the work group as a whole. In the event an agent is not performing satisfactorily, management can then take whatever corrective actions are necessary.

If abandoned calls are escalating because the agents are on the phone too long, several other activities might result. These could include such things as new training, analysis of the call type and the information requested, or simply determining the morale and productivity within the work group.

Beyond the call processing and the call monitoring capabilities, CTI can be implemented to integrate the facilities and capabilities of the PBX as well as the computer systems into one homogenous unit. The typical PBX today has features that are rarely used. There can be as many as 300 to 400 of these features, which are designed to either improve productivity or make the job simpler. Even though the average user typically activates only three or four of the normal features, many of the functions are available but never used. The CTI applications can customize features and functions for individual users within a work group and allow for more powerful interface using point-and-click GUIs. The screen on a PC through a CTI application can be used for dialing, activating features, conference calling, call transfer, or any other feature necessary. Using the mouse to click on a feature resembling a telephone set on a PC screen removes the risk and the unfamiliarity of the PBX features from the end user. This would prevent cutoffs or lost calls and facilitate better utilization of the PBX features.

Taking this one step further, each individual user within a group supported by a CTI application can customize his or her own features and functions in personal folders on the PC. These can either be stored on the individual PCs, or alternatively on the server. Therefore, using the CTI application, individuals may select different forms of call screening, call forwarding, or call answering according to the applications and the individualized services they most prefer.

Technology Enhancements

When CTI was first routed in the industry, it was done through two very large computing platforms, either the mainframe or the high-end midrange computing systems. These included such things as the IBM mainframe, AS/400, DECVax, or HP-type systems. Although they worked, they were very expensive and very sophisticated, requiring extensive investment as well as application programming interfaces that made the service available only to very large organizations.

With the implementation of the server-based platforms, or LAN-based platforms, the technology has trickled down to the very small organizations. No longer can one determine or assume that a company using CTI applications is very large. As a matter of fact, many of the CTI applications are now rolling out on PC-based platforms for the very small organization. Companies with 3 or 5 to as many as 10 call answering or telemarketing positions have implemented CTI very effectively. These lower-cost solutions have made CTI a reality within organizations around the world. It was through LAN technology and not PBX technology that CTI actually got a foothold within organizations. Imagine if we had waited for the PBX manufacturers and the telephone companies/long-distance companies to roll out CTI integration for us. Unfortunately, we would probably still be waiting. In the past few years the computer manufacturers have provided the push and the software developers the innovations to make CTI a reality. This relatively new yet rapidly accepted approach to using the server as the instrument to provide CTI has thrust CTI into the forefront of telecommunications technology. As mentioned earlier, the voice server on a LAN is designed to connect directly to the public switch telephone network and handle calls coming into the group, then process those calls directly to the desktop. Priority customers and special handling arrangements allow specific users to work around the high-end PBX and Centrex platforms and go directly to the individual department or customer service group without proceeding through the corporate platform. This in turn changes the architecture because of the CTI applications that can work on a server platform. PBXs, once known for their large investments and proprietary nature, can now remain single-line telephony service providers. When higher-end features and functions are necessary, the end user merely has to buy computer-based software as opposed to high-end PBX architectural software. Of course, the PBX manufacturers have recognized this shift in the technology implementation and all of the PBX manufacturers are now developing the CTI interfaces or the software with third-party developers to reclaim customers. Because the PBX need only be an uncomplicated telephone system for the masses within the organization, the technology can last significantly longer. In the old days of the PBX, the plan was to keep the system for a period of about 10 years. However, reality said that the PBX was changed on a basis of about 5–7 years. This involved major investment and changes within an architecture and caused a significant amount of corporate stress. As a quick side note, whenever a new PBX was installed, it usually meant that the telecom manager within a corporation would be leaving soon. Regardless of how many technological advancements or enhancements were installed, users' expectations were never met. The frustrations of the users and the complaints made to management usually led to the telecom manager's demise. Now, with the features and functions moved to a PC-based platform, the telecom manager can breathe easier. Without the need to upgrade the PBX or change an entire infrastructure, the telecom manager can implement on a department-by-department basis whatever features or functions are necessary and available. On a larger scale, all PBX features would have been available to all users. Using the CTI implementation on a server, features can be purchased on a department-by-department basis and subsequently be less expensive. This has been the boon of the 1990s.

Other Technologies

Because of the innovations and enhancements in the telecommunications environment, and the server marriage, many other applications and features can be made available by other producers. For example, through the use of the telecom server on a LAN, the automated attendant, voice messaging, ACD, and integrated voice recognition and response (IVR) functions can all be united in a single server-based platform.

The developers of voice messaging systems automated attendants recognized this opportunity several years ago. They leapfrogged the market, bypassing the PBX manufacturers, and developed single-card processing systems that could use high-end digital trunking capabilities directly into the servers. Using microprocessor control devices, these companies were able to write the necessary software that would provide the capabilities of all of these features. No longer would an organization have to buy a room-sized voice messaging system; this function can now be performed on a PC-based platform. When voice messaging was first introduced, the size, heat, and cost of systems were exorbitant. Now, using PC-based systems, just about every vendor offers the capability of allowing several hundred to several thousand active users on a voice messaging system or a call processing system at a much lower cost. Actually, it's becoming much more difficult to tell a PBX performing CTI applications, from a CTI server performing PBX capabilities. This convergence and marriage is blurring the lines between the various departments. Many of the organizations that now produce systems with these capabilities may have once been niche market providers but are now moving across the border that once separated these two technologies. Just about every feature, function, and capability can now be had using a very low-end server platform at a very reasonable price.

The integration includes such features and functions as:

- Voice messaging
- Automated attendant
- IVR
- Text to speech
- Speech to text
- Directory services
- Fax services
- Fax-back services
- Intranet access for catalogs

Taking these one at a time, we'll see how they can all play together and provide unified messaging and integration capability. The capabilities of the integrated messaging and unified messaging services allow the desktop user to functionally perform all day-to-day operations at a single interface device, now the desktop PC.

Automated Attendant

With the technology moving as quickly as it is, the use of single processing cards in a PC can deliver a combination of voice messaging and automated attendant functions directly to the CTI application. Literally, a digital signal processing capability can compress voice calls so that they can be conveniently stored on a hard disk drive. The voice is already in a digital form when it arrives from a digital trunk or digital line card; therefore storage is a relatively simple technique. The application software used in a voice mail system is basically a file service, where storage and retrieval can be easily accommodated.

Integrated Voice Recognition and Response

Integrated voice recognition and response systems are based primarily on the same type of technology as the auto attendant and voice mail. Using a single digital processing card, the capability now allows users to arrange for prescripted calling capabilities that will actually walk a caller through a menu. The IVR will play digitally stored messages and solicit a response from the caller at each step, usually in the form of a touch tone from a telephone set. The response from that tone will then cause the next step of the message to be played in accordance with the script. This is useful when a user is trying to access information from a host-based system, for example, and a played-back message will allow the user to retrieve any form of information. This IVR capability has been used by several medical providers or insurance providers to enable a caller to dial in and access information regarding payment or processing of claims by merely using the Touch-Tone phone. When dialing into the IVR, the user is prompted each step of the way by the system. As the user enters his or her ID number, a query is sent to a database in a host computing platform. This appropriate information is then retrieved and provided in a played-back message: "Payment to your provider was mailed on this date." Using this CTI application saves an immense amount of time for an organization, because this normally labor-intensive activity can now be achieved through technology. One can just imagine how much savings could be achieved if we no longer needed humans to answer these routine and repetitive types of normal questions.

Fax-Back and Fax Processing

The digital signal processor (DSP) card can also be programmed to function as a fax modem that can provide for the sharing of fax services within a single server environment. A fax image can be downloaded across a LAN, then converted by a fax card and transmitted out across the network over a digital trunking facility. In the reverse direction, if an incoming fax is received from the network, it is then converted back to a file format that is easily usable within the PC environment. This can then either be stored in a fax server file for later retrieval by the individual recipient of the message or redirected by a fax operator. Some of these systems and services take more effort to implement and facilitate, but they may well be worth the effort in terms of the organization's needs. Fax-back capability means that when a user dials into an organization equipped with CTI applications, that user can be directed to a fax server that has a numerical listing of specific documents that the user can retrieve by keying in a telephone number. When the user enters the telephone number, the server retrieves the fax from the file and then automatically transmits it to the designated telephone number the user has just entered. Through this application, catalog information or specific customer information can be retrieved without the use of human intervention. One can see how much could be saved by using these types of services.

Text to Speech and Speech to Text

In speech to text applications, a pre-stored pattern of words can be used through the CTI application to enable a highly mobile workforce to dial into a server-based platform and literally speak to the machine as opposed to using touch tones. Speech from the callers, whose voice patterns are already stored in the computer, can then be converted into usable text using a server-based CTI application. This is instrumental when the user cannot access a Touch-Tone phone (yes, there are many places around the country that do not yet have Touch-Tone services, such as hotels, pay phones, and public telephones). Without the Touch-Tone phone, the user would have to carry a portable touch tone pad generator, which is very inconvenient. Inevitably, the batteries on these devices die at the very moment when the user needs access to information. Consequently, the use of voice patterns or speech patterns that have been pre-recorded with a series of words such as *get, save, retrieve, file,* and so on, can be used to facilitate and walk through a computing system.

The text to speech applications are comparable in that when a user accesses a particular file, again without a terminal device, for example, the system can convert the text into a speech pattern. What this effectively means is that e-mail and files can literally be read back to us no matter where we are. This is exciting because an end user might well dial into the CTI application while traveling on the road and learn that he or she has six voice messages and four e-mail messages waiting. Rather than that user being forced to log on with a different form of terminal device, these e-mail messages can be read right down the telephone line to the end user, facilitating the easy retrieval, storage or redirection of messages. It is through these types of services that the CTI applications are drawing so much excitement.

Optical Character Recognition (OCR)

Another form of digital-signal-processor-based technology can convert scanned images into text. When used with fax machines or fax images, OCR can change an incoming fax into a document that can easily be edited or incorporated into other types of applications. This would include editing a fax and plugging it into a word processing document for easy edit capabilities in word processing. Then the additional storage capabilities convert that OCR, which usually would be a file of significant size, into a text-based document, which would be much smaller. Furthermore, using the OCR by scanning a pre-typed document, for example, the application can then incorporate the scanned document into a text to speech application, where it can be read aloud. One can just imagine the uses and applications for some of these technologies as they start to come together.

Summary

Hopefully, this discussion of CTI will give the reader some appreciation of the capabilities and features that can now be ported into the merger of both computing and PBX architectures. The use of an on-screen interface at a desktop PC allows users to manage and maintain their mailboxes for voice messaging as well as e-mail. Beyond that, a visual display can be received directly to the desktop outlining the number of faxes, e-mails, or voice messages waiting to be retrieved. With the integration of the voice and text applications, the user can also see who the messages are from and prioritize the receipt of each of these messages on the basis of some pre-conditioned arrangement. All of this facilitates the integration of the computer and telephony capabilities onto a single, simple platform that empowers the end user to access information more readily. Moreover, with the implementation of CTI as a front-end processor for the organization's telemarketing or order processing departments, customers have the ability to retrieve information at will. This use of Touch-Tone or voice response systems allows a customer to literally walk through catalogs, check status on orders, check inventories, or even check the process of billing information, all without human involvement. It is not the intent of this discussion to rule out the use of all humans, but rather show how humans can be more productive in performing the functions for which they were initially hired. By taking the repetitive, "look up" type applications and data applications and moving them to something that is controlled by the end user (or customer), the organization can save a significant amount of time and money and better utilize the human resources they do have. The industry is now facing a severe shortfall of skill sets and talents to be able to facilitate some of these functions. With the use of the CTI application, this human resources shortfall can easily be supplemented through technology. As things progress even more, additional applications such as video servers may well be added to this architecture and allow callers to view displays on a downloadable file so that catalog information can be easily retrieved with a video clip that would show exactly what the customer is ordering or buying. Moreover, as the video clips and the fax services and voice messaging capabilities all become integrated into one tightly coupled architecture, customers could see the article, place the order, and literally "construct" the order customized to their needs. One can only imagine some of the possibilities of these features and functions that will be available in the future. However, as with anything else, the first steps must be implemented. An organization must recognize the potential savings or benefits that can be derived from a CTI application. These things are exciting because of all of the different ways the CTI's capabilities can be used. With a graphical user interface (GUI)-based system, the ability to use the single point-and-click mouse-driven application at a desktop allows end users as well as customers to literally walk their way through all catalogs and information.

13
Data
Communications

Everyone who has ever had to deal with data communications has shuddered at least once. Telecommunications engineers, data processing personnel, and vendors alike all throw data communication terms around as though they were going out of style. The interesting point is that many of them really don't understand what they are talking about. Many people think: "If I learn the buzzwords, everyone will think I know what I'm talking about." Nothing could be further from the truth; these folks make complete fools of themselves in front of knowledgeable professionals. However, there is really no a mystique associated with the use of data communications. Although some complexities do exist in this technology, the basics are fairly straightforward. If you can surmount the initial hurdle of setting up a data transmission, the rest can be fairly well assimilated.

In 1997, for the first time in the history of the telecommunications industry, data was carried across the networks in an equal share as voice. This showed the heavy emphasis on the growth of data communications. As we near the end of the millennium, the growth of voice on the networks is averaging between 3 and 4 percent per year. However, data is growing at a rate of 30 percent per year. It will take 12 years to double the amount of voice carried on the network today at the current growth rate, whereas the data is doubling approximately every 90 days. It is important to understand that the data world grew out of the voice world. Voice traditionally paid for the data transmissions on the network, and still today, 90 percent of all the revenues generated across the wide area networks are the result of voice usage. But that will all change quickly as we enter the new millennium. The use of the analog dial-up network is where it all started. In order to communicate from a terminal, computer, or other piece of equipment, you merely have to put the pieces together in the proper order:

- Select and deal with the transmission media
- Use communicating devices that will present the proper signal to the line (the communicating device is called the DCE)
- Add a device called the data terminal equipment (DTE)
- Set up or abide by accepted rules (protocols)
- Use a pre-established alphabet that the devices understand
- Ensure the integrity of information before, during, and after transmission
- Deliver the information to the receiving device

This chapter demystifies the elements involved in all data communications processes. Later chapters focus on specific technologies, using the concepts and terminology introduced here.

Concepts

Like learning computer programming, learning data communications technology is a nonlinear process. That is, whatever starting point one chooses, one almost has to use terms that will be defined elsewhere. The usual solution to this problem is *iterative teaching:* teach a basic set of concepts, then go back and both use and expand on those concepts, refining them along the way. Our basic set begins with a discussion of some important concepts that permeate the world of data communications. Those concepts include:

- Standards
- Architectures
- Protocols
- Error detection
- Plexes
- Multiplexing
- Compression
- Standards

Standards

A *standard* is a definition or description of a technology. The purpose of developing standards is to help vendors build components that will function together or that will facilitate use by providing consistency with other products. This section discusses what standards are, why they exist, and some of the ways they could affect you. Specific standards are mentioned in other sections where applicable.

There are two kinds of standards: de facto and de jure.

De facto means *in fact.* If more than one vendor "builds to" or complies with a particular technology, one can reasonably refer to that technology as a standard. An excellent example of such a standard in data communications is IBM's Systems Network Architecture (SNA). No independent standards organization has ever "blessed" SNA as an "official," or de jure, standard. But dozens, if not hundreds, of other vendors have built products that successfully interact with SNA devices and networks.

Note that de facto standards rarely become standards overnight. SNA was available for some time before vendors other than IBM could or would provide products that supported it.

Moreover, some technologies become standards because the creating vendors intend them to become standards (e.g., Ethernet), while others become standards despite the creating vendors (e.g., Lotus 1-2-3 menu structure).

De jure means *in law,* although standards do not generally have the force of law. In some parts of the world, when a standard is set, it in fact becomes law. If a user or vendor violates the rules, the penalties can be quite severe. A user who installs a nonstandard piece of equipment on the links could be subject to steep fines and up to one year in prison. These countries take their standards seriously. In the United States, no such penalties exist; we are more relaxed in this area. But a standard is a de jure standard if an independent standards body (i.e., one not solely sponsored by vendors) successfully carries it through a more or less public standards-making procedure and announces that it is now a standard.

You might well ask, "What's the difference? And who cares, anyway?" But understanding which technologies fall into which category, if either, can be a factor in deciding what to buy. Generally, de facto standards are controlled by the vendors that introduced them. For example, Microsoft Windows is a de facto standard; many vendors provide programs that comply with and operate in this environment. But if it decides to change the way a new version of Windows works, Microsoft in theory can obsolete all of those programs. (In practice, Microsoft is most unlikely to do this, at least intentionally, because much of its market power stems from the fact that all those other products are built to its standard. Were Microsoft to make such a change, it is likely that those other software providers would look elsewhere for a target operating system. Microsoft's stock would drop precipitously, to say the least!)

One reason that vendors often build to de facto standards is that the standards-making process tends to be somewhat lengthy. A minimum of a 4- to 12-year period to come out with a new or revised standard is not at all uncommon in the industry. With product cycles under one year in some areas of the communications industry, waiting for finalization of a standard before introducing a product could result in corporate suicide. Ethernet is a good example of a standard that started as a de facto standard. Intel, Xerox, and Digital Equipment Corporation (DEC) introduced Ethernet with the intent of making it a de jure standard. But that process took years, and the final result was slightly different than the technology originally created by the three vendors. Nonetheless, many networks were created based on Ethernet before the 802.3 de jure standard was finalized, bringing profit to its creators and operating environments to their customers.

In the real world, the standards-making bodies rely in large part on vendors to develop the details of new and revised standards. In fact, most standards bodies have vendor representatives as full participants. It is a fascinating political process with much pushing and pulling to gain advantage in the market. The vendors participate for several reasons, not the least of which is to get the jump on competitors that are not as close to the process. Other reasons include the ability to state in marketing materials that they contributed to or were involved in testing of a new standard, as well as the opportunity to influence the actual details of a standard to favor technology that the vendors are most familiar with.

To be fair, it should be stated that the primary goal of most of those involved in the standards process is to define a good and useful standard. But when the process produces a dual standard, as in the case of the Ethernet and Token Ring local area network standards, one can presume that the "best" was compromised somewhat in favor of what could be agreed on.

Even de jure standards (usually identifiable by virtue of their unintelligible alphanumeric designations, such as X.25, V.35, V.42 bis, etc.) change in ways that significantly affect the market—and you. One set of standards that affects thousands of users is the set of modem standards, discussed later. But beware a vendor that trumpets compliance with a "new standard!" The vendor's claim might be legitimate, but if no other vendors have products available in the same space, the company might simply be hyping its own product in hopes that it eventually will become a standard. Or there might be a standard under development but not yet approved. In the latter case, if that standard changes before final approval, the vendor's current products will instantly become nonstandard without changing in any way!

In the recent past, the standards committees were working on a new modulation technique to speed up data communications. The standards committees were locked in discussion about the rules to be applied. Yet, at the same time, every modem manufacturer began producing a new modem that was advertised and sold as compatible or compliant with the new V.Fast or V.34 standard. It was not yet a completed standard, but the manufacturers wanted to get their products on the shelves as quickly as possible and corral their piece of the market. So they produced a product with a disclaimer that offered a free or minimal cost upgrade to the V.34 standards if the standard changed. This is a classic example of industry leaders setting the pace before the standards were completed.

Architectures

As with constructing a building, an overall design is needed when planning a communications environment. For a building, that design is described by architectural drawings. A communications architecture is a coordinated set of design guidelines that together constitute a complete description of one approach to building a communications environment.

Several communications architectures have been developed. Some of the best-known include IBM's SNA and DEC's DNA. Architectures are covered in much greater detail in Chap. 15. But for those already familiar with the open systems interconnect (OSI) model described in Chap. 15, most of this chapter (excluding codes, which reside at the presentation layer) addresses the physical and link layers. The newest architecture to run away with the industry and the fancy of all developers is the Internet architecture using TCP and IP protocols. Every day new applications and protocols are being developed to run on the Internet architecture. This includes voice, data, streaming audio and video, and multimedia applications.

The data communications architectures were modeled after the voice architectures. This is understandable, because data is merely a logical extension of the dial-up voice network. Devices are therefore constructed to fit into the overall voice network operation. Data equipment is designed and built to mimic the characteristics of a human speech pattern. Now, however, we see that voice is data and data is, too! This paradigm shift marks the true convergence of voice and data onto a single architecture. The world appears ready to embrace the technologies that will fall out from this convergence.

Protocols

Protocols are key components of communications architectures. Architectures are guidelines on how environments connecting two or more devices can be constructed, so most components of a given architecture in a network will be found on each communicating computer in that network. Protocols provide the rules for communications between counterpart components on different devices.

More detail, with examples, is provided in the discussion of the OSI model in Chap. 15. However, there is one aspect of protocols that also applies to hardware: whether they are synchronous or asynchronous. These key characteristics are covered in the following section.

Transmission Protocols (Synchronous vs. Asynchronous)

All lower-level data communications protocols fall into one of the two following categories: synchronous or asynchronous. The words themselves are based on Greek roots indicating that they either are "in" or "with" time (*synchronous*) or "out of" or "separated from" time (*asynchronous*). The underlying meanings are quite accurate, so long as one understands to what they must be applied.

All data communications depend on precise timing, or clocking. The discussion of analog versus digital transmission in Chap. 5 covers how voltage levels are sampled in the middle of a bit time in order to maximize the odds that the sample value will be clearly distinguishable as a 1 or a 0. But how does the equipment determine precisely when the middle of a bit time occurs? The answer is clocking; equipment at both ends of a circuit must be synchronized during transmission so that the receiver and the sender agree regarding beginnings, middles, and ends of bits during transmissions. There are two fundamentally different ways to do this clocking: asynchronously and synchronously.

Simply put, asynchronous transmissions are clocked (or synchronized) one byte at a time. Synchronous transmissions are clocked in groups of bytes. But the differences in how these two approaches work go beyond the differences between individual bytes and groups of bytes.

Asynchronous communications is also called start/stop communications and has the following characteristics.

Every byte has added to it one bit signaling the beginning of the byte (the *start bit*) and at least one bit (possibly two) added at the end of the byte (the *stop bits*). Bytes with 7 data bits typically also include a parity bit (which is an error-checking bit), whereas 8-data-bit bytes usually do not. Thus, generally speaking, 10 or 11 total bits are actually transmitted for every asynchronous byte. To get 7 usable data bits, we must transmit approximately 10 to 11; strictly speaking, we use a 30 to 35 percent overhead. Fig. 13.1 shows the layout of a data byte in an asynchronous form. This was a special concern when data communications was initially used in the late 1950s and early 1960s. Back then, the cost per minute of a dial-up line was $0.60 to $0.65. Using that value, 30 cents of every dollar were spent just to provide the timing for the line. This amount of waste concerned everyone.

This also makes nominal speed calculations for such connections easy: dividing the rated speed of the circuit (e.g., 9600 bits per second) by 10 bits

Figure 13.1
To send 7 usable bits
of data, we must use
1 start, 1 parity, and
1 or 2 stop bits.

```
S                  S  S
T   1 000 001 P    T  T
A                  O  O
R                  P  P
T
```

per byte gives a transmission speed in characters per second (e.g., 960 cps). As a rule, we divide the bits per second by 10 to get the nominal speed of an asynchronous circuit. (*Nominal* here means best-case; in the real world, circuits rarely deliver 100 percent of their nominal capacity. But it's a starting point for capacity calculations.)

The bytes are sent out without regard to the timing of previous and succeeding bytes. That means that none of the components in a circuit ever assume that just because one byte just went by, another will follow in any particular period of time. Think of a person banging away on a keyboard. The speed and number of characters sent in a given period does not indicate in any way how many or how quickly characters can be sent in the succeeding similar period.

Clocking is controlled by data terminal equipment (DTE). For example, when a personal computer is used to dial into CompuServe, clocking on bytes going toward the service is generated by the sending PC (see Fig. 13.2). That first start bit reaching the modem begins the sequence, with all succeeding bits in the same byte arriving in lockstep at the agreed-on rate until the stop bit is received. Then clocking stops until the beginning of the next byte arrives. Any intervening devices (especially modems) between the communicating DTEs take the clocking from the data sent by the originating DTE for any given byte.

Figure 13.2
The data terminal equipment controls the timing to the data communications equipment (modem), as the bits are sent in an asynchronous protocol—a start bit and 1 or 2 stop bits help to set the timing.

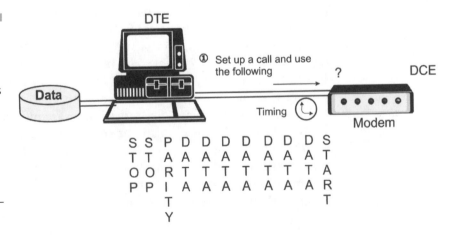

Most PC and minicomputer terminal communications employ asynchronous techniques. The default communications ports on PCs (the "serial" or COM ports) only support asynchronous communications. To use synchronous communications on a PC, a special circuit board is required.

Synchronous communications have the following characteristics:

- Blocks of data rather than individual bytes (characters) are transmitted.

- Individual bytes do not have any additional bits added to them on a byte-by-byte basis, except for parity.

However, bytes are sent and clocked in contiguous groups of one or more bytes. Each group is immediately (with no intervening time) preceded by a minimum of two consecutive synchronization bytes (a special character defined by the specific synchronous protocol, of which there are many) that begin the clocking. All succeeding bits in the group are sent in lockstep until the last bit of the last byte is sent, followed (still in lockstep) by an end-of-block byte. This layout of the synchronous characters (SYN) is shown in Fig. 13.3.

Clocking is controlled by data communications equipment (DCE). Specifically, on any given circuit one specific DCE component is optioned (i.e., configured) at installation time as the master device. When the circuit is otherwise idle, the master generates the same synchronization character mentioned above on a periodic basis to all other DCE devices so that all DCE clocks on the circuit are maintained in continuous synchronization.

Except in cases where smaller numbers of bytes (fewer than about 20) are sent at a time, synchronous communications makes more efficient use of a circuit, as can be seen from Table 13.1.

Generally speaking, all circuits running at greater than 2400 bits per second actually operate in synchronous mode over the wire. This is simply because building modems to reliably operate asynchronously at higher speeds over analog circuits is much more difficult than taking this approach. Asynchronous modems that run faster than 2400 bits per second

Figure 13.3
The layout of data transmitted synchronously. In this case, the block size is 512 bytes.

S Y N	S Y N	S O H	S T X	Data up to 512 Bytes	E X T	E O T	S Y N
8	8	8	8	4096	8	8	8

TABLE 13.1

Comparison of the
Utilization of the
Circuit

Data Bytes	Asynchronous Bits	Synchronous Bits	Synchronous Savings
1	10	32	−220%
5	50	64	−28%
10	100	104	−4%
20	200	184	8%
30	300	264	12%
100	1000	824	18%
1000	10,000	8024	20%
10,000	100,000	80,024	20%

actually incorporate asynchronous-to-synchronous converters; they communicate asynchronously to their respective DTEs but synchronously between the modems as shown in Fig. 13.4. This doesn't normally impact performance: when smaller groups of characters are sent, there is time to include the additional overhead for synchronous transmission. When larger groups are sent, the reduced overhead of synchronous transmission comes into play. In practice, these higher-speed modems actually communicate with other modems at even higher than their rated speeds; the extra bandwidth is used for overhead functions between the modems.

Error Detection

We mentioned at the beginning of the chapter that ensuring the integrity of the information was one of the key responsibilities of a data communications environment. This does not mean that the data must be kept honest. Rather, it means that we must somehow guarantee with an extremely high degree of probability that the information is received in exactly the same form as it was sent.

More precisely, there are two tasks required: detecting when transmission errors occur and triggering retransmissions in the event that an error is detected. It is the responsibility of protocols (see Chap. 15) to trigger and manage retransmissions. Here we discuss some of the various approaches that have been developed to detect errors in the first place.

All of the code sets used in data communications (see the following) are designed to use all of their bits to represent characters (letters, num-

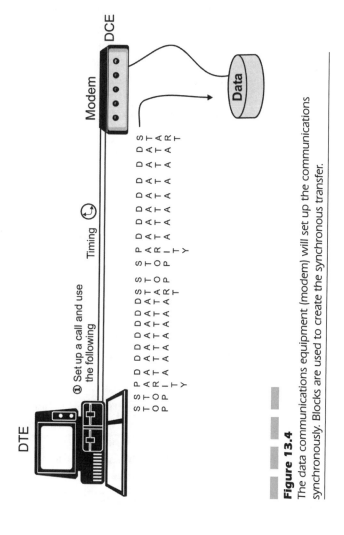

Figure 13.4

The data communications equipment (modem) will set up the communications synchronously. Blocks are used to create the synchronous transfer.

bers, other special characters). A not-so-obvious implication of this fact is that every byte received in such a code set is by definition a valid code. How can we detect whether the received code is the code that was sent?

The solution is to somehow send some additional information, some data about data (sometimes described as *metadata*) along with the primary data. All error-checking approaches depend on sending some additional data besides the original application-related data. The additional data is created during the communications process, used to check the underlying data when it is received, and then discarded before the information is passed to its final destination.

In order of increasing reliability, the major methods used to detect data communications errors include:

- Parity bit, or vertical redundancy checking (VRC)
- Longitudinal redundancy checking (LRC)
- Cyclic redundancy checking (CRC)

Suppose you are my rich aunt and that I am living in Paris (to further my cultural education, of course) and have run out of money. I've called you (collect, of course) to request that you electronically transfer some money into my account at the Banque de Paris. In a fit of generosity, you have decided to send me $1000.00. If the network used is not perfectly reliable and appropriate error detection methods are not applied by the transmitting financial service, a change in a single character—for example, changing the period after the first three 0s to another 0—could result in your sending me considerably more money than you intended: a total of $1,000,000.

What are the chances of my getting my inheritance early in this way? Not very high, given the odds that only the decimal point would change, and only to a 0 (out of either 126 or 254 other possibilities). But consider the probability of detecting the error, assuming that it has occurred. Using parity bits, the likelihood of detecting this kind of error (which requires several bits to be wrong at one time to change an entire character) is about 65 percent. A somewhat better method, longitudinal redundancy checking, would up the odds of detection to about 85 percent. But the cyclic redundancy checking method improves the odds of detecting and correcting such a multibit error to 99.99995 percent.

Because the networks used to send monetary amounts generally use CRC techniques, it doesn't look as though I'm going to get rich because of their errors. But let's examine these methods in a bit more detail anyway.

Parity Bit/Vertical Redundancy Checking (VRC)

The parity bit approach to error detection simply adds a single "parity" bit to every character (or byte) sent. Whether the parity bit is set to 0 or 1 (the only two possibilities, of course) is calculated by the sending digital device and recalculated by the receiving device. If the calculations match, the associated character is considered good. Otherwise, an error is detected.

This is much simpler than it sounds. Two approaches are typically used: even and odd parity. Other forms of parity exist, such as mark or space parity. It makes no difference which is used; the only requirement is that the sending and receiving devices use the same approach. To illustrate even parity, consider the ASCII bit sequence representing a lowercase letter *a:* 1100001. Because we are using even parity, we require that the total number of 1 bits transmitted to the receiver to send this *a*, including the eighth parity bit, be equal to an even number. Their position in the underlying byte is irrelevant. If we count the 1s in the 7-bit pattern, we get 3, an odd number. Therefore we set the parity bit to 1, resulting in 11100001, an 8-bit pattern with an even number of (i.e., four) 1s. (The parity bit is sent last. In our illustrations the bits farthest to the right are sent first, so we show the parity bit being added at the left.) Figure 13.5 shows an ASCII illustration of the word *hello*, complete with even parity bits. As you can see from the illustration, the bytes are represented as vertical sets of numbers, thus vertical redundancy checking: if we orient the digits vertically, we add a vertical bit that is redundant to help check the correctness of the underlying byte. The vertical orientation is arbitrary, of course. However, when we illustrate longitudinal redundancy checking, you will see that there is a reason for this display approach.

Having gone to the trouble of describing parity bits, we must confess that they are of limited use in data communications. Parity checking will catch 100 percent of errors where the number of bits in error is odd (1, 3, 5, etc.)...and none of the errors where the number of bits in error is even. Put another way, if an error occurs (and communications errors rarely affect only a single bit), there is only about a 65 percent chance that parity checking will detect it. (The probability is better than 50 percent because there are somewhat more one-bit errors than any one type of multibit error, whether the numbers of the latter are odd or even.)

Parity checking is used extensively inside computers. There it makes sense because it is entirely plausible that errors would occur one at a time (if they occur at all). Parity checking does well in this environment. Also,

Figure 13.5
The vertical redundancy check shows the flow of data. The parity bit is added after each character is generated.

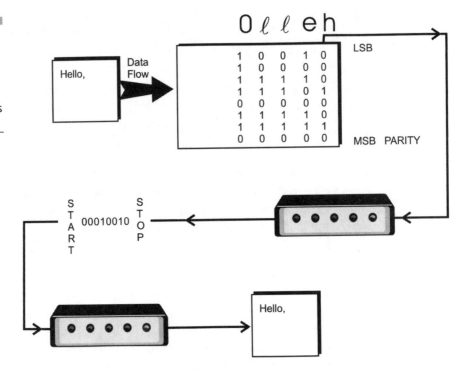

some networks (e.g., CompuServe) still have users set their communications software to use parity checking. But even CompuServe uses a more sophisticated protocol for file transfers. Some more sophisticated error detection protocols are described in the following paragraphs.

Longitudinal Redundancy Checking (LRC)

The concept of LRC follows directly from VRC, taking VRC a step further. This example uses 8-bit rather than 7-bit bytes. But LRC checking needs to operate on a group of bytes, rather than on one byte at a time. For this example, it doesn't really matter what the bits represent, so let us create a set of eight 8-bit bytes. As you will see, although the bit patterns do not matter for the example, the number of bytes used does (see Table 13.2).

If we only use VRC (odd parity), as described above, we produce the pattern shown in Table 13.3. But of course, we said that VRC only catches about 65 percent of errors, hardly acceptable. But what if we apply odd

TABLE 13.2

Setting Up for Vertical Parity Checking

1	0	1	0	1	0	1	0
0	0	1	0	0	0	1	0
1	1	1	0	1	1	1	0
1	0	1	0	0	0	1	0
0	1	1	0	0	1	1	0
1	1	1	0	1	1	1	0
0	1	1	0	0	1	1	0
1	0	1	0	1	0	1	0

parity checking across the bytes in addition to vertically? In that case, the completely filled-in Table 13.4 would be generated.

Adding both the horizontal and vertical checking, together referred to as longitudinal redundancy checking, improves the odds of detecting errors to about 85 percent. Not bad, although we wouldn't want to trust our money to such a transmission. But there is another disadvantage to LRC. Using 8-bit bytes for every eight data bytes, an additional two LRC bytes must be transmitted. That works out to 20 percent added overhead for error checking (2 LRC bytes divided by the 10 total bytes transmitted in the set), not counting any degradation due to time required to compute the check bytes. This is not an efficient error-checking mechanism. In fact, considering that error checking is only one of several sources of transmission overhead, it is abysmal.

LRC does have one advantage over cyclic redundancy checking (CRC), the approach discussed next: far fewer computational resources are required for calculating the LRC bytes than for calculating a CRC (unless

TABLE 13.3

The Parity Bit Is Inserted in VRC

1	0	1	0	1	0	1	0
0	0	1	0	0	0	1	0
1	1	1	0	1	1	1	0
1	0	1	0	0	0	1	0
0	1	1	0	0	1	1	0
1	1	1	0	1	1	1	0
0	1	1	0	0	1	1	0
1	1	0	1	0	1	0	1

TABLE 13.4

The Vertical and Longitudinal Redundancy Checks Are Inserted Here

1	0	1	0	1	0	1	0	1
0	0	1	0	0	0	1	0	1
1	1	1	0	1	1	1	0	1
1	0	1	0	0	0	1	0	0
0	1	1	0	0	1	1	0	1
1	1	1	0	1	1	1	0	1
0	1	1	0	0	1	1	0	1
1	1	0	1	0	1	0	1	0

the CRC is implemented with hardware). In fact, until recent generations of PCs became available, with their vastly more powerful CPUs, CRC checking for asynchronous data communications in the PC environment was not practical because of its computationally intensive nature. Now, however, it is routinely used. Read on to see why.

Cyclic Redundancy Checking (CRC)

Although no practical error-checking algorithm can guarantee detection of every possible error pattern, CRC comes close. A complete explanation with examples of how CRC works would (and does, in several data communications textbooks) require several pages of somewhat hairy binary algebra. Rather than put you through that, we'll describe some of the method's key characteristics and indicate how this method is used.

Like the previously described approaches to error detection, CRC relies on on-the-fly calculation of an additional bit pattern (referred to as a *frame check sequence,* or FCS) that is sent immediately after the original block of data bits. The length of the FCS is chosen in advance by a software or hardware designer on the basis of how high a confidence level is required in the error-detection capability of the given transmission. All *burst errors,* or groups of bits randomized by transmission problems, with a length less than that of the FCS will be detected. Frequently used FCS lengths include 12, 16, and 32 bits. Obviously, the longer the FCS, the more errors will be detected.

The FCS is computed by first taking the original data block bit pattern (treated as a single huge binary number) and adding to its end (after the low-order bits) some additional binary 0s. The exact number of added 0s will be the same as the number of bits in the desired FCS. (The FCS, once calculated, will overlay those 0s.) Then, the resulting binary number, including the trailing 0s, is divided by a special previously selected divisor (often referred to in descriptions of the algorithm as P).

P has certain required characteristics:

- It is always 1 bit longer than the desired FCS.

- Its first and last bits are always 1.

- It is chosen to be "relatively prime" to the FCS; that is, P divided by the FCS would always give a non-0 remainder. In practice, that means P is normally a prime number.

- The division uses binary division, a much quicker and simpler process than decimal division. The remainder of the division becomes the FCS.

Specific implementations of CRC use specific divisors; thus the CRC-32 error-checking protocol on one system should be able to cooperate with the CRC-32 protocol on another system; the CRC-CCITT protocol (which uses a 17-bit pattern, generating a 16-bit FCS) likewise should talk to other implementations of CRC-CCITT. Selection of a specific P can be tuned to the types of errors most likely to occur in a specific environment. But, unless you are planning on engineering a new protocol, you needn't worry about the selection process; it already has been done for you by the designers of your hardware or your communications software. Figure 13.6 illustrates the CRC creation process.

CRC is typically used on blocks or frames of data rather than on individual bytes. Depending on the protocol being used, the size of the blocks can be as high as several thousand bytes. Thus, in terms of bits of error-checking information required for a given number of bytes of data, CRC requires far less transmission overhead (e.g., CRC-32 sends four 8-bit bytes' worth of error-checking bits to check thousands of data bytes) than any of the parity-based approaches.

Although binary division is very efficient, having to perform such a calculation on every block transmitted does have the potential to add significantly to transmission times. Fortunately, CPUs developed in the last few years are up to the challenge. Also, unlike with most other check-digit types of error correction, the receiving device or software does not have to recalculate the FCS in order to check for an error. Instead, the original data plus the FCS are concatenated together to form a longer pattern, then divided by the same P used as a divisor during the FCS cre-

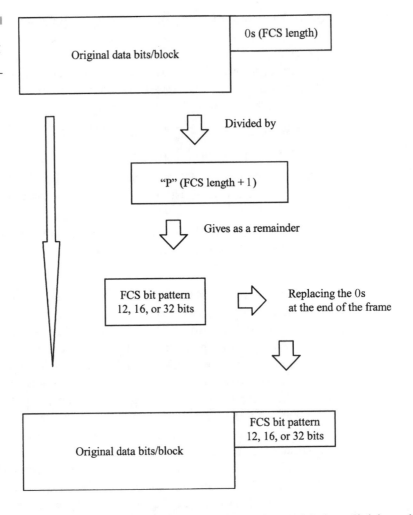

Figure 13.6
The creation of a CRC
cycle.

ation process. If there is no remainder from this last division, then the CRC algorithm assumes that there are no errors. And, 99.99995 percent of the time, there aren't.

Plexes—Communications Channel Directions

The next area of discussion is the directional nature of your communications channel. Three basic forms of communications channels exist.

One-Way (Simplex)

This is a service that is one way and only one way. You can use it to either transmit or to receive. This is not a common channel for telephony (voice), because there are very few occasions where one person speaks and everyone else listens. Feedback, one of the capabilities that we prize in our communications, would be eliminated in a one-way conversation. Broadcast television is an example of simplex communications.

Designing an efficient data communications application using a simplex channel can require quite a bit of ingenuity. A good example is stock ticker tape radio signals. Bear in mind that in a true simplex system (such as this one) there is absolutely no feedback possible from the receiver to the transmitter. How then does someone using such a signal get useful, timely information? After all, unless one is simply gawking at the symbols as they go by, it is not practical to wait for on average half of the symbols to go by in order to find out that the particular stock in which you are interested just went up or down a bit.

The answer is a combination of communications and computer technology. The applications that implement this technique memorize locally (on a PC) the entire repeating communications stream once, then accept each new symbol/price combination received as an update to the local "database." The user inquires against that local database, getting what appears to be instant information, even though it might have been received several minutes ago. Naturally, the computer must be set to continually receive; otherwise, the user will have no assurance that the data is even remotely current.

Another approach requires the user to specify in advance to the software a set of symbols to collect. As that stock information goes by, the program snags only the specified information for retention and local query. This is not any faster than the previous approach, but it does require less local storage capability.

Two-Way Alternating (Half Duplex)

This is the normal channel that is used in conversations. We speak to a listener, then we listen while someone else speaks. The telephone conversations we engage in are normally half duplex. Although the line or medium (air, in this case) is capable of handling a transmission in each direction, most human brains can't deal well with simultaneous transmission and reception.

Many computer and communications configurations use half duplex technology. Until a few years ago, most leased-line multidrop modems were half duplex. One of the key differentiators among such modems was their turnaround time, that is, how quickly a pair of such modems could reverse the channel direction. This was measured in milliseconds—the fewer, the better. Entire communications protocols were built around this technology (e.g., IBM's Bisynchronous Communications or BSC). All-block mode (e.g., IBM 3270s) terminals still operate in half duplex mode only, even if full duplex facilities are available. This simply means that at any given time, the terminal is either sending to its associated computer or receiving from it, not both at once. This does not cause a problem because the entire system is designed around this behavior, and it works quite well. Of course, almost half of any given circuit's raw capacity (if the capacities of the two directions are added together) is wasted. (IBM's more recent SDLC-protocol-based front end processors can talk to one terminal and receive from another at the same time, minimizing this waste, but the individual terminals are still functioning as half duplex devices.)

With some technologies, half duplex can be used so effectively that the one-way-at-a-time characteristic of the circuit is invisible; it appears to be full duplex (see below). An excellent example of such an approach is local area network communications. Most actual LAN technologies, including both Ethernet and Token Ring, are actually half duplex on the wire. But the information moves so quickly, and the responses are so fast, that the path appears to an observer to be full duplex.

Two-Way Simultaneous (Duplex) or Full Duplex

True full duplex communications make maximum use of a circuit's capacity—if the nature of the communications on that circuit takes advantage of it. In data communications, a circuit is implemented and used in full duplex mode if a device can send to a computer and receive from the computer at the same time. Although we mentioned that human conversation is typically half duplex, there is an exception to this: conversation among teenagers, who seem to be able to speak and listen (somewhat) at exactly the same time.

A common example of full duplex communications is that seen when one dials into an on-line information system such as CompuServe or Prodigy. Most such systems allow users to continue typing at a keyboard even while the service is sending information to the user for display on the screen. Were the connection not a full duplex one, this simultaneous bidirectional communication would not be possible.

One of the points that confuses some people is that the terms *simplex, half duplex,* and *full duplex* can refer to varying levels of a communications architecture. If the three levels are considered to be three points on an increasing scale of capability, one can say that a given level of a communications architecture must rely on lower levels with at least the capability of that given level, as presented in Table 13.5.

What does this table mean? It means that a wide-area analog data circuit built to handle full duplex communications (requiring the telephone company to support simultaneous communications in both directions, and use of full duplex modems) can fully support half duplex or simplex communications. But if a similar circuit is implemented with half duplex modems, then full duplex communications on that circuit will not work, although simplex will. A citizens band radio provides a half duplex channel—two directions, but only one way at a time. Simplex would work— one sender could lock down a key and just keep sending—but full duplex would be impossible.

Air is a full duplex channel. But a simplex signal such as the output from a stereo speaker system has no trouble traveling this full duplex channel.

Compression

Although compression is not exactly a modulation technique, it does (usually) produce faster transmissions. To understand how compression works, consider first how human beings communicate. Most human communication is inherently redundant. This does not imply waste; rather,

TABLE 13.5

Comparing the Capability and Directionality of the Circuit

This Capability Level	Requires All Lower Levels to Have at Least the Following Capability	But Can Also Function Without Impairment on Top of Levels with the Following Capabilities
Simplex	Simplex	Half or full duplex
Half duplex	Half duplex	Full duplex
Full duplex	Full duplex	(no additional levels)

human beings use that redundancy as a continual cross-check on what information is really being sent and meant. For example, in face-to-face conversation much more information is being sent than just the words. Facial expressions, tones of voice, limb positions and movement, overall carriage of the body, and other less obvious cues all contribute to the information stream flowing between two people having a conversation. But much of the information is duplicated. For example, anger can be communicated by the words themselves; but it can also be conveyed by tone of voice, facial expression, involuntary changes in the color of one's complexion, the stress in the voice, arm movement, and other cues. If some of these items were removed, the message received might be just as clear but the total amount of raw information might be reduced.

In data communications, compression is a technique applied either in advance to information to be transmitted or dynamically to an information stream being transmitted. The underlying technology is essentially the same in both cases: removal of redundant information, or expression of the information in a more compact form, is used to reduce the total number of bytes that must pass over the communications medium in order to reduce the time the medium is occupied by a given transmission to a minimum.

A detailed discussion of compression techniques is beyond the scope of this book. But in this section we describe two very basic approaches in order to elucidate the fundamentals of the technology. Later, during the discussion of modems, we will briefly identify and describe the power of some compression techniques that are often built into such devices.

The simplest form of compression is the identification and encoding of repeating characters into fewer characters. For example, consider the transmission of printed output across a network to a printer. A typical report contains a very high number of blank characters, often occurring consecutively. Suppose every such string of four or more consecutive blanks (which are of course themselves ASCII characters) is detected and replaced with a 3-byte special character sequence encoded as follows:

- The first character is a special character (one of the nonprint characters in the ASCII code set, for example) indicating that this is a special sequence.

- The second character is one occurrence of the character that is to be repeated, in this case a blank.

- The third character is a 1-byte binary number indicating how many times the character is to be repeated. With 1 byte (using binary format), we can count up to 255, high enough to get some real savings!

How much can we save? Look at Table 13.6. We'll assume that on average, blanks occur in 10-byte consecutive streams, a very pessimistic assumption.

As can be seen from the table, the savings depends on the number of occurrences of the character to be repeated. In practice, this is a reasonably powerful technique. In the example, we addressed only blanks. In practice, any character except the special character would be fair game.

But what if we actually want to send that special character? After all, unlike print jobs, many transmissions must be able to handle every possible code; there are none left over that are "special." No problem—we add the following rules:

- We'll never try to compress multiple occurrences of the special character.

- Every time we encounter the special character as input during the encoding process, we'll simply send it twice. If the receiving hardware sees this character twice, it drops one of the occurrences.

With this approach, we only have to select a special character that is unlikely to occur frequently. If it then does occur frequently in a particular transmission, our compression algorithm doesn't break; it will just be very inefficient for that one transmission.

Note how the overall redundancy is squeezed out of a transmission using this approach. But, just as in human communications, eliminating redundancy increases the risk that some information will be misinterpreted. In human communication, if the reddening of an angry person's face and other visual cues were not visible (e.g., if the conversation were on

TABLE 13.6

Quick Analysis of Compression Benefits

Total Characters in Print Stream Before Compression	Blanks in Print Before Compression	Total Characters in Streamprint Stream After Compression	Percentage Savings
1000	10	993	1
1000	100	930	7
1000	500	650	35
50,000	2000	48,600	3
50,000	5000	46,500	7
50,000	20,000	36,000	28

the telephone) and the speaker was otherwise very self-controlled, the listener might misinterpret angry words as being a joke; after all, many American subcultures routinely use affectionate insults without anger. Unrecognized anger is a very serious loss of information. In data communications without compression, omission of a single space in a series of spaces might cause a slight misalignment on a report, but will most likely not seriously distort its meaning. If compression is used and the binary count field is damaged—i.e., changed to another binary digit—dozens or even hundreds of spaces or other repetitive characters might be either deleted or added to the report, seriously compromising its appearance and perhaps distorting its meaning. Consider the havoc that could be wrought on a horizontal bar chart! The error-checking techniques discussed earlier become much more important in a system that uses compression!

Another more sophisticated method of compression requires pattern recognition analysis of the raw data rather than just detection of repeating characters. Again, some special character must be designated, but now it precedes a special short code that represents some repetitive pattern detected during the analysis. For example, in graphics displays capable of showing 64K (65,536) colors, every screen pixel has associated with it two 8-bit bytes (which together can represent 65K different values) indicating the color assigned to that pixel. If someone sets the screen to display white on a blue background, the 2-byte code for blue is going to appear thousands of times in the data stream associated with that display. The repeating 1-byte compression algorithm described earlier will not detect anything to compress. But if analysis shows that a 2-byte pattern occurs many times in succession, a more sophisticated approach might assign a specific character (preceded by the special character) to represent precisely 20 (or some other specific number of) consecutive occurrences of that 2-byte sequence. The savings can be considerable, but they again depend on the characteristics of the data being transmitted.

A third, very computationally intensive approach to compression has been designed especially for live transmission of digitized video signals. Unlike most other compression methods, this approach does not involve movement of representations of the entire digitized data stream from one point to another. Video signals consist of a number of still frames composed each second (visualize 30 photographs per second, in the highest-quality case). Although the first picture must, of course, be sent in its entirety, special equipment and algorithms must then continuously examine succeeding video frames to be transmitted, identifying which pixels have changed since the last "picture" was taken. Then, information addressing just the changed pixels is sent to the receiver, rather than the entire new frame. The receiving equipment uses this change information combined with its "memory" of the previous frame to continuously, locally build new versions of the picture for display. This approach is particularly fruitful for pictures that in large part remain static; for example, video conferencing. In video conferencing, usually the only moving features of the picture are the human beings. The table(s), walls, and other room fixtures stay still, therefore requiring transmission only once. Frequently, only the lips move for long periods of time.

One other compression-related concept is worth mentioning: lossy vs. lossless compression. "What?" I hear you ask. Does lossy mean what it sounds like? Would we ever tolerate transmission that loses information? The answer, for some applications, is yes. Moreover, you have probably settled for information loss when working daily with computers, and it caused you no hardship at all. If you use a personal computer with a video graphics adapter (VGA) screen, but display any type of graphic that inherently has Super VGA (SVGA) level resolution, your VGA screen loses the additional definition in the image that is visible only when an SVGA controller card and monitor are used. And in fact, this example, while not involving compression as such, demonstrates precisely the type of situation where lossy compression would be tolerated: transmission of video images. Some compression algorithms used for transmission of video images lose some of the resolution of those images. However, if the received image is acceptably precise, the maintenance of the speed of the moving image might be more important.

Multiplexing

The paths available for moving electronic information vary considerably in their respective capacities or bandwidths. If a company requires many paths over the same route (for example, many terminals each requiring a connection to one distant computer), it often makes sense to configure one large-capacity circuit and bundle all the smaller requirements into that one big path. The process of combining two or more communications paths into one path is referred to as *multiplexing*. There are three fundamental types of multiplexing, all of which have significant variations. These main types are:

- Space-division multiplexing (SDM)
- Frequency-division multiplexing (FDM)
- Time-division multiplexing (TDM)

SDM

SDM is the easiest multiplexing technology to understand. In fact, it is so simple it would hardly rate its own special term, except that it is the primary method by which literally millions of telephone signals reach private homes. With SDM, signals are placed on physically different media. Then those media are combined into larger groups and connected to the desired end points.

For example, telephone wire pairs (which of course are also used for data communications), each of which can carry a voice conversation, are aggregated into cables with hundreds or even thousands of pairs (see Fig. 13.7). The latter are run as units from telco COs out to wiring center locations, from there splitting out to individual customer buildings.

The biggest advantage of SDM is also its biggest disadvantage: the physical separation of the media carrying each signal. It is an advantage because of the simplicity of managing the bandwidth; one only must label each end point of the medium appropriately. No failures (other than a break in the medium) can affect the bandwidth allocation scheme. But because there is a direct physical correlation between the physical link and the individual communications channel, a provider has some difficulty in electronically manipulating the path to achieve efficiencies of technology or scale. For example, probably the largest single factor blocking the conversion of the overall telephone network to all-digital technology is the embedded base of copper wire (and analog amplifiers) supplying telephone service to millions of homes and businesses.

Figure 13.7
Individual pairs are bundled together into much larger cables on a 1-to-1 ratio.

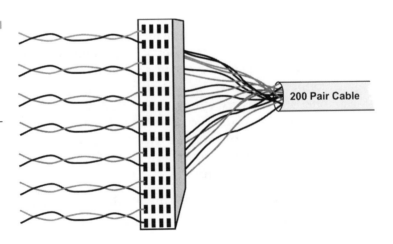

200 Pair Cable

FDM

FDM is inherently an analog technology. It achieves the combining of several digital signals onto (or into) one medium by sending signals in several distinct frequency ranges over that medium.

One of FDM's most common applications is cable television. Only one cable reaches a customer's home, but the service provider can nevertheless send multiple television channels or signals simultaneously over that cable to all subscribers. Receivers must tune to the appropriate frequency in order to access the desired signal. Figure 13.8 demonstrates the combining of signals on a cable television coaxial cable.

Certain modems have built-in FDM capabilities. Users can modify, with controls on the modem, how much bandwidth each connected digital device will be provided.

TDM

Time-division multiplexing is a digital technology. It involves sequencing groups of a few bits or bytes from each individual input stream one after the other and in such a way that they can be associated with the appropriate receiver. If this is done sufficiently quickly, the receiving devices will not realize or care that some of the circuit time was used to serve another logical communication path. The really high-speed communications technologies covered in later chapters all, without exception, use some form of TDM, so it is worthwhile to try to understand it.

Consider an application requiring four terminals at an airport to reach a central computer. Each terminal communicates at 2400 bits per second, so, rather than acquiring four individual circuits to carry such a low-speed transmission, the airline has installed a pair of multiplexers (see

Figure 13.8

Frequency-division multiplexing breaks the whole bandwidth into separate full-time channels.

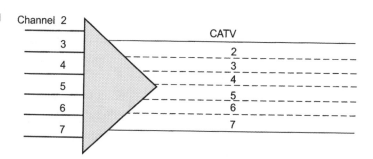

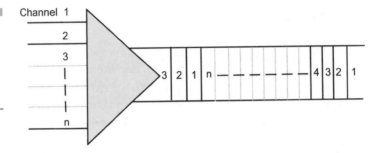

Figure 13.9
In time-division
multiplexing, the
entire channel is
allocated for short
periods of time.

below), a pair of 9600-bit-per-second modems, and one dedicated analog communications circuit from the airport ticket desk back to the airline data center, as illustrated in Fig. 13.9.

The time-division multiplexers work together to merge the data streams onto the 9600-bit-per-second circuit in such a way that each terminal appears to have a dedicated 2400-bit-per-second circuit. The multiplexer has enough buffer (or storage) space so that as any of the four clerks presses a key, that keystroke is stored locally (in the multiplexer or *mux*) until the time slot assigned to that clerk's terminal comes along. The flow of information for four hypothetical clerks named Ellen, Joe, Susan, and Allan is shown in Table 13.7.

As an example, we'll assume that Joe is inquiring as to the available of space on XXXXX Airlines flight 243. As part of this inquiry, he types the letters *XA243*. At the same time, Susan is seeing where there is space at the Compris hotel in Chicago, so she types *Compris*. The other two clerks are out to lunch; their terminals are inactive. Those letters will be interspersed among others typed by their colleagues. TDM assigns time slots

TABLE 13.7	**Byte**	**From**
The Flow of Information for the Various Time Slots Used	1	Ellen
	2	Joe
	3	Susan
	4	Allan
	5	Ellen
	6	Joe
	7	Susan
	8	Allan

to each configured device, whether or not the slots are used. If the data stream is to be interleaved one byte at a time, the data stream toward the computer resulting from this typing might look as shown in Fig. 13.10 (the information on the right goes toward the computer first).

This is not really a form of compression, because the same number of bytes is sent as is typed. In fact, as can been seen from the figure, there is a certain inefficiency here. Those 4-byte blocks are being pumped out by the multiplexer no matter whether anyone types a character or not. The first position in each 4-byte block is reserved for Ellen's characters, while the fourth position is reserved for Allan. Because Ellen and Allan are both at lunch, their blocks are empty—and wasted. Joe is a slower typist than Susan, so even though he types fewer characters, his last character goes at the same time as Susan's last; some of his slots are empty. He "missed the train."

In real life, many more slots would be empty because the capacity of the line is 240 characters per second—per clerk! No one can type that quickly. Also, in real life, multiplexing is more commonly done at the bit level rather than at the byte level. This would be harder to illustrate understandably, so we use bytes in the example. But the principle is identical.

Next, we modify the example to assume that the characters typed (*XA243* and *Compris*) are coming back from the computer instead of being typed. In such a case, they will be coming at "wire speed"; the computer can pump them out far faster than the line can absorb them, so we are now talking about wasting some time (Fig. 13.11).

Were the multiplexer configured for only two clerks' terminals, the information from the computer could appear much more quickly (Fig. 13.12).

Figure 13.10
The use of TDM and the data time slots for two operators.

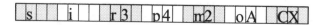

Figure 13.11
The time wasted on the circuit can be compensated for using TDM techniques.

This, in effect, is what a variation on a TDM called a statistical time-division multiplexer (STDM), does. The sending STDM analyzes the data stream on the fly to determine which ports or "tail circuits" are active—that is, how much service they require. If they are inactive, or less active (slow typists?), they are provided fewer of the time slots (or bandwidth). This allocation changes dynamically depending on the traffic pattern. Again, this is not compression, because all the information is sent. Nonetheless, it seems like compression because extending this technique allows overcommitting the line. If 10 terminals are connected to a 9600 circuit via an STDM, each terminal can be set to 2400 bits per second. So long as not all terminals are busy at full speed at every moment, the STDM can make it look as though each terminal has its own 2400-bit-per-second circuit, even though the aggregate bandwidth required to support these settings without using an STDM would be 24,000 bits per second!

Paradoxically, STDM techniques are used more at lower speeds than on the really fast multimegabit circuits. At very high speeds, the equipment is so busy just performing TDM functions that too much extra computer power would be required to do the on-the-fly analysis for STDM.

Codes

Chapter 5 covered digital transmission. The concept of a *bit*—an electronic expression of a 1 or 0—should now be clear. However, how do you get from 1s and 0s to transmitting your resume over wires?

The alphabet must be built up from sets of 1s and 0s. Specifically, we employ one or more sets of *codes* or *alphabets*, standard definitions of patterns of 1s and 0s that we will agree to use to represent letters, numbers, and other symbols that we wish to transmit and receive.

The alphabets most frequently used are either American Standard Code for Information Interchange (ASCII), which is fairly universal, or Extended Binary Coded Decimal Interchange Code (EBCDIC), which is an IBM alphabet. These two code sets, or alphabets, are used to convert a series of 1s and 0s into an alphabetic or numeric character.

ASCII

One character (also known as a byte or octet) must be represented as a consistent bit pattern by both sender and receiver. As we use a keyboard (standard typewriter keyboards are known as *QWERTY keyboards* because of the sequence of the first row of alphabetic keys), we create a stream of combination of letters, numbers, and symbols.

When ASCII (usually pronounced *ass-key* with a mild accent on the first syllable) was originally defined for use by the United States government, the bit pattern was defined to be seven data bits long. With 7 bits, it is possible to differentiate 128 different patterns. So, to recreate these typed characters with 1s and 0s, we can use a combination of up to 128 possible ASCII characters (Fig. 13.13).

Using the combinations in this table, we should be able to transmit just about everything we presently understand in our vocabulary. And, in fact, for many years, the 7-bit ASCII was used for most non-IBM mainframe communications. But there were two factors that caused this form of ASCII to become less popular.

First, most computers handle data in 8-bit chunks (rather than 7) to represent characters. The terms *byte* and *octet* almost always refer to 8-bit, not 7-bit, patterns. Second, while 7-bit ASCII can indeed represent all the English letters and numbers, with some symbols left over for special characters and control information, there are many other characters used in written communication that cannot easily be expressed in a 128-character code set. Accented characters in the Romance languages, character graphic drawing symbols, and typographical indications in word processors (bolding, underlining, etc.) are just a few examples of symbols difficult to handle with standard ASCII.

Extended ASCII

Extended ASCII is a superset of ASCII. Extended ASCII is the code used inside virtually all non-IBM computers, including personal computers. It is an 8-bit code set, doubling the possible distinguishable characters to 256. The 7-bit ASCII codes are present in extended ASCII in their original form with a 0 prefixed to the base 7 bits. Another 128 characters are also available with the same base 7 bits as the original ASCII, but with a 1 prefixed instead.

But whereas ASCII is a standard, extended ASCII is…well, not quite standard. While the original 128 characters communicate well from ven-

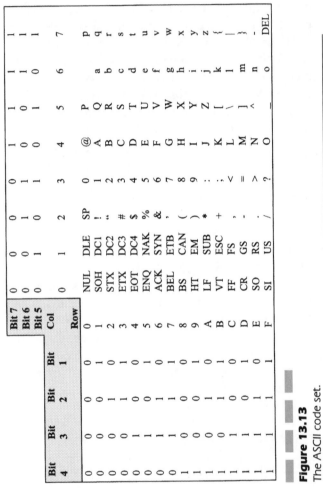

Figure 13.13
The ASCII code set.

dor to vendor, even in extended ASCII, every application defines its own use of the additional 128 characters. For example, you can easily write out ASCII text from most word processing programs, with the result being readable by most other word processors. But if you attempt to read a document created by a word processor in its native form with another, different word processor, you will only be successful if the latter specifically contains a translation module for material created by the first.

Nonetheless, the lion's share of non-IBM communications is now conducted using extended ASCII. Virtually all personal computers use it, including IBM's. If you set your communications protocol to *N,8,1* (no parity, eight data bits, one parity bit), those eight data bits are encoded in using extended ASCII.

EBCDIC

IBM, not a company to follow the herd, realized early on that 128-code ASCII did not contain enough patterns for its requirements. So, IBM created an entirely different code set, one twice as large as the original ASCII code set. IBM's 8-bit, 256-character code set, used on all of its computers except personal computers, is called Extended Binary Coded Decimal Interchange Code or EBCDIC. (Most people pronounce it *eb-suh-dick* with a mild accent on the first syllable.)

As with ASCII, certain of the characters are consistent wherever EBCDIC is used. But other characters vary depending on the specific communicating devices. In Fig. 13.14, the white space can be used differently depending on the EBCDIC dialect in use.

Unicode

Two hundred fifty-six codes might seem to be all anyone would need. But consider the requirements of Chinese, which has thousands of characters. Or the Cyrillic alphabet, which, although it does not have a terribly large number of characters, does not overlap any of those defined in ASCII or EBCDIC. Another code set, called Unicode, is now being implemented in some products. Unlike the 8-bit extended ASCII and EBCDIC code sets, Unicode uses 16 bits, or 2 bytes, per character. While only 1s and 0s are used, this allows up to 65,536 (2 to the 16th power) separate character definitions. Of course, each character takes up as much storage and transmission time as two eight-bit characters. But Unicode is a truly international code set, allowing all peoples to use their own alphabets if they wish.

Figure 13.14 The EBCDIC code set.

Bits 4 3 2 1 identify the columns; bits 8 7 6 5 identify the rows.

Bits 8765	0000	0001	0010	0011	0100	0101	0110	0111	1000	1001	1010	1011	1100	1101	1110	1111
0000	NUL	SOH	STX	ETX	PF	HT	LC	DEL			SM/M	VT	FF	CR	SO	SI
0001	DLE	DC₁	DC₂	DC₃	RES	N;	BS	IL	CAN	EM	CC		IFS	IGS	IRS	IUS
0010	DS	SOS	FS		BYP	LF	EOB	PRE			SM			ENQ	ACK	BEL
0011			SYN		PN	RS	UC	EOT					DC₄	NAK		SUB
0100	SP										¢	.	<	(	+	|
0101	&										!	$	*	)	;	¬
0110	-	/										,	%	_	>	?
0111											:	#	@	'	=	"
1000		a	b	c	d	e	f	g	h	i						
1001		j	k	l	m	n	o	p	q	r						
1010			s	t	u	v	w	x	y	z						
1011																
1100		A	B	C	D	E	F	G	H	I						
1101		J	K	L	M	N	O	P	Q	R						
1110			S	T	U	V	W	X	Y	Z						
1111	0	1	2	3	4	5	6	7	8	9						

Modulation

How does the transmission process work? How does the data get onto the voice dial-up telephone line? We use a device to change the data. This device, known as a modem, changes the data from something a computer understands (numeric bits of information) into something the telephone network understands (analog sine waves, or sound). A modem generates a continuous tone, or carrier, and then modifies or modulates it in ways that will be recognized by its partner modem at the other end of the telephone circuit. The modems available to do this come in variations, each one creating a change in a different way. Remember that the word *modem* is a contraction for *modulation/demodulation*. In order for the communications process to work, we need the same types of modems at each end of the line operating at the same speeds. These modems can use the following types of modulation schemes (or change methods):

Amplitude Modulation (AM)

Amplitude modulation represents the bits of information (the 1s and 0s) by changing a continuous carrier tone. Figure 13.15 illustrates amplitude modulation. Because there are only two stages of the data, 1 or 0, you can let the continuous carrier tone represent the 0 and the modulated tone represent the 1. This type of modem changes the amplitude (think of amplitude as the height or loudness of the signal). Each change represents a 1 or a 0. Because this is a 3-kHz analog dial-up telephone line, the maximum amount of changes that can be represented and still be discrete enough to be recognizable to the line and the equipment is about 2400 per second. This cycle of 2400 changes per second is called the *baud rate*. Therefore, the maximum amount of data bits that can be transmitted across the telephone line with AM modulation is 2400 bits per second. Most amplitude modulation modems were designed to transmit 300–1200 bits per second, although others have been made to go faster.

As we look at the different ways to change the 1s and 0s generated by the computer into their analog equivalents, we have another choice in the process. Voice communications (or human speech itself) is the continuous variation of amplitude and frequencies, so we could choose to use a modem that modulates the frequency instead of the amplitude. An explanation of this type of modem follows.

Figure 13.15
The concept of
amplitude
modulation.

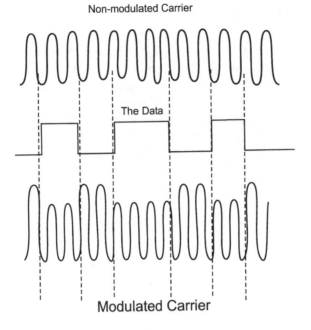

Non-modulated Carrier

The Data

Modulated Carrier

Frequency Modulation (FM)

Frequency modulation is provided by an FM modem. This modem represents the 1s and 0s as changes in the frequency of a continuous carrier tone. Because there are only two states to deal with, we can represent the normal frequency as being a 0 and slow down the continuous carrier frequency when we want to represent a 1 to the telephone line (Fig. 13.16). The modem uses the same baud rate on the telephone line as the amplitude modulation technique, that being 2400 baud or discrete changes per second. These modems modulate 1 bit of information per cycle change per second, or a maximum of 2400 bits per second. Note that the baud rate and the bits per second rate are somewhat symmetrical. Although both AM and FM modems are designed around what was once considered pretty fast transmission rates, we have continually been unsatisfied with any rate of speed developed. We want more and more, faster and faster.

Because we (as humans) and our creations (the computers) are never satisfied with the speed of transmission over the telephone line, we demanded faster. Throughput was expensive under the old dial-up telephone network. Therefore we asked for additional speed to get more throughput and less cost. The engineers came up with a new process that modulates on the basis of phases.

Figure 13.16
The concept of
frequency
modulation (FM) for
data modems.

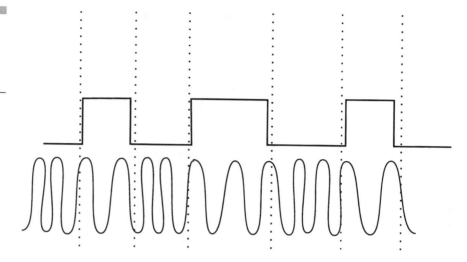

Phase Modulation

If we can change the phase of the sine wave as it is introduced to the line, at positions of 0, 90, 180, and 270 degrees, we can encode the data with more than 1 bit of information at a time. A phase modulation technique allows us to transmit a di-bit of information per signaling-state change. This gives us 4800 bits per second of throughput.

The di-bit represents the information as shown in Table 13.8.

As you can see, two bits of information at 2400 baud gives us the 4800 bits per second. This was a step in the right direction. But we wanted more; so a combination of phase modulation and amplitude modulation was developed.

QAM

Called quadrature with amplitude modulation (QAM), this combination allows us to use up to 16 possible steps of phase and amplitude modula-

TABLE 13.8

The Four Phases in
Phase Modulation
Cause a 90° Shift
When a Di-Bit Is
Received

Bits	Phase
00	0
01	90
10	180
11	270

tion. QAM is mostly used with 4 bits of information per baud rate, thereby producing 9600 bits per second of throughput across an analog telephone line. Theoretically, the system should be able to produce 38,400 bits per second of information (4 phases at 4 bits per phase at 2400 baud = 38,400 bits per second).

However, in practice, line rates can rarely support sustained throughput across the telephone network at this speed. Also, the telco limits the bandwidth to 3 kHz, and we use 2400 baud. If we try to send more data at a higher baud rate, the band limitation (bandpass filters) will strip the frequencies that go beyond the filters. This is a function of the telco equipment on the line at the CO. Thus we usually have to settle for analog data transmission at slower speeds.

The driving force behind improving modulation techniques has always been to increase the speed possible over an analog circuit. Why? In addition to the obvious reason (i.e., accomplishing the task more quickly improves productivity), the cost of communications over dial circuits is directly proportional to the amount of time those circuits are in use; going faster saves money. But, when the absolute best available modulation technology is in use, one has not yet necessarily squeezed the absolute best transmission volumes out of a circuit. One can push it even further by using compression.

V.90 Modems

Practically all new PCs and laptop computers sold today have a new modem integral to them. The use of modem technology has become commonplace. However, as modems have dropped in price, the ability to go faster has increased. Today, the new modem technology is asymmetrical in that it transmits at one speed and receives data back at a different rate. The world was looking for better ways of moving data, but had to deal with the limits of the wires in the local loop. Over the decades, the cost of modems has dropped while the ability to move the data was going in an upward fashion. Something had to be done because the amount of data we were moving was escalating exponentially. Enter the 56 Kbps modem (V.90 standard), which operates at a 33.6-Kbps transmit rate and a 56 Kbps receive rate

The rated speeds can be misnomers because in North America, the FCC and the Canadian CRTC limit the speed that a telco can offer to a customer to no greater than 53 Kbps. The rate was selected years ago when we were still trying to drive modem communications at 9.6 Kbps. So the arbitrary decision was something that no one ever thought would occur. Alas, here we are with a modem that can achieve 56-Kbps and the regulators have capped it. Moreover, the asymmetrical rate of speed facilitates access to one of the most commonly used networks today— the Internet. The 56 Kbps modem works well because the user has little data to send (typing <www.tcic.com>, for example, requires very little data) but lots to receive (a Web page can be millions of bytes large). So the different speeds allow for dial-up communications on a telephone company circuit (a voice channel) in one direction and a digital-access method for the return path at a much higher rate of speed.

Typically, however, when a user accesses the network with these modems, they will transmit at between 28.8 and 33.6 Kbps upline and receive approximately between 38 and 45 Kbps downstream. One can get a slightly better response from the 56-Kbps modem, but there are still limitations that must be dealt with.

A note that puts this all in perspective though—in the early 1980s we were transmitting data at between 4800 and 7200 Bps, and the modems were expensive (hundreds to thousands of dollars at the time). Here, in a mere decade, we have achieved quantum leaps in technological advancements, and we have seen the cost drop exponentially to a few dollars. What we can expect is faster and better, but cheaper, through the new millennium.

Devices

DTE versus DCE

A basic of data communications is that every communicating device is either a terminal-type device (data terminal equipment, or DTE; also sometimes referred to as data circuit terminating equipment, or DCTE) or a communications-type device (data communications equipment, or DCE). DTEs use communications facilities; their primary functions lie elsewhere. DCEs provide access or even implement communications facilities. Their only role is moving information.

You might presume that some cosmic requirement is satisfied by this overall categorization. But in fact the nitty-gritty reality of cabled communications is the primary cause of this division. Although we will get into cabling standards later, consider the following situation:

Two devices, A and B, must be connected. The connection will be via two wires, numbered one and two. Let us assign wire number one to transmit the data, and number two to receive. But wait! If A transmits on number one, and B also transmits on number one, and if they both receive on number two, neither will listen to what the other is sending. It would be as though two people each spoke into opposite ends of the same tube at the same time, with neither putting the end to his or her ear.

The solution seems simple: have A transmit on wire one, B listen on wire one, A listen on wire two, B transmit on wire two. But now let us introduce device C. How should C be built? If it is configured as is A, it cannot communicate with A; if it is configured as is B, it cannot communicate with B.

But devices are not configured at random. For example, terminals do not usually connect directly to terminals; printers are never connected directly to printers; and so on. Perhaps if one broad category of devices usually connects not to another in that category but rather to a device in another category, we could standardize on only two default configurations. This is what was done with DTEs and DCEs.

Categories and Examples of DTEs

The grouping is not perfect. Some devices routinely connect to both DTEs and DCEs (e.g., multiplexers). Such devices do not clearly fall into either category, and must be configured depending on the specific installation requirements. But for the most part, any data processing device that can communicate falls cleanly into one of these two categories. Most of the devices covered in this book are DCE devices. Here are some examples of the DTE devices to which we provide data communications services:

- Computers (mainframe, mini, midi)
- Terminals (CRTs, VDT, teletype)
- Printers (laser, line, dot matrix)
- Specialized (bar code readers, optical character recognition)
- Transactional (point of sale equipment, automated tellers)
- Intelligent (personal computers)

The remainder of this section will focus on DCEs and ways of connecting components.

Modems

As mentioned earlier, *modem* is a contraction of two words (*modulator* and *demodulator*). The role of the modem is to change information arriving in digital form into an analog format suitable for transmission over the normal telephone network. Naturally, modems work in pairs (or sets of at least two), and a second modem at the other end of the communications path must return the analog signal to a digital format useful to terminals and computers.

Dozens of manufacturers make modems, and hundreds of models exist. The market can be divided in many ways:

- Speed
- Supported standards
- Leased line versus dial-up
- Two-wire versus four-wire
- Point-to-point versus multipoint (multidrop)
- With or without compression capability
- With or without error correction capability
- Manageable or nonmanageable

Entire texts could be (and no doubt have been) written just on this topic.

14

T1 and the
T-Carrier System

Evolution of the T-Carrier System

In the early 1960s, the Bell System began to introduce and use a new digital technology in the network. This was necessary because the older carrier and cabling systems were rapidly becoming strained for capacity. The demand for newer and higher-speed communications facilities was building among customers as well as within the systems themselves.

When this digital technology was being introduced, it was deployed in the public network as a means of increasing the traffic capacity within the telephone company on the existing wire-pair cable facilities as interoffice trunks. The older systems, including the N-carrier system, used a two- or four-wire connection through an analog multiplexing device to deliver either 12 or 24 analog channels. This was still an inefficient use of the line capacity, and the analog service was noisy and required expensive line-treatment equipment. Therefore, the telephone company introduced its newer technology to overcome the limitations of the existing plant and transmission services.

Some of the problems with the analog systems were also related to the circuit quality. Anyone who can remember the older analog network knows that a call placed across the country from East to West Coast was significantly different than such a call made today. The static and noise on the line made the call sound more like it went to the moon and back. Because that's all that was available, that's what the user became accustomed to.

The use of older analog systems was ending within the telephone company networks. The telcos had to find a way of improving the utilization of the cable plant on an interoffice basis to overcome problems with underutilized pairs of wires; and the continued installation of inefficient systems was expensive and bulky. The average length of the wires between the telcos' offices was approximately 6.5 miles. As calling requirements continued to grow, the telcos needed to increase the traffic-handling capacity on these interoffice routes. They were in a quandary, however. First, they didn't want to continue running bulky cables between offices because there simply wasn't enough space. Second, costs for maintaining the cable plant were escalating. Something had to be done quickly. Figure 14.1 shows the use of wires between the telco offices. These wires provide the interoffice communications from telco to telco. The end user or customer was kept on the old analog twisted pairs of wire. From a user's perspective the changes were invisible, except that some improvement in call quality was evident.

Figure 14.1
Bell used high-capacity services between its offices. Users were still kept on twisted-pair analog.

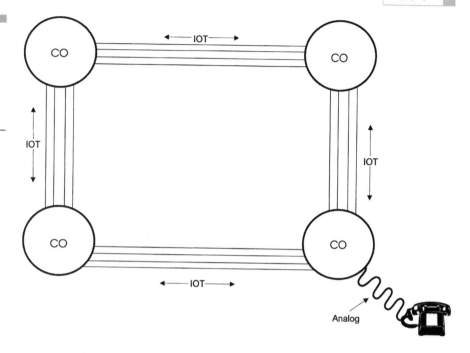

Analog

Analog Transmission Basics

Before going much further, it would be prudent to furnish a brief description of the dial-up telephone network to aid in understanding the need for the digital architecture that was introduced through the use of the T-carrier system. The telephone system was designed around providing analog dial-up voice telephony. Everything was based on voice communications services on a switched (non-dedicated) basis. A user could connect his or her own telephone to that of another user on the network through either an operator-connected or dial-up (a later evolution) addressing scheme.

Because voice is the primary service provided, the telephone set (Fig. 14.2) has evolved into a device that takes the sound wave from the human vocal cords and converts that sound into an electrical current, represented by its analog equivalent. The human voice produces constantly changing variables of both amplitude (the height of the wave) and frequency (the number of cycle changes per second). As these constantly changing variables of amplitude and frequency are produced, the telephone set converts the sound pressure into an equivalent electrical wave. This electrical energy will be carried down a pair of wires. As the electrical energy is introduced into the wires (using twisted pairs of copper for this reference)

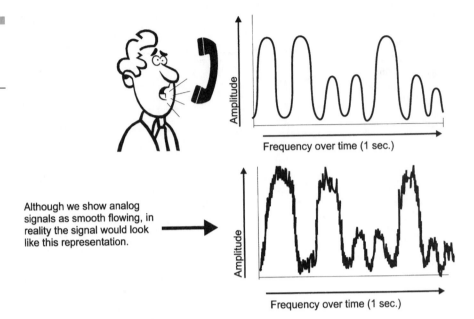

Figure 14.2
A representation of the analog wave being created.

Although we show analog signals as smooth flowing, in reality the signal would look like this representation.

certain characteristics begin to work on it. First, resistance occurs on the wire, impeding the flow of electricity and reducing the strength of the created signal. Second, the wires act as an antenna, drawing in noise (such as static, cross talk, or electricity from other conversations on wire pairs adjacent to yours, etc.). This loss of signal strength coupled with the introduced noise continues to distort the signal.

To overcome these problems, analog amplifiers are used to boost the signal strength back up to its original value. The amplifiers are normally placed on circuits greater than 18,000 to 20,000 feet in length. Figure 14.3

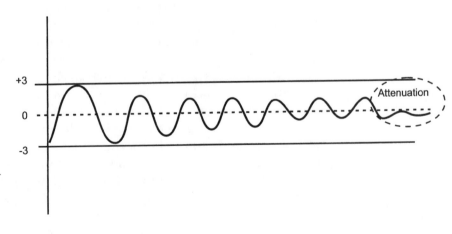

Figure 14.3
As the electrical signal is introduced, attenuation begins immediately. If the signal must travel great distances, it could run out of strength and disappear.

shows how the electrical signal on the wires is attenuated or weakened. As the signal is boosted back to its original strength, the noise begins to accumulate, causing further distortion of the call. Figure 14.4 represents the analog amplifier on the line boosting up the signal. To overcome the loss on the line and to ensure that the signal gets to the other end, the telcos used analog amplifiers. These amplifiers were used on circuits to increase the strength of the analog transmission. However, they were prone to failures and had other qualities that were undesirable. Figure 14.5 is a representation of the problem encountered with the amplification process. As the signal is amplified, the signal and all the noise on the wires are amplified together. This does allow the signal to travel further, but after each amplification process, noise also accumulates. A good deal of the problem stems from the fact that the amplifier can't distinguish the analog signal from the noise. Therefore it amplifies everything.

The Evolution to Digital

The T-carrier system was developed to cope with these amplification problems. The telcos were looking to enhance the quality of calls and

Figure 14.4
As the signal gets weak, the telco uses analog amplifiers to boost the signal and keep going.

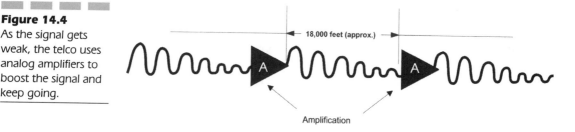

18,000 feet (approx.)

Amplification

Figure 14.5
As the signal and noise are amplified, the noise takes on a cumulative effect until the transmission becomes virtually useless.

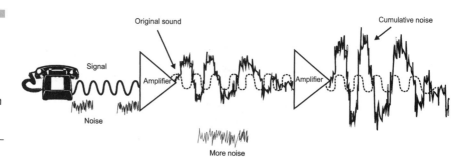

Original sound

Cumulative noise

Signal

Amplifier

Amplifier

Noise

More noise

better utilize the cable facilities. The T-carrier system allowed them to increase the call-carrying capacity while taking advantage of the unused transmission capacity of their existing wire pair facilities and to improve the quality of transmission by migrating away from the older analog techniques. The evolution to T-carrier was important for a number of reasons:

- It was the first successful system designed to utilize digitized voice transmission.

- It identified many of the standards that are employed today for digital switching and digital transmission, including a modulation technique.

- The transmission rate was established at 1,544,000 bits per second.

- The T-carrier technology defined many of the rules, or protocols, and constraints in use today for other types of communications.

As mentioned earlier, the analog transmission capabilities were inefficient. In many cases, the telephone companies used a single pair of twisted wires to carry a phone call. This meant that as uses of the interoffice facilities grew, the cable plant grew exponentially. As more central offices (or end offices) were added, the need for meshing these together grew. Interconnecting each office with enough pairs of wire to service user demands was becoming a nightmare. Figure 14.6 depicts the need for cabling or interoffice trunking to support the growing demand for services. As each new end office was added, the wiring systems had to grow to serve customer demands.

Remember that the telephone companies saw this carrier system as a telco service only. Even as the end user population grew, the telcos still held back on deploying this digital capability to the end user. The telcos used the higher efficiencies to support the end user digitally, from end office to end office or through the network. The user was still relegated to a single conversation on a twisted pair using analog transmission. Thus this system operated in an analog format on the local loop, but digitally on the interoffice trunks (IOTs). In metropolitan areas, the telcos were reaping the benefits of the T-carrier system. Figure 14.7 is a representation of the telephone company deployment of analog/digital transmission capacities.

Some immediate benefits were achieved: The quality of transmission improved dramatically, and the utilization of existing wire facilities increased. A single four-wire facility on twisted wires could now carry 24 simultaneous conversations digitally at an aggregated rate of 1.544 Mbps.

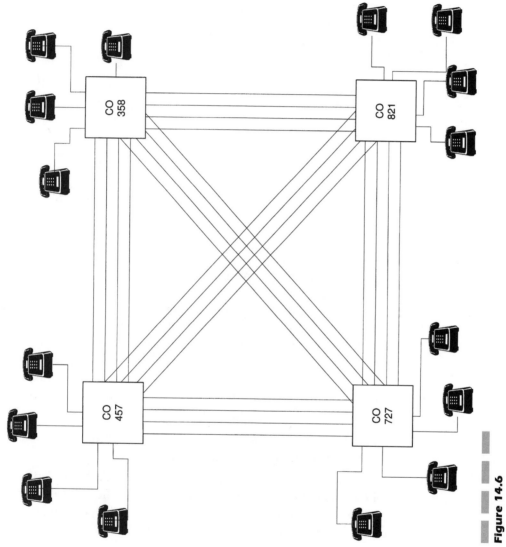

Figure 14.6
As more central offices were added to the network, the need for wires grew exponentially.

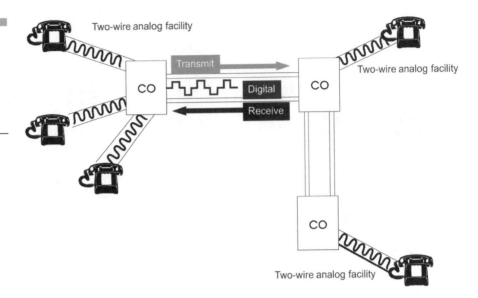

Figure 14.7
The telephone companies reaped the benefits of interoffice trunking using digital transmission.

As digital switching systems were introduced into the dial telephone network, they were designed around the same techniques employed in the T-carrier system. The digital switching matrices of the #4 and #5 electronic switching systems (ESSs) developed by Western Electric (now called Lucent Technologies) and the DMS systems (10, 100, 250) developed by Northern Telecom, Inc. (now called Nortel) utilize pulse code modulation (PCM) internally, so that digital carrier systems and channels can be interfaced directly into the digital switching systems. By adhering to the standards set forth, the operating telephone companies and the long-haul carriers could build integrated digital switching and digital transmission systems that:

■ Eliminated the need to terminate the digital channels or equipment to provide analog interfaces to digital switching architectures.

■ Avoided the addition of quantizing noise that would be introduced by another digital-to-analog conversion process. Whenever a conversion of the signal from analog to digital or from digital to analog is necessary, the risk of errors and noise increases.

These techniques were used to provide lower-cost, better-quality dial-up telephone services. However, this same technology underlies the idea of full end-to-end digital networking services that is the basis of integrated services digital networks (ISDN), the future and the higher-end broadband services that are emerging.

Analog-to-Digital Conversion

Prior to covering the T-carrier fundamentals, it would be appropriate to discuss the analog-to-digital conversion process. Digital architecture dictates the use of a digital bit stream; therefore the analog wave must be converted into a useable format. Digital pulses are represented by a 0 or a 1. As the analog signal is being transmitted to the network, it must be converted from a wave of varying amplitudes and frequencies to a digital format of 1s and 0s (or presence and absence of voltage).

The analog wave must be sampled often enough and converted to create a stream of 1s and 0s that is precise enough to be recreated at the distant end, producing a signal that sounds like the original conversation. According to the Nyquist rule, a digital signal should be created by sampling the analog wave at twice the highest frequency on the line. For an analog circuit, delivered by the local telephone company, the frequency range on a twisted pair of wires is represented in hertz (Hz). To maximize the utilization of the older carrier systems, the telcos delivered a usable bandwidth of 3000 Hz (3 kHz), as compared to the 4000 Hz (4 kHz) capable of being carried on the line. The difference between the 3-kHz and 4-kHz services is in how the line is filtered. The human voice produces understandable information in the range of 300 to 3300 frequency changes per second. Therefore, the telcos knew that all they had to provide to the user was 3 kHz of usable bandwidth. But for separation between conversations and to minimize cross talk and other interference, they allocated 4 kHz per analog line. Using Nyquist's rule, the sampling of an analog wave at twice the maximum frequency of the line meant that a minimum of 6600 samples per second should be sufficient to recreate the wave for the digital-to-analog and analog-to-digital conversion. However, to produce the quality of the higher range of frequencies of the human voice, the number of samples is rounded up to 8000 per second.

The sample now had to be created into a bit stream of 1s and 0s, as already stated. To represent the true tone and inflection of the human voice, enough bits need to be used to create a digital word. Using an 8-bit word creates enough different points on a wave to do just that. The wave is divided into 256 possible amplitude combinations at the moment of the sample itself. Figure 14.8 shows how the sampling is accomplished using two states and 8 bits to create 2^8 or 256 points on the analog wave. As each sample is taken, the amplitude of the wave is sampled.

Once the sample has been taken and the digital equivalent created into an 8-bit word, the digital (or square) wave can be transmitted. The 1 repre-

Figure 14.8
With 256 possible points being samples, the quality of the digital transmission, when converted back to analog, should be exact.

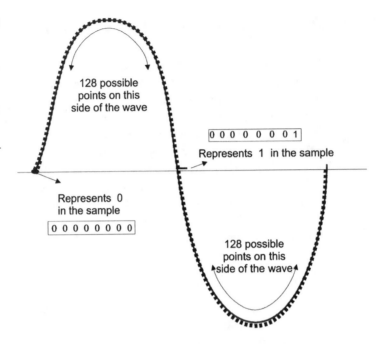

128 possible points on this side of the wave

|0 0 0 0 0 0 0 1|

Represents 1 in the sample

Represents 0 in the sample

|0 0 0 0 0 0 0|

128 possible points on this side of the wave

sents the presence of a voltage and the 0 represents the absence of a voltage (Figure 14.9).

This conversion uses pulse code modulation (PCM) to create the sample. Using PCM and the rules established, the transmission of the digital equivalent of the analog wave results in 8000 samples per second × 8 bits per sample = 64,000 bits per second as the basis for a voice transmission in PCM mode. Whether the end user transmitted voice, data, or facsimile, the copper wires were used to carry the analog wave to the central office, where the wave was converted to a digital signal for transport across the telco network. At the far end, the receiving telco converted the digital sig-

Figure 14.9
A sampling of the analog wave yields a transmission of 8 bits in a stream. Two pulses (voltages) are required in this sample in the sixth and eighth positions.

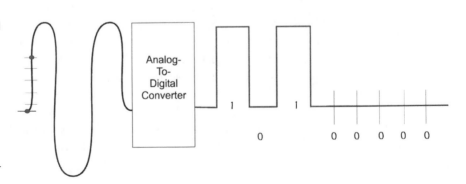

Analog-To-Digital Converter

1   1

0   0 0 0 0 0

nal back to its original analog equivalent for delivery to the customer. All this served to improve transmission quality for the telcos, but left the customer, who still received analog transmission to his or her premises, out in the cold (Fig. 14.10).

As the telcos were reaping the benefits of their new transmission capability, the users began to request this service. More user demands created some upheaval within the Bell system. This technology was perceived to be a telco-only service. But users also wanted (demanded) end-to-end digital transmission to improve their throughput and quality. Because analog transmission was noisy, there was a limit to the types of signals users could exchange across the network. Specifically, the constant growth in computer technology and the deployment of terminals and printers around the country left the end user with limitations and restrictions that were becoming intolerable. Analog leased line or dial-up services left the user with data communications capabilities that were limited to 9.6 to 19.2 Kbps (leased) transmission, as described in Chap. 13. Not only were these services too slow for most users, they were expensive.

The Movement to End Users

Ultimately, the telcos had to install digital transmission to the customer premises. Originally this took the form of lower-speed digital data services (DDSs) at speeds of 2.4, 4.8, 9.6, and 56 Kbps digital. Newer introductions added 19.2–64 Kbps and high-capacity (HICAP) T1 services at 1.544 Mbps.

In order to deliver this digital capability to the customer, a special assembly or individual-case-basis (ICB) tariff arrangement was initially used. This emergence didn't occur until the mid-1970s. At first, this was done with some reluctance, but as the movement caught on the deployment was both quick and dynamic. The telcos had to modify the outside plant to accommodate the end user needs. To do this, they first had to engineer the circuit. Next they had to remove all of the analog transmission equipment from the line. Then the circuit (now a four-wire circuit) had to be checked for splices, taps, bridges, and other problems that would introduce loss and noise and thereby impair transmission. These had to be removed or cleaned up. The next step was to provide digital equipment, called regenerators or repeaters, on the line.

With analog transmission, amplifiers are used on circuits over 18,000 feet in length. However, with digital transmission, repeaters are used at whatever intervals are necessary to ensure quality (typically 2000 to 6000

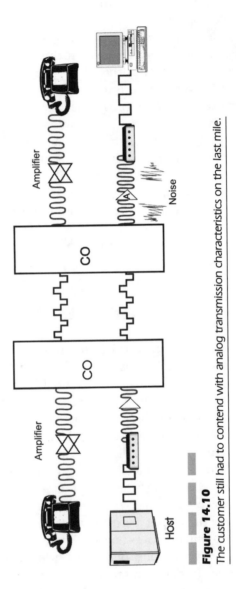

Figure 14.10

The customer still had to contend with analog transmission characteristics on the last mile.

feet). In digital transmission—the presence or absence of a voltage—the signal can deteriorate very quickly. Therefore the telco needed four times as many repeaters as it had analog amplifiers. These repeaters or regenerators ensure that the proper signal is moved across the line. The signal is repeated to prevent it from falling below a threshold where it can't be distinguished from noise. Figure 14.11 is a representation of how the signal is actually recreated rather than amplified to produce the quality we have come to expect. As the signal moves across the wires, the same problems are inherent; resistance on the line depletes the signal strength, and noise is prevalent. The digital regenerator is placed close enough to get around this problem. Therefore the end user could move to an effective use of a four-wire circuit at 64 Kbps. The first deployment of this service was for voice, but data quickly became the dominant rider of the new digital transmission. It should also be noted that as T1 became more readily available, the primary use (75%) was to consolidate WATS lines onto a single digital trunk. This digital transmission capability was designed around the voice dial-up network; therefore, the service was a natural fit.

T1 Basics

T1 is characterized by the following operating characteristics.

A Four-Wire Circuit

Because this technology evolved from the old twisted-pair environment, four wires were used. Two wires are used to transmit and two are used to receive. Other facilities can be used, but for now, think of it this way.

Figure 14.11
A representation of how the signal is recreated.

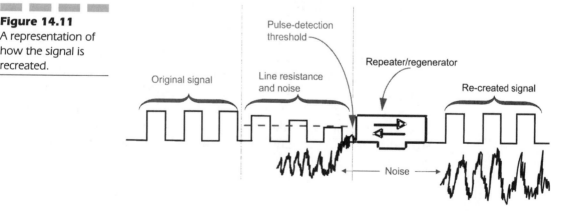

Pulse-detection threshold

Repeater/regenerator

Original signal

Line resistance and noise

Re-created signal

Noise

Full Duplex

Transmission and reception can take place simultaneously. Many customers derive other uses, such as one way only for remote printing, file transfer, and so on, or two way for alternate service such as voice communications.

Digital

This is an all-digital service. Data, analog facsimile, analog voice and the like are all converted to digital pulses (1s and 0s) for transmission on the line.

Time-Division Multiplexing

The digital stream is capable of carrying a standard 64-Kbps channel; 24 channels are multiplexed to create an aggregate of 1.536 Mbps. Time division allows a channel to use a slot $1/24$ of the time. These can be fixed time slots made available to the channel.

Pulse Code Modulation

Using the example, the analog voice or other signal is sampled 8000 times per second; an 8-bit word is used to represent each sample, thus yielding the 64-Kbps channel capacity.

Framed Format

As the pulse code modulation scheme is used, the 24 channels are time-division multiplexed into a frame to be carried along the line. Each frame represents 1 sample of 8 bits from each of the 24 channels. Added to this is a framing bit. The net result is a frame of 193 bits. There are 8000 frames per second; therefore, a frame is 125 microseconds long. Framing accounts for 8 Kbps overhead (1 bit × 8000 frames). Adding this 8 Kbps to the 1.536 Mbps discussed in the paragraph on time-division multiplexing yields an aggregate of 1.544 Mbps.

Bipolar Format

T1 uses electrical voltage across the line to represent the pulses (1s). The bipolar format serves two purposes. It reduces the required bandwidth from 1.5 MHz to 772 kHz, which increases repeater spacing. And the signal voltage averages out to zero, allowing direct current (dc) power to be simplexed on the line to power intermediate regenerators. Think of it as an alternating current (ac) version of a dc line. Every other pulse will be represented by the negative equivalent of the pulse. For example: the first pulse will be represented by a positive 3 V (+3 V), the next pulse will be represented by a negative 3 V (–3 V), and so on. This effectively yields a 0 voltage on the line, because the +–+– equalizes the current. This bipolar format is also called alternate mark inversion (AMI). The mark is a digital 1. Alternate marks are inverted in polarity (+,–).

Byte-Synchronous Transmission

Each sample is made up of 8 bits from each channel. Timing for the channels is derived from the pulses that appear within the samples. This timing keeps everything in sequence. If the devices on both ends of the line do not see any pulses, they lose track of where they were (temporary amnesia). This means that the T1 will be synchronous unto itself, but not to others. When using the T1 as the basis of other higher multiplexing schemes (like T2 and T3) this is important to remember. The timing is derived from the pulses on the specific link, so the link is timed to itself. If one needs to multiplex into higher-speed services, additional stuffing bits may have to be added to get the individual T1s timed to each other.

Channelized or Nonchannelized

Generically, T1 is 24 channels of 64 Kbps each, plus 8 Kbps of overhead. This is considered channelized service. However, the newer multiplexing equipment can be configured in a number of ways. For example, the T1 can be used as a single channel of 1.536 Mbps (such as for a router connection to the WAN); or two high-speed data channels at 384 Kbps each and a video channel at 768 Kbps. These examples can be mixed into a variety of offerings. The point is that the service does not have to be configured in 24 channels, but can be nonchannelized into any usable data stream needed (equipment allowing, of course).

Any other suitable medium (fiber, digital microwave, coax, etc.) can also be used. The T1 is still treated as a four-wire circuit. When the other media forms are used, the T1 will be suitably taken from the transmission mode and converted back to the appropriate interface. For a four-wire circuit, the carrier will normally terminate the four wires into a demarcation point (DEMARC) or network interface unit (NIU). Individual circuits can be terminated into a recommended jack—RJ-48 (sometimes called dumb jacks) or RJ-68, (now called smart jacks). These jacks serve as the interface to the four wires. In a larger environment or where multiple circuits are involved, other methods of termination can be used.

For example, an RJ-2IX or a BIX block can be used for terminating a T1 on a main distribution frame in a customer location. From the RJ-48, a 15-cable (using a subminiature 15-pin DIN connector) will be extended to the customer premises equipment (CPE). Usually, the CPE is a channel service unit (CSU). Figure 14.12 is a representation of the connection from central office to CPE.

The customer premises equipment uses time-division multiplexing to carry the multiple voice and data conversation across the line. Time-division multiplexing is somewhat efficient in that it allows time slots to be dedicated to each of the conversations to be carried on the line. Using this fixed time slot, a conversation will be in the same bucket (or slot) at each sample time.

Remember that the equipment using the Nyquist rule operates at twice the highest frequency of the line. Therefore each individual con-

Figure 14.12
A representation of a four-wire circuit terminating into an RJ-48. This is probably the most common method of delivery. The RJ-48 is an eight-conductor modular jack.

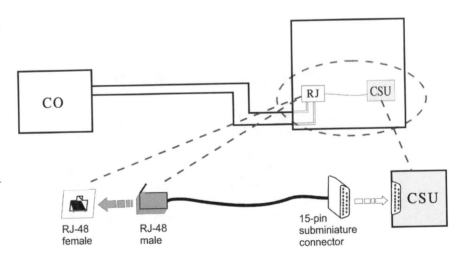

versation is sampled 8000 times per second. There are 24 (typical) paths being simultaneously multiplexed onto the T1. Each of these paths therefore carries one sample ($\frac{1}{8000}$ of a second) of voice, data, fax, video, etc.

The time slot is always present, and if no traffic is being generated between points across the line the time slot goes unused. The slot is empty. For this reason, time-division multiplexing is inefficient. Attempts to get the most amount of service from a T1 would be thwarted because of the fixed time slot problem. Figure 14.13 is a representation of this time-division multiplexing (TDM or Mux) scheme.

Framed Format

T1 uses some very specific conventions to transmit information between both ends. One of these is a framing sequence that formats the samples of voice or data transmission. It's easy to think of a frame in terms of the 8-bit samples of the 24 channels being strung together in a logical sequence. After the 192 bits of information are compiled, a framing bit is added, creating a frame of 193 bits of information.

Figure 14.13
The time-division multiplexing scheme.

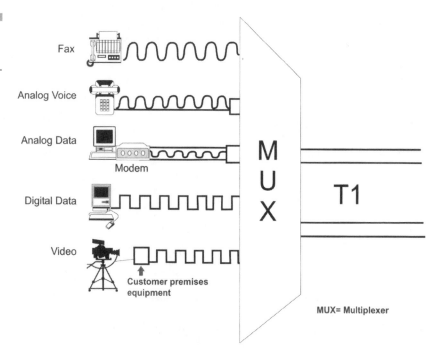

Fax

Analog Voice

Analog Data

Modem

Digital Data

Video

Customer premises equipment

M U X

T1

MUX= Multiplexer

The framing bit can be equated to a pointer or address. Because the line is moving bits of information at 1,544,000 per second, it would be very easy to skew left or right and get information out of sequence. Therefore the extra 8000 bits of information (1 bit per frame, 8000 frames per second) create a locator for the equipment to lock in on. This pointer allows the devices to read a pattern of bits in order to know which frame is being received or transmitted and the location of each channel thereafter.

Framing has undergone several evolutions over the years. The first use of T1 service was for voice, so the information was easier to use. The evolution of framing followed a sequence as outlined in Table 14.1.

The D4 superframe uses a framing pattern for voice and data. The framing pattern in the D4 superframe is a repeating 12-bit sequence (1000 1101 1100) that allows signaling bits to be "robbed" in the 6th and 12th frames of the superframe. The 8th bit from each sample in frames 6 and 12 is used to provide signaling. This pattern of the framing format is shown in Fig. 14.14.

Table 14.2 shows the actual framing of a data stream in the D4 framing format. The framing bit pattern repeats every 12 frames. Twelve frames make up a superframe.

Once again, this requires each of the 24 channels to be robbed of the least significant bit (LCB) in the 6th and 12th frames. This has an impact on the data capacity that always limits the user to 56 Kbps of effective throughput (resulting in a net overhead of 192 Kbps given up to provide signaling).

TABLE 14.1

Framing Evolution

Framing	Use	Pattern/Use
D1	Voice or analog data	Alternating 1s & 0
D2	Voice or analog data	12-frame sequence Superframe
D1D	Voice or analog data	Upgrade compatibility for D1
D3	Voice or analog data	Superframe format and sequence bits
D4	Voice & data (digital)	Superframe
D5*	Voice & data (digital)	Superframe
ESF	Voice & data plus maintenance	Extended Superframe (ESF)

*D5 really doesn't exist as a standard. However, in the electronic switching systems this framing was termed a software version of the D4 format.

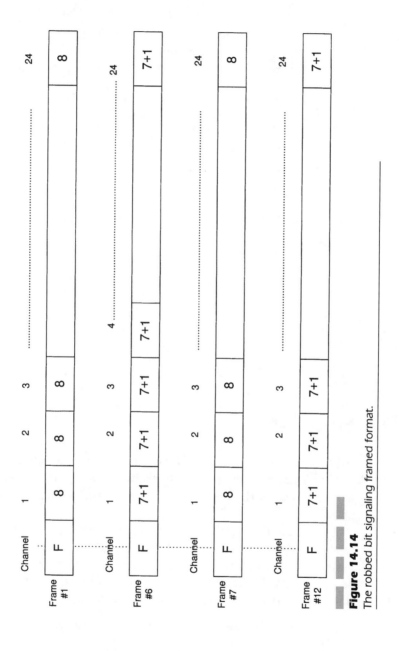

Figure 14.14
The robbed bit signaling framed format.

TABLE 14.2

Superframe Format, 6th and 12th Frames Used to Denote the Presence of Signaling Bits

Frame	Bit	Framing Bits		Bit Use in Each Time Slot		Signaling Bit Use Options	
		Term Frame F1	Sig Frame F1	Traffic (All Channels)	Sig	T	Signaling Channel
1	0	1	—	1–8	—		
2	193	—	0	1–8	—		
3	386	0	—	1–8	—		
4	579	—	0	1–8	—		
5	772	1	—	1–8	—		
6	965	—	1	1–7	8	—	A
7	1158	0	—	1–8	—		
8	1351	—	1	1–8	—		
9	1544	1	—	1–8	—		
10	1737	—	1	1–8	—		
11	1930	0	—	1–8	—		
12	2123	—	0	1–7	8	—	B

Bipolar

When encoding the voice and data streams onto a digital circuit (the T1), as mentioned earlier, up to 1,544,000 bits of information can be transmitted. The 1s show the presence of a voltage, and are represented by 3 V per pulse.

As a result, the bipolar concept was introduced. The first pulse (a 1) is represented by 3 V positive, followed by the next pulse, which is represented by 3 V negative. This would serve to bring the overall line voltage down to zero. Figure 14.15 is a representation of the voltage/pulses on the line. Bipolar encoding does the following things:

- Reduces the bandwidth necessary for transmission from 1.5 to 0.772 MHz.

- Allows the use of isolation transformers for inexpensive connections.

- Allows simplexed dc voltage to power intermediate repeaters. This can be as much as + and –130 volts.

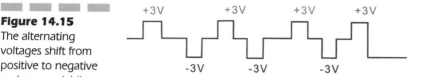

Figure 14.15
The alternating voltages shift from positive to negative and so on, yielding an average of 0 volts on the line.

Byte Synchronous

When transmitting on the T1, timing and synchronization come from the bits transmitted. Because each sample is made up of 8 bits, a byte is formed. The transmitters and receivers synchronize on the basis of the pulses in the byte format. If a string of 0s is transmitted in several frames, then synchronization is lost; the devices get amnesia as to where they were. Consequently, conventions on the transmission of the byte were established. These conventions are called the *1s density rule*. What this means is that in order to maintain synchronization, a certain amount of 1s must be present.

For voice communication this doesn't pose a problem, because the voice generates a continuous change in amplitude voltages which, when encoded into an 8-bit word (byte), displays the presence of voltage (1s). However, in data transmission, strings of 0s are possible (even probable) when refreshing or painting screens of information. Hence the 1s density rule comes into play.

Simply put, the 1s density rule states that in every 24 bits of information to be transmitted, there must be at least 3 pulses (1s), and that no more than 15 0s can be transmitted consecutively. A more stringent requirement set by AT&T in the implementation of their digital transmission was that in every 8 bits of information, at least one pulse (1) must be present.

Think of how this affects data transmission. If your data must have at least one pulse per byte, you have to change the way you deliver data to the line. A technique known as pulse stuffing is used to meet these conventions. The 8th bit in every byte was stuffed with a 1. This limited the data transmission rate to 56 Kbps, because 7 usable bits plus a stuff bit had to be transmitted. The end receiving equipment will receive 00000001 and not be able to tell if we sent 00000001 or if we sent 00000000 and the last bit was forced to a 1 for timing purposes. Because of this rule, we cannot trust the 8th bit, so we relegate ourselves to using 7 bits for data and the 8th bit for timing, all the time. Figure 14.16 is a representation of the 1s density rule and pulse stuffing.

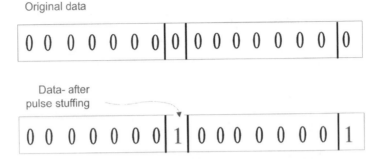

Figure 14.16
When strings of 0s must be transmitted, a 1 is stuffed into the eighth position (LSB). This yields a 56-Kbps throughput on the line.

To overcome this bit stuffing, yet meet the one's density requirement, a technique was developed. This is known as bipolar eight 0 substitution (B8ZS—pronounced *bates*). At the customer location, B8ZS is implemented in the channel service unit (CSU). As data bits are delivered to the CSU for transmission across the line, the CSU (a microprocessor-controlled device) reads the 8-bit format. Immediately recognizing that a string of 8/16/24 zeros will cause problems, it strips off the 8-bit byte and substitutes a fictitious byte.

- Eight 0s are stripped off the data stream and discarded.
- The CSU then inserts a substitute word (byte) of 00011011.

However, to let the receiving end know this is a substitute word and not real data, two violations to the bipolar or AMI convention are created. Remember that the bipolar convention states that alternating voltages will be used for the pulses.

In the fourth and seventh positions, violations will be created that act as flags to the receiver that something is wrong. See Fig. 14.17 for a graphic representation of this substitution.

Channelized versus Nonchannelized

Up to now, discussions have focused on using the T1 for 24-channel capacity at 64 Kbps per channel. For the average user, this was the norm. However, as newer uses for the T1 service became evident, it became possible to configure the capacity to meet the need:

- A single channel of 1.536 Mbps is possible for point-to-point video conferencing.

Figure 14.17
The receiving CSU, noting that two bipolar violations occurred in the fourth and seventh positions, recognized this as a fictitious word. It strips off the substitute word and replaces it with all 0s for delivery to the computer, terminal, printer, whatever. Thus the ability to transmit all usable data at a 64-Kbps clear channel becomes a reality.

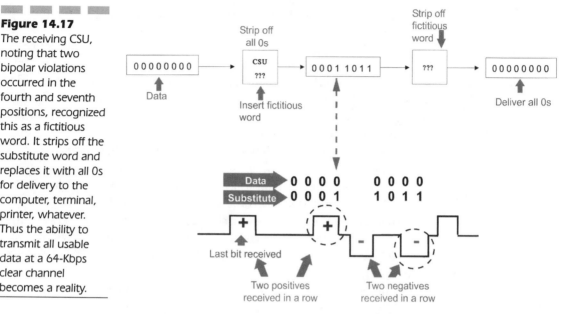

- High-speed data on a channel of 512 Kbps, plus 16 channels of lower-speed data and voice at 64 Kbps each, are possible needs.
- Any other mix of services might be required.

To accommodate the mix of needs, the end user (CPE) equipment can be superrate or subrate multiplexed. This is fine, except that you must ensure that the provider (LEC, AAC, and IEC) knows you are using the service for something other than the conventional 24 channels at 64 Kbps.

If confusion exists, the carrier or provider could do some reconfiguring in the network (at its end) that could literally disrupt service for the throughput being used. Consider what would happen if a 512-Kbps data channel were somehow rerouted through a telco office and then brought back and forced into a single 64-Kbps channel. This obviously would not work. Hence all parties should know what uses are being made of the transmission capacity.

Digital Capacities

One of the by-products of the T-carrier system was that the transmission rate employed became the standard building block for a multiplexing hierarchy. In high-speed digital transmission systems, a carrier combines many lower-speed signals into an aggregate signal for transport. To simplify this process and to hold the line on the costs of equipment, standard rates were defined.

Each of the transmission rates was assigned a number that identified the rate and configuration for the signal. Each time a higher speed is created, the lower speeds are combined and extra bits (bit stuffing) are added to come up to the signaling rate. These extra bits allow the equipment (multiplexers) to compensate for distances, clocking, and so forth, which cause transmission delay cycles.

The designations for the digital transmission are called digital signal X. Many times people will transpose or intersperse these designators. For example, a T1 is often referred to as a DS1 in the industry. To clarify this point, T1 is the first level of the T-carrier system. It is physical components such as the wires, plugs and jacks, repeaters, and so forth. These physical devices combine to create the T1. DS1 is the multiplexed digital signal, first level, carried inside the T-carrier. The DS1 is the electrical signal (the pulses) running on the T1.

The digital hierarchy for North America includes the following.

DS0

Although originally not a formally defined rate, the DS0 is a 64-Kbps signal that makes up the basis for the DS1. 24 DS0s combined produce the DS1. The standard 64-Kbps pulse-coded modulation (PCM) signal is the basis for all of the future networks employing digital signaling. Included in this is the integrated service digital network (ISDN).

DS1

The digital signal level 1 is a time-division multiplexed (TDM) pulse-code modulation aggregate of 1.544 Mbps, regardless of the medium used to carry the signal. The physical side of this capacity is called the T1. Comparative costing for a T1 between the East and West Coasts is approximately $19–20,000 per month.

DS1C

This is a digital signal equivalent to 2 DS1s, with extra (stuff) bits to conform to a signaling standard of 3.152 Mbps. Few (if any) of these circuit capacities are still in use today. In the early days of digital and data transmission, the 3 Mbps data rate was used to link mainframes together. The physical side of this circuit is called T1C.

DS2

This is a composite of four DS1s multiplexed together yielding an aggregate rate of 6.312 Mbps. The Bell system used a DS2 capability to deliver subscriber loop carrier (SLC-96) to customers. A total of 96 DS0s could be carried across the DS2. This service was used more by the LECs in their outside plant world. However, the 6 Mbps data rate is now becoming repopularized with the promise of ADSL technologies in the local loop. If we can get 6 to 8 Mbps downloads from the Internet on our local wires, the carriers (ILEC and CLEC) stand to gain from not having to replace their infrastructure. This is also called a T2 by the ILECs.

DS3

This 44.736-Mbps aggregate-multiplexed signal is equivalent to 28 DS1s or 672 DS0s. The T3 service is typically used by high-end data and voice customers who can afford the cost of this channel capacity. A price used for comparative purposes is from the East Coast to the West Coast, for which the T3 costs approximately $103,000 per month. One can see where smaller companies could not financially justify the use of the T3.

DS4/NA

This 139.264-Mbps aggregate-multiplexed signal is equivalent to 3 DS3s or 2016 DS0s. The high-end and long-haul telecommunications carriers use this size of channel capacity. It is unlikely that a DS4 would ever be installed at a customer location.

DS4

This 274.176-Mbps aggregate-multiplexed signal is equivalent to 6 DS3s or 4032 DS0s.

See Table 14.3 for a summary of these bandwidths. These rates are based on the ANSI T1.107 guidelines. From an international perspective, the

TABLE 14.3

ANSI T1.107 Rates

Designator	Capacity (Kbps)	Equiv. DS1	Equiv. DS0	Stuff Bits
DS0	64	—	1	
DS1	1544	1	24	8000
DS1C	3152	2	48	64,000
DS2	6312	4	96	136,000
DS3	44,736	28	672	1,504,000
DS34/NA	139,264	84	2016	5,056,000
DS4	274,176	168	4032	14,784,000

ITU has set a hierarchy of rates that differs from the North American standard. A comparison of these rates is shown in Table 14.4. As can be seen, differences between the North American standards and the international standards exist.

However, other rates are evolving for transmission capacities. The standard 64 Kbps is derived from using pulse-code modulation. In several vendor products, 64 Kbps for the transmission of voice and/or analog data is considered too much bandwidth to carry traditional voice or analog data. Therefore lower speed capacities are derived at 32 Kbps.

TABLE 14.4

International CCITT G.702

Designator	Capacity (bps)	Equiv. DS1	Equiv. DS0
E1	2,048,000	1	32
E2	8,448,000	4	128
E3	34,368,000	16	512
E4	139,264,000	64	2048
DS0	64,000	—	1
DS1	1,544,000	1	24
DS2	6,312,000	4	96
J1*	32,064,000		
DS3	44,736,000	28	672
J1*	97,728,000		
DS4	139,264,000	84	2016

*J designators are for Japan's network/also these capacities will be used for the BISDN rate H21.

Signaling

Signaling comes into play when dealing with voice and dial-up data services. Traditionally signaling is provided on a dial-up telephone line, across the talk path. This is referred to as in-band signaling. You might recall that the need to find bits to send between transmitter and receiver was accomplished in the 6th and 12th frames. Bit robbing, or stealing the 8th bit in each of the channels (1–24) in these two frames, allows for enough bits to signal between the transmit and receive ends. These ends can be customer premises equipment (CPE) to the central office for switched services, or CPE to CPE for PBX-PBX connections, and so on.

The most common form of signaling on a T1 line is four-wire E&M signaling of type I, II, or III. It would be safe to say that the easiest implementation, acceptable to all vendors and carriers alike, is four-wire E&M type I.

Signaling is used to tell the receiver where the call or route is destined. The signal is sent through switches along the route to a distant end.

The common types of signals are:

- On hook
- Off hook
- Dial tone
- Dialed digits
- Ringing cycle
- Busy tone

Four-wire E&M is used for tie lines between switches. Occasionally, other services are bundled onto a T1 circuit. These could include:

- Direct inward dialing (DID)
- Direct outward dialing (DOD)
- Two-way circuit
- Off-premises extension (OPX)
- Foreign exchange service (FX)

With these other services, the type of signaling might differ. DID/DOD on a PBX might use ground-start trunks, which requires a ground to be placed on the individual circuit to alert the central office that service is requested or there is some other change on the line. Regardless of the type signaling or the services used, signaling requires bits of information. There is a way to overcome the robbed-bit signaling that limits data to 56 Kbps. *Common-channel signaling,* or CCS, is a method to get the clear channel capability back.

Remember that there are constant demands on the use of T1 for clear channel capacity. The use of bit stuffing to conform to the 1s density rule for timing was overcome by using B8ZS. Now, to overcome bit robbing for signaling, CCS can apply.

If the 24th channel in the digital signal is dedicated, 23 clear channels can pass 64,000 bits of information. The choice is how much to give up. Table 14.5 is a decision table to help decide this process.

This common-channel signaling technique is also called the transparent mode for signaling. A single 64-Kbps channel (#24) is given up rather than 8 Kbps per channel [[×]] 24 (192 Kbps) for robbed bit. Carriers now use CCS7 (or SS7 as it is called in the United States) in their newer digital dial-up services under the auspices of Intelligent Networks and ISDN. In the T1 world, this means the primary rate interface (PRI) at 23B+D—simply stated, 23 bearer (B) channels at 64 Kbps clear channel, plus a data channel for signaling at 64 Kbps (D). The choices are not always obvious, but understanding the requirements helps to steer the decision better.

Clocking (Network Synchronization)

Any digital network synchronization between sender and receiver must be maintained. As the DS1 (the digital signal level 1 of 1.544 Mbps) is delivered to the network, it is likely that it will be multiplexed with other digital streams from many other users. All of these signals will then be transported as a single signal over high-capacity digital links (DS3 and above). Further, if the line is directly interfaced to a digital switching

TABLE 14.5

Decision Matrix for Deciding on Robbed Bit or Common Channel Signaling

If (Switched)	Then	O/H Given Up (Kbps)
Voice only	Robbed bit	192
Voice and analog data	Robbed bit	192
Voice, analog data, and 56 Kbps digital data	Robbed bit	192
Digital voice or digital data @64 Kbps	CCS	64
Digital data @64 Kbps or greater	CCS	64

device, then synchronization must be maintained between the customer's transmitter and the switch.

Synchronization is imperative in a digital transmission system. If the timing of arrival or transmission is off, then the information will be distorted. Regardless of whether voice, data, video, or image traffic is present, the presentation of a digital stream of 1s and 0s is contingent on a timed arrival between the two ends.

There are a number of ways to synchronize a digital network, but the issue must definitely be addressed. Some of the ways to address the synchronization deal with levels of synchronization.

The levels are:

- Bit
- Time slots for time-division multiplexing
- Frame

Bit Synchronization

As a digital stream of 1s and 0s is delivered to the line, the timing (or clocking) of the bit is important. The transmitter should be sending bits at the same rate the receiver can take them in. Any difference—faster or slower—could result in lost bits. Therefore the bits must occur at a fixed time interval. This is a bit technical, but the timing for the bits is set up in a window of 648 nanoseconds (that is, 648 billionths of a second) and the pulse must occur in $1/2$ of that window. Therefore, the pulse is 324 nanoseconds in duration. When we only have a 3-volt pulse, and it occurs for only 324 ns, one can see why the timing is crucial. A slight delay and the timing is off by a lot. If the timing is off enough, the pulse can wind up in the wrong time slot. This is called "jitter."

Time Slot

Whenever several links are connected or routed through a network processor, switching system, or end mode, the potential for lost bits or degradation of the link increases exponentially. Using the pulse-coded modulation technique, 8 bits are encoded from each sample of information. These 8 bits are then assembled and placed into a time slot. As signals and links are processed through a network, it is the 8-bit pattern that is routed from time slot to time slot. Should a slippage or mismatch occur, multiple streams of information will be lost.

Frame Synchronization

After the data stream of 192 bits of information is assembled (8 bits × 24 channels), an extra overhead bit is added to let both transmitter and receiver know the boundaries of the frame (think of it as a start/stop bit sequence). When dealing with a T1, the bit sequencing is easier to derive. However, at multiplexing schemes above the T1 rate, additional bits are inserted in the data stream to maintain a constant clocking reference. The use of these overhead (or stuff bits) brings both transmitter and receiver up to a common signaling speed to maintain frame synchronization.

Potential Synchronization Problems

When a digital system is scheduled to receive a bit, it expects to do just that. However, clocking or timing differences between the transmitter and receiver can exist. Therefore, while the receiver is expecting a bit that the transmitter hasn't sent, a slip occurs. There will most likely be slips present because of multiple factors in any network. These can result from the two clocks at the ends being off or from problems that can occur along the link.

Along the link, problems can be accommodated. The use of pulse stuffing helps, but other methods also can help. Each device along the link has a buffer capability. This buffer creates a simple means of maintaining synchronization. Pulse stuffing is done independently for each multiplexer along the way, enhancing overall reliability of the network. However, pulse stuffing has its negatives too. The overhead at each multiplexer is basically a penalty. Further, at both ends the location and timing of each stuff bit must be determined, then signaled to the receiver to enable it to locate and remove the stuff bits. When de-stuffing occurs, a timing problem known as *jitter* can cause degradation of the signal. When passing through multiple switching sites, the signal must be de-stuffed from a received signal, then re-stuffed to a newly transmitted signal. This is expensive in both equipment and overhead.

Obviously, when slippage occurs or if a problem exists in the network buffers, the retransmission of a frame or frames of information will be required. For voice, this isn't too bad, but for data transmission it can result in errors that can render the data unusable. The result will be retransmission requests (NAKS) that will affect the throughput of the link and potentially increase the burden on other systems to detect and correct the errors.

Performance Issues

Once a decision is made to use digital transmission capabilities and to consolidate 24 or more voice, data, or video circuits on a single T1, the performance of the T1 becomes an issue. The placing of all the eggs in a single basket is an issue to be dealt with.

The first users of this technology were risk takers. If 24 analog voice and data lines are being used and 1 fails, 23 are still working. However, when 24 lines are digitally encoded on a single circuit and 1 fails, they all fail. This is a concern to many users. Further, it should be noted that failures do occur. A T1 is based on the four-wire circuit in an error-prone network. Equipment failures, cable cuts, flooded cable vaults and the like all contribute to circuit failures. If total circuit failures were the only problem, the risks would be minimized. But cases can occur that cause distortion, high bit-error rates, timing slips, and loss of framing sequence (to name a few), which can impair or disrupt transmission. The thought of all these situations can leave the potential user's knees weak.

A further statistic that causes concern is that 90 percent of all errors/failures will likely occur at the local loop (the last mile from central office to customer premises). This is not to imply that the local exchange carriers (LECs) are doing a poor job of provisioning these circuits. It is merely a statement of fact, because the local loop is exposed to more of the external problems mentioned above.

Considering average performances, the use of T1 technology is based on 7 days per week × 24 hours per day. Table 14.6 is a summary of total hours and percentage comparisons at availability times. With these statistics you can see the impact of a few percentage points in terms of downtime and your exposure to lost time.

TABLE 14.6

Total Hours
Available
@24 × 30 = 720

Performance (Percent)	Hours Down/Month	Hours of Yearly Downtime
99.5	4	48
99.0	8	96
98.5	11	132
98.0	15	180
97.0	18	216
95.0	36	432
90.0	72	864

Although these numbers are dramatic, the actual performance is completely different on the basis of the problem, customer, location, and so on. Situations have been recorded where services have been out for six to seven consecutive days because of variable circumstances. However, such extreme outages are rare.

To ensure availability of the circuit and the call-carrying capacity of the channels, carriers, equipment vendors, and the like have all adopted to a new standard. This new standard is known as *extended superframe format* or ESF. Before delving into ESF, however, an understanding of the older format is appropriate.

D3/D4 Framing

First, the D3 and D4 framed formats were designed to format the channelized information of a T1. Because D3 was a voice-only (analog data input) and fixed format for input, the use of T1 was designed around tie lines, WATS lines, and dial-up analog data. The evolution to the D4 frame was to accommodate voice, data, and image. Both of these formats had their limitations. The superframe (SF) format used provisions to allow for signaling (robbed bit) and 1s density for timing (stuff bits). These two capabilities, combined with the framing bits, were constantly taking channel capacity from the T1. All of the user information was designed around a 56-Kbps data stream. The robbed or stuffed bits were designed around control. Yet, when looking at the potential for maintenance and diagnostic capabilities, no spare capacity exists for the data channels to do this, short of giving up additional overhead. Given the pricing structure and the other forms of overhead, this option was not acceptable.

Maintenance Issues

Whenever a user of T1 with D3/D4 framing experienced problems on the line, the sequence of events became elongated and disruptive. Things would occur in the following order:

- The user would experience some hits on the line (or downtime) that affected data throughput. These hits would be in the form of loss of framing, loss of synchronization, bipolar violations, and such. Regardless of the problem, disruptions on the circuit for seconds or longer occurred. The user could see lights flashing on the channel service unit (CSU).

▪ The user would notify the appropriate carrier (either local exchange or interexchange carrier) of the problem. Further discussions would evolve regarding the errors and so on.

▪ To test the line, the carrier would ask the customer to get all users off the circuit and give the line to the carrier for testing. In many cases, this might be an immediate situation; in other cases, it was a scheduled event. Whether or not this was scheduled for a later time, the issue at hand was the need to give up 24 channels of voice, data, or image for a couple of hours for testing.

▪ The carrier would then test the line. A loopback arrangement would be used in some cases. The carrier would loop the circuit from the exchange office to the CSU or demarcation point. A pattern of 1s and 0s was transmitted to the customer premises and immediately shipped back around to the originating office. As the data was shipped out, a comparison was made with what was received back. If any errors were evident, the necessary fixes could be either done on the spot or scheduled.

▪ However, often the carrier conducted the tests and detected no problems. A technician might then be dispatched to the customer's premises to provide further testing. Equipment using test patterns (*quasi-random signal source* or QRSS) was used. Again, the possibility existed that no errors would be detected.

▪ The carrier would then give the circuit back to the customer. When challenged as to the nature of the problem, the carrier's response was *no trouble found* (NTF). Obviously, the customer had problems accepting this answer. The feeling that the carrier was either hiding something or just didn't know what was going on prevailed.

▪ Once the circuit was placed back in service (two hours later), the customer might see errors again. At this point the cycle would start all over again, usually with the same end result—NTF. Consequently, the working relationships and confidence levels between the customer and provider would become very strained.

Error Detection

The easiest way to describe errors on a T1 is as the process by which any logical 1 is changed to a 0 or a logical 0 is changed to a 1. These logic errors are the source of the corrupted data on the line. The errors can be caused by noise from a spike of electrical energy on the line (i.e., lightning) or induced by electromechanical interference (i.e., a large motor in a building). Further types of errors can be caused by cable losses, flooded cables, rodent damage, or cross talk on the line, to name a few. Equipment can introduce errors from the digital repeaters, multiplexers, or DACS on the line. Unfortunately, the process of detecting these errors can be tenuous. Any pulse (a 1) or absence of a pulse (a 0) transmitted along the line is subject to noise that can change the actual data. When traveling through equipment such as multiplexers, repeaters, DACS, CSU, DSU, and the like, the pulses are accepted as valid. The equipment therefore cleans up the valid pulses, but because the electronic systems can't differentiate between valid and erroneous pulses, it cleans up the entire data stream. Consequently, the receiver at the distant end accepts the clean, well-timed input as correct information. Hence errors are introduced and undetected.

You might think that an error on the line could be easily detected by the equipment because certain format problems might also be evident. For example, an error (an extra pulse on the line) might violate the bipolar format convention, or a dropped pulse might violate the 1s density rule. Clearly these would be easy to detect. However, if the signal must pass through equipment along the way, that equipment might clean up the errors. The equipment does have the ability to log an error on the inbound side, then reformat the signal on the outbound side and make it appear to be valid.

To further understand this process, look at both errors—where a pulse is inserted and where a pulse is dropped.

Errors of Omission/Commission

The errors are either dropped pulses (omitted) or inserted pulses (committed). An error where a pulse is omitted should in turn create a bipolar violation. If the number of pulses omitted is even, the bipolar violation won't be created. Only an odd number of errors causes this violation. Figure 14.18 is an example of how this works with an even number of removed pulses.

A error occurring with an odd number of pulses being dropped would be detected. Figure 14.19 is a representation where three pulses are dropped, causing a bipolar error.

Figure 14.18
An even number of pulses being dropped will not be detected from a bipolar convention.

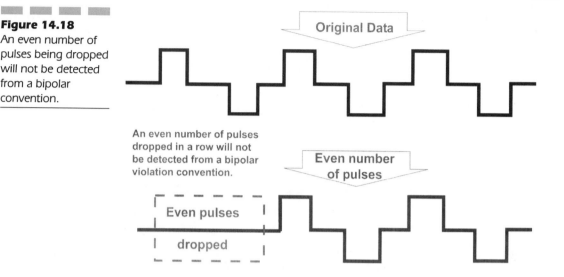

An even number of pulses dropped in a row will not be detected from a bipolar violation convention.

When pulses are introduced (committed), a similar problem occurs in detecting the errors. For example, when an odd number of pulses is introduced, a bipolar error will occur, but when an even number of errors occurs, a bipolar violation might be created. This is not guaranteed, however. Figure 14.20 shows the commission of an odd number of pulses, which will create a bipolar violation.

However, as mentioned earlier, when an erroneous signal passes through equipment along the link, the signal is cleared up and put into a correct format state. Figure 14.21 is a sample of how this would work.

Figure 14.19
Because an odd number of pulses changes, a bipolar error occurs. This error can be detected.

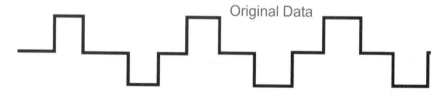

Since an odd number of pulses changes, a bi-polar error occurs. This error can be detected.

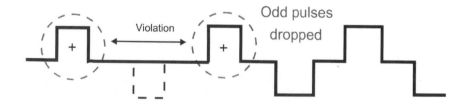

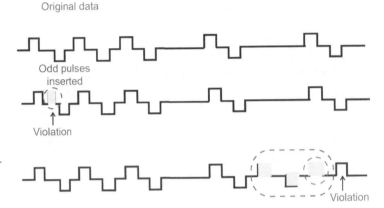

Figure 14.20
When an odd number of pulses is inserted, a bipolar violation will occur, either with the first pulse inserted or with the first valid pulse after the commission of the pulse.

The detection of errors, therefore, becomes more difficult in the transmitted data stream due to these problems. A format error (bipolar violation) implies that a logical error has occurred; thus the data is erroneous.

ESF—A Step to Correct the Problem

In the early 1980s, AT&T suggested the implementation of the extended superframe format as a means to provide nondisruptive error detection and perform nonservice affecting diagnostics on T1 circuits. This represented a step toward correcting correct the problem of lost customer confidence and the disruptive performance testing. As service providers, the

Figure 14.21
As the data with errors and bipolar violations passes through the MUX, CSU, DACS, DSU, and so on, the electronics cleans up the signal so that it comes out properly formatted.

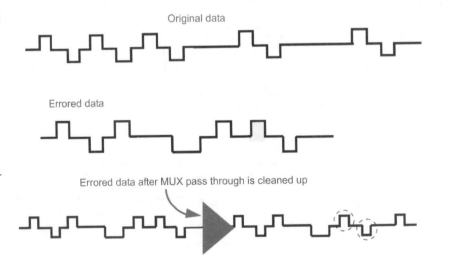

interexchange carriers write tariffs to provide service within certain guidelines of availability. ESF helps to address this problem.

Essentially, because overhead on the circuit is at issue, customers are unwilling to give up more channel capacity for maintenance and diagnostics. ESF uses the 8 Kbps of overhead normally assigned for the framing format and therefore does not affect the data bandwidth.

What Does ESF Do?

Extended superframe format, as its name implies, extends the superframe from 12 consecutive frames of information to 24 frames of information. The 8 Kbps of overhead, originally used strictly for framing, is now subdivided into three functions. Table 14.7 summarizes how the overhead is shared to support ESF.

ESF takes advantage of newer technologies to use the 8 Kbps of framing overhead and uses only 6 bits (2 Kbps) for framing synchronization. Six additional bits are used for error detection by means of a cyclic redundancy check 6 (CRC-6). This leaves 12 of the framing bits for a facility data link communications channel (4 Kbps). Table 14.8 is a comparison framing bit use for SF and the newer ESF.

Using this shared facility, the carrier has the option of performing the three functions necessary to support the customer needs:

■ Framing synchronization

■ Error detection

■ In-service monitoring and diagnostics

Framing

As already stated, the fundamental purpose of the bit pattern is framing synchronization. Framing is important because its loss can impact per-

TABLE 14.7

The Original 8 Kbps of Overhead in SF Format Is Now Subdivided to Provide Three Separate Functions

Original SF Format	Revised ESF Format
8 Kbps for framing *only*	2 Kbps for framing
	2 Kbps for error detection (CRC)
	4 Kbps for facility data link

TABLE 14.8

Frame Bit
Definitions

Frame	Bit	FPS	FDL	CRC
1	0	—	m	—
2	193	—	—	C1
3	386	—	m	—
4	579	0	—	—
5	772	—	m	—
6	963	—	—	C2
7	1158	—	m	—
8	1351	0	—	—
9	1544	—	m	—
10	1737	—	—	C3
11	1930	—	m	—
12	2123	1	—	—
13	2316	—	m	—
14	2509	—	—	C4
15	2702	—	m	—
16	2895	0	—	—
17	3088	—	m	—
18	3281	—	—	C5
19	3474	—	m	—
20	3667	1	—	—
21	3860	—	m	—
22	4053	—	—	C6
23	4246	—	m	—
24	4439	1	—	—

formance through incorrect synchronization. (Channel 1 could wind up connected to channel X [2–24] if the framing is incorrect.) Errors might cause a loss of frame and disrupt the circuit, and far-end and/or intermediate equipment could be thrown into a loss-of-frame condition and thereby a loss of data. While the devices along the circuit are reframing data, throughput is disrupted for a period of time. Depending on the equipment in use, this loss could be quite substantial (approximately 200 ms). Newer equipment performs a reframing operation in less time (approximately 10–20 ms). This helps to improve performance. With ESF using only 1 of 4 bits, the delay could increase significantly unless the proper equipment is purchased.

CRC-6

Cyclic redundancy checking is designed to detect errors that occur on the line. Because the network is error prone, it is a foregone conclusion that errors will occur. Using CRC-6, an entire ESF (24 frames of information or 4632 bits) is checked for accuracy. This check is a fictitious number created by the CSU as it performs a mathematical computation on the 4632 data bits. When the math is calculated using a prime number polynomial, a 6-bit pattern is the end result. In the next ESF (24 frames), the result of this mathematics is transmitted to the distant end.

At the distant end, the receiving CSU calculates exactly the same mathematical operations on the ESF data and then computes an answer. When the next ESF is received, the result of the first calculation is contained in the CRC bits (6 bits) of the overhead. When the two results are compared, the answers should be the same. If, however, they differ, an error has occurred. The use of CRC-6 allows for an error detection efficiency of 98.4 percent, or, simply stated, detection of 63 of 64 errors and 99.9995 percent of bursts of errors less than 16.

The Facility Data Link

The 4 Kbps of overhead set aside for facility data link control is a synchronous communication channel. This channel can serve multiple purposes, one of which is the exchange of information between equipment devices along the circuit. AT&T publication 54016 specifies a set of standard message formats for communicating across the data link to the storage devices. The standard conforms to a basic X.25 (BX.25) link procedure and uses an AT&T standard known as telemetry asynchronous block serial protocol (TABS). The carrier or end user can communicate with the remote equipment (CSU) using a maintenance message format. Some of the maintenance messages allowed under the AT&T technical reference (54016) include:

- *Send 1-hour performance data reports.* This includes the existing status of the link; time since the last check (up to 24 hours); errored seconds; and failed seconds in the present 15-minute cycle, the overall 24-hour cycle, and the past four 15-minute cycle performances.

- *Send 24-hour errored seconds (ES) performance data.* The CSU will send specific error events logged within the past 24 hours in 15-minute increments.

- *Send 24-hour failed seconds (FS) performance data.* The CSU sends the 24-hour historical information of errored events logged in 15-minute increments.

- *Reset registers.* Empty the buffers and start counting errors all over again.

- *Send errored ESF.* The CSU will send the current count of errors accumulated (up to a maximum of 65,535).

- *Reset ESF register.* Empty the buffer and start at zero.

Benefits of ESF

The reasons to use ESF are to increase circuit availability and improve performance. Once again, putting 24 channels (circuits) on one T1 puts the user at risk if performance is poor. Just about all of the major carriers (interexchange) have implemented ESF. Because all of the interexchange carriers internetwork—that is, they connect to each other—the use of the AT&T standard (54016) is the norm.

Vendors and carriers can see the improvements brought about by ESF. When a burst of errors occurs, the carrier need only access the CSU from the facility data link and query for status. The problem of old, the "no trouble found" situation, tends to fade into the background because the CSU has buffers to collect the errors for later retrieval. Further, with the facility data link, the carrier can monitor real customer data traveling across the line nondisruptively. Errors occurring on the line can be seen and diagnosed in real time. This all leads to a better track record. In fact, many of the carriers will contractually guarantee circuit availability between 98.4 and 99.5 percent because of the proactive capability gathered in ESF.

This seemed to solve the problem for users and carriers alike. Despite the passive collection capability and the on-line diagnostics, the parties all began to create a working rather than an adversarial role.

Problems with ESF

Unfortunately, one player that is not mentioned in this scenario and yet is an essential ingredient in the total circuit is the local exchange carrier

(LEC). An integral piece of the circuit is the "last mile," or the circuit running from the LEC to the customer premises. ESF was designed to function as a part of the customer premises equipment (CPE). However, the LEC by mandate cannot poll the CSU that stores the error information. Under the guidelines of the modified final judgment (MFJ), or what we know as divestiture, the LEC cannot go beyond a clearly defined demarcation point. This demarcation is the line termination in the RJ-48, RJ-68, or smart jack—where the LEC stops and customer takes over. Some of these limitations are falling aside with the deregulation of the telecommunications industry and the Telecommunications Act of 1996. Yet the local utilities commissions across North America have still held some of the barriers intact. Over time, the change will be finalized, but it is a phased project.

This point, therefore, excludes the LEC from benefits derived from ESF as defined by the AT&T technical reference (54016). To overcome this problem, the American National Standards Institute (ANSI) came up with a proactive approach to ESF: ANSI's standard T1.403 uses the CSU to monitor performance on the T1 line, then puts together an activity report every second and transmits this ESF report back out across the network. This solves the problem but leaves the door open to two sets of standards. As a result, all North American carriers have been migrating to the ANSI T1.403 specification and will have their networks converted by the year 2000.

15

The Open Systems Interconnect Model (OSI)

The three-letter acronym *OSI* (open systems interconnect) appears in several places in this book. No one concept in this industry has been more misunderstood, but that is what happens when several different operations are merged into one.

We first had a telephone industry. Later we started to call it a telecommunications network—especially as newer services were being supported on this one infrastructure. The merger started when we introduced data processing into the telecommunications industry. Now we have the convergence network. The data processing industry had already become entrenched in its own jargon and concepts. As long as the data processing function was localized to the mainframe and all transactions were performed in the computer room, no problem existed. However, like all things, this industry evolved. Soon the movement was from the computer room to the desktop, where users wanted connectivity to the mainframe. This was accomplished through the use of specialized wiring systems. Nothing is new here, other than that these special wires were very expensive. The wires had to carry the data from the terminal to the computer and also do the reverse. This was accomplished with the use of traditional communications techniques—by converting the data into electrical energy and carrying it down the wire—and the data was represented in digital form without notice.

Then it happened: not only did we need connectivity from the desktop to the computer room, we also needed it from a desktop across town or across the country. Now the process was going to change the way we did business. The data communications industry became the next battleground. The data processing people all thought that data processing and data communications should be similar enough so that they would control the deployment. The telephony people felt that this computer stuff was a pain anyway, so they were content to let it reside in the data processing arena.

But some problems existed with this whole situation. To send the data across town or across the entire country, it would be necessary to use a device called a modem. Recall that we talked about this in an earlier chapter, but there the modem was used to make the data look like voice. As long as the data looked like voice, the data folks wanted to send it back to the telephony department, especially when they didn't understand or care to understand how this analog network functioned.

So the marriage of the three techniques began back in the late 1950s and into the 1960s. The evolution was slow because the telephone monopoly controlled the delivery of the transport system. The data processors were looking for a way to make the network digital to avoid the digital-to-analog and the analog-to-digital conversions. This seemed like too many steps, and as each step was taken there was a risk of introducing errors into the data stream.

Things rolled along nicely, but the data processing evolution kept going from the mainframe environment to one leaning toward distributed computing. Organizations were cropping up that would offer smaller-scale computers (midi and mini) that could handle very specific functions and off-load this process from the mainframe. Once again, this started the ball rolling. Many of the computers in use at the time were not compatible with each other. This led to a new set of predicaments. As the smaller computer companies were competing with the mainframe world, two major players were doing battle—IBM and DEC. Each had its own hardware and software platforms. Now organizations began to move computers from different manufacturers into their buildings. No problem: these were very specific application processors, so there need be no conflict. That was true, but a user who needed access to the mainframe and the specialty machine now required two separate sets of wires run to the desk and two different terminals on top of the desk. This is reminiscent of the situation portrayed the historical section of this book, where the interconnection of various telephone companies would not work and users needed two or three phones on their desks to intercommunicate. The battle was being fought over compatibility rather than technology. The problems were just starting.

First, let's take another look at this from the perspective of the managers of the computing systems. In Fig. 15.1, the organization had a mainframe that was based on architecture created by the manufacturer—in this case, IBM. The hierarchy of this architecture allowed various connection arrangements based on the device's level on the planes of the architecture. Figure 15.1 shows how the transfer of information among and between users would take place. Things worked pretty well because IBM had everything in place: they delivered the hardware, the software, the connections, and the talent to make everything work together. This was all based on a proprietary and closed architecture called systems network architecture (SNA). Figure 15.2 shows the architecture in its stack. This is a seven-layered architecture that IBM created to make everything work in harmony. Using this structure, IBM would fix something that did not work—for a fee, of course.

Later, as the specialty machines appeared, a separate approach was used by the provider. This provider was created as a result of spin-offs from IBM. The founders of this particular company (DEC) were upset with the strategy and direction that IBM used. So they created their own offshoot

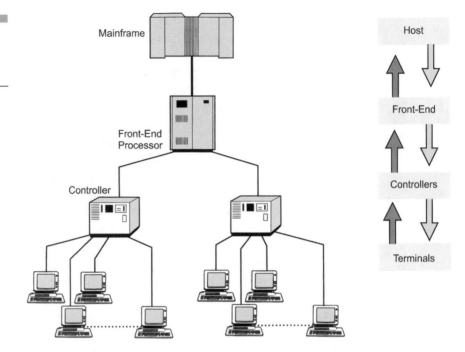

Figure 15.1
The hierarchy in a
mainframe
environment.

Figure 15.2
The SNA architecture.

Transaction Services
Presentation Services
Data Flow Control
Transmission Control
Path Control
Data Link Control
Physical Control

company to offer lower-end machines to the customer, but at more reasonable prices and in a more open environment. DEC, of course, being an offshoot of IBM, used a lot of similar approaches. Let's face it: when an engineer or manager leaves one place of employment, it's tough to leave behind all learned information, so the cross-pollination of ideas and goals took root. In the design and rollout of their products, DEC introduced their own closed and proprietary architecture (shown in Fig. 15.3), called digital network architecture (DNA), that just happened to have seven layers. These seven layers were based on DEC hardware, software, connections, and talent to make everything work in harmony. Sounds familiar, doesn't it?

Progressive end users who had embarked on a multivendor approach in meeting their computing needs would be faced with two separate architectures to work with. The third aspect of this is that many users wanted to be connected to both systems. Ultimately the end user wanted to be able to access these two platforms, hardware, software, applications shared data sets, or whatever from a single device. Moreover, many of the specialty applications that the users wanted required the data that was stored in the mainframe. This meant that the users were re-keying the data from printed reports that came from the mainframe. This left a new set of problems:

- The timeliness of the data was suspect. Because it had to be re-keyed, it could be obsolete before it was ever entered.

- The accuracy of the information was questionable because data entry errors are not uncommon.

- The cost of the information was now more expensive because the input was done at least twice.

- The frequency of the input was becoming exponential because as soon as the data was printed out from the mainframe, it could be obsolete as a result of changes made by other users. Unfortunately, this was changed in the mainframe, but the update in the specialty machine was not automatic.

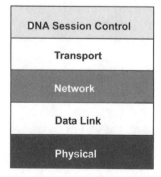

Figure 15.3
DEC's DNA architecture.

DNA

| DNA Session Control |
| Transport |
| Network |
| Data Link |
| Physical |

Therefore, the scenario became almost comical. Here's an illustration to make this concept easier to understand. This is a hypothetical situation—all names of people and companies are used only for the purposes of this example. Do not try this yourself, as it was tested by professionals who are skilled at ducking the issues.

Let's use an example of an organization that has two suppliers of computing platforms. One is an ABC mainframe, the other is a DEF BAX. The company has users attached to both machines, and has received complaints about the connections. Users are getting confused when they try to log on for sessions on both machines. The MIS manager decides to go to the vendors and ask whether there is a way to provide transparent connectivity. The dialogue goes something like this:

IS Mgr.:   Hello, ABC? I've got this ABC mainframe where the users are attached using a 333X terminal. I also have a DEF BAX system and my users are using a VL2000/3000 type terminal to connect to the BAX.

ABC: Yes, what can we do for you?

IS Mgr.:   Well, what I'm trying to do is let my BAX users log onto the mainframe and view or manipulate the data on the ABC host. When the need dictates, I'd also like for my users to pull the data from the host and move it over to the BAX where the user can modify or delete the data. Then, when they are done, they should be able to save it back to the original data set on the host.

ABC: So you want your users to access the data on the host, use it, modify it, or whatever. Then you want that data saved on the host so that others can see it and use it. Is that correct?

IS Mgr.:   That's it exactly. But I also want the user to be able to use the same keystrokes on their terminal that my 333X terminals use.

ABC: OK! There's no problem, we have the perfect solution for you.

IS Mgr.:   You do! That's great! this is exactly what I was hoping to hear. What do I have to do?

ABC: It's simple. All you have to do is sell all those DEF terminals and the BAX, then buy all ABC terminals and hosts. You'll have full transparency and everyone will enter the exact same keystrokes.

That's not exactly what the IS manager was hoping to hear. So off to DEF. The dialogue will be somewhat the same.

IS Mgr.:   Hello, DEF? I have these two systems, one a BAX using VL2000/3000 terminals serving a special application. I also have this ABC mainframe that is serving some of my back office functions like accounting, billing, inventory control and so on. What I'd like to do is provide seamless and transparent connectivity between the two computers and let any user access any data from the terminals at their desk. This can be a VL2000/3000 or a 333X. Do you have a way of handling this for me?

DEF: You have a BAX and an AB who?

IS Mgr.: An ABC host.

DEF: Oh, this is easy. We have these requests all the time and we help our customers through this arrangement very easily. As a matter of fact, we can do it all for you.

IS Mgr.: You can? This is great. I knew you folks would come through for me. What do I have to do?

DEF: It's really quite simple. Sell all those other devices, buy more BAXs and VL2000/3000s. It will work transparently and everyone will love it!

Once again the dilemma—what was the IS manager to do? Fortunately others had already gone through this problem and a new supplier had emerged: the third-party manufacturers of "black boxes."* The box was not black; the name referred to the fact that no one really knew what it did. It was a protocol converter. All the user knew was that if somebody on one system typed the normal way at a terminal, the commands and data typed went into the black box as *DEF* and came out the other side as *ABC*, and vice versa. These protocol converters were not inexpensive, but they were certainly a more attractive option than the others that the IS manager got. So, the IS manager bought a third-party converter and attached it to both machines. With a connection to each machine, the converter sat in the middle. This is shown in Fig. 15.4, where the two hosts are connected to the converter. All is solved. Or is it?

After the installation, things appear to be working fairly well. Then one day they start to go awry. The IS manager calls ABC and asks if they anything different has been done with the system or code. The converter worked fine yesterday, but today nothing seems to be functioning. ABC, of course, responds that it has done nothing. It would, therefore, suggest it must be those other folks. Now, the IS manager goes to DEF and gets the exact same response. So, it must be those others. Finally, off to the third-party supplier, who of course suggests that it has done nothing different so it must be those others. The result of all this is shown in Fig. 15.5. This is called the finger-pointing routine. Point and hide is really what happened.

---

*The reference here is generic to a device that converts the information from one system and format to another. The use of the black box in no way refers to The Black Box Company, an organization in Pittsburgh, PA. These are highly skilled and qualified personnel that do an excellent job for their customers.

Figure 15.4
A black box was used
to create
transparency.

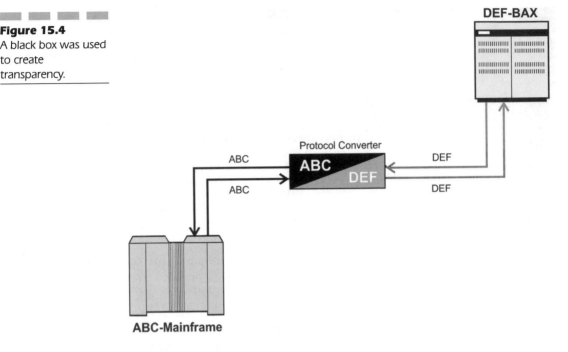

Even if nothing went wrong, things might not be exactly the same as what the end user accesses at the various machines through protocol converters. The follow-up to this is that the code sets and command lines might be different. Specifically, in a 333X world, the end user might go through the following keystrokes to perform a function:

- To repaint a screen with a blank accounting form, press the CMD key, which is located at the upper left corner of the keyboard, using the left index finger.

- At the same time, press the F9 key using the right index finger.

- Voilà! The new screen appears with the blank form ready to go.

To perform the same function from a VL2000/3000 device in the same host, the user might do the following:

- Press the ESC key, located in the upper left corner of the keyboard, with the left index finger. Because there is no CMD key, a remapping of the keyboard must take place.

- At the same time, press the F9 key with the right index finger.

- At the same time, press the shift key with the left thumb. The shift key is on the lower left side of the keyboard.

Figure 15.5
When a problem occurred, the finger-pointing started, leaving the IS manager at a loss.

- At the same time, press the ALT key with the right thumb. The ALT is located at the lower right side of the keyboard.

- At the same time, press the space bar with your left big toe. The space bar is located at the bottom center of the keyboard.

- Voilà! You now have the screen repainted with the blank form ready to be filled in! Figure 15.6 shows what this might look like.

Obviously, by now you have recognized that this is strictly in jest. There are past ills in the industry, but they have not been this bad. The issue was that transparency in connectivity, communications, and use of the data was not to be had easily. The result was that the users put pressure on their coordinating committees in the industry to come up with a resolution. In 1978, the International Organization for Standards (ISO) was asked to come up with a solution that would allow the transparent communications and data transfer between and among systems regardless of manufacturer. This was originally thought of as the task of the data communications devices. Thus, a new committee was formed to evaluate how this might happen. Everyone in the industry held their breath. What form of solution would be achieved? Because this was a committee made up of users, vendors, and manufacturers of the systems as well as standards members, what could be expected? The result was to come several years later, with what we now have come to know as the ISO's OSI reference model. No, they did not just reverse the letters of the organization; this stands for the open systems

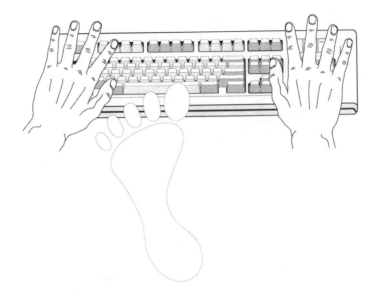

Figure 15.6
Using protocol
converters was not
that transparent.

interconnect reference model. It is a seven-layer architecture, just like the other two architectures that already existed (SNA and DNA)—two of the member companies of the committee were IBM and DEC. This seven-layered architecture is shown in Fig. 15.7. The seven layers will be covered in more detail later, but are listed here:

- *The application layer* sits at the very top of the model. This is the direct interface with the application used when requesting a service. This layer provides communications services to the end user.

- *The presentation layer* sits below the application layer and provides a service to the application layer. This is where formatting, code conversions, data representation, compression, and encryption are handled for the application.

- *The session layer*, located under the presentation layer, is responsible for the establishment and maintenance of connections to a process between two different users or systems. The control of the direction of data transfer is handled here.

- *The transport layer* is right below the session layer. The transport layer controls the connection for error recovery and flow control. It is responsible for assuring error-free data delivery end to end in a cost-efficient manner.

The next three layers deal with the network-specific issues of the communications process. Whereas the upper four layers deal with the

Figure 15.7
The OSI reference model.

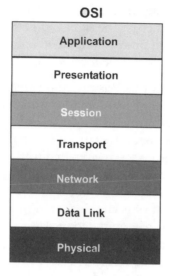

end-to-end communications and data transfer, the bottom three layers are concerned with the node-to-node connection. This portion of the network may be provided by a single entity (such as a carrier, PTT, or third-party communications company).

■ *The network layer* sits just under the transport layer and is responsible for the switching and routing of the connection. Also, this layer is responsible for taking the data that is to be shipped from node to node, not end user to end user, and breaking it into smaller pieces to accommodate the transmission system. Congestion control, alternate routing, and other such connectivity services are handled here. The network layer provides establishment of the connection, transfer of the data, releasing the connection, and in some cases the transfer of data in a connectionless service.

■ *The data link layer* is responsible for the delivery of the information to the medium. It will create frames of information, if required, and send the frames along to the wires. Additionally, this layer is responsible for the reliability of the information and error checking on a node-to-node basis.

■ *The physical layer* that sits below the data link layer is the actual electrical or mechanical interface to the physical medium. In this layer we have the physical pieces and the necessary components based on the dependency of the medium. This layer is responsible for transmitting the bits onto the guided medium (wires). It specifies various physical portions such as the voltage levels, the pins to use in placing the voltage onto the link, and whether the signal will be electrical or photonic (fiber) or modulated onto a nonguided (radio)-based service. We will see how all of these pieces tie together in later text.

Even with all of the effort that was placed on creating this set of protocols and the entire architecture, many industry and end user groups were somewhat frustrated. They wanted a single solution to provide the transparent communications between systems of different manufacturers, and what they got was still another seven-layer architecture. They were amazed that the standards committees would do such a thing. In reality, this was the best way to provide the solutions, but it took a while to sink in. The seven layers are independent of each other; even though they provide services to the upper layer and rely on the lower layer, there is still some freedom. This is important because in the event that any protocol suite in one of the seven layers needs to be changed, the rest of the architecture will remain intact. A single solution would not have done that.

Figure 15.8 compares the three seven-layer architectures—the OSI as a base, the SNA, and the DNA—to see how they line up against each other. In fairness to both DEC and IBM, their architectures were in place before the OSI model was created; therefore they cannot be expected to be the same. DEC, over the years, has changed the architecture of its DECNet protocols to be more consistent with the OSI. However, IBM has basically done nothing over the past 18 to 20 years to change. IBM felt that it already

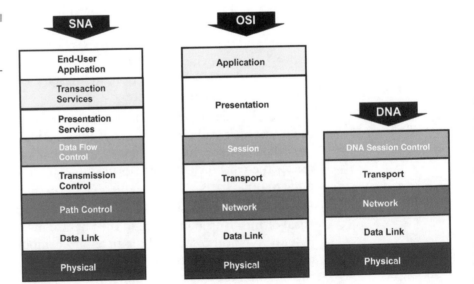

Figure 15.8
Comparing OSI, DECNet, and SNA.

had the single largest and most widely accepted architecture installed at all of the Fortune 1000 companies, so therefore there was no need to change what already worked. So, even though the architecture was developed around 1978, it was not totally accepted until 1983, and there have been very few implementations of it.

Part of the reason was the cost of moving over to a new architecture. The U.S. Government wanted a standard applied to all of their purchases in the past. In an effort to initiate OSI acceptance and implementation, the GSA specified a modified version of the OSI reference model. If you wanted to sell something of a computing or data communications nature to the government, you had to comply with their modified version. It was called the Government Open Systems Interconnect Profile (GOSIP).

This model has been the base reference on which many of the changes of the past 20 years have been founded. Although very few manufacturers have attained fully open systems, they have attempted to use the OSI as a model for their future compliance. Again, this will become clear as later in this chapter. It must be remembered that there was little incentive for any manufacturer to comply fully with the OSI model because the openness could potentially jeopardize future revenue streams. Not to pick on any vendor, but when you bought a computer system using SNA from a large computer manufacturer, it was also an unwritten rule that most if not all future software and hardware purchases would be from that same vendor. If you bought from that vendor, you were assured that your purchases would fit seamlessly into your existing hardware and software platforms. If you went the third-party route and something did not work, you could expect little support from IBM. Thus IBM's future revenue streams flowed from the proprietary nature of their architecture. If a completely open attitude and platform were introduced, it would work with any hardware you acquired.

Now back to this complex model. The seven layers always confused even those with a rigorous background in the data processing or data communications industry. Furthermore, these folks took 5- and 10-day workshops, only to leave even more confused than when they went in. How can we show you in one chapter of text how this works and keep it simple? We assume that you are likely a novice or getting into some new portion of the industry, and that this is your reference manual. So, let's try it with our usual method: a story. We can hope that by the time this works through, you will have a basic understanding of the function of the model and each of the layers as they play into the industry. Here goes!

Figure 15.9 shows a situation where two people (Don and Bud) are thinking about moving to a new area in a suburb of a major city. There is

Figure 15.9
Building on new
subdevelopment
areas.

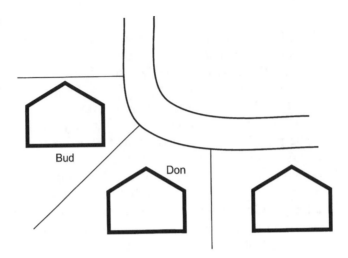

no subdivision yet, but there is a lot of property that can easily be broken down into perfectly sized lots, all with a great view.

These two players meet with a builder who owns the land and ask for their respective lots, which just happen to be right next to each other. The builder agrees to build both Bud and Don's houses at a certain price.

Now that the agreements have been reached, both parties decide that they'd like their houses rather quickly. So they both go back to the builder and ask that the process be accelerated. The builder, of course, will balk at the idea of doing the job too quickly. But perseverance wins out and the two potential homeowners get the builder to cave in and speed up the construction.

The result is that in two weeks the houses are ready to go. Now Bud and Don can move in. Immediately, a couple of things become apparent. Because the houses were built so quickly, the builder did not have time to notify the local incumbent local exchange carrier (ILEC) or competitive local exchange carrier (CLEC) that the owners were moving in. So, the two owners take possession, but they have no form of telephony service. Here are Don and Bud, living next door to each other without a means of communicating. What options do they have?

Immediately, they both call the local telephone company (ILEC or CLEC) and are told that the installation of facilities in their new area will take 6 to 12 months. This is too long for next-door neighbors to be without a communications method. So they consider their options and come up with the choices shown in Table 15.1. The option is shown in the first column, and the pros and cons are in the other two columns. These are not all-inclusive; we have limited our list to a few options.

TABLE 15.1

Summary of
Communications
Options

Option	Benefits	Deterrents
Use a megaphone and yell out the window from one house to the other (see Fig. 15.10).	• Easy to install • Relatively short distances • Can be implemented immediately	• What if one person isn't home? • What if Bud wants to talk to Don at 2:00 a.m.? • Loss of transmission due to distances. • Noisy for other neighbors in the area. • Total lack of privacy.
Throw pebbles at the window, when the receiver hears the pebbles on the window, open the window and yell out (see Fig. 15.11).	• Pebbles are readily available • The method of signaling each other is straightforward	• Given the distances, the throw will take some effort. This can cause damage. • In inclement weather, this can be inconvenient and awkward.
Use a flashlight and send the signals in Morse code.	• Easy to use • Quick communications	• Don doesn't know Morse Code. • During daylight hours, the sun may diminish the light. • Batteries can fail.
Go next door and visit.	• Easy to accommodate	• Weather may make this difficult.
Have our conversations during the visit.	• Doesn't cost anything	• What about those 2:00 a.m. conversations? • The inconvenience will outweigh the benefits.

In the figures illustrating the table results, there are benefits and losses associated with the communications options. So Don and Bud decide to model their communications device after the service that they really wanted—the telephone. The two neighbors aren't going to go out and build a central office. After all, they really just want the convenience of speaking with each other. So they decide to build on a model. Let's use the OSI model and make sure that this all ties in while we provide transparent communications.

Bud comes up with the idea that he and Don need to send their information across a medium. The earlier options used the airwaves as the medium, but this was limited by weather conditions and other factors. Therefore both neighbors agree to use a physical medium. They decide to install a piece of everyday common household string between both of their houses. This is shown in Fig. 15.12, where the string is run in a direct

Figure 15.10
A megaphone isn't a good choice, but it could work.

Figure 15.11
Throwing pebbles (or rocks) may not be a good alternative.

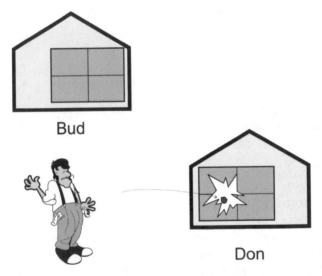

line of sight between both parties. The string is laid out and the two parties are connected with the medium.

Given that the physical string is now attached, can the two parties communicate? It is somewhat obvious that there is still something missing. So the neighbors decide that they have to use the model to handle their next step. Don comes up with the idea that they need to address the physical medium with some electrical or mechanical interface (a la layer 1 of the OSI). They decide that an electrical interface is not applicable because

Figure 15.12
A piece of ordinary string produces a physical connection (medium).

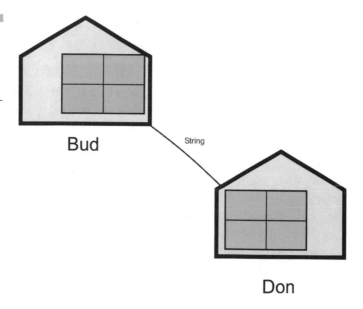

they do not have an electrical conductor in the string. So all they need is a mechanical interface. Don suggests that each neighbor attach a can on his end of the link to build this interface. Using everyday household tin cans on both ends of the string, Don and Bud have now addressed the bottom layer of the OSI model (Fig. 15.13).

A problem has arisen with this arrangement. To make this all work properly and transparently, everyone must build the connection in the same way. Bud has attached the string through the side of the can, whereas Don has attached the string to the bottom of the can. Although this might work, when you pull the string tight, the can at Bud's end is in an awkward position. This means that the two have to agree how they will connect the two cans to the string to make this work more efficiently. The agreement is that they will both take an empty can, punch a hole through the bottom, and then attach the string from the outside of the

Figure 15.13
Two tin cans are used to satisfy the mechanical interface (layer 1 OSI).

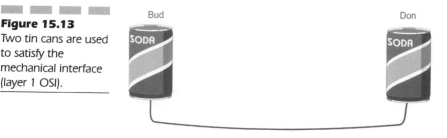

can, through the hole in the bottom, and knot it off inside the can. Then they will pull the string tight to get a good connection. Now the bottom layer of the OSI is handled.

The strings are now attached the same way, so can the parties now communicate? Well, maybe! Bud decides to test it out, so he picks up the tin can on his end and starts to talk to Don. Don is preoccupied with the television set and does not hear that the communications system is being used. The whole conversation is wasted thus far. So the parties need a means of using the link. They build a protocol that will solve this problem. First, the two parties agree that they should have an alerting mechanism to let the other party know that a conversation is requested. They decide that when one party wants to talk to the other, the originating party will grab the can on his end and yank on it three times. This will cause the can on the other end to rattle around. When the called party sees the can bouncing around, he will know that the party on the other end wants to talk. To use the data link (layer 2), Bud and Don establish a protocol called the *yank and rattle*. One end yanks the string, causing the can on the other end to rattle. Now they have overcome this obstacle.

We can skip by layer 3 of the model for now. Is it obvious why? Look at Fig. 15.14 and it should become fairly clear. The strings are attached to both parties directly. The connection is a point-to-point private line. There are no switching or routing decisions to be made, because the connection always goes to the same location. Therefore the network layer doesn't apply per se, so it will be transparent. But now the neighbors have the necessary

Figure 15.14
The point-to-point connection doesn't need a network layer. No switching decisions need to be made.

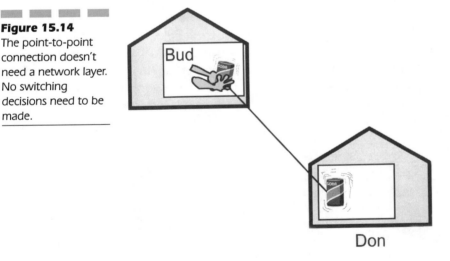

connection and the alerting mechanism, so they should be able to communicate without much ado. Correct? Maybe not, but see what happens.

Bud now decides that he wants to talk to Don. So, Bud picks up the can and yanks three times. This causes Don's can to bounce around on the other end, which alerts Don that Bud wants to talk to him. Kind of like a ringing phone, isn't it? So Don jumps up and grabs the can, picks it up, places it to his mouth and says, "Hello." At the same time, Bud is yelling into the can on the other end of this connection for Don to pick up: "Don. Hello, Don, are you there?"

Are these two parties communicating? No, they are not. Both have the can to their mouths, and no one is listening. If both parties are talking and no one is listening, then the information transfer is nil. So a new set of protocols is required. Let's introduce the transport layer in the OSI model now. Because this whole idea is Bud's, the neighbors agree that whenever Bud wants to speak to Don, or whenever Don wants to speak to Bud, after the can has been rattled a couple of times, Don will always pick the can up and place it to his ear. Bud, on the other hand, will always start the conversation off by placing the can to his mouth. Now the problems already encountered are handled. So we should be able to communicate without any further complications. Right? Here is the transmission process as it has been defined.

Bud: (Yanks the string).
Don: (Picks up the can on his end and places it to his ear because that's the rule).
Bud: "Hi, Don! Can I borrow your lawn mower today?"
Don: (Takes the can from his ear and places it to his mouth and begins responding) "No way, you told me last week that you were going to finally buy your own."
Bud: (At the same time Don is responding) "I know I told you last week that I was going to buy my own, but I haven't had a chance to get to the store yet."

Now both parties are talking at the same time. After a given amount of time, they will probably both be listening at the same time. The communications flow is not working properly. So they have to add a new set of protocols to allow the orderly transfer of the information. The problem stems from the half duplex nature of the link. There is only one transmitter/receiver on each end. There are two possible solutions. These are as follows.

First, a new string could be run between both parties. On this new string, Don and Bud will attach two new cans, just as they did for the first connection. Each can will be labeled *T* or *R*. The *T* can on Don's end will be connected to the *R* can at Bud's end, and vice versa. Now when an alert comes in, each party will pick up the two cans at his end. The *T* can will be placed in front of each party's mouth, and the *R* can will be placed in front of each party's ear (Fig. 15.15). Now there are two separate transmission paths, one for transmitting and one for receiving. This is probably the best option, but Bud and Don are reluctant to do this because it adds the complexity of maintaining two strings between their houses and requires twice as many cans on each end. This will take up too much space in their homes.

The alternative to this communications dilemma is the use of a set of rules (protocol) that will define the session layer (the control of the data transfer between the two ends). The session layer will work like this. Because this is Bud's idea, whenever Bud or Don yanks on the string, Don always listens first and Bud always talks first. The problem occurs when the transmitting end is done sending and wants to await the reply: the two parties need to know how to time this out properly. So Bud will talk

Figure 15.15
Two separate cans allow simultaneous transmission or reception, not the converse.

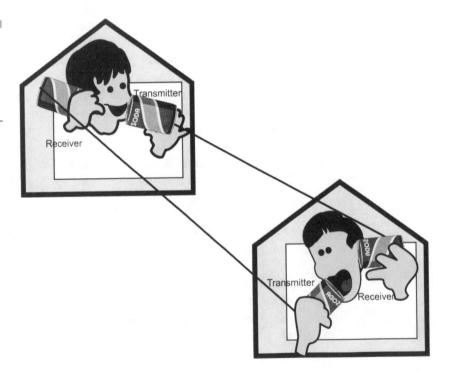

first (have you noticed that Bud talks a lot?). When Bud finishes his statement, he will use an "over" protocol. When Don hears the word *over*, he will take the can from his ear and place it to his mouth. At the same time, Bud will take the can from his mouth and place it to his ear. Then Don finally gets a chance to talk. When Don finishes, he will say "over," and the process will reverse back to the beginning. And so on…

Now the two parties have the ability to communicate with each other. Essentially, they can send their information back and forth as they had planned. Now that this whole scenario is established, you have probably noticed that a couple of layers were missed in the OSI model. These would be handled transparently by Don and Bud because they are working with the same communications protocols. However, let's go back and look at these layers to see how they will play out under our scenario. The three we have yet to address are the network, presentation, and application layers. So we shall complicate the issue by introducing the following add-on needs.

Just as Don and Bud get the communications working by the rules established, a new neighbor, Helen, moves in next door to Don. This is going to happen all around the neighborhood, and we should be prepared. Helen is neighborly and decides that she might have a need to speak to Don and Bud from time to time. The rules already in place can be used to provide the connectivity between all three parties: Fig. 15.16 shows this. To set up communications between Helen and Don, a string can be run between the two houses and cans attached to the string. The rule will be as follows: because Don was here first, Don will always speak first and Helen will listen, regardless of who rattles the can. The "over" protocol can also be used to provide the session layer, and so on. Now Helen also needs to talk to Bud, but, as the figure shows, in order for this to happen they need to run the string from Helen to Bud. This is not a problem for Helen or Bud. As a matter of fact, we can even color-code the cans (red for Don, blue for Bud).

Figure 15.16
Adding a third party may complicate the process (for Don).

Bud   Don   Helen

But there is a problem for Don in this scenario. In order for Bud and Helen to connect the string to each location, the string must go in one side of Don's house, pass through and exit the other side to connect to Bud. Don will therefore have to leave his windows open on both sides of his house at all times for this to work. Don doesn't like this idea, so he offers to become a network layer relay point. The process, as shown in Fig. 15.17, will work this way. When Bud wants to speak to Helen, he will pick up the one can that is associated with this communications system. He will then yank the can, causing Don to pick up and immediately listen. Bud will then say, "Don, this is a call for Helen." On hearing this, Don, who has two cans and strings, will grab the other can that is connected to the string to Helen's. Don will yank the can, causing Helen to pick up the can and listen. Don speaks first on this connection, so this is fine; he has a can to his ear from Bud and the other can to his mouth leading out to Helen. When Don knows that Helen is on the link, he will say, "Stand by for a message from Bud." At this point, Bud will speak to Don, who in turn will automatically relay everything he hears to Helen. Now the routing of the information is accommodated. The network layer is met, of sorts.

A problem exists with this whole scenario that isn't as obvious. Helen only speaks German. Bud only speaks English. Now the problem: if Bud speaks English into the can connected to Don, how do we make the format or language conversion into something that is usable and understandable to Helen? Luckily, when Don was in school he took German 101. He didn't excel, but he has a working knowledge of the language. Now, when Don hears Bud say, "Good morning, Helen," in English, Don will relay the information out through the other can in German as: "Guten Tag!" And when Bud finishes and says, "over," the process switches. Bud takes the can from his mouth and moves it to his ear. Don takes the

Figure 15.17
By using the relay point, we satisfy the bottom three layers of the model.

Bud   Don   Helen can connected to Bud and moves it from his ear to his mouth, at the same time he takes the can connected to Helen and moves that one from his mouth to his ear.

Helen in turn moves the can from her ear to her mouth and begins to speak or respond. When Helen responds with "Morgen, wie geht's?" Don hears it and converts it to Bud as "Morning. How are you?" And so it goes. The presentation layer has just been addressed to accommodate the format or protocol (language) conversion so that transparent communications can take place. The only layer in this case is the applications layer that has not been addressed. What is the application in this regard? Chitchat: "I just called to say hello!"

By now, you should have an understanding from this simplistic comparison of how the process works through the OSI model. Furthermore, the importance of this all taking place transparently should be obvious. If we had to go through the trials and tribulations of information transfer every time we placed a call, this would be far too cumbersome. The next time a connection to another neighbor is needed, the same rules can be applied.

We already have a simplistic network here, so let's draw it to a conclusion by introducing a couple of other players. Mary moves in behind Bud. There will be a need for Mary to communicate with Bud, Don, and Helen. Rather than trying to run strings all around the neighborhood we'll just use our relays and rules already in place. So Mary connects a string to Bud. Communication to Bud is simple. To reach Don, the call will go through Bud to Don; to reach Helen, the call will go through Bud and Don to ultimately get to Helen. One added complication is that Mary only speaks French. Therefore, when Mary says "Bonjour" to Bud, he repeats the information to Don as "Good morning." Don then finally sends it to Helen as "Guten Morgen." Now Roy moves in behind Don. The neighbors run a connection between Don and Roy—no problem. Roy only speaks pig latin, but that's okay because Don can deal with it. Now when Mary wants to speak to Roy she calls through Bud in French, which Bud converts to English and sends on to Don. Don receives the message in English and converts it to pig latin for Roy. The connection is sent through and routed/switched depending on the intended target location. Helen, as you recognize, is left out of this conversation, but she has no need to be connected. The result is shown in Fig. 15.18: the connection is routed and switched transparently between and among the neighbors. Don has also become very similar to a central switching system and service provider. Now for voice or data, the tele-

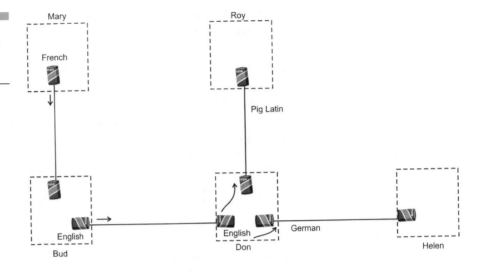

Figure 15.18
The layers can all be met through this network.

communications network is built to handle communication from any user to any other, regardless of the language (format and protocol) or origin of creation (manufacturer). This is what the OSI model was designed to do for us.

Is it here? Partially! Work continues on the standards, especially in how this can all be achieved. Many manufacturers are implementing OSI compatibility in different ways. That's okay, but it might lead to some kludges. The issue is not how the arrangement is created, but the opportunity to have a universal set of rules and protocols that everyone can build to. That is what the OSI reference provides.

That description is probably more than most of us need to know about this. However, we should at least be aware of just what the reference was meant to do for the industry as a whole. If we substitute our players in the example above, then we can have an organization using a computer platform manufactured by Company Bud. This is connected to or through a computer manufactured by Company Don and ultimately sharing and swapping information with a computer manufactured by Company Helen, and so on. . . . The communications network might well be used as a point-to-point, point-to-multipoint, or switched dial-up service. Protocols and services such as ISDN, X.25, switched 64 Kbits/s and leased-line T1 have been applied, so that the connectivity arrangements adhere to those of the tin cans and string. Yes, some of the cans might be larger or rounder than others. As long as the same rules apply for electrically and mechanically attaching to the string (wires, light beams, radio waves), then this should all work.

Other Network Architectures

Using the OSI as the reference is fine. It will be the basis of all future development. But there are still other architectures that were in existence or that sprung up during the wait for the ubiquitous implementation of OSI. We might never see the end, but we will see variations and improvements over the OSI as it applies to vendor-specific architectures or proprietary applications. Here are some of the more common models:

- SNA, which is now over 20 years old and has undergone very few modifications from the original architecture.
- DEC Net. As covered, this is also 20+ years old, but has gone through several revisions in an attempt to create openness.
- TCP/IP, a set of protocols that works on a different architecture developed back in the late 1960s and deployed more in the early 1970s for the government. Now it is the protocol for the Internet and the primary set of protocols that LAN, metropolitan area network (MAN), and wide area network (WAN) users are implementing for openness and robustness in their networking needs.

SNA

We have already discussed the SNA world earlier in this chapter. Therefore the continued discussion of this architecture will concentrate on the subtleties of the SNA to OSI comparison. In 1974, IBM introduced its architecture called Systems Network Architecture (SNA). SNA defines a structure and all of the protocols required to implement a network in which a wide variety of computers, software, and terminal devices can interact and provide a very high degree of network efficiency. Some of the characteristics that were introduced with the SNA concept are as follows:

- Application software uses a standard interface for communications, leaving the software network and device transparent and independent.
- All of the data links can use the same set of rules, called *link control procedures*.
- Applications can exchange data with each other.
- All applications can be intertwined to allow multiprocessing capabilities.
- Regardless of the location, different devices can access any application on any host.

Various techniques exist for local area data transport through recommendations and implementations by the vendors. Remember that the installations were merely recommendations to move data across a cable system between any two stations on the localized data transmission facility. Networking, whether local area, metropolitan area, or wide area, requires higher-level protocols to manage the transfer of information and make it usable and understandable. Without these higher-level protocols, nothing is guaranteed.

The function of SNA is to augment and supplement the local area network. Imagine that the LAN actually existed under a different name prior to industry acceptance of the term LAN. In the earlier days, this was implemented in the hierarchical networking format provided by SNA, which was introduced in 1974 and changed on a limited basis only. IBM felt that there existed an infrastructure in the major organizations that did not require change. The goal of SNA was to introduce the structure and protocols required to implement a network in which a wide variety of computer hardware, software, and terminal devices could interact and provide a high degree of efficiency and utilization of the network. SNA represents several characteristics of the architecture; for example:

- Applications software uses a standard communications interface and maintains network and device independence.
- All data links in SNA use the same link control procedures.
- Applications can exchange data between and among each other, allowing the use of multiprocessing services.
- Different types of devices can access applications in different hosts, regardless of the location.

These characteristics of SNA function to work cohesively in any implementation from IBM. SNA, therefore, details the specifications for the devices and nodes on a network and the logical path between these nodes. Both paths and nodes are arranged in the hierarchy using several layers of protocols. This is organized to facilitate the distribution of processing under a common control for the nodes. Furthermore, it is structured to provide flexibility and robustness in routing on the paths.

SNA Components

Each device in an SNA network, from the individual dumb terminal to the host computer, functions with a certain level of control depending on its position in the hierarchy. It will then operate under the control of the device at the next level up. Terminals, for example, function under the control of the cluster controller. The controller functions under control of the front-end processor and the front-end functions under control of the host. This is the true hierarchical concept IBM introduced 20 years ago.

The *network addressable unit* (NAU) is any segment or code that emphasizes the device to a network, program, or device. There are basically three different types of NAUs, as follows:

- The systems services control part (SSCP)
- The physical unit (PU)
- The logical unit (LU)

The SSCP

The SSCP functionally resides in the communications access method of a mainframe computer, or the control program in a midrange computer. Figure 15.19 shows a representation of this control part. This function controls the addressing and routing tables, the name service to address translation services, the routing tables for the network, and all instruction sets that deal with these entities. The SSCP establishes the communica-

Figure 15.19
The SSCP works as the communications software and manages the network. It will control the setup and teardown of connections between users.

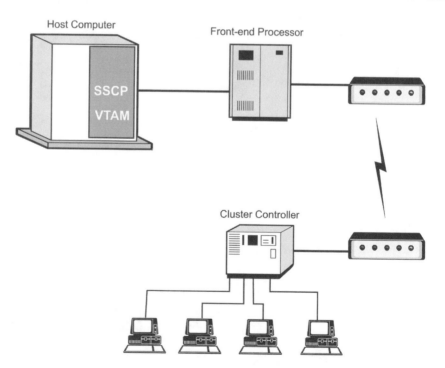

tions connection between nodes in the network, provides the informational flow control and queuing services for network efficiency, and selects the route for the nodes to communicate with each other.

The Physical Unit

IBM defines the physical unit as a single device on the network. Figure 15.20 is a representation of the PU. In a host computer and a front-end communications processor, this is implemented in software. In a terminal device (less intelligence applied in these), the physical unit is implemented in firmware.

IBM defines the PUs as follows:

- Type 1 is a dumb terminal or a printer.
- Type 2 is a cluster controller (3274) for the terminals or a batch terminal.
- Type 3 is under study.

Figure 15.20
The physical unit is a supervisory function controlled by the SSCP.

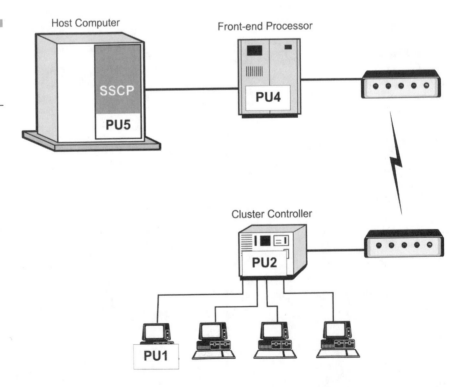

- Type 4 is a communications controller such as the front end processor (37X5).

- Type 5 is a host computer with a system services control part (SSCP).

In SNA, the SSCP controls the physical unit. Each PU is treated as a physical entry point on the network between the network and one or more logical units.

The Logical Unit

The logical unit is the basic communications entity in SNA. The LU is a port where end users access the SNA network. Figure 15.21 shows this access method. Two LUs communicate through what is called an LU-LU session, a temporary connection where data can be exchanged on the basis of some mutually agreed upon protocols. The layers of SNA, shown a little differently in Fig. 15.22 than in previous versions of the architecture (not to change but to clarify the architecture), build on each other from bottom up. Every SNA node contains the bottom two layers shown in this figure. These two layers—the data link control (DLC) and the path control (PC)—combine as what is called the path control subnetwork.

This is the basic method for the movement of information from a source to a destination node. Sometimes this can involve passing through intermediate nodes. The LU, which resides in the SNA nodes, consists of the upper three layers in Fig. 15.22 and uses the data transport capability of the path control subnetwork.

The different types of LU are:

- LU0 are sessions for special applications where IBM or the end user defines the parameters.

Figure 15.21
The logical unit is a port where end users access the SNA network.

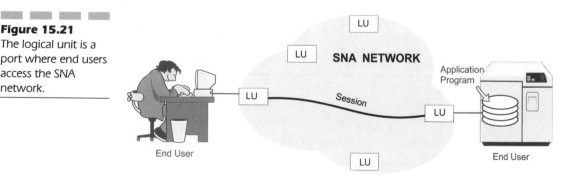

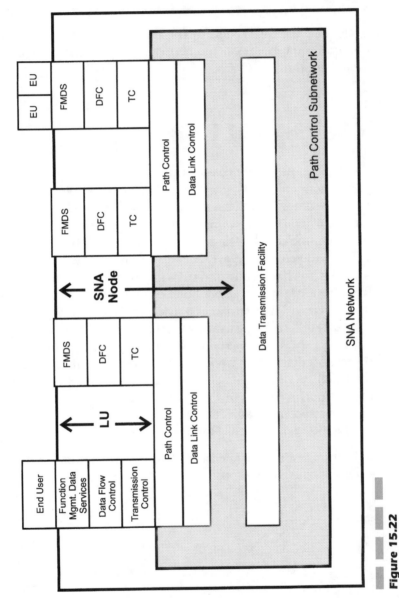

Figure 15.22

The SNA architecture shown differently.

- LU1 is a session between a host application and a remote batch terminal.

- LU2 is a session between the host application and a 3270 display terminal.

- LU3 is a session between a host application and a printer in the 3270 display family.

- LU4 is a session between a host application and a word processor, or between two terminals.

- LU5 is under study.

- LU6 is an intersystem communication, a session between two or more different applications, usually in two or more different systems. The more common type of this LU is advanced program-to-program communications (APPC), a session between an application program in a host and an application program residing in an intelligent workstation or terminal. This is also called the LU6.2.

The first SNA networks were simple tree networks, using the hierarchy shown. All terminals and controllers were attached to a single host computer. LUs in the terminal and controller communicated only with LUs in the host, but never with each other. All of these arrangements were purely hierarchical. Later versions of SNA allowed multiple hosts to reside on an SNA network and communicate as peers on the network. However, LUs in the terminals and controllers still only communicated with host LUs, never with each other. The difference was that they could communicate with LUs in multiple hosts instead of a single host.

Later advances in very-large-scale integration (VLSI) changed the computer networking strategies. Dumb terminals became intelligent workstations. Personal computers and midrange computers were introduced that changed the complexion of the network. Because these devices contained their own processing and storage capability, it was a natural evolution to allow for distributed processing. SNA evolved with the market and adopted more of a peer-to-peer processing capability and the advanced peer-to-peer networking strategy.

Digital Network Architecture (DNA)

Digital Network Architecture, by Digital Equipment Corporation (now acquired by Compaq Computer Corporation), is the model and architecture for the functions of DECNet. DecNet is a system of communications software and hardware that enables DEC's operating systems and components to function in a network with other DEC systems and is open enough to communicate with systems manufactured by others. DECNet is a group of computers with equivalent DEC software and communications hardware that connect with physical channels or lines. Each computer implemented in a network with DECNet software is called a *system*.

DNA defines standard protocols, interfaces, and functions that allow DECNet systems to share data and access various resources, programs, and functions. DNA currently has two separate stacks of protocols, one supporting proprietary protocols and interfaces on a DEC environment and the other supporting the protocols defined in an open systems environment (OSI model). Using a layered approach, DNA is functionally grouped into services and functions based on the layers associated with DECNet. Functionally, DNA is broken down into the following:

- User functions
- Network functions
- Communications functions

Using a slightly different view of DECNet compared to the OSI model than was shown earlier, DNA is shown in Fig. 15.23. Two separate stacks are shown here: the first is DEC's version; next to that is the function served by this layer. The labeling on DNA is different than that on the OSI, indicating that DECNet is not 100 percent compatible with the OSI reference. This was

Figure 15.23
The DECNet comparison shown differently.

OSI	DNA	Function		
Application	User	• File Transfer • Down-Line Loading • Virtual Terminal • Remote Resource Access • Remote Command File Submission		
	Network Mgmt.			
Presentation	Network Application			
Session	Session Control	Task to Task		
Transport	End Communications			
Network	Routing	Adaptive Routing		
Data Link	Data Link	DDCMP Point-to-Point Multiport	X.25	Ethernet
Physical	Physical			

demonstrated when DEC introduced its DECNet Phase V. This met with only moderate interest and limited implementation in the industry.

DEC has introduced a family of products and services over the years to allow access to and from DEC computing platforms and other manufacturers' platforms.

Internet Protocols (TCP/IP)

While the industry and standards bodies wrestled with openness in the communications and computing arenas, another evolving set of protocols emerged. This is also a layered architecture composed essentially of four layers. A result of a government contract to look at the internetworking of computers in the event of a national disaster, the protocols emerged from the original Advanced Research Projects Agency Network (ARPANET). Developed by Bolt, Beraneck, and Newman of Cambridge, this simple but robust set of protocols was geared toward linking systems in an open architecture. The transmission control protocol with Internet protocol (TCP/IP) continued its evolution from the original ARPANET to what we know as the Internet. The TCP/IP architecture is shown in Fig. 15.24. This is a four-layer stack that deals with the equivalent seven-layer architecture of OSI, SNA, and DNA.

Actually, the TCP/IP is fast becoming the most widely accepted set of protocols in the industry. Included in all forms of the UNIX operating systems and now the protocol of choice in many LAN-to-WAN environ-

Figure 15.24
The TCP/IP protocol
stack (architecture).

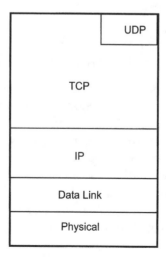

ments, it is the "middleware" for interconnectivity. The two most prevalent features of this protocol stack are its ability to work with just about any environment because of it can use application program interfaces to emulate most services on other architectures and its ability to packetize the data and send it out.

IP

Internet Protocol, abbreviated IP, performs a packetization (called *datagrams*) of the user data to be sent between and among the various systems on a network. When a large file is sent down the protocol stack, the IP function is responsible for the segmentation and packetization of this data. Then a header is placed on the datagram for delivery to the data link. The routing and switching of this data is handled at the IP (network) layer. To be somewhat simple in our analogy, IP is also a dumb protocol. When a datagram is prepared for transmission across the medium, IP does not specifically route the call across a specific channel. Rather, it just puts the header on the datagram and lets the network deal with it. Therefore the outward-bound datagrams can take various routes to get from point A to point B. This means that the datagrams are not sequentially numbered as they are in other protocols. IP makes its best attempt to deliver the datagrams to the destination network interface; however, it makes no assurances that:

■ The data will arrive.
■ The data will be error free.
■ The nodes along the way will concern themselves with the accuracy of the data and sequencing, or come back and alert the originator that something is wrong in the delivery mechanism.

This might sound strange in a networking environment, but it allows robustness to be achieved. The nodes along the network merely read the header information and route the datagram to the next logical downstream neighbor. This means that if anything gets corrupted on the network, the node will not know where to send the datagram, so it will throw it away. The network will not send a message back to the originator and let it know that the datagram did not get delivered. Another possibility is that the nodes along the network can send the datagram out on the basis of the best route (based on some costing algorithm or the number of points to pass through). Yet, in the transfer of the datagrams, if one of these routes is busy or broken, the datagrams will be rerouted around the problem to another node. This has great possibilities, but it can lead to a problem.

It is possible in the IP routing of a datagram that it can be sent along the network in a loop (Fig. 15.25). The routes might set up in a loop fashion, and the datagram will be circulating on the network like a spinning top. This doesn't cause concern if it is only one datagram, but if it were a general occurrence, the network could get quite bogged down with datagrams spinning around. Therefore IP has a mechanism in its header information that allows a certain amount of "hops" or what is called *time to live* (TTL) on the network. Rather than let an undeliverable datagram spin around on the network, IP has a counter mechanism that decrements every time the datagram passes through a network node. If the counter (usually 16 hops or less) expires or gets set to 0, the node will discard the datagram. As you might imagine, if the network can throw away datagrams, we should be alerted when this happens. This is not part of the IP protocol; it does not tell the originator that a datagram has been discarded. Additionally, datagram number four (just an example) was spinning around the network for several hops and was finally discarded. Yet numbers five, six, and seven go through the network without ado. Thus the later datagrams arrive at the destination, but datagram four is

Figure 15.25
IP can send data into a loop, meaning it will not reach the destination computer. The switches could send the data around and around.

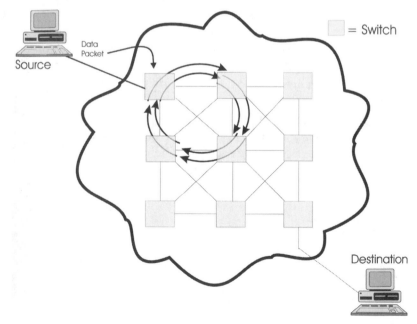

gone. IP does not have a mechanism to know that this has happened, and no numbering sequence to put everything back in order. Why would anyone favor such a dumb protocol? It is this robustness to deliver the datagrams across the network and throw away bad or undeliverable datagrams that helps to improve the efficiency of network utilization.

TCP

No network manager or administrator would accept not knowing whether data gets delivered across the network. Therefore working together with IP is the transmission control protocol (TCP). TCP provides the smarts to overcome the limitations of IP. Here the controls will be put in place to ensure that the reliable data stream is sent and delivered. At the sending end, TCP puts a byte count header on the information that will be delivered to the IP protocol layer. This is encapsulated as part of the data in the IP datagram. The receiving end, when it gets the datagrams, is responsible for putting the data back into its proper sequence and ensuring its accuracy. If things are not correct, the byte count acknowledgment (ACK) or nonacknowledged (NAK) message is sent back to the sending end. The sending end receiving a NAK will resend the bytes necessary to fill in the blanks. Furthermore, TCP holds all of the later received bytes (datagrams five, six, and seven from the previous example) until it gets datagram four resent. Thus it buffers the data at the receiving end. This makes data reception and accuracy the responsibility of the end user (node), not the network responsibility. This is to prevent the network from getting bogged down if errors or loops are occurring. The network is, after all, a transport system, not a computer processing function.

Using TCP/IP, the network operates efficiently, and improvements are being seen every day as the quality of the circuits is improved with all of the fiber-optic backbones being installed. Therefore the movement in the past few years has been heavily toward TCP/IP.

16
Internet

Introduction

Are you on the Net? For an increasing percentage of the world's population, the answer to this question is yes. The Internet is a social as well as a technological phenomenon. Huge numbers of people otherwise unfamiliar with data processing nonetheless have gone out and bought computers, modems, and other required paraphernalia—and then proceeded to hook up to and actually use a resource based on technologies that, if asked about, they could describe in only the most general terms if at all. With some qualifications, the Internet ultimately may be one of the few innovations that actually justifies the hype associated with it. Often described as a "network of networks," it provides a path to incredible amounts of information, much of it free for the taking (free if you don't count the basic network access costs). The continuing price of access is reasonable (about $19.95 per month per household). So many people have figured this out, in fact, that the Internet has quickly become overloaded to the point that it is sometimes essentially unusable. This chapter attempts to provide you with a basic understanding of just what the Internet really is; why you should care; the main types of capabilities the Internet provides; how to go about acquiring and using those capabilities; and a discussion of where the Internet may be going.

A Little History

Back during the cold war, the United States government instigated the construction of DARPAnet, the Defense Advanced Research Projects Agency network. Initially, it only connected a few key government research laboratories and a few universities under contract to the government for various research projects. Over time, additional universities were added. In addition, companies dealing with the government were connected (Fig. 16.1). Some overseas (i.e., outside the United States) institutions were also added, making this network international. The bandwidths connecting all these institutions were continuously increased to support all kinds of communications (see below for more details). In reality the Internet (as we now know it) started out as a 56 Kbps data communications network used to connect these sites together. When the rollout started, the backbone was immediately increased from 56 Kbps to a T1 (1.544 Mbps) transmission system. Later, in the late 1980s, it evolved again to a T3 backbone.

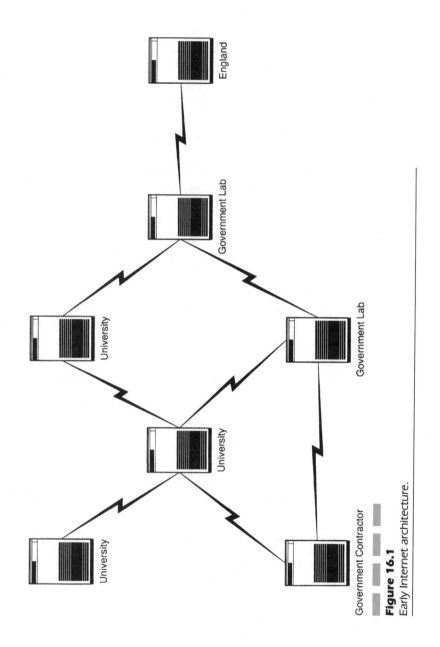

Figure 16.1
Early Internet architecture.

In theory, all of these institutions were connected only to support research associated with the United States government. In reality, most provided access to all members of their organizations. Moreover, in the universities' case, that included students. Those students took the knowledge of how this service worked out into the world when they graduated. Somewhere along the way, the *D* in *DARPAnet* was dropped—when people realized that the types of research and other activities on the network went well beyond "defense" and supporting military research.

The service that was most used at that time was electronic mail—e-mail. It was much less sophisticated than the capabilities used today. In fact, most implementations of the time sent only the text of a message, the address of the recipient, and a text subject string. No attachments of any kind, no separate names of the addressees (vs. their addresses), no copy-to's. Nevertheless, this mail capability was based on the same underlying protocol used for most mail on the Internet today: simple mail transfer protocol, or SMTP. If one were to describe SMTP in terms of the OSI model, it is a layer 7 protocol. It rides on top of TCP/IP, just as other protocols mentioned later in this chapter (Fig. 16.2). The SMTP mail protocol was developed around an ASCII text transmission system. This poses an interesting situation when a user wants to transmit information to a colleague. Has the following ever happened to you?

You spend the whole day preparing a report. Great pains are taken to build out a table with four columns of text. Background shading is

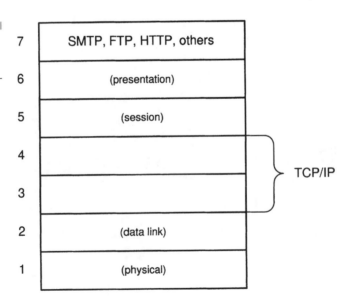

Figure 16.2
SMTP is a layer 7 protocol.

7	SMTP, FTP, HTTP, others
6	(presentation)
5	(session)
4	
3	
2	(data link)
1	(physical)

TCP/IP

applied in your word processor to highlight certain fields in the file. The headers on each of the columns are bolded and italicized in the word processing formatting service. Finally, you lay this all out with both reverse text for emphases (white text on a black background for important fields, black text on white background for normal text). Eureka! You have finally finished the document. Now you want to e-mail it to your colleague for review and discussion. A few compliments are also expected as the fruits of your labor. So, you type up an e-mail message announcing your accomplishment, paste the document to the mail message, and let it fly through the net. Soon you call your colleague and ask what he/she thinks about it. Alas, you are told it is garbage! No, the text is not garbage but the format of the document that arrives is garbage. The formatting has all disappeared. No longer is the table and column format intact. The bolding and shading have also disappeared. Moreover, the text is all over the place, indented where it shouldn't be, and so forth. You are crushed! What happened? After all that work, the other end thinks the document is trash. Well, let's analyze what happened.

Word processing systems use an extended ASCII code set these days. That means that an 8-bit code set is used to format the document in current packages. The net, on the other hand, was built on a true ASCII code set (7-bit). As a result, when you transmitted the document, the receiving device was looking at the seven bits (not eight) and interpreting the information using a different language. The result is what showed up. To solve this problem, other protocols like file attachments (MIME and BINAR) are used in the mail programs on the market today. Relief at last!

The other major layer 7 protocol in use both then and now is file transfer protocol, or FTP. As its name implies, FTP is used to move files from one system to another across a TCP/IP-based network. FTP can be used not only by humans, but also by programs; that is what makes it a layer 7 protocol rather than just an application program. (This is also true of SMTP. If a program sends mail, the program is sometimes referred to as "mail enabled.") But whereas now most Internet access software provides GUI-based FTP built in, systems in the early days had only a command-line interface (similar to DOS, although generally on UNIX-based machines). Calling it user hostile would be charitable. Incidentally, most modern PC-based operating systems (Windows 9x, Windows NT, OS/2) still provide command-line FTP as part of their base package (along with TCP/IP). However, because few people know how to use it (or need to), the vendors do not promote this feature in their marketing materials. To see it, simply start a DOS window, type *FTP*, and press the Enter key. Or another way in a Windows 9X system, use the "Start" button, then click on the "Run" command. When the run window pops up, type "FTP" in and press the Enter key. This will launch the core protocol and you then enter the destination address you want to transfer the files to. It is actually simpler than it seems, but we have become addicted to the GUI interfaces, so we do not like all that complicated typing and command line interfacing. *HELP* should list the available commands; *HELP command*, where *command* is one of the choices, should provide you more information on individual commands. *QUIT* will get you back to DOS. There are several additional layer 7 applications in common use until the World Wide Web became popular—which are still probably used by those not enamored of the Web. Among these are the picturesquely named Gopher and Finger. Gopher is a text-based ancestor of the Web; a gopher client, like a Web browser, jumps from Gopher site ("server") to site retrieving text-based information at the command of its user. Finger is a method of identifying users of UNIX systems. Since most Web users do not have or need UNIX accounts, its usefulness is less now than in the past. Other search capabilities include Ph, WAIS, and Veronica.

All of the above capabilities and services were used by thousands of people for many years before anyone used the words *the "Web"*. Nevertheless, you should not feel left out. Compared to the Web, none of these capabilities were packaged in a user-friendly fashion. Moreover, the types of information available via these services were not in general of interest to typical users. After all, the *R* in *ARPAnet* did stand for *research*.

Structure (Who's at the Center—and Why)

Obviously, the backbone is critical—but it's just a start. The vast majority of Internet users (both companies and individuals) are not part of the backbone. Moreover, the backbone providers in general are not in the business of providing end-user (or end-business) connections—other than to their own personnel and customers. Instead, a group of intermediary companies called Internet service providers (ISPs) has sprung into being. It is your friendly local ISP that you go to get a connection to the Internet. ISPs themselves connect to the backbone directly, usually for a fee (which is recouped, month by month, from customers like you). Note that underlying all this is the assumption that actual physical interlocation circuits are still provided by either local telephone companies or

interexchange carriers. Where a backbone provider, or ISP, *is* an interexchange carrier, it acts in a dual role. Otherwise, such providers and ISPs
construct rooms full of routers, modems, DSUs, etc. and use their local
telco to connect up, both among their own locations and to the Internet
backbone itself or to other backbone providers in the case of the central
organizations. We generally divide ISPs into two categories: regional (sometimes called local) and national (Fig. 16.3). These backbone providers are
now called national access providers (NAPs) and they have national access
points (NAP), which they call POPs. Confused? Actually, the national category should really be called international. See the section on gaining
access for more on these providers. Current 1999—2000 statistics point out
that there are over 4500 registered ISPs in North America. Unfortunately,
none of these ISPs is making any money.

Why the Hype?

It is reasonable to describe the Internet as a peer-to-peer network. This
means that, as long as things are set up properly, any computer on the Net
can communicate "directly" with any other computer. What do we mean
by *any computer?* Any computer in the world with a connection to the
Internet and a properly configured suite of TCP/IP software can participate fully in any service for which it and its users have authorized access.
Of course, in practice performance considerations may limit what one
can actually do with any particular computer. Essentially, by buying computer equipment costing under $1000, ordering a dial-up telephone line
(the extra line can cost as much as $30 a month), and signing up with a
local Internet service provider for about $19.95 per month, an individual
or an organization can gain full access to the Internet. The ISPs and computer manufacturers have teamed together and now offer free computers
(not real powerful, but they do the job) if you subscribe with the ISP for
three years.

And not only access. In addition to pulling information *from* the network, individuals can publish information *to* the network. To do this
effectively, one must have access to a computer that is connected to the
network full-time (i.e., 24 hours per day). However, there are many network providers (ISPs and others) that are more than happy to provide
shared access to such full-time-connected computers for a nominal fee (as
little as $10 per month specifically for this publishing capability), or in
some cases as part of a basic Internet access fee (Fig. 16.4).

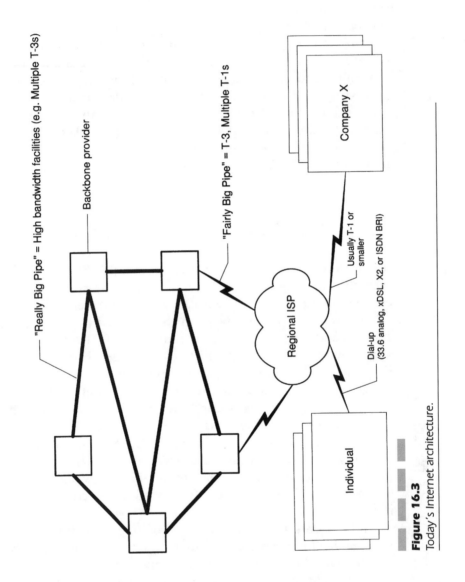

"Really Big Pipe" = High bandwidth facilities (e.g. Multiple T-3s)

Backbone provider

"Fairly Big Pipe" = T-3, Multiple T-1s

Regional ISP

Company X

Usually T-1 or smaller

Dial-up
(33.6 analog, xDSL, X2, or ISDN BRI)

Individual

Figure 16.3
Today's Internet architecture.

374

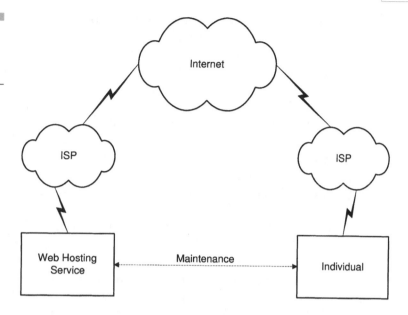

Figure 16.4
Web hosting using a
remote Web hosting
service.

Everybody as a Publisher

The invention of the printing press made it possible for large numbers of individuals and organizations to publish their ideas and information to the world at large. The "free press" is an ideal in the United States, and mostly one that still is a reality. However, with few exceptions, practical and economic limitations usually constrain the geographic areas to which print media spread. It is an axiom in the newspaper business that "all news is local." Only a tiny percentage of the world's population ever sees much detail on what is happening outside the borders of its own country—even in countries without significant limitations on newspapers. After all, how many people in the United States read the *Times* (London)? And setting up an actual printing press is an expensive proposition, not to mention the logistical might required to actually distribute a paper worldwide on a regular basis. There are a few papers that have done this (the *New York Times*, the *International Herald Tribune* and the *Wall Street Journal* come to mind), but the total number of truly international publications is quite limited.

However, on the Internet, there is no particular reason why the shared publishing service mentioned earlier must reside in the country of the individual contracting for the service. Also, to someone accessing that service, it looks the same as a local service and is just as likely to be accessed—so long, of course, as it is in the same language! It is almost as easy to maintain a Web site (later we discuss more details on the Web) from 10,000 miles away as it is from 5 miles away. The only real difference between the two is that the 10,000-mile connection may be a bit slower. This remote maintenance capability is currently causing heartburn for a great many bureaucrats around the world. Most countries have some kind of limitation on the types of content that can be made available, or published, on the Internet. In most cases, these limitations parallel those placed on print media. But the rules vary significantly from country to country. Some countries focus on limiting political expression, others on controlling dissemination of pornographic materials (the definitions of which also vary widely). Still others prohibit the propagation of hate materials. In addition, of course, many attempt to block more than one of these.

The problem is, of course, that any one individual or organization that wishes to publish information banned in its home country can simply select a network provider in a country that does not block the particular type of information desired. Thus, since the United States does not block most forms of political expression, opposition parties from around the world and of all political persuasions can publish from the United States to the world with impunity. Likewise, at least one Scandinavian country places few limits on what might elsewhere be described as pornography. There need be no evidence in the home country to betray the originators to local authorities.

To prevent such uncontrolled international communications from circumventing local rules, some countries are considering attempting the monitoring of all electronic traffic across their borders. But the Internet is a phenomenon that reaches down to the level of individuals, rather than just companies and other organizations. Unless a country is willing to give up the benefits of having the individuals in its population gain reasonable access to the Internet, monitoring all such traffic is a manifestly impossible scenario, even for the most automated countries in the world (such as the United States). Of course, the threat of enforcement can sometimes be an effective deterrent if the penalties of being caught are harsh, even if the probability of being caught is small. After all, the United States Internal Revenue Service can audit only a tiny percentage of the taxes due to the United States government, but by and large people pay taxes anyway.

One can see, however, that the ability of individuals and organizations to communicate with the world has been increased by orders of magnitude by the ubiquity of the Internet.

Commercial Opportunity?

Where do you shop? Chances are that most of your shopping is done at physical "storefronts" or malls within a few miles of your home. Perhaps you also shop from catalogs that reach you by *snail mail* (that is, the post office). What chance does a small vendor located thousands of miles away from you have of marketing products to you? Until the Internet became available, very little. That vendor could publish nice color catalogs and mail them out. Many small companies have done just that, and have become large companies by doing so. A good example of this is L.L. Bean. However, this is an expensive route to take; the production of a good catalog is a costly effort, and postage is very expensive if large numbers of pieces are to be distributed. The point here is that the cost of even trying is high.

Enter the Internet. Literally tens of millions of people either have or will shortly have access. And, with minimal investment (a few thousand dollars), a vendor can in theory reach all of them. A nice potential market, wouldn't you say? Many think the answer to this question is a resounding *yes!* However, there are a few flies in the ointment, depending on your point of view. . . .

There currently are several ways to attempt to reach potential customers via the Internet:

- Send out large numbers of electronic messages (the electronic equivalent of junk mail, referred to as *spam* by those who would prefer not to receive such transmissions).
- Place advertising for products and services on others' Web sites.
- Create a Web site (see below) to describe products and services, and perhaps to actually take orders for those same products and services.

Spam

What is your definition of *junk mail?* Do you like to receive it? Do you often purchase products as a result of receiving it? We didn't think so. While people who receive small amounts of postal mail might like receiving anything at all in their mailbox, most would prefer it to be a bit more focused than typical junk mail. Likewise, while new users of the Internet (*newbies*) may initially be excited by messages in their electronic inboxes with titles like *""""!!!!Make big money stuffing envelopes!!!!""""*, after the fifth, tenth, or hundredth such message, things get tedious very quickly. There is always the risk that something worthwhile is buried among the garbage. While the delete button is always there, it does take some time to filter out the spam. And time is usually in short supply.

The good news is that with a small expenditure of money and a little applied intelligence, one can greatly reduce the amount of spam that one actually sees. There are products, both for regular e-mail and for newsgroups, that are capable of filtering out unwanted messages (Fig. 16.5). For example, Qualcomm's Eudora Pro e-mail program can select e-mail and delete before reading on the basis of an unlimited set of user-specified character sequences, whether those sequences appear in a message's subject, in the main body of text, or as part of the sender's name. If you never want to see a message containing the word *toner* in the subject (many scam artists try to sell copier toner via the Net—we have no idea why), you don't have to. Anawave's Gravity, a sophisticated newsgroup reader program, has similar filtering capabilities. On the other hand, as P. T. Barnum said, "there's a sucker born every minute"—and the law of large numbers says that if you send out thousands of such messages, at least a few suckers are going to read them and respond. Because the cost of sending out spam is so much smaller than the costs associated with paper mail, it is likely that people will continue to send it out.

A number of organizations have attempted to limit spam in various ways. Some ISPs (for example, CompuServe and America Online) have rules against their own users generating spam; if discovered, a spam sender on those networks will find him- or herself without an ISP. This is helpful, but only partially successful. There are literally hundreds of other ISPs, at least a few of which exist primarily for the purpose of supporting spam senders. It's not illegal, merely greatly annoying (except to those senders, of course). The United States Congress itself has considered

Figure 16.5
Filtering programs dump spam in a trash can.

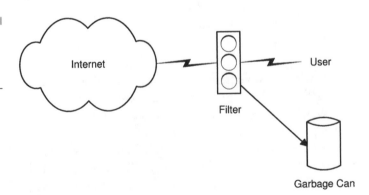

legislating against spam. Unfortunately, this is one of the areas where the social aspects of the Internet phenomenon come to the fore. Spam is just a form of communication. One person's junk e-mail is another's shining opportunity, or so it is said. Moreover, in the United States, at least, there are strong constitutional protections against most kinds of attempts to limit communication, particularly via legislation. If one regards Internet mail as a faster alternative to the post office, then spam can be considered unavoidable. At least in the case of the electronic version, there are methods to automatically handle a portion of it, no recycling required.

Web Advertising

One of the distinguishing characteristics of computers is that, if programmed correctly, they can aid in focusing transmission of commercial messages (i.e., advertising) to only those people most likely to be receptive to those messages. The Internet embodies a number of ways to advertise to its users. One, mentioned above, is spam. But spam is a sledgehammer approach; by definition, it is not focused. There are other, much more subtle approaches. One is "Banner ads"—colorful, sometimes animated rectangles on Web pages, similar to magazine ads but with a significant difference: when clicked on, they normally change the user's current Web page to one on the advertiser's Web site. This is referred to as a *link*. Banner ads are usually placed (for a fee, of course) on sites with a lot of traffic; they are thus the Net equivalent of putting up a poster for some kind of product or service in the window of a supermarket.

Web Site

Another way one can advertise one's wares on the network is to create a Web site. Creating a Web site is more than just a way to advertise (Fig. 16.6). A Web site puts your corporate or personal image up close and personal in front of any Web user who wants to access it. Unlike spam, however, creation of a Web site can be very expensive. On a per-user basis, it may not be expensive compared to print advertising; however, the skills as well as the resources involved in creating a Web site are completely different. What does it take to create a Web site? Here are just some of the requirements. For one thing, you need a computer; an ISP may provide

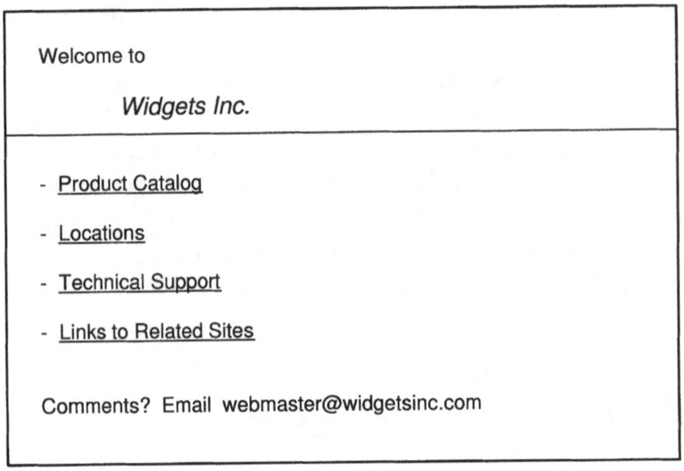

Figure 16.6
Sample home page
for a company.

```
Welcome to

        Widgets Inc.
─────────────────────────────────────────────

  - Product Catalog

  - Locations

  - Technical Support

  - Links to Related Sites


  Comments?  Email  webmaster@widgetsinc.com
```

one for you if that's the way you'd like to go. An ISP can provide either a dedicated computer—one that is just going to be your own Web site—or a shared (sometimes referred to as a virtual Web site) computer, in which case your Web pages will share the computer, the disk space, and so on of a number of other Web sites. Two categories of computers, or rather operating systems, tend to be used for Web sites. The two are UNIX computers and Windows NT computers. (We realize that these don't really specify computers, but rather operating systems.) In practice, any computer that runs UNIX, such as a Sun, IBM RS 6000, or any other computer capable of running some variant of UNIX can be used. Windows NT computers are becoming increasingly popular because they are somewhat easier to manage; however, the size of the Web site you can create tends to be larger with a UNIX computer. Now the implementation of Linux-based systems is becoming the norm, because of the simplicity and cost of the software. In addition, the movement against Microsoft,* the NT platform, is now greeted with mixed results. Linux seems to be a movement all of its own as we approach the new millennium. At least it seems easier to create larger sites with a UNIX/Linux computer than with a Windows NT—based machine.

You need a lot more than just a computer, however. To create a Web site, you need connections to the Internet that require an ISP (discussed elsewhere in this book) and a way to create the pages. Initially, when the Web first came into being, the way one created Web pages was using a text edi-

---

*Many people are anti-Microsoft. Therefore, they go out of their way to avoid using the products and software produced by that company.

tor where pages were defined in hypertext markup language (HTML).
Markup languages have been around for quite some time. They are distinguished from WYSIWYG (what you see is what you get) systems by the
fact that they are text. Markup languages use what are called *tags*. A tag is
a word surrounded by angle brackets (<tag-goes-here>) that tells the system processing the markup language what to do with text that immediately preceeds or follows the tag. Markup languages are particularly useful
for manipulating truly large amounts of text in a consistent way. If you
are going to use a WYSIWYG system, then everybody working on a project has to have a whole series of standards defined in excruciating detail.
For example, normal paragraphs have to be a certain point size, headers
have to be a certain point size, and so on, and every individual has to
know how those work and set their WYSIWYG editors accordingly. With
a markup language, instead of saying exact point size for heading, you
would say "heading 1." That will be a tag. Of course, while the WYSIWYG
editor requires a great deal of specification of detail, a markup language
requires the user to know all of the tags. Since HTML is a relatively new
language, not many people initially knew those tags. That's one of the
main reasons why so-called Web masters command such high salaries.
The creation of a Web page is not inherently difficult. However, it
requires a skill set that is completely new and therefore rare—at least so
far. There are a fair number of businesses that have had Web pages created,
then examined what was created and realized that maintenance is not all
that difficult, at least at a conceptual level. In any case, there are now a
large number of tools that allow Web maintainers to take at least one large
step away from the HTML coding that was required until recently. In
addition, there are a number of editors that focus on the HTML coding
but have point-and-shoot selections for the various tags so that you don't
necessarily have to memorize all of them.

There is a major additional element here. For example, it is very easy to
create a basic newsletter document; however, a really good newsletter
requires skills far beyond just the use of a word processor, no matter how
sophisticated that word processor is. Page layout skills, an understanding
of design elements, and of course an understanding of subject matter are
all required. In the case of a Web page or site, a significantly larger number of factors needs to be taken into account. For example, there is always
a temptation with modern tools to create strikingly beautiful Web pages
with all kinds of wonderful graphics. However, graphics of any size
require a considerable amount of bandwidth or network capacity to be
moved out to the people who are going to view them. Most people still
access the network via dial-up lines. For the most part, those lines at the

moment max out at 33.6 Kbps. Some faster speeds are available (see else-where in this book), but it is a mistake to design a Web page for popular use and then not take into account the vast majority of people who do not have truly fast connections to the Web. The 56 Kbps modems discussed earlier will not produce fast enough connections for very busy Web sites or dense graphic images. Large files will serve only to frustrate visitors to a Web site and chase them away forever. This problem is sometimes aggravated by the fact that companies typically have dedicated, higher-speed connections to the Internet and the Web. By *higher-speed* in this case we mean as fast as a T1 circuit or 1.544 million bits per second. Really large companies have T3s or 45 million bits per second. Connect these to a local area network in a developer's area and the people who are doing the testing on the Web site will see all those lovely pictures positively snap onto their screens. But those same graphics will take a long time to reach a typical dial-in user's screen. The Web is not referred to as *"World Wide Wait"* for nothing.

The cost of a Web site varies radically. At the low end, it is possible to get an account on America Online and create a Web site for no extra cost. We are talking about $20 per month or so. But for this, the amount of storage you get is minimal—2 or 3 megabytes—and you do all the work yourself. All you have is a facility. Moreover, anyone who is serious about providing a Web site for access by many other people will probably not want to put it on America Online or CompuServe. The focus of these providers is serving individual users: they're not really set up to provide excellent Web site support. They are changing, but are not yet there. These companies (AOL, Compuserve, and Prodigy) are charted as on-line service providers, not ISPs. There are many other companies that have gone into business to do just that—notably many ISPs. The real cost of setting up a Web site is not the network; it's the computers themselves, which for high-capacity machines could easily cost in the tens of thousands of dollars, possibly even the low hundreds of thousands for a single high-capability server. Some of the more frequently accessed sites are *hit* (accessed) literally millions of times per day. One doesn't want to use a low-end PC to handle something like that. The other major cost is people time. We mentioned above that the skills aren't all that rarified: but they are still rare. Because of that, people who really know how to do a good job on a Web site still command premium rates. A typical high-end Web site will not come into being for less than $50,000 to 100,000 up front; it can easily cost several times that. Also, there are continuing maintenance costs. One of the most common phenomena seen on the Web today is that a company, possibly very well funded, spends the bucks to create a

Web site, puts it on-line, and gets buried by the number of accesses coming in initially. But those accesses will taper off unless at least two things happen:

1. The Web site is kept up to date. That doesn't mean just current information; it means that there is perceptible change in the content of the Web site, whether it's content or layout, on a continuing basis. People won't come back frequently if, when they come back, they don't see anything new.

2. Enough bandwidth is ensured for the site to handle the actual number of users accessing it. Bandwidth in this context is not just the network access; it's also the capability of the server. As with many other environments, everyone typically blames the network itself when access is slow. More often than not, it turns out that the slowness of the access is due to the undersizing of the server's capabilities.

Advertising, of course, is not the only reason to create a Web site. Many companies, for example, Amazon.com, have created a Web site that is an entire enterprise. Of course there's advertising, but also there is a great deal of content regarding products or services for sale. In the most complete cases, there are actually products and services for sale directly via the Web.

Sounds like a major opportunity, right? It could be. There are billions of people who now routinely access the Web on a daily basis. That's a large pool of potential customers. However, people are still in some cases reluctant to put the necessary information on the network to make real purchases. Both of the major vendors of Web browsers, Netscape and Microsoft, have addressed this concern by incorporating something called secure sockets layer (SSL) into their browsers, as have the people who created Web sites that actually sell products and services (Fig. 16.7). SSL is pretty good; but since very few people understand that the encryption is present, we believe that it will take some time for Web sales to really take off. It doesn't help that the Web is really slow at some times. Often, it's just quicker to pick up the telephone and call a vendor than it is to try to go through all the text entry to purchase products and services on the Web.

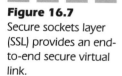

Figure 16.7
Secure sockets layer (SSL) provides an end-to-end secure virtual link.

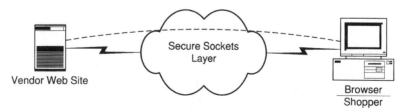

One should be aware that the Web is almost a science fiction kind of environment. One of the characteristics is that things change amazingly rapidly. Web pages come and go, companies appear and disappear, and the features of individual Web pages, as mentioned above, change rapidly. There are some ways to try to keep on top of this. One is to use searchers. There are a few services or Web sites that specialize in helping other users find Web sites. Paradoxically, these services or searchers are free. How do they do it? They accept a great deal of advertising. And this advertising is focused in a way that occurs in no other medium. The Web sites that provide searchers pay attention to what individuals search for, and, based on what those individuals request, different advertising appears along with the responses to the searches. Two of the major Web services that provide searching are AltaVista (*http://www.altavista.com*) and Excite (*http://www.excite.com*). The technology used by these companies is amazing. They have enormous databases that are kept up to date on a continuing basis via automated searchers that themselves go out and collect information from everywhere on the net. Those databases are then indexed and used for searches for users. And the searches are amazingly fast. Of course, to an individual who does not realize what's going on behind the scenes, the searches may not be so amazing. But a typical search may return 20,000 results. That would be daunting, but, at least in Excite's case, the results are sorted in a descending order according to the likelihood of satisfying the request.

Protocols

We're going to get a little technical here for a moment. TCP/IP was defined elsewhere in this book. TCP/IP is a collection of protocols. The base protocols described by the letters TCP/IP refer to the lower three or four layers of the OSI model. There are layer 7 protocols that ride on top of TCP/IP. Among these are FTP, Telnet, simple mail transfer protocol (SMTP), and simple network management protocol (SNMP). Newer protocols like the real time protocols (RTP) and network time stamp protocols (NTP) ride at layer 7.

FTP stands for file transfer protocol. When you use a browser to access a list of files or updates to programs on a Web site, when you actually click on an icon or link to retrieve a file, there's a better-than-even chance that file is being retrieved using FTP. FTP is not the protocol used for routine browser operations. But the major browsers do incorporate the capability of moving files around using FTP. FTP existed long before the World Wide Web. Most operating systems come with software that allows you to use FTP directly without using a browser. However, FTP typically has a command line interface in such an environment. If you are comfortable with the *C:* prompt, then you might want to use FTP in cases where you have problems getting the browser to work properly. There are also a few GUI-based FTP implementations that are stand-alones rather than integrated into a browser. That could be very useful in cases where you're going to spend some time doing nothing but retrieving files, for example, from a vendor's support site.

Telnet (not Telenet) is a means to sign onto a remote system network directly as a user of that system (Fig. 16.8). This is not the same as using the World Wide Web, although the physical infrastructure is the same. When you use a browser to access a Web page, every time you click on a link you send a request for a specific set of information to a specific Web site. That Web site, if it can, sends back just the information you requested. Generally speaking, it then completely forgets it had anything to do with you. There is no "conversation" taking place: each exchange is a stand-alone transaction. In the case of Telnet, however, you sign onto a remote machine. To do this you have to have an account on that machine, although some machines, notably in university environments, may allow you to sign on anonymously or even create your own account. Usually, if you are trying to sign onto a system anonymously, you use the user ID *anonymous* and a password consisting of your complete e-mail address. The latter is not normally checked by the software; rather, it's just a courtesy and a convention that you comply with in order to get the free service that you're about to use. With the popularity of the Web, Telnet is not a facility you are likely to need on a regular basis. However, if a vendor gives you access to a system for some reason, you may need to access it using this approach. To use it, get to the prompt on your operating system and type *Telnet* [*space*] and the name of the system you wish to access. You

Figure 16.8
Telnet allows signing onto a remote computer.

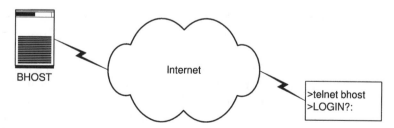

BHOST

Internet

>telnet bhost
>LOGIN?:

can also tell Telnet a specific IP address of the form *nnn.nnn.nnn.nnn*. FTP can also be used to access a remote system in the same way, with the same address and convention.

SMTP stands for simple mail transfer protocol. You will probably not use this protocol explicitly; however, if you use Internet mail (for example, the mail that is provided as part of Netscape), then you are using the SMTP protocol. As mail protocols go, SMTP is pretty simple. That has caused some problems, since it does not in and of itself incorporate ways to send attachments, particularly not binary or program-type attachments. But a number of vendors have gotten together and built software and established conventions that allow these capabilities.

While among the least sophisticated mail environments in the world, SMTP mail is sort of taking over the world. Why? Because SMTP is built into pretty much all systems that use the Internet. It is the kind of mail that you get when you sign up with an ISP.

The form of an SMTP or Internet mail address is *name@node.extension*. For example, the Internet mail address of one of the authors of this book is *bud@tcic.com*. Until recently, the list of extensions was very short; there were only about eight or so. There have since been some agreements made that will probably significantly increase the number of suffixes. There is a whole structure of these that is beyond the scope of this book.

One of the major successes in the area of intervendor cooperation was the creation of a couple of standards for including attachments with e-mail. The preferred one of these is called Multipurpose Internet Mail Extensions (MIME). A MIME attachment can include just about anything that can be represented in digital form. Moreover, most of the more sophisticated e-mail software packages can automatically create MIME inclusions without users having to go through several steps. MIME extends the format of the Internet mail to allow non-U.S.-ASCII textual messages, nontextual messages, multipart message bodies, and non-U.S.-ASCII information in message headers.

SNMP stands for simple network management protocol. Although this protocol is integral to the management of the Internet, most users will never be involved with it since it is used by people operating the Internet rather than by end users. We will not go further into it in this book.

The World Wide Web is just a virtual structure running on top of the basic Internet itself. All of the things that were there before the World Wide Web are still there; however, the vast majority of Internet users are not aware of them because the Web itself is so much easier to user. Browsers (and therefore the Web, which really just consists of browsers and Web sites talking back and forth) use a protocol called hypertext transfer protocol (HTTP). One of its key characteristics was mentioned earlier, but bears repeating: exchanges (or transactions) based on HTTP are "stateless." This means that unless special measures are taken, Web sites do not maintain true "conversations" with users of those Web sites. Every time you access a Web site—that is, every time you get a page from a Web site—that Web site recognizes the request from you, responds to it, and forgets about it. You may be thinking, "But many Web sites that I interact with do seem to keep track of what's going on from one screen to the next." You are correct; they have figured out ways to do this. But those ways don't use the basic HTTP capability; they use some additional capabilities, for example, that are built into browsers. The developers of the major browsers realized that this was a problem that had to be solved. What happens is that when a Web site wants to retain some contacts to an exchange, it causes the browser to create something called a *cookie* on the user's local disk drive, somewhere on a directory below the browser. Cookies can either be line entries in a file of cookies, or they can be one per file; Netscape maintains a file of cookies. A cookie is just a sequence of information that the Web site places on the local disk to retain some information. It may have the user's name and ID, and possibly some information that the user filled in on a form on the Web site. One very common use of a cookie is to retain an ID and password that is granted to the user as part of the exchange with the Web site in order to allow the user to get into the Web site in the future without having to go through an elaborate sign-on procedure. Thus cookies can be very beneficial to efficiency and user satisfaction. However, there has been a fair amount of concern that cookies can also be used to compromise the privacy of individuals. Some people do not want a Web site to retain information about them, especially without telling them about it. Discussions are under way to address this. By the time you read this, there may be enhancements to browsers that allow them to tell the user when information from a cookie is about to be used. In practice, we don't believe that users will make heavy use of this capability, since it will really slow down accessing Web site services.

Naming Standards

Many things exist on the Internet. For example, there are files or Web pages. There is a general format for things on the Internet. That naming format is *protocol://node/path/filename.type*. Breaking this down, *protocol* will be something like HTTP or FTP. *Node* can be a number of different things. In the case of a Web site, the node name often starts with *www*, but it does not have to. An example of a node name would be *www.excite.com*. In the case of a Web site, there may not be anything after that. In fact, there is an implied file name, since the file name *index.html* is the default file name if one is not specified. But it is routine to try and locate a company's Web site by typing *http://www.companyname.com*. More often than not, you will be successful with this syntax. A more elaborate example for the name of a file on the Internet is Netscape's search location. That address is: *http://home.netscape.com/search/index.html*. Note a couple of things here: Netscape does not use the *www* syntax at the beginning of its node name. It doesn't have to; it's not a requirement. There is a single-level path below the node name: *search*. This could be multiple levels, but in this case, it is not. Finally, Netscape has specified the default file name: *index.html*. Again, this isn't required, but it is sometimes nice to show what the defaults are.

There is another service that is integral to the Internet, although not obvious to Web browser users. It is, however, a critical component at this point, and is used routinely and frequently by Web browsers. That service is a distributed naming service called Domain Name Services (DNS). Like many of the other protocols mentioned, it has been around for much longer than the Web. DNS is the protocol or service that allows you to specify the name of a Web site instead of the Internet address of that site. There is a whole complex infrastructure of DNS servers or computers that is maintained on a continuing basis by the Internet providers. Web browser users access DNS service frequently during a session. Every time a user clicks on a Web link or pointer, the browser makes a request to a DNS server that is defined as part of the TCP/IP software on the user's machine. That DNS client accesses the DNS server—possibly several servers—in order to get the TCP/IP address that corresponds to that name. It then passes that address back to the rest of the IP software in order to allow it to make a direct access out to that Web site server. Not only that, but once a Web page is being retrieved, there may be many links within that Web page that constitute the page. When you retrieve a page, you don't just get a bunch of stuff in one file; rather, in most cases, there are intrasite links built into the page that retrieve images and tables and other objects that are part of the page. There may be multiple retrievals and accesses to DNS servers during the retrieval of a single Web page. This can really slow things down if the network itself is not moving quickly. But if it weren't there at all, maintenance of these sites would be much more difficult, since everything would have to be coded with Internet addresses instead of readable names. That's why, when you set up your computer, you may or may not have to give it an IP address (it gets one anyway, but sometimes it's dynamically assigned); however, you absolutely have to have a DNS address, usually two, in order to provide a backup. There is a way to assign DNS addresses dynamically, but not everyone uses it yet.

Gaining Access

To gain access to the Internet, you need an Internet service provider (ISP). ISPs come in two flavors, regional and national; also, one might consider an international category as well. National and international carriers tend to be very well-known companies. Such organizations as AT&T, CompuServe, and IBM fall into this category. They have access points, or points of presence (POPs), all over the world. Interestingly, CompuServe is probably the most pervasive service. But the others are also up there. It would be worth your while if you travel to check into the dial-in services for these services before choosing an ISP. It is a major factor. All have an 800 number access, but it tends not to work terribly well over truly long distances because you are going through the dial-up network—and, as has been mentioned elsewhere, the dial-up network was not really designed for data.

Regional ISPs tend to have slightly lower access charges than the international ones. This is probably because of the additional cost of those international circuits. Another possible advantage of a regional ISP is that, in the region it focuses on, it probably has more dial-in points than the national carriers. So if one of the dial-in points (or numbers) has a problem, you can use an alternative one in your area. The national and international ISPs are really gunning for your business. They have in most cases brought their prices down to very close to those of the regional carriers. Moreover, they do have the advantage of providing dial-in points across the country and even around the world.

Another factor in selecting an ISP is less obvious than the number of dial-in points. That factor is the ISP's connection to the backbone of the Internet itself. The number of simultaneous users an ISP can effectively handle is directly related to the speed of its connection into the Internet. We discuss high-speed circuits elsewhere in this book. An ISP that has only a T1 connection to the Internet really does not have enough capacity to handle significant numbers of people. This is not normally a problem with the national and international ISPs; however, it is a real consideration when selecting among regional carriers, of which there are often several in any given region. Normally, you probably want to use an ISP that has at least a T3 connection and perhaps several T3 connections to the Internet. This ISP connection to the Internet factor is particularly important if you are selecting an ISP on behalf of a medium to large company, as opposed to just an individual. A company with hundreds or thousands of individuals connecting to the Internet simultaneously can put a really big load on an ISP. It may increase the ISP's total traffic by a double-digit percentage if it's just a regional ISP. The moral of the story: do your homework.

Another consideration in selecting an ISP is the variety of connection alternatives for the user to the ISP. Obviously, dial-up will be supported. But straight dial-up, at the writing of this book, only reaches speeds of 33.6 Kbps. That's a lot faster than speeds in the past, but it isn't anywhere near as fast as some available alternatives. Not all ISPs for example, support ISDN—and those that do, do not all support bonded (two ISDN B channels bonded together to give 128 Kbps) circuits. Of course, you may not be able to get ISDN in your area anyway; however, for those that can get it, it is a very desirable means of connecting to the Internet. Another alternative now available is xDSL. xDSL, sometimes referred to as just DSL (this stands for *digital subscriber line*), represents a variety of technologies that, as a class, deliver megabit-per-second speeds to homeowners. But an ISP has to have a special connection to your local telephone company in order for you to access the ISP using xDSL. As a practical matter, xDSL will only be available in limited locations for the next few years. It may not take off if other services become generally available like ISDN. But xDSL does have the potential to take off, since it provides higher bandwidth than just about any other service to individual homes at a tiny fraction of the cost of equivalent services. Other means of connecting to ISPs involve dedicated circuits, switched 56, or even frame relay. You have to decide what you need and what you are willing to pay for—but you also have to check with the ISP to find out what it is willing to accept.

If you are selecting an ISP for a business, two ISPs may be better than one (Fig. 16.9). We mention diverse routing elsewhere in this book. Several ISPs over the past couple of years have had total failures for hours at a time. If you are going to purchase a service like this for your business users, then, as with any other service, you need to consider what happens if the provider fails. One way to handle this is to purchase service from two different ISPs. Another is to use an ISP that can provide redundant dedicated connections into your business. Don't forget to arrange for diverse routing from your location to the ISPs as well. That may involve, for example, getting circuits to two different central offices, something

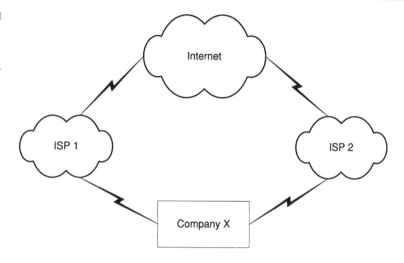

Figure 16.9
Two ISPs provide backup.

that the telephone companies tend to resist but that still is a good idea if you can swing it.

Some of the other miscellaneous things to consider when selecting an ISP include factors that you might consider when selecting any vendor. Among these are optional services, prices for those services, what support and coaching is available for setting up the use of the ISP, and a hidden gotcha that you really should consider. The key question is, will the ISP still be in business a few months from now? The regional ISPs are often surprisingly small businesses operating on a shoestring. It doesn't take very much equipment to set up an Internet ISP—and many individuals have taken advantage of this fact. We are not recommending that you automatically discriminate against a small ISP; in fact, some of the best ones out there are small organizations. Nonetheless, you may not be comfortable dealing with a very small company, if only because you may have concerns about its ability to handle growth. Again, do your homework.

Internet Futures

There has been a great deal of concern that the Internet will experience spontaneous self-destruction due to its speed of growth. We mentioned earlier that change happens rapidly on the Internet. In fact, there is in common usage a concept called *Internet time*. It is difficult to quantify Internet time, but it means that things that one would expect in other environments to take years take only months or even days on the Internet.

A good example of this is the speed at which, for a while, Internet browsers were being revised by their producers. In fact even now, Netscape releases new beta versions of its Netscape browser on a very frequent basis, sometimes as frequently as once every couple of months. It says something that a high percentage of the people who use browsers on the Internet are actually using beta versions of those browsers just to get the latest features being taken advantage of by the sites that they access.

However, we were talking about the growth of the Internet. There are two major concerns regarding the growth of the Internet:

1. The traffic levels will grow to the point where things will bog down completely.

2. The number of nodes or computers on the network will exceed the number that can be handled by the addressing scheme.

Actually, in many respects the Internet has already run out of gas as far as performance is concerned. For good reason, the Web is described as the *"World Wide Wait"* rather than the *World Wide Web*. All the major backbone providers are adding capacity at a very rapid clip. It remains to be seen whether they can keep up with demand. After all, the growth eventually has to taper off a little bit, since it is unlikely that more than one computer can be used by every man, woman, and child. Once everybody has a computer, we're limited to the growth of the population. Moreover, the population has not been growing nearly as quickly as the Internet has. Of course, that point may not be reached for a while—so in the meantime, we may experience a fair amount of pain. We can't predict whether the Net will self-destruct. We suspect that it will go through phases of slowing and speeding as new capacity comes on-line and new facilities become available on the network, bringing in new users and increasing the use by existing users.

As far as the network address availability is concerned, we're happy to report that the powers that be on the Internet have defined a new standard, IPv6. Going into the details of the IPv6 protocol is really beyond the scope of this book. Suffice it to say that there will not be any problems with the number of nodes that can be addressed by IPv6 any time in the next millennium. Not only that, but the designers have done a really good job of stripping out unnecessary overhead in order to make this protocol inherently slightly faster than the existing IPv4 protocol now in use on the network (version 5 was a specialty protocol that never received general usage). The main question is not whether IPv6 can handle the number of addresses that it will have to handle, but rather how quickly can it be brought on-line. Naturally, there are migration issues, since there are millions of people running software that does not support IPv6 at this time. There is a certain amount of backward compatibility built into the new protocol, but we suspect that the people operating the network will have to dance very quickly to get from where they are now to where they want to be in a smooth fashion.

Voice Over the Internet

Lately, the hype in the industry is the convergence of the voice and data transmission systems. This is nothing new; we have been trying to integrate voice and data for years. However, a subtle difference exists with today's convergence scenarios:

- In the beginning, the convergence operation mandated that data be made to look like voice; then it could be transmitted on the same circuitry. Circuit-switched voice and data used the age-old techniques of a network founded in voice techniques.

- Now, the convergence states that voice will look like data and the two can reside in packetized form on the same networks and circuitry. Circuit-switching technologies are making way for the packet switching technologies.

These changes take advantage of the idle space in voice conversations, where it has been determined that during a conversation, only about 10 to 25 percent of the circuit time is actually utilized to carry the voice. The rest of the time, we are in idle condition by either:

1. Listening to the other end

2. Thinking of a response to a question

3. Breathing between our words

In this idle capacity, the compression of voice stream can facilitate less circuit usage, and encourage the use of a packetized form of voice. Data networking is more efficient because we have been using data packeting for years through packets, frames, or cells.

The use of a packet-switching transmission system allows us to interleave voice and data packets (video too) where there is idle space. As long as a mechanism exists to recoup the information and reassemble it on the receiving end, it can be a more efficient use of bandwidth. It is just this bandwidth utilization and effective saving expectations that has driven the world into a frenzy over packetizing voice and interleaving it on a data network, especially the Internet.

A note on the cost of voice and data communications is probably in order here. *Currently, the drive is to get free voice on a data network.* In the early years of data communications, data always was given a free ride on the voice networks. Telecommunications managers diligently fine-tuned their voice networks and allowed the data to run over the voice networks during the off-hours (after hours). This use of the circuitry was paid for through the dial-up voice communications.

This method of providing data over the voice networks crept into some of the business hours when real-time communications were needed. However, many times, the voice tie-lines and leased lines (point-to-point circuits) usually had some reserve capacity. Therefore, the data was placed on the point-to-point circuits that were justified by voice. As the competition in the telecommunications market heated up, we saw the costs for a minute of long distance drop from $.50–.60 per minute in North America down to an average cost today of about $.05–.10 per minute. This significant drop in cost has been the result of competition and technological enhancements. At the same time, the data convergence took advantage of this falling cost factor. Meanwhile, the data was migrating from a point-to-point technique on a dial-up connection to a more robust packet form of transmission using X.25 or IP packet-switching techniques. The cost per bit of data transmission was dropping rapidly. Then the introduction of the Internet came to the commercial world. With the commercialization of the Internet in the early 1990s, the culture and the cost factors changed at an escalating pace. The cost of data transmission has been touted as being free on the Internet, the only cost being the access fees. However, the cost of data networking has been rapidly declining on private line (Intranets), drawing the attention from both a public and a private networking focus. This convergence has everyone looking for free voice by interleaving voice on the data networks. I agree with the scenario and would not attempt to detract from that goal. However, the other side of the equation is the circuit-switched networks are continually driving the cost per minute for voice down. Closely aligned to this paradigm is the fact that the North American long distance companies are now offering specials in their service. For example, in 1999 many of the carriers offer $.05 per minute on nights and weekends; another carrier offers Fridays for free; and so on. Looking closely at their advertisements, we see a culture change slowly moving its way into the industry. More of the ads state "not quite free, but almost," or "free weekends," or "free Fridays." It is this shift that makes one wonder if free voice on the data networks is as much of a deal as we think. Alternatively, it may be the movement to draw the crowd in. I believe that voice will be free after the turn of the century, with the strategy being "give them the voice, they will buy the data!" However, we have to wait and see how this plays out.

Voice over the Internet Protocol

The public telephone network and the circuit-switching systems are usually taken for granted. Over the past few decades, they have grown to be accepted as almost 100 percent reliable. Manufacturers built in all the necessary stopgaps to prevent downtime and increase lifeline availability of telephony. It was even assumed that when the commercial power failed, the telephony business continued to operate. This did not happen accidentally, but through a very concerted development cycle jointly by the carriers and the manufacturers alike.

Access to a low-cost, high-quality worldwide network is considered essential in today's world. Anything that would jeopardize this access is treated with suspicion. A new paradigm is beginning to occur because more of our basic communications occurs in digital form, transported via packet networks such as IP, ATM, and frame relay. Packet data networking has matured over the same period of time that the voice technologies were maturing. The old basic voice and basic data networks have been replaced with highly reliable networks that carry voice, data, video, and multimedia transmissions. Proprietary solutions manufactured by various providers have fallen to the side, opening the industry to a more open and standards-based environment. In 1996, there was as much data traffic running on the networks as there was voice traffic. Admittedly, industry pundits are all still saying that 90 percent of the revenue in this industry is generated by voice applications. This may be an accounting problem, because on average, 57 percent of all international calls originating in North America going to Europe and Asia are actually carrying fax, not voice. Yet they are considered dial-up voice communications transmissions because of the methodology used. Moreover, the voice market is growing at approximately 3 to 4 percent per year, whereas data is growing at approximately 30 percent per year. Since data traffic is growing much faster than telephone traffic, there has been considerable interest in transporting voice over data networks (as opposed to the more traditional data over voice networks). Support for voice communications using the Internet Protocol (IP), which is usually just called *Voice over IP* or *VoIP*, has become especially attractive given the low-cost, flat-rate pricing of the public Internet. In fact, toll-quality telephony over IP has now become one of the important steps leading to the convergence of the voice, video, and data communications industries. The feasibility of carrying voice and call-signaling messages over the Internet has already been demonstrated. Delivering high-quality commercial products, establishing public services, and convincing users to buy into the vision are all still in their infancy. The evolution of all networks begins this way, so there is no mystique in it.

IP telephony will also have to change, somewhat. We will expect it to deliver interpersonal communications that end users are already accustomed to using. These added capabilities will include (but not be limited to):

- Calling line ID (CLID)
- Three-way calling
- Call transfer
- Voice Mail
- Voice-to-text conversions

Users are very comfortable with the services and capabilities delivered by the telephone companies on the standard dial-up telephone set using the touch-tone pad. IP telephony will have to match these services and ease-of-use functions in order to be successful.

IP telephony will not replace the circuit switched telephone networks overnight; this will be a coexistence for the near future. Analysts expect that in 2003 the amount of IP telephony will amount to 3 percent of all voice traffic domestically, and approximately 10 to 15 percent of international traffic. This amounts to 50 billion minutes of traffic, so it is consequential. One must be prepared for both alternatives to carrying voice in the next decade. Thus, the differences between the two opposing network strategies will be ironed out and the world may shift into a packet-switched voice network over the next decade.

Quality of Service

One of the arguments against IP-based telephony today is the lack of quality of service (QoS). The manufacturers and developers will have to overcome the objections by producing transmission systems that will assure a quality of service for lifeline voice communications. Mission-critical applications in the corporate world will also demand the ability to have a specified grade of service available. The CTI applications with call centers being web-enabled, interactive voice recognition, response, and other speech activated technologies will demand a quality of service to facilitate the use of these systems. Each will demand the grade and quality of service expected in the telephone industry.

Another critical application for IP telephony will be the results of quality of voice transmission. Noisy lines, delays in voice delivery, and clicking and chipping all tend to frustrate users on a voice network.

	Strategy	Description
TABLE 16.1 Different Approaches to QoS	Integrated Services Architectures (Int-Serv)	Int-Serv includes the specifications to reserve network resources in support of a specific application. Using the RSVP protocol, the application or user can request and allocate sufficient bandwidth to support the short- or long-term connection. This is a partial solution because Int-Serv does not scale well, because each networking device (routers and switches) must maintain and manage the information for each flow established across their path.
	Differentiated Services (Diff-Serv)	Easier to use than Int-Serv, Diff-Serv uses a different mechanism to handle the flow across the network. Instead of trying to manage individual flows and per-flow signaling needs, Diff-Serv uses DS bits in the header to recognize the flow and the need for QoS on a particular datagram-by-datagram basis. This is more scalable than Int-Serv and does not rely solely on RSVP to control flows.
	802.1p Prioritization	The IEEE standard specifies a prority scheme for the layer 2 switching in a switched-LAN. When a packet leaves a subnetwork or a domain, the 802.1p priority can be mapped to Diff-Serv to satisfy the layer 2 switching demands across the network.

TABLE 16.2

QoS Requirements
for IP Telephony

Layer Addressed	Technique	Variable
1	Physical Port	Variations of port definitions, or the prioritization of port interfaces based on application
2	IEEE 802.1p Bits	Dedicated paths or ports for high bandwidth applications, but very expensive to maintain
3	IP addressing	RSVP protocol (Int-Serv) DS bits in the IP header (Diff-Serv)

Packet data networks carrying voice services today may produce the same results. Therefore, overcoming these pitfalls is essential to the success and acceptance of Voice over IP telephony applications. Merely installing more capacity (bandwidth) is not a solution to the problem; it is a temporary fix. Instead, developers must concentrate on delivering several solution sets to the industry such as those shown in Table 16.1.

The QoS requirements for IP telephony can therefore be summarized as shown in Table 16.2, which considers the layered approach that vendors will be aggressively pursuing. IP telephony datagrams entering the network will be treated with a priority to deliver the QoS expected by the end user. The routers and switches in the network will assign a high priority marking on each datagram carrying voice, and treat these datagrams specially. Queues throughout the network will be established with variable treatments to handle the voice datagrams first, followed by the data datagrams.

Given the differences, why should we even consider the use of IP telephony? If the IP telephony world is only going to account for 3 percent of domestic traffic by 2003, is it worth all the hassles? The answer is a mixed bag, but overall the efficiencies of VoIP will outweigh the need to develop better control mechanisms to satisfy the telephony industry.

Intranet

Introduction

With all of the hype and fanfare about the Internet that has surfaced since the early 1990s, a lot of organizations have spent inordinate amounts of money to use and facilitate access to the wide area through the World Wide Web (WWW). These organizations saw the benefits of providing the capabilities of remote access and catalog information access for customers to log onto their Internet-based server. Through the use of the graphical user interface (GUI), organizations have allowed customers to pick up information and dispense or print it locally. This eliminates a lot of the costs associated with mailing flyers and catalogs. Customers can literally print down their own catalogs as opposed to waiting for information to come through the postal system. Beyond the savings of the catalog printing, there's also the postage savings for the companies using this distribution method.

With the emphasis on electronic access to document information and delivery through an architecture called the *Internet*, large corporations began to view this as an opportunity to support their internal client/server architectures. Therefore these larger organizations began to use the same tools and techniques consistent with those already available on the Internet to build an *Intranet*. The Intranet becomes an internal Internet, so to speak. As more and more users within organizations moved away from the terminal device (dumb terminal) and became more PC savvy, they began looking for the ability to conduct and transact business on-line. The Internet allowed this as a transport vehicle for most of the consumer access capabilities. The Intranet provides the same capabilities and features inside the organization. The desktop device, a la PC, can be used with a Web browser to scan for information that will reside on a large Web-based server that is for exclusive use by internal organizations. Moreover, organizations can blend together the combination of both Intranet and Internet services on a single server-based platform, as long as isolation and separation can be maintained. It is here where much of the current spending and emphasis is now placed. Users are becoming pros on the Internet, so they now are becoming internal pros at managing, maintaining, and using their Intranets.

Organizations have been impressed with the ease of use and deployment capabilities of the Internet and have seen it as an opportunity to change the way the internal infrastructure works. By using Web servers as mentioned, Web browsers, and hypertext-based applications, they are employing the same technology as their own internal business communications tool. These organizations have discovered that the same features that facilitate the Internet architecture and make it attractive for inter-enterprise communications also make it an excellent technology for intra-enterprise communications. In particular, very large organizations that have various application servers, hardware platforms, and multiple sources of information (such as on mainframes) can now use this Intranet server capability to provide for a single user interface on a hardware independent platform. With the various categories of operating systems and hardware systems (PCs and Macs), organizations can now take advantage of the ability to merge the hardware platforms into a single homogenous network.

This movement toward the Intranet technology is heavily supported by other industry trends. For example, TCP/IP—the protocol used throughout the Internet (and developed for the Internet)—has become the de facto standard as a multivendor protocol in many corporate networks because of its robustness and its simplicity. In the area of e-mail, simple mail transfer protocol (SMTP) is rapidly becoming the popular standard for internal e-mail systems. Moving away from the proprietary architectures of the past, organizations now use the SMTP and POP server capabilities to facilitate and to allow access for users who want to "surf" an internal net as well as navigate and access all organizational data housed on differing computers and operating system software. The browsers used to do this allow users to get away from using the old command line interface, which was cumbersome and platform-specific. Users were faced the challenge of remembering varying keystrokes and organizational needs to be able to talk to different computing systems such as IBM3270 platforms and UNIX. Beyond the TCP/IP platform and protocols, users have simplified access to e-mail systems through the use of the point-to-point protocol (PPP) that allows them to access Internet resources as well as internal mail system information, using a remote serial communications connection. These two simplification protocols have really expanded much of what takes place in an office today.

With the use of a single point-and-click arrangement within an organization, the information residing on an IBM mainframe, a UNIX server, a Novell or NT server, or a PC-based LAN or any other computing platform can be accessed through the same easy-to-use arrangement. This allows universal access to data on the different systems as well as different applications. In a sense, Internet and Web technology is playing much the same role as the proprietary solutions such as the "groupware" technology used on Lotus Notes or Novell's GroupWise services. The Web server's ability to provide access to a variety of information—even the multimedia applications, files, and video streams—enables it to act much like a

Lotus Notes database, yet it can cost much less when an open interface as opposed to the proprietary interface is used. In the future, the Intranet technology can be expanded to incorporate all collaborative functions (such as those offered in Notes or in Netscape's Collabra software) and broader support for the multimedia technology (such as desktop video communications). A significant amount of work has been done in most organizations to provide collaborative computing capabilities between various departments. For example, in an advertising or marketing department, much has been done to interface and to integrate the dynamic work group capabilities between artists, freelancers, copywriters, and final editors. Through the use of a single GUI-based environment, these organizations and departments will have access to each other's information or any work in progress. Each of the users will have access to the data so that they can manipulate or view status on a project under way. This allows for a much more dynamic environment within the organization without extensive retraining and application-specific programs on everybody's desk. Much of the time and cost associated with the development of a new campaign or advertising literature is accrued in the back-and-forth information flow between these various groups, both externally as well as internally. By deploying the information through an Intranet, that information can be made readily accessible and available to anyone at any time. The delays in getting the information back and forth between the responsible persons leads to market delivery being continually extended from inception to final product. The use of Intranet technologies will pare down the development cycle by days if not weeks. The cost of transmitting or faxing documents, or expressing packages both domestically and internationally, can be reduced significantly through the use of an Intranet.

Issues

The use of the Internet technology inside the organization on the Intranet therefore gives credence to the fact that electronic access and delivery services should be deployed using these types of capabilities. From an external communications perspective, the benefits are very clear. Because many parties with whom organizations will deal use the Internet, it provides a common method of handling external connectivity. Because it is also widely used by organizations and individuals throughout the world, it provides an excellent platform for future growth within any organization. The costs associated with transmitting information from the Intranet to the Internet can easily be justified on the basis of the simple-to-use dial-up communications technology. In many cases, a local call will allow access to either the Internet or Intranet and therefore allow users to share information regardless of where they are in the world. Most of the commercial on-line services provide links to the Internet. Internet services are typically inexpensive. Over time, it would be safe to say that the organization would be able to reach virtually any computer with a modem from an Intranet to the Internet.

Another of the key advantages of this technology is the use of the Web browser. Web browsers tend to be somewhat platform independent and offer the same look and feel to a user regardless of the hardware or operating system. These can typically be purchased inexpensively at the local computer stores or downloaded free to individual users. They are typically built into many of the operating systems on the market today (such as the Windows 9X Internet Explorer). Based on a hypertext technology, Web software is generally very simple to use. This makes it a natural platform for internal user applications such as directories, employee information bulletin boards, news and highlights within the organization, and access to the data warehouse. The costs associated with a training program and with introducing users to a proprietary operating system versus the cost of learning to use a Web browser are significantly different. The Web browser software is much less expensive to implement and requires much less user training regardless of whether the software is on a PC, Mac, or now the newer Net PCs.

For Intranet users, an internal Web site used to distribute corporate information to employees to accommodate day-to-day business operations makes sense. The organization can develop a team concept for production methods from inception to final product development. By using the Intranet, these team members can easily be kept up to date on product status, project meeting notes, and other incremental information based on a project at hand. This will empower employees to share business data using the same friendly Web browsers that they use to access their external information over the Internet. Moreover, the Intranet will enable linkage between various business units in an organization so that those employees can easily navigate throughout the available data for more complete information at any time.

This approach also facilitates another important element within the organization: off-site access by employees. By building a common set of Internet-based access and delivery mechanisms, the organization can allow remote users (either road warriors or employees working at home) to easily access the information resources. Off-site employees could use the Internet connections to pull down company data and information, send e-mail, or upload new business information at a moment's notice. In other cases, mobile employees and home-based employees can access faster, more highly secured links directly to the internal Intranet while still having the Internet accessible to them.

The Architecture

The ideal access mechanism concerns itself primarily with an infrastructure, which includes the following areas:

- External access
- E-mail
- Directory services
- Security
- Archival systems
- Web services
- Fax servers

In Fig. 17.1 a graphic representation of how the Internet and Intranet might work is shown. In this figure, the LAN within the organization is shown at left. Many different organizational departments can be attached through these different LAN servers. A high-speed backbone infrastructure would be used to link all the servers and the departments together. Just to the right of the server backbone are several different components, called *Intranet servers*, that would now be used to interlink these different departments. The Intranet servers can be one or many servers plugged into the network for access by various departments. These servers function like warehouses, storing much of the corporate information and allowing all users with the rights and privileges assigned to access corporate information with the graphical user interface. The browser would allow for the connection to the various hosts based on an IP addressing scheme used internally within the organization. Beyond the Intranet servers would be a mail server that serves the entire enterprise. This mail server would run the standard SMTP and POP protocols to supply electronic mail internally as well as to provide a gateway capability to the external world. The enterprise mail server as well as the LAN itself would be attached through a high-speed router connected to a device called a

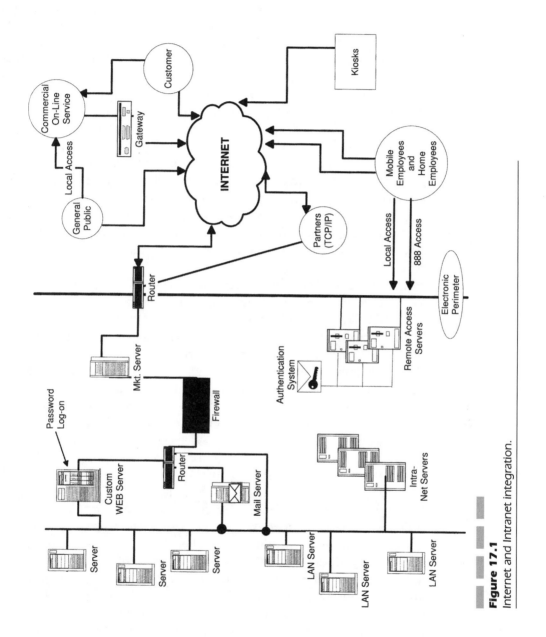

Figure 17.1
Internet and Intranet integration.

firewall. Note that extending off the router is another connection called a customer Web server. This web server would be internal to an organization but allow customers to access electronic document interchange (EDI), order entry, or inventory and status on projects under way. This would be an internally protected Web server using password and log-on capabilities to secure and protect the information as well as provide a barrier to keep users from getting onto the LAN itself. In the center of the Intranet is the firewall, and all access through the firewall is to the router.

Just to the right of the firewall we see a marketing server, which would be optional. This marketing server might be something that allows for customers to dial in or contact the organization from a front-end perspective. This marketing server would provide the general public with access to catalog information that the organization offers to the consuming public. In the event an organization or an employee accessing the network remotely needed to reach any of the other servers, the marketing server would in turn pass the information through to the firewall for access to the backbone network.

Beyond the marketing server is another router. This router is the external router, as opposed to the internal router as shown on the left of Fig. 17.1. This external router is also the barrier between the outside world and the Intranet. On this router, specific access is made on the basis of different types of services. Note that hanging from the bottom of that router is a LAN operating at 10 or 100 Mbps and on that LAN are several remote access servers for the remote dial-in users. These would provide authentication services from a security standpoint to protect the network from allowing just anyone to get through. These authentication servers and remote access servers would provide for telecommuters and home office workers to access all the information on the Intranet as needed. One can see the depth and complexity this infrastructure set up. To the right side of Fig. 17.1, we have the external communications world with the access to this router on dedicated facilities (such as leased line and T1 access) into the Internet. Several other methods are being used to access both the Internet and, passing through the Internet, the Intranet. This can include mobile users who are traveling and using the Internet as the access method, or general public users and connected on-line services. Several other access methods are provided either on dial-up or leased line connections into the remote access servers. Access to the router can also be provided through either the Internet or the public switched telephone network, allowing additional types of access. One can see that boundaries and separations between the external and internal worlds are a must. However, through these different access methods, the user who has authorization will be able to access just about any service that is necessary. For those not authorized, the combination of routers and firewalls will deny the access and protect the Intranet, and therefore the company data.

Accessing the Network

At this organization's main location, external connectivity for end users is already improved. Individual modems that normally would have been used at the user's desktop to provide external dial-up connection to outside resources would now be replaced with a server providing a modem pool arrangement. This step has to be taken, because all modems must be protected in an environment like this. By using a centralized modem pooling arrangement (such as a modem server), the number and type of modems can be minimized. Modems are only occasionally used by end users, so the pooling allows the resources to be available as needed but eliminates the need for a one-to-one ratio. This approach will provide improved price performance as well as the authentication via some token-based access security mechanism.

Because this approach eliminates the need for employee traffic to pass through the Internet firewall, greater accessibility and flexibility for the organization's resources is provided. The employee can use the Internet if desired, but will be limited to e-mail and access to a small number of pre-defined services. Traveling (nomadic) employees outside the organizational headquarters can be given an 800/888/877 number to remotely access the server, or they can use a local Internet point to presence (POP) if they so desire. By taking this structured approach, the organization will reduce its reliance on costly and disparate means of access. This approach is also more secure because it consolidates external network access via only one of two methods: the firewall or the remote access authentication server. Consolidating access should make external links easier to monitor and manage. Using the Internet will bring along reduced costs.

Another benefit to this approach is consolidation and centrally manageable IP address translations. If the organization chooses to use a high-end addressing system internally (such as a class A address) but at the Internet use a class C externally, address translation will be required at the Internet firewall. Remote access servers will dynamically assign internal class A addresses to individual users when they dial into the organization, so that no address translation will be necessary for these devices. A standard procedure to follow when approaching this translation need is the use of RFC1597.* In Fig. 17.2 is a general description of how this address translation can work. With RFC1597, the entire organization will use a private network with a Class A address (10.XXX.XXX.XXX, which has not been assigned or registered to any organization). However, the only access to the Internet would be through the firewall hosts that perform the address translation between the internal network address and one or more of the registered class C addresses (e.g., YYY.YYY.YYY.XXX). To the outside world, every host on this network will appear to belong to one or more of the organization's class C networks.

The firewall host will maintain an address translation table with an entry for each of the internal hosts that requires the Internet access. One alternative for each host or user is to preregister its internal address with the firewall to be assigned an external compatible address; adding an entry to the address table will permit up to 256 hosts to have an inbound and outbound Internet access. Alternatively, the address translation function in the host or firewall could be dynamically assigned 1 of the 256

*RFC is a request for comment issued by the Internet Engineering Task Force (IETF) describing how services should work.

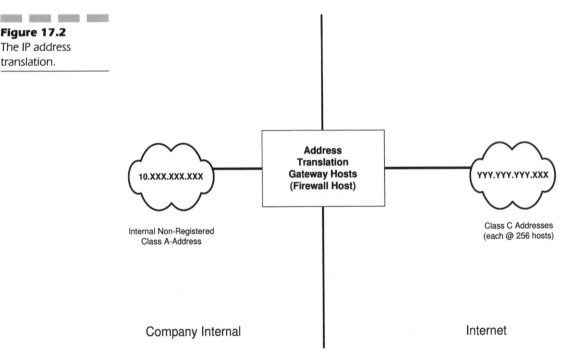

Figure 17.2
The IP address translation.

class C addresses at the time of connection through the firewall. This is through a dynamic host controlled protocol (DHCP) that allows this capability. Now, any IP messages outbound from the Intranet to the Internet will pass through the gateway host function. The gateway host will swap the source address (from the internal to the external Class C address) but leave the destination address alone. For any messages inbound from the Internet to the organization, the gateway host will swap the destination address from the registered external class C number to an internal private class A address but leave the source address alone. In this approach, there is no ambiguity regarding the destination address. This will permit standardized packet-level gateway access to the Internet as opposed to only application-level gateway functionality or use of a cumbersome dual log-on procedure. This strategy enables the organization to migrate gradually to the newer IP version 6 addressing standard, converting internal addresses before changing external addresses (or vice versa).

Intranet: The Logical Evolution

One can imagine an organization with hundreds of sites located and dispersed around the country or the world. Thousands of employees need access to each and every database for timely information on policy changes, personnel guidelines, training materials, company directories and phone books, product and pricing information for sales organizations, and general dissemination of information. Where individual work groups and departmental requirements provide for secure, unlimited access to their data, the organization now has a need to share this information with all employees on either a wide-open or a limited basis. Some information must remain confidential, but that can also be facilitated. Each geographic region within an organization typically has its own separate databases and servers and provides its own reporting functions that should be shared with the corporate headquarters. In the past, these reports and databases would be transmitted via electronic means. During the transmission, a conversion process would take place that might eliminate formatting. Manual conversion and translation processes for reformatting are labor intensive and time consuming. Beyond that, if the information is not electronically transmitted, then a hard paper copy must be transmitted via fax or through the postal service. In either case, this hard paper copy would then have to be reentered into the computer database at the corporate headquarters. One can imagine the amount of time and money that is being lost in this manual processing of information when disparate systems exist. Also included is the risk of error when data is re-keyed. The disparity in systems occurs because each regional or geographic location may well have installed its own hardware and software operating systems on a site-by-site basis, as opposed to a global infrastructure on a corporate-wide basis.

Traditionally, organizations have always had multiple policy and procedure handbooks, printed materials, newsletters, and pricing catalogs produced at some centralized location. All of the printed materials are time consuming and expensive to produce and do not necessarily generate revenue in their form and format. Trying to guarantee that this information is transmitted or shipped across the country (or the world) in a timely and reasonable fashion has always been a headache. Consider all the costs associated with this, such as the data gathering coming from multiple sites with all of the associated feedback cycles necessary to ensure accuracy, the production process in terms of printing and typesetting and final production, and then the distribution, whereby boxes of materials are packaged and shipped at great expense using surface or air transportation. If all of this could be done electronically on a common, user-friendly interface, it is natural that the organization would want to consider this.

When viewing today's cost-cutting environment and the need to do "more with less", one can hardly afford to eliminate all of the communications necessary. However, the staff (being limited) will not have the time to manage and administer all of the various informational tools necessary to run the organization. In today's competitive environment, timely access to the information is probably more mission critical than ever in the past. A solution to this would be just to have a broadcast e-mail system and send the information out to all employees. However, this method is inefficient, because users and employees will view the mail at differing times or they will scan the mail and not necessarily note the changes that are being introduced. As the mail is being sent out, the information is concurrently generating multiple stored messages in everybody's mailbox, consuming hard disk space as well as producing a directory or mailbox that is being jam-stuffed (comparing it to the in box in a person's office). If, in fact, the mail is starting to stuff the boxes, users can only access, approach, and view a certain amount of information when needed. When information is not needed, the mail will either be discarded or filed for later retrieval. This, however, creates the other scenario: "Where did I file this one?"

As a means of overcoming some of these problems and pitfalls, the Intranet was the logical solution to produce a cost-efficient and timely approach to the distribution of information. Using the World Wide Web (WWW) technology, the *intraprise* has become the global perspective. At the foundation of the Internet is the server providing the WWW interface.

Organizations that built their WWW sites store and update information electronically on this WWW server attached to a LAN. As information changes, the server content can be updated with the new or revised data. This allows the organizational employees to access information in a timely and consistent manner and provides accurate information to all employees without the expense of typesetting, printing, distributing, and mailing all the changes. In a very large organization, multiple servers may well exist on the differing LANs spread across the country. However, if employees need a price list from a group that is located halfway around the world, they need only access the world wide server to research and gather the data using the GUI. Using the point-and-click method, the employee will find the host (the server) and access the information for immediate viewing. In the event a printed copy is needed, the file can be downloaded to the local printer on the LAN and thus be immediately available to the user in hard copy. This prevents the need for having multiple manuals, price lists, and catalogs sitting on people's desks, rarely accessed or used. When the information is needed, it is there; however, when it's not, it doesn't need to be stored on the bookshelves.

For highly confidential information, security on the World Wide Web site offers numerous advantages for internal exchange of information. As an example of this, an organization can restrict information to certain users or departments, such as human resources and financial data, allowing users to access the information securely. This can be done using the Internet as an *external* backbone and the Intranet as the *internal* backbone linking the two sites together. In this scenario, the Human Resources department of a large organization can share ladder plan strategies, salary planning information, personnel reviews and files, or any other type of data in a secure method. In any case, the server would be far easier to configure, use, and manage. Individual home pages can be set up for each of the departments or functional areas. This would allow for an easy-to-use interface on a home-page-by-home-page basis. Additionally, when corporate information needs to be distributed, the home page can allow for video and audio downloads to discuss notes or special considerations by providing separate interface information.

Is It for Everyone?

Clearly, organizations that have multiple locations spread across the country and hundreds or thousands of employees provide a good fit for this Internet/Intranet technology. Because of the disparity and the distribution necessary to keep information flowing, the Intranet becomes a logical choice. However, more and more applications are being developed and Intranets are being deployed for smaller organizations with a handful of sites and possibly 50 or less employees. This is particularly true when dealing with the *virtual office*. The virtual office allows for the small office and home office* locations to access information through either dial-up access or Internet access into the main headquarters infrastructure. This prevents home office users from having books stacked all over the place. Instead, when the information is needed, the home office users need only access it through their Intranet. Several other applications may tie into the Intranet. An example of this would be the CTI, as described in an earlier chapter, linked together with the home page or the Intranet so that a home office worker can call a specific telephone number and, if the home user's telephone number matches an approved access number, readily access information available on the Intranet. This would imply that CTI is used both at the home office and the corporate office. Combining features and services through these telecommunications techniques allows for accessibility of information regardless of where the worker is.

What this means, then, is that the Intranet, once viewed as the large corporate infrastructure, can now move down into the smaller organizations in much the same way it is used in the larger ones. The real name of the game is communications capability. By linking services together through a communications infrastructure, organizations have the ability to provide the access whenever, wherever, and however they choose. Printing resources can be consolidated and locally provided based on Intranet technology and the dial-up or remote access services. Therefore, the Intranet is a logical evolution and a good fit for the both the large and the small organization that needs to provide timely, accurate, and efficient information regardless of the employee location. One can see the benefits and the challenges of building an Intranet within an organization, comparing it to some of the higher-end cost solutions provided by the proprietary vendors already offering services. Is it any wonder that many manufacturers of these proprietary solutions have now adopted the name *Intranet* as part of their business solution packages? Intranet will be one of the major communications infrastructures that will roll through the turn of the century. If an organization has not yet begun to deploy this architecture, it may well be in a position of not having the information readily available and lose its competitive edge. It is through the competitive edge that organizations prosper and grow, and it is through technology that this is accomplished.

---

*SOHO—small office and home office

The Full-Service Intranet

Companies like Microsoft and Netscape have a vision of how organizations can use standard Internet technologies to deploy fully functional ubiquitous communications access for information sharing and applications built on top of open networking technologies and on an open network-based application platform.*

The full-service Intranet is a concept that was first envisioned in early 1995 and that was talked about throughout 1996, ultimately becoming a reality in late 1996 and early 1997. A full-service Intranet is supposed to include the ability to run a TCP/IP network inside an organization that provides connectivity to all the people and all the information. This ideally increases productivity and allows navigational services through the computing environment on a seamless-based network. This full-service Intranet takes advantage of all the open standards and protocols that emerged as a result of the Internet. When an open standards interface is used, applications and services like e-mail, groupware, security systems, archival systems, database access, management, and directory services can all be more powerful than they are when run on traditional proprietary systems. Because of the openness in this architecture, end users can reap all the benefits of the cost, platform support, flexibility, and independence from operating systems and vendors and gain the ability to leverage their products and data as a single entity.

Actually, the model is really described in terms of services rather than protocols and standards. Services are provided as a direct result of the interface to all the software and applications. Because Intranet software runs across a client and server architecture, a common network environment spans even heterogeneous networks. Therefore the services provide users with the capabilities of looking up any information, sending and receiving mail, searching directories, using facsimile servers, and providing access to the outside world via the Internet. They also allow for custom application development from a myriad of third-party providers for sales automation, help desk functions, and financial applications (such as loan calculators and mortgage calculators), and take advantage of all of the capabilities of replication and security. The intent of the Intranet is to make integrated technology (IT) simple and easy to use technology and to simplify the process of building and managing a network for the IT professionals. A quick note here: there is a potential shortage of qualified

---

*From Netscape Communications, Inc.

and talented IT professionals facing all industries. By the year 2000, it is anticipated that a 35 percent shortfall in IT skills will be evident. If, in fact, we can use the capabilities of the Intranet to manage technologies far more simply, we can overcome the limitations and shortfalls in talents in the future.

Companies like Netscape envision their services of an Intranet in four basic categories:

- Information sharing and management
- Communications and collaboration
- Navigation
- Application access

Beyond the user services, the full-service Intranet as portrayed by Netscape also includes four major network services:

- Directory
- Replication
- Security
- Management

The services provided by a full-service Intranet will drive the network to a focal point of business computing. The days of old, where the desktop PC was dominated by a command line interface and proprietary operating software, are just that—old. Using an open architecture and the navigational browsers, an end user sitting at a desktop PC will have the basis of being able to access any piece of information. As a result, what are called network-centric environments can be created. Some of the features and capabilities of these environments are as follows:

- Easy to use and implement; avoid the need to replace all hardware systems and operating software.
- Encompass all of the existing investments on desktop computers, mainframes, applications, and databases that already exist.
- Provide a network that allows for interoperability and Internetworking capabilities. This allows users to directly address and access any associated equipment existing in the Intranet.
- Write once—read many: the ability to write a document or information and distribute or replicate it across the network to multiple servers so that it can be read and accessed by many different locations either simultaneously or on an ad hoc basis.

- Centrally managed, so that ease of use and operational stability can be accomplished within an Intranet.

- Generally lower in cost; using a simple, robust set of applications and the Web browsers provides for less expensive communications infrastructure. This avoids the need to buy separate proprietary and domain-based software and operating systems that can be very expensive.

- Flexible operational capabilities; the ability to use individual vendors and operating systems on an open standards-based platform.

- Proven and mature, as evidenced by the several major Fortune 100 companies that have deployed an Intranet on an enterprise or an intraprise environment.

Applications

When you really think about it, it would be difficult to list all of the applications that could fit on an Intranet. However, the obvious applications that most folks will attempt to integrate and deploy would include such things as:

- Database access
- On-line financial transactions
- Inventory analysis
- Order entry
- Company newsletters
- Job posting and opportunities
- Corporate-wide public relations announcements

Although these are the primary applications that have been implemented and used extensively by organizations, a wealth of other opportunities could exist, including the routine day-to-day operations that an organization may also need to provide resource sharing. When one thinks of the day-to-day operation, it takes just a little imagination to realize how the integration of the databases can supplement these operations. But taking that a step further, on-line forms provide for things such as:

- Expensive report submission and printing
- Help desk functions

- Frequently asked questions about technology

- Questions regarding medical expenses or coverage

- Educational opportunities or training classes available within the organization

- Outside educational courses that are approved and accepted by the organization

- Supply ordering and forms

- Maintenance requests for internal or external facilities-based applications

- Security issues, such as application for a new security card

- Automotive expenses

- Travel requests

- Overtime production and scheduling

When one thinks of the applications that could be put on an on-line form, whereby an end user within an organization using the Intranet can access the forms and fill them out in a user-friendly fashion then submit them directly to the corresponding department, the possibilities become endless. Updating directories, maintenance of policies and procedures, suggested changes in operations, or any other application could all become part of the Intranet capability. For example, think of how, if a user notices an operation inefficiency, an on-line suggestion box could simply be directed to the quality control group or to senior management for administration and action. All of the possibilities that today require employees to fill out extensive forms and store paper everywhere within the organization could literally become electronic applications on the Intranet. Using these capabilities, the organization could save a fortune in printing and distribution costs alone. Mail handling through an organization for interoffice communications is an expensive proposition. By using the electronic distribution, the amount of mail handling personnel might be equally reduced, supporting the Intranet in its truest distribution form. This book is not trying to encourage the reduction of operating personnel, but rather the use of people already on board in a more productive manner. An inordinate amount of waste in an organization surrounds filling out and submitting the appropriate forms. Studies have been done around the world showing that the cost and preparation for documentation and administrative handling, postage requests, and things of that nature, is excessive and redundant. It is this multiple handling of paper and documents that impedes truly getting the job done. By routing the appropriate form electronically throughout an organization, the handling can be minimized and the work can be done efficiently. This is the goal of the Intranet.

Getting There from Here

Creating an internal World Wide Web application and an Intranet site, although low in cost and a minimal investment, is still a challenge regardless of who is chosen to do it. Although it will be easy to implement, with little training and equipment needed, there are still decisions to be made before the product is rolled out. The basic system should consist of a server with an operating system and some Web server software. Assuming that the organization already has PCs in place, then the client investment is probably going to be relatively insignificant. One can imagine that most users will either have an Internet explorer or Netscape browser already on their PCs, as this has become quite common. In many cases, browser software is already included in the operating system on PCs purchased today. That being the case, client software is probably simple enough to deploy and implement. As a matter of fact, one would hope that in developing the client access to the Intranet, multiple browsers could be used, depending on what is already on the end user's desktop device.

At the server, the rule of thumb is that server hardware with sufficient memory and disk space to run one of the operating systems such as Windows NT, Windows 95, and/or UNIX will depend particularly on the preference and in-house expertise. The department responsible for the Intranet will need to configure the hardware with LAN cards, if not already installed, for a TCP/IP connection over the network to the clients. In many cases, more organizations are opting for either the Windows NT or Windows 95 interface because of ease of use and their open architecture. This is not an endorsement or a directional statement, merely an observation.

The Web server software will enable the organization to manage the internal presence on the Intranet. The right server software will provide the functionality required to set up and manage home pages and develop the content base in hypertext markup language (HTML), as already discussed in the section on the Internet. Further, the Web server software will also use the ability to provide search engine capabilities and integrate with internal corporate databases and back office functionality.

The typical interface at the end user device will likely involve the PC running a 486 or 586 (Pentium) and a minimum of 8 to 16 megabytes of random access memory (RAM) to run NT or Windows client browsers. Typically, a commercial browser would cost around $40, although freeware versions are always available for downloading off the Internet. The client browser can launch a variety of applications, access the databases, and retrieve information from across the Internet or Intranet. Client software is also required to generate the HTML code so that it can be added and converted on a document-handling basis. It is very easy to develop the content for a Web using one of the inexpensive authoring tools, including Microsoft's Internet Assistant (a freebie). Depending on the organization's requirements, you can also take advantage of numerous other tools available, including the graphics, software, and packages to convert frame maker documents into HTML as well as integrating text retrieval and indexing software, links to database management systems, and server configurations or management tools.

Therefore the decisions appear to be fairly straightforward and simple. However, decisions still required include such things as:

■ Should the server run a UNIX/LINUX, NT, or Windows 9X operating environment?

■ Do employees require access to all information or just certain subsets of it?

■ Do employees need access to the databases such as SQL?

■ Should information be made available to all employees, or just certain employees or departments? If only a small percentage of the employees require the access, how will data be secured from others reaching or accessing it?

■ What form of security will be used to secure the Intranet from the Internet?

■ Does the firewall exist, or will one need to be purchased? In either case, is the expertise available to install and maintain the firewall database?

■ Will multiple browsers be used? If so, what types of access and support, such as HTTP, gopher, FTP, or Telnet, will be required? Does the server support these various protocols?

■ Will multiple home pages be installed on the same server? If so, how easy will they be to manage and maintain?

■ Is remote administration a requirement?

■ Who will be installing the server? Will this be IS, IT, or LAN administration? How easy will it be to install?

- What type of interface will the server use? Will it be intuitive enough and easy to use?

- What type of search engine and text retrieval is supported on this server?

- What addressing schemes does the server support?

- Who will set up the home page and maintain it? Is a Web Master available or is training required?

- Who has responsibility for managing the content? Will this be the marketing group, technical support, or a separate Web Master?

- Are there special configuration requirements on your hardware platform?

- Will on-line help and support be available? Who will script it, support it, and maintain it?

- Under what group or department will the Intranet reside? Is this IS/IT?

- How will users request updates or access to information that may not already be available to them? Is this an application of the Intranet because of its on-line form?

- How does management support play in this whole scenario? Is management convinced that an Intranet will have benefit?

One can clearly see that the decisions and the implementation of an Intranet, although simple, still require some thought before installation is attempted. Management must support the idea and make the decision on how the process will take place.

Other Issues

Clearly, as the organization places all of its data on an Intranet, accessible to all, there are other issues that must be dealt with. When users can access the data, will they be able to edit and delete, or just to read? What would be the scenario if a disgruntled employee decided to go onto the home pages and literally destroy them? These issues are a little more sensitive, but they must be addressed. By placing all of the corporate data and access to that data on an Intranet, productivity will be improved; however, one must not go into this with rose-colored glasses and think that all problems can be solved on a single platform. The issues of security, manageability, and recoverability in the event of loss or destruction of information must also be addressed. The appropriate groups must be assigned and selected before a deployment is started. As users progress and roll out the access to the data, a flexible approach must also be used. If things aren't exactly the way they should be, then the infrastructure has to be put in place to react and adapt very quickly. If the form or format of the data is not readable by all browsers (there are many versions of browsers available on the market today), or if in fact the data is not capable of being downloaded due to incompatibility of some systems, then issues must be resolved quickly. The Intranet is an exciting opportunity for many organizations to quickly link all of their databases and information on a single hardware-independent platform, but still poses problems and frustrations if not approached correctly.

Still other issues might address the bandwidth available on the LAN or WAN. Normal word processing documents that might be significantly smaller in a word processing system may increase in size when all of the GUI information is included. Any graphical user interface will clearly increase the size of the files, including graphic interchange format (GIF) or Joint Photographic Experts Group (JPEG) files, tagged image file format (TIFF) or even bit map images. As the size and the complexity of the document increases, the decision on how the bandwidth will be handled is a must. Will this involve an upgrade to the entire infrastructure on the LAN? Clearly, opportunities exist to provide for higher throughput capabilities on the LAN by going from 10 to 100 Mbps. Another opportunity is to leave the 10 Mbps at the desktop and use a switched hub inside a telephone closet. It involves some hardware implementation and deployment, but this can be kept to a minimum if planned properly. The worst-case scenario would be to deploy the Intranet and discover that the capacity and throughput is not there. Users will not accept the same throughput that they see on the Internet. When dialing into the Internet with modem speed communications in the Kbps range (28.8–33.6 Kbps), users tend to accept slow response and slow downloads. However, on the local area network, if that same throughput and response were made available to the users, they would burn down the MIS department. Therefore, on the higher speed LANs at 4, 10, or 16 Mbps, the capacity assumes that the LAN is not already bogged down. If, in fact, the GUIs and pictures increase the network demands, the response could be slowed down significantly. Therefore it behooves the administrator to analyze and understand in advance the capacity and actual throughput on the LAN today. If, in fact, an Intranet will be deployed, then the necessary upgrades to the LAN will have to be considered as part of the budgeted items.

This again is not a condemnation of the Intranet, but merely a caution that tools do exist and capabilities can be provided so long as the appropriate bandwidth is there to support them. Refer to the sections on LANs, Ethernet, and token ring for more details on the throughput capabilities and the response times for the LAN.

Using an Intranet can be an exciting opportunity to provide high-speed throughput capabilities and easy-to-use information and to eliminate the amount of training necessary at the desktop user device. This is one area that will be a hot button at the turn of the century.

18

Packet Switching Technologies (X.25)

We have spent a considerable amount of time in past chapters covering the pros and cons of the dial-up telephone network. Keep in mind that when the network was originally fashioned, it was designed as an analog voice telephone network. Restating this position, it was a network that was designed to carry voice communications only.

In the late 1950s, the thought of carrying data communication began to blossom. New needs to move information from a user to a mainframe computer system rapidly emerged. Therefore the industry was challenged to create a device that would allow a terminal in a remote office (across town or across the country) to communicate with a central computer system, in other words the mainframe. This was not a major problem because many other systems (Telex, TWX, etc.) were already in place to carry low-speed data communications. Dial-up modem communications on the telephone network were already moving about rapidly. Most users were transmitting their data across the dial-up telephone network at relatively low speeds, such as 2400 or 4800 bits per second. Only in rare cases did speeds of 9600 bits per second come into play. The data transfers and host access applications being used on the network were rudimentary.

A problem existed, however, with data transfer. In the chapter on data communications, we discussed the problems that could be experienced on the dial-up telephone network. Because the use of a modem implies that an analog data transmission is used, all of the problems inherent in analog communications were present. For voice, the analog systems posed no major problem. If a human does not understand part of a conversation, the problem is easily rectified. We merely shout, "What?" into the phone, and the conversation is immediately retransmitted—that is, the person on the other end repeats what was just said. If the line is too bad (e.g., full of static, snap, crackle, and other noises), then the two humans agree to hang up and replace the call, hoping for a better connection.

Data Communications Problems

In the data communications world, things were different. Back in the early days of data transmission, transmission protocols were not as sophisticated as they are today. Consequently, if a data transfer was attempted, the overhead was significant. Refer back to the section on overhead in the chapter on data communications and you will see that in the case of an asynchronous transmission, employing an ASCII code sent over the dial-up telephone network used approximately 40 percent overhead. This was a tremendous burden in data transfer. When you apply a 2400-bit-per-second transmit mode with 40 percent overhead, the net data rate is only around 1440 bits per second.

Again, if you think back to the earlier chapters discussing voice communications, the cost per minute for a voice (or data) call was $0.50 to $0.60. This overhead on the call meant that approximately $0.20 to $0.24 per minute was automatically lost and transmitted nothing usable. This problem became more pronounced as the data needs grew. Adding to the problem, the network was highly unreliable for data transmission. You can only imagine how frustrating it was to set up a file transfer of 2 megabytes of information.

Transmission protocols that were rudimentary in their approach to handling data transfers were still limited. During the data file transfer, if the end user got 1.5 megabytes transmitted over the network, timing was critical. Transfer of a 1.5-megabyte file at 2400 bits per second would take a significant amount of time. As a matter of averaging this out, the connection would take approximately 83 minutes (without the associated overhead). In reality, the time could be 116 minutes (close to 2 hours). Now imagine this: the first 1.5 megabytes of this 2-megabyte file got through in 116 minutes, then the inevitable happened. The line failed for any number of reasons. Whether a poor connection existed or the network degraded over time, the result was the same. The call disconnected three-quarters of the way through the transmission. Clearly, this would pose no problem today. By redialing the far end and establishing the connection, one could pick up the transmission where it left off. Not so back then. If the call disconnected, the protocols could not pick up in the middle of a data set and begin where it left off. As a matter of fact, everything had to be scrapped and the whole file had to be retransmitted. This was not only frustrating, but expensive too! After 116 minutes of transmission at $0.50 per minute, you would have just scrapped $58.00.

The Data Communications Review

This scenario was more the norm than the exception. Users were easily frustrated with the data communications system and dreaded using a modem. Clearly, the industry had to do something about the problem. In the late 1960s, the CCITT (now called the ITU) commissioned a study group to look for a resolution to this and many other problems. The study group arrived at several initial findings:

- The network was unreliable. Further, the use of data transmission across the analog network was at risk.

- The rollout of digital services was still very limited, and the service was primarily a carrier service.

- Other systems, such as message store and forward, although more reliable than straight dial-up connections, still became congested delivering messages at customer demand. Therefore the network could delay delivery for hours if not days because of the congestion factor.

- Other services did not offer any reliability, regardless of the delay.

The response to this whole study was the recommendation to use a packet switching transport system, which would guarantee the reliable delivery of data, break the data down into more manageable pieces, deliver the data at the convenience of the network, sequentially deliver the information, and recover from any failures that might occur on the link without requiring retransmission of the entire file.

Packet Switching Defined

Packet switching was born to accomplish these goals. To define packet switching, then, we will use the following guidelines:

Packet switching is a means of taking a very large file of information (data) and delivering it to a piece of hardware or software. From this interface, the hardware or software breaks the information down into smaller, more manageable pieces. As these pieces are broken down, additional overhead is applied to the original segment of data. This overhead is used for control of the information. Because the information is segmented, the packet service inserts the telephone number of the addressee, along with the segmentation number (i.e., packet #1, packet #2, packet #3, etc.), so that the data can be reassembled at the receiving end. Once the overhead is attached to the segmented data (now called a packet), the packet is transmitted across a physical link to a switching system that reads the address information (telephone number) and routes the packet accordingly. This establishes a virtual connection to the distant end, and each packet is sent along the same route as the first packet. The system uses a connection-oriented transport based on a virtual circuit.

That is quite a definition to work from. From this as a starting point, the rest should be easy to break down and discuss. Actually, we can "packetize" the definition and work at explaining it piece by piece.

What Is Packet Switching?

As already mentioned, packet switching is a means of breaking down the larger data files into more manageable pieces. Here's an example that might make the whole thing a little easier to comprehend.

The Packet Switching Analogy

For the sake of describing this concept, visualize a meeting with all of your peers in the industry. You have all been called into a local meeting to discuss a matter of extreme importance. The U.S. postmaster general asked that you meet regarding a serious problem. The problem is that more and more people (like you) are using other types of delivery services for various articles (i.e., mail, packages, etc.). This, of course, detracts from the revenues of the U.S. Postal Service (USPS). Therefore, the postmaster general has an offer to get you back as a customer for your routine deliveries. The scenario plays like this:

Postmaster General: We know that you have some concerns about using the U.S. Postal Service. Therefore, in order to get you back as a customer, we are willing to hold the price of a postage stamp at $0.32! Yes the rumors are true that the cost of a stamp will rise again, but if you'll sign an agreement here today, we will hold the rate at the $0.32.

Author: How many of you would sign an agreement to stay with the Postal Service for all of your delivery needs? Just as we thought, this was not enough to move you.

PG: OK, let's sweeten the pot. Not only will we hold the price of the stamp at $0.32; if you agree to give the USPS 100% of your mail, we'll lower the cost of a stamp to $0.25 and hold that rate for 5 years. This I guarantee.

AU: Some people will be moved to sign this agreement now! Many of us think in terms of business needs and money is always a decision factor. Only a few of the audience will bite this time. This leaves a bunch more that have not committed. Now for the PG's next move....

PG: Okay, folks, here's my last offer. I will drop the price of a stamp to $0.25 and hold it there for 5 years. In addition, I guarantee that any letter you drop in a USPS mailbox will be delivered anywhere in the country overnight! Now how many of you will sign the agreement?

AU: By now you are worn down and will sign on the bottom line. This is an attractive offer. For 5 years, you get cheaper postage and overnight delivery guaranteed. How can you lose? So you go ahead and sign.

PG: Now that you've all signed on the bottom line, we have a contract that you will use the USPS for all of your mailing and delivery needs. However...

AU: Watch out! Here it comes.

PG: In order to do all of the things we contracted to do, I have to give you the conditions of this agreement. You never asked, so I can assume that overnight delivery at \$0.25 per letter is what concerns you the most. To meet the terms of our agreement, the conditions are:

All mail that you wish to send through the USPS must be placed in a standard #10 business envelope. [This is shown in Fig. 18.1.] The

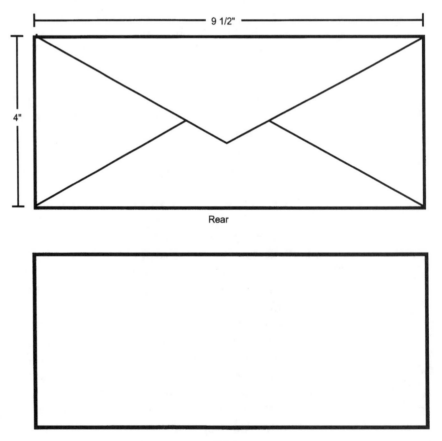

Figure 18.1
The standard business envelope will be used.

9 1/2"

4"

Rear

Front

envelope will be a white $9\frac{1}{2} \times 4$-inch solid (no windows) business envelope. There will be no other envelopes used. We don't want to see the holiday red or green; no Mother's Day orchid color; no big seasonal envelopes for birthdays and Valentine's Day, and so on.

In each of these business envelopes, you can only put one sheet of paper (standard U.S. business page of $8\frac{1}{2} \times 11$ inches). You can print on it front and back, but you must still limit the contents to only one page per envelope. [See Fig. 18.2 for the standard page letter.]

For those of you who have a major pamphlet, book, or other document to mail, the solution is easy. If the pamphlet has 200 pages, you merely have to send 200 envelopes. This means that at each end you will hire a person to handle the process. On the sending end, the postal clerk in your organization will take the pamphlet apart and stuff one envelope at a time with one of the pages. This will require a total of 200 envelopes and stamps ($50.00). The mail clerk will then add some overhead to the envelope, such as:

Figure 18.2
The standard page for letters shall be $8\frac{1}{2} \times 11$ inches. You can print on both sides of the page.

- A "from" address, to be placed in the upper left corner of the envelope.
- A "to" address, to be placed in the center of the envelope.
- A $0.25 postage stamp or meter imprint, to be placed in the upper right corner of the envelope.
- A page number (1 of 200, 2 of 200, 3 of 200, etc.), to be placed in the bottom right corner of the envelope, to assure that the pages stay in order.

AU: How are you doing? Are you still with us? Does this sound like a workable solution just to save a few dollars? Will you really save that money? Probably not! But let's continue, there's more.

PG: Once the letter is ready for delivery to the USPS [see Fig. 18.3 for the final envelope], your mail clerk has to drop these into the mailbox or at the post office. Your job is partially done. At the mailbox, a new set of sequences will begin:

- When the postal employee comes along, the envelopes will be stacked up and checked (counted) to make sure that all of the envelopes are there. If the postal clerks find that an envelope is missing, they will search around for it in the mailbox. If this proves futile, they will contact the mail clerk in your organization and ask that a new copy of the missing envelopes be generated. This of course means that the mail clerk must keep a copy of every page until we accept the envelopes and determine that none are missing.
- If none are missing or if one was missing but has been replaced, the mail clerk will receive an acknowledgment from the postal

Figure 18.3
The completed business envelope has all the necessary information. Note the numbering system on the bottom right corner.

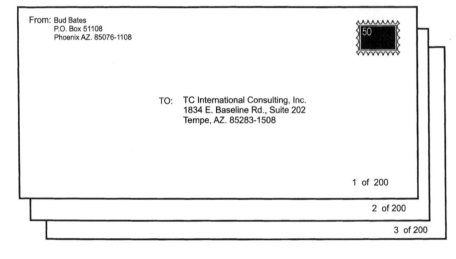

From: Bud Bates
     P.O. Box 51108
     Phoenix AZ. 85076-1108

50

TO: TC International Consulting, Inc.
      1834 E. Baseline Rd., Suite 202
      Tempe, AZ. 85283-1508

1 of 200

2 of 200

3 of 200

employee. The mail clerk can now discard all of the spares that were held [called buffering].

- Before the acknowledgment is given to the mail clerk, the postal employee will do a couple of things. The envelope count confirmed, the postal employee will open each and every envelope and read all of the information to ensure that the information contained inside is intact. If something has happened to the page [as shown in Fig. 18.4], we merely alert the mail clerk to give us a new copy. Also, as the information is read and validated, the Postal employee will make a duplicate of each page of information and store it in a holding bin.

- After these steps have been completed, the acknowledgment will be sent to the company mail clerk, who empties the holding bin at the originating location [flushes the buffer]. Now the USPS is responsible for the letter and the integrity of the information. The letter will be passed on to the next location in the chain.

- At each step along the delivery trail (e.g., mailbox to post office to plane to receiving post office, etc.) the information will be counted, opened and read, copied and accepted. The previous holder of the information will be given an acceptance before

Figure 18.4
Many things can happen to destroy the integrity of the printed page.

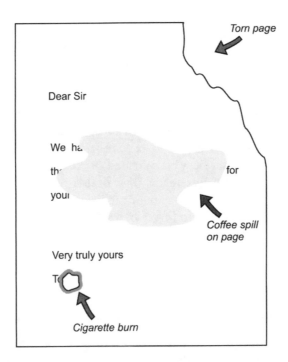

destroying the copies. If the system breaks down anywhere along the line, the holder of the stack need only go to the previous location and request a new copy.

■ At the end point (your location), the process will end. A mail clerk at the receiving end will count all the envelopes, open and read them, and accept them one at a time. But in this case, the mail clerk will finalize the process by putting the page in each envelope back in the original order until the pamphlet has been reassembled.

■ Your organization has just received overnight delivery of the letters, with guaranteed integrity of the information and sequential arrival.

AU: By now you are probably saying to yourself, "These authors have been sniffing too much laser printer toner while preparing this book." This whole scenario sounds too complex to be real, doesn't it? But this is exactly is how X.25 packet switching works. The process is a little simpler, but the concept is the same. How does this sound now?

The Packet Concept

Packet switching works like the scenario described above. If an organization has large amounts of data to send, then the data can be delivered to a packet assembler/disassembler (PAD). The PAD can be a software package or a piece of hardware outboard of the computer system. The PAD acts as the originating mail clerk in that the originating PAD receives the data and breaks it down into manageable pieces or packets. In the data communications arena, a packet can be a variable length of information, usually up to 128 bytes of data (one page in the example). Other implementers of X.25 services have created packets up to 512 bytes, but the average is 128. The 128-byte capability is also referred to as a *fast select*. The packet switching system can immediately route the packet to a distant end and pass data of up to 128 bytes (1024 bits).

Overhead

To this packet, the PAD then applies some overhead as follows:

■ An opening flag that is made up of 8 bits of information. Using a standard high-level data link control (HDLC) framing format covered in the chapters on data and LANs, the opening flag is a sequence of 8 bits that should not be construed as real data. (This is an envelope.)

- A 16-bit address sequence that is a binary description of the end points (the "from" address).

Control information consists of 8 bits of data describing the type of HDLC frame that is traversing the network (this is a notation on the envelope that describes the information inside). These can be supervisory, unnumbered, or information fields. To be more specific, these break down as follows:

- Information (I) is used to transfer data across the link at a rate determined by the receiver and with error detection and correction.

- Supervisory (S) is used to determine the ready state of the devices— receiver is ready (RR), receiver is not ready (RNR), or reject (REJ).

- Unnumbered is used to dictate parameters, such as set modes, disconnect, and so on.

Packet-specific information follows the HDLC information. The packet information will consist of the following information (this information is similar to the "to" address and the designation of the routing that will be used, such as first class, book rate, etc.):

- General format identifier (GFI)—4 bits of information that describes how the data in the packet is being used; from/to an end user, from/to a device controlling the end user device, and so on.

- Logical channel group number (LGN)—4 bits that describe the grouping of channels. Because only 4 bits are available, only eight combinations are used.

- Logical channel number (LCN)—an 8-bit description of the actual channel being used. The theoretical number of channels (ports) available is 2048. The logical channels are broken down by channel groups so that the numbers play out (Table 18.1). Although the number of logical channels can be 2048, most organizations implement significantly fewer ports or channels.

The next overhead elements consist of the packet type identifier (PTI), an 8-bit sequence that describes the type of packet being sent across the network. Six different packet types are used in an X.25 switching network. These packet types define what is expected of the devices across the network. The packet types are shown in Table 18.2. Note that the packet types have different uses, so their binary equivalent is shown for purposes of differentiation.

TABLE 18.1

Summary of Logical
Groups and
Channel Numbers.
Although the
Number of Logical
Channels Can Be
2048, Most
Organizations
Implement
Significantly Less
Ports or Channels.

Type	Logical Group	Logical Channel	Number of Channels
PVC	0	1–255	0–255
PVC	1	256–511	0–255
Incoming-only	2	512–767	0–255
SVC	3	768–1023	0–255
Two-way	4	1024–1279	0–255
SVC	5	1280–1535	0–255
Outgoing-only	6	1536–1791	0–255
SVC	7	1792–2047	0–255

TABLE 18.2

Summary of the
Packet Types and
Identifiers

Packet Type	Description	Packet Type Identifier
Incoming call	Call request	00001011
Call connected	Call accepted	00001111
Clear indication	Clear request	00010011
Clear confirmed		00010111
Data		xxxxxxx0
Receiver ready		xxx00001
Receiver not ready		xxx00101
Reject		xxx01001
Interrupt indication	Interrupt request	00100011
Interrupt confirmation		00100111
Reset indication	Reset request	00011011
Reset confirmation		00011111
Restart indication	Restart request	11111011
Restart confirmation		11111111

The variable data field is now inserted. This is where the 128 bytes of information are contained in the packet. The 128-byte field is the standard implementation, but as mentioned, it can be larger (as much as 512 bytes).

Following the information field is the CRC, a 16-bit sequence that will be used for error detection and/or correction. Using a CRC-16, the error detection capability will be approximately 99.999995 percent. This is to ensure the integrity of the data; rather than having to deliver information over and over again, the concept is to deliver reliable data to the far end.

The closure of the packet is the end of frame flag. In the HDLC frame format, this denotes the end of the frame, so the switches and related equipment know that nothing follows. The switches then calculate all of the error detection and accept or reject the packet on the basis of the data integrity.

Summary of Packet Format

Figure 18.5 is the actual HDLC frame format used for the packet of information. In this case, one can look at the fields and recheck through the sequencing information of what each field represents. Users intimidated by this process can relax. This will be done by the equipment that generates the packets across the network.

The Packet Network

Figure 18.6 shows a typical network layout. Here the cloud in the center of the drawing represents the network provided by one of the carriers. From our discussions in earlier chapters, the obvious network configuration is that each of the designated carriers has its own cloud. Actually, each cloud interconnects to the clouds provided by other carriers so that transparent communications can take place. Once the cloud is established, the next step is to provide the packet switching systems (called packet switching exchanges or PSEs). These are nothing more than computers that are capable of reading the address and framing information. The PSEs then route the packet to an appropriate outgoing port to the next downstream neighbor. In many cases, these PSEs are connected to several other PSEs.

A meshed network is thus provided by the carrier, as shown in Fig. 18.6. The packet switches can select the outbound route to the next downstream neighbor on the basis of several variables. The selection process can be the circuit used least, the most direct, the most reliable, or some other predefined variable. This again is the magic that takes place inside the cloud. Now that the network cloud and the packet exchanges are in place, the next step is to connect a user.

The User Connection

Users sign up with the carrier of choice and let the carrier worry about the physical connection. As shown in Fig. 18.7, the user connection into the cloud is through a dedicated or leased line. The carrier notifies the

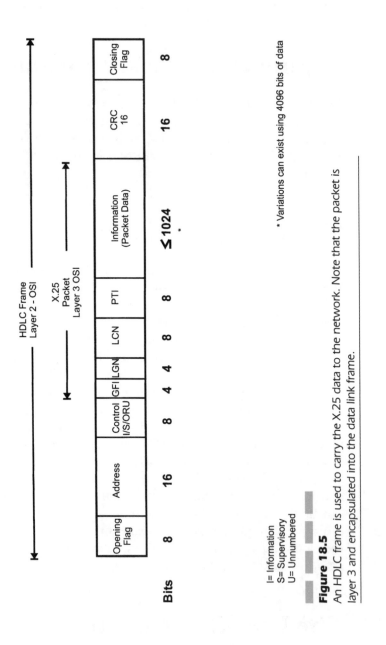

Figure 18.5
An HDLC frame is used to carry the X.25 data to the network. Note that the packet is layer 3 and encapsulated into the data link frame.

436

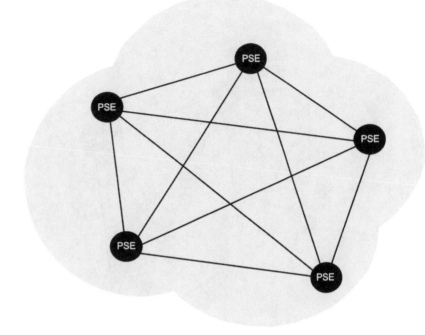

Figure 18.6
The typical X.25 network is a meshed network. The cloud represents the network.

PSE — Packet Switching Exchange

LEC and orders a leased line at the appropriate speed (in the leased line, this can be up to 64 Kbps on digital circuits). The original network connections back in the initial rollout of X.25 services were on analog lines at up to 9.6 Kbps. Either a modem was provided by the carrier at the customer end, or the customer purchased and provided it. Now that the modem is attached at the customer end, the circuit is terminated in a port on the computer (PSE). This is the incoming port that can be used as a permanent virtual connection or as an incoming-only channel. This is part of the addressing mechanism inside the packet where the PSE reads the packets' originating address. The incoming-only channels are the channel numbers assigned to each customer.

The next step follows the connection where the customer must initiate the PAD function as shown in Fig. 18.8. Remember that the PAD is the hardware or software installed to break the data into smaller pieces, attach the overhead, and forward the packetized data to the network. In the reverse order, the PAD is responsible for receiving the packets, peeling off the overhead, and reassembling the data into a serial data stream to the

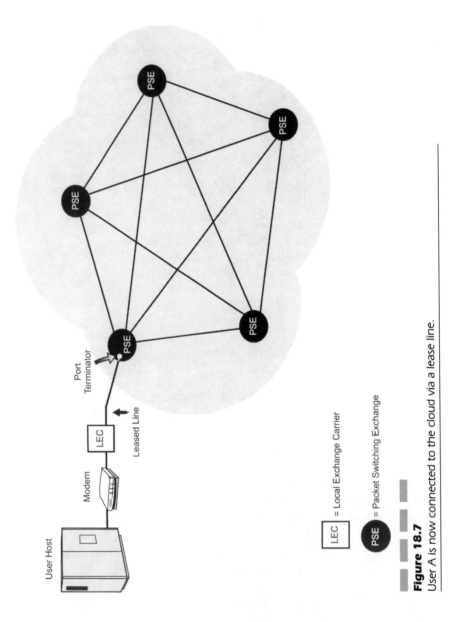

Figure 18.7
User A is now connected to the cloud via a lease line.

LEC = Local Exchange Carrier

PSE = Packet Switching Exchange

User Host

Modem

Port
Terminator

Leased Line

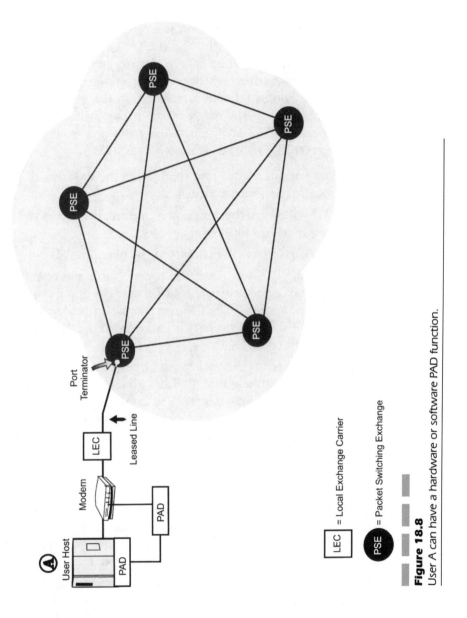

Figure 18.8

User A can have a hardware or software PAD function.

LEC = Local Exchange Carrier

PSE = Packet Switching Exchange

data terminal equipment, whether it is a terminal or host computer. So the PAD's function is crucial. If a software package is performing the PAD function, then it is the customer who must purchase (or license) the software and install it. If the solution is a piece of hardware, the options are different. The customer might buy the PAD and install it, or the carrier might provide and install it. This hardware can be rented, leased, or sold by the carrier. Now the connection exists on one end of the cloud.

In Fig. 18.9, another device is attached on the other end of the cloud. What happens now is the magic of the packet switching world. As packets are generated through the network from user A to B across the network, several things happen, as shown in Fig. 18.10.

- The data is sent serially from the DTE to the PAD (which acts as the DCE for the computer terminal).

- The PAD will break the data down into smaller pieces and add the necessary overhead for delivery.

- The PAD then routes the packets to the network PSE.

- The receiving PSE (PSE 1) sees the packet coming in on a logical channel, so it remembers where the packets are coming from. After analyzing the data (performing a CRC on the packet) and verifying that it is alright, the PSE will send back an acknowledgment to the originating device (which is the DCE).

- The PSE then sends the packet out across an outgoing channel to the next downstream neighbor (PSE 2). This establishes a logical connection between the two devices (PSEs) from the out channel to an in channel at the other end. The logical channel is already there; it is used for the transfer of these specific packets. A virtual connection is also created, allocating the time slots for the packets from A to B to run on the virtual circuit.

- Back at PSE 1, the next packet is sent down from the originating PAD. This again is analyzed and acknowledged, then passed along. At the same time this is happening, PSE 2 is sending the first packet to PSE 3. Again, at each step of the way the packets are opened and a CRC is performed before a packet is actually accepted.

- At each PSE along the way, the packet is buffered (at PSE 1) until the next receiving PSE accepts the packet and acknowledges it (PSE 2). Only after the packet is acknowledged, does PSE 1 flush it away. Prior to that, the network node (PSE 1) stored it just in case something went wrong.

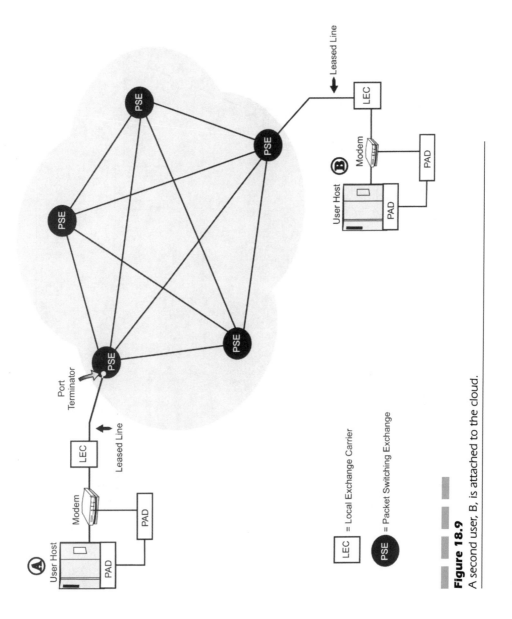

Figure 18.9

A second user, B, is attached to the cloud.

LEC = Local Exchange Carrier

PSE = Packet Switching Exchange

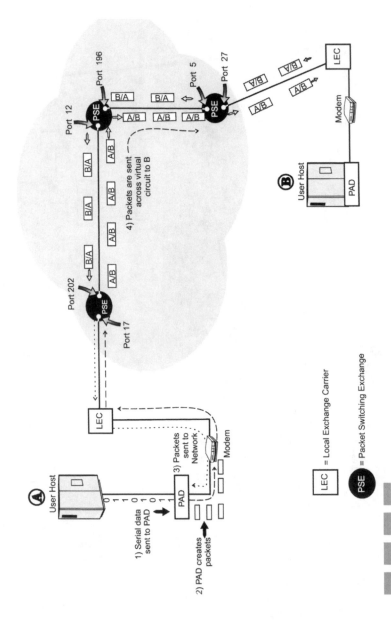

Figure 18.10

Serial data sent to the PAD get packetized (A/B) and then sent across a virtual circuit to location B. B can return responses (B/A) on a duplex circuit.

LEC = Local Exchange Carrier

PSE = Packet Switching Exchange

- Don't forget what happens at the receiving end (device B), where the packets arrive in sequential order and are checked and acknowledged. Then the overhead is peeled away so that a serial data stream is delivered to the receiving DTE.

- This process continues from device A through the network to device B until all data packets are received. Every packet along every step of the link is sent, accepted, acknowledged, and forwarded until all get through. Here is the guaranteed delivery of reliable data properly sequenced. The logical link that is established between devices A and B is full duplex. The two devices can be sending and receiving simultaneously.

You can see from the packet delivery process that there is a benefit to the packet switching process. However, nothing is perfect. The overhead on the packet, the buffering of multiple packets along the route, the CRC performance at each node along the network, and the final sequenced data delivery all combine to present the risk of serious delays. What you receive in integrity and reliability can be offset in delays across the network. You must always weigh the possibilities and choose the best service.

Benefits of Packets

The real benefit to this method of data delivery ties into the scenario presented in the beginning of this chapter. Remember the problems with the dial-up telephone network and the risk of sending three-quarters of a file transfer only to have a glitch in the transfer? Using the packetized effort, if a glitch occurs the network might have from 7 to 128 outstanding packets traversing the links. Therefore, instead of scrapping the entire file, the network automatically recovers and resends the packets that were lost or corrupted. This means that the users would save time and money on an error-prone network. Again, the risk of congestion and delay on the network might cause others to look for alternative solutions.

Other Benefits

Beyond these benefits, other benefits can be achieved from the use of the dedicated link into the network. As Fig. 18.11 shows, the link is not solely

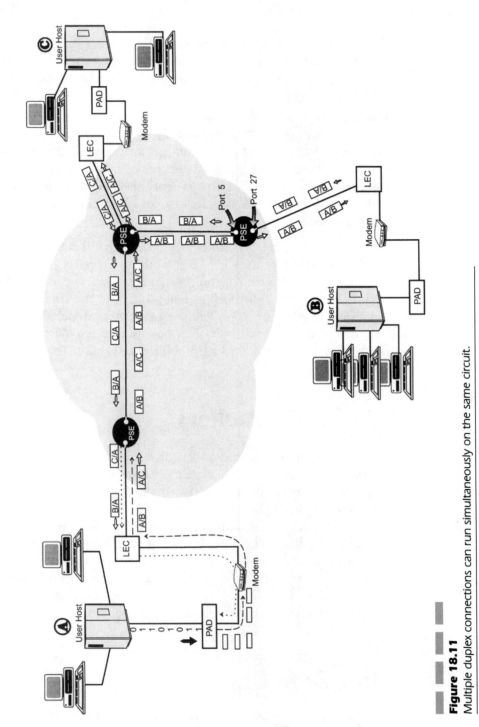

Figure 18.11

Multiple duplex connections can run simultaneously on the same circuit.

for one user at a time, nor is it for two specific locations. When the organization uses a packet switching network, the users might need to have multiple simultaneous connections up and running. Therefore, the PAD will act similarly to a statistical time division multiplexer (*statmux*). The statmux capability allows multiple connections into the single device, and will sample each of the ports in a sequential mode to determine whether the port has anything to send. If a packet has been prepared, the statmux will generate the call request (initiate a call) to the network. The connection will be created and the data will flow. This assumes that no problems are being experienced on the network.

As a new user logs on and generates a request to send data, the PAD will then set up this connection. Packets will then be interleaved across the physical link between or among the various users. One can imagine that not all devices will transmit at the same rate of speed or have the same amount of two-way interactive traffic. Therefore, the statmux function of the PAD interleaves the packets based on an algorithm that allows each device to appear as though a dedicated link is available to it. The use of a statmux will be a benefit because it will enable users to employ the expensive leased link to the maximum benefit of the organization. This requires fewer physical links and takes advantage of the dead time between transmissions, and so on.

Figure 18.12 shows a series of sessions running on a single link, all interleaved. This also shows that packets are not specifically interleaved in the order of A,B,C.... Instead, they can be interleaved on the basis of the flow or delivery method used, such as A,B,A,B,C,B,A.... Herein lies the added benefit of packet switching—the user achieves the data throughput necessary without having a dedicated resource that is only periodically used.

Advantages of Packet Switching

Packet switching is considered by many to be the most efficient means of sharing both public and private network facilities among multiple users. Each packet contains all of the necessary control and routing information to deliver the packet across the network. Packets can be routed independently or as a series that must be maintained and preserved as an entity. The major advantages of using this type of transport system are:

■ Shared access among multiple users, either in a single company or in multiple organizations.

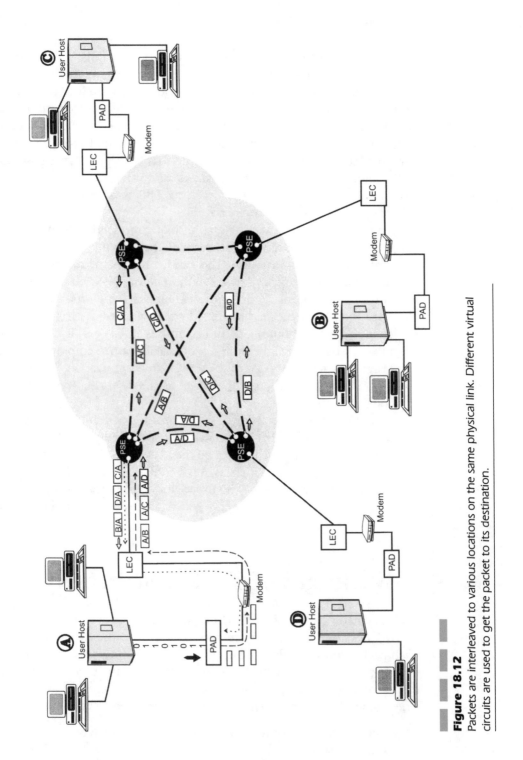

Figure 18.12

Packets are interleaved to various locations on the same physical link. Different virtual circuits are used to get the packet to its destination.

- Full error control and flow control in place to allow for the smooth and efficient transfer of data.

- Transparency of the user data.

- Speed and code conversion capabilities.

- Protection from an intermediate node or link failure.

- Pricing advantages; because the network is used by many, the prices are on a per-packet, rather than per-minute, basis.

Other Components of Packet Switching

Although the primary components of packet switching have already been discussed, there are others. As the evolution of the X.25 standard continued in the early 1970s, several different implementations were enhanced. This could include devices or access methods that you should understand. These are summarized so that you can gain an appreciation of how these pieces all work together. Some of these added pieces include:

1. The ability to dial into the packet switching network from an asynchronous modem communication. Although packet switching (a la X.25) is a synchronous transfer system, the need for remote dial-up communications exists. Therefore, the standards bodies included this capability in one of the enhanced versions of the network. As shown in Fig. 18.13, a dial-up connection can be made from a user. In this case, the asynchronous communication will be a serial data transfer to a network-based PAD. The PAD will accept the serial asynchronous data, collect it, segment it into packets, and establish the connection to the remote host desired. In this case, the connection is now across the X.25 world, synchronously moving packets across the network. This uses an X.28 protocol to establish the connection between the asynchronous terminal and the PAD.

2. The PAD-to-host arrangement is controlled by the X.29 protocol. Control information is exchanged between a PAD (X.3) and a packet mode DTE or another PAD (X.3). This is shown in Fig. 18.14. In this case, as the communication is established between these devices, the X.3 PAD will set the parameters of the remote device. This could be the speed, format, control parameters, or anything that would be appropriate in the file transfer. The X.29 parameters can also provide keyboard conversion into network-usable information.

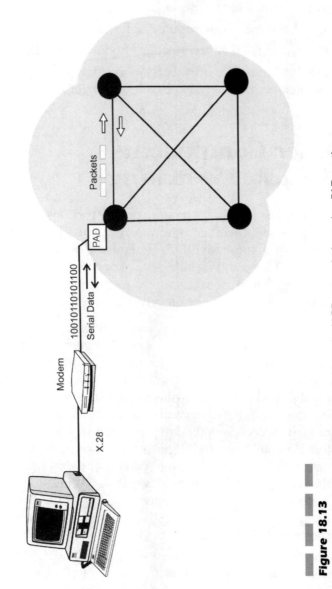

Figure 18.13

Asynchronous dial-up terminals can dial into the X.25 network through a PAD on the network.

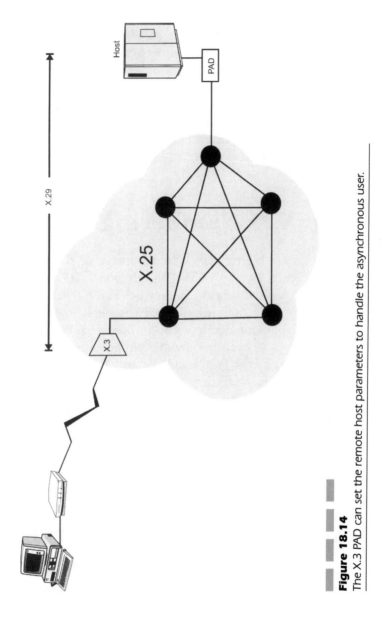

Figure 18.14
The X.3 PAD can set the remote host parameters to handle the asynchronous user.

3. The internetworking capability of an X.25 network uses a protocol called X.75. Although this should be user transparent, the network needs the X.75 parameters to provide a gateway between two different packet networks or between networks in different countries. In each case, the gateway function is something that should only concern the network carrier. However, as more organizations install their own private network switching systems, the need to interconnect to the public data networks rises. Therefore, the internetworking capability is moving closer to the end user's door. A representation of an X.75 interconnection is shown in Fig. 18.15.

The X.25 Numbering Plan

The X.25 numbering plan takes advantage of the worldwide numbering plan designed by the CCITT International Telecommunications Union Telephony Standardization Sector (ITU-TSS). It is known as the X.121 worldwide network numbering plan for the public data networks. This was a design to allow for public networks only. However, in most of the world this is acceptable because only one public network exists that is controlled and operated by the Post Telephone and Telegraph (PTT) organizations.

In the United States, however, the proliferation of both public and private networks is completely different. Looking at all of the players involved today in the carrier community, you can see that easily 300+ networks could exist publicly. Couple that with the private organizations that can add packet switching to their networks and want to access public networks, and the number can expand exponentially.

Applications for X.25 Services

There are no single solutions that fit any transport system. Therefore, the variations of the X.25 capabilities are as numerous as the organizations using them. However, if you think in terms of the overall use, the following can be applied.

If a user organization has data to transfer between or among a limited number of locations and the data transfer will be continuous (such as 8–12–24 hours per day) or the files are constantly very large, the use of X.25 might or might not apply. In this case, the connection will likely be

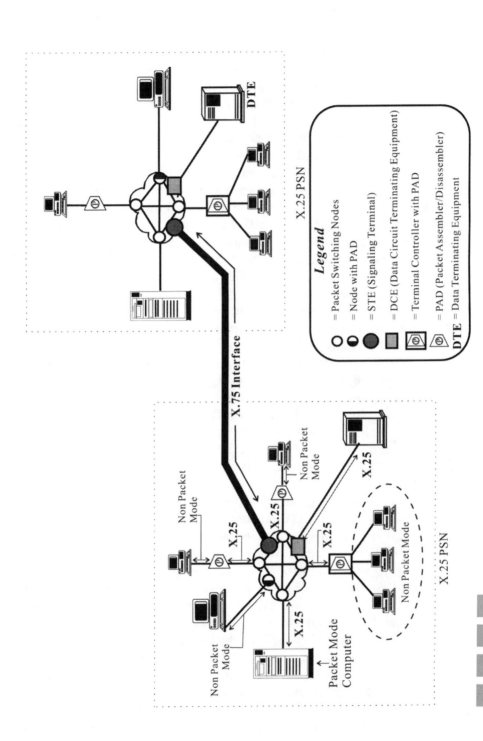

Figure 18.15

X.75 interconnectivity protocols allow different networks to communicate.

451

a leased line between two hosts (for example) and the data transfer will be constant. The leased line is an expensive proposition unless the volume of data justifies the connection full time. A data transfer will use very little time to initiate a session between hosts or end-to-end devices, because they are physically connected together.

If, on the other hand, a user organization has multiple locations (10, 20, 30, etc.) and only needs to establish a connection occasionally during the day (twice or three times, perhaps), and the data is batched in a large transfer that will take 1 hour, then a dial-up circuit might be appropriate. In this case, the cost associated with a leased line would likely be unjustified. Therefore, a dial-up modem communications will be a better solution. The time it takes to generate a connection and handshake between the modems can be approximately 1 to 2 minutes. As a portion of the data session, the call setup is insignificant. Again, the dial-up communications will be a possible solution.

If, however, an organization has many (50, 100, 200, 1000) locations that must pass information to a central computing platform, and the amount of data is relatively trivial, then packet switching might be the most appropriate solution. In this case, let's use a department store that has 100 locations spread around the country. The stores use a point-of-sale (POS) system to run the registers and perform inventory and other associated data processing functions. Each night, the store must communicate back to the home office three pieces of information: amount of daily sales, amount of money deposited in the local bank for transfer back to corporate, and items totally out of stock and in need of immediate replenishment. The entire data transfer here will be approximately 10 seconds.

It would not be prudent to spend the money to lease data lines to each of the 100 locations. The expense could never be justified. Even in a multipoint circuit environment, this will be too expensive.

A dial-up connection would be inappropriate, because at 1–2 minutes of setup and handshake it will take longer to establish the connection than to actually transmit the information, since only 10 seconds' worth of data will be transferred. Also, the amount of time lost in sequentially dialing 100 locations with the associated timing and handshakes would be far too great to get all the calls through in a reasonable amount of time. Therefore, the packet switching arrangement would be more appropriate.

Using the packetized data, several devices can be simultaneously generating packets of information to the host and transmitting the nightly information across the network. Different logical connections will be established and the packet reassembly process will take its due course to deliver the required information in the appropriate form.

An organization that has nightly or periodical information to transfer is a prime candidate for packet switching. Other forms of information transfer that could be performed by a packet switching service are:

- Transactional processing, such as for automated teller machines (ATMs).
- Process control information, such as for electrical distribution, oil drilling, and chemical production facilities.
- Database inquiries.
- Credit card approvals.
- Motor vehicle registration information.
- Signaling system seven (SS7) messages for call setup and teardown.
- File transfers where reliable and guaranteed data delivery is required.
- Money transfers, such as on the SWIFTE network.

One can see that many of these organizations can take full advantage of the packetization of the data streams and still conduct the normal day-to-day business activities.

Other Forms of Packets

As an industry accepted standard, X.25 is the most widely installed and accepted form of data transmission. This is a worldwide standard, and the interconnectivity solutions defined have created a usable form and format throughout the world. In later chapters of this book, we discuss other forms of data transmission. However, these are emerging forms of packet switching (fast packets) that are not ubiquitous. Other forms of transmission might or might not be in use in third-world countries. Speeds are limited and facilities to transmit at various other means might not yet exist. Therefore, the X.25 standard has become the industry norm.

Still, other forms or variations of packet switching exist today. One is called fast select, which is a single packet transfer (rather than a sequentially numbered series of packets) where only one packet needs to be sent. This can be applicable in a UNIX world where a single character is sent across the network to a UNIX host. It would be inappropriate to have a group of packets sent across the network when all that is needed is a single packet. Another use of such a system is in credit card approval. The single packet that gets generated is limited by the data needs, such as:

- The credit card number
- The expiration date of the card (optional)
- The ID number of the vendor requesting the approval
- The dollar amount of the transaction

In this case, because only a very limited amount of information is required, it can all be contained in a single packet. The network does not use the sequential transfer as it does in X.25, nor does the reliability check take place on the network. Instead, the packet is acknowledged by the POS equipment when the main credit clearinghouse sends back an acknowledgment. If something is corrupted in the transfer of the packet, the vendor must retransmit the entire transaction. This is a sporadic type of data transfer and works well for the most part. Delays do occur, but mainly as a result of the clearinghouse computers being busy.

Still another form of packet switching, called a Datagram, is used in the industry. A datagram transfer is a form of packet switching. Typically, the datagram does not follow the same virtual circuit concept. Every packet is sent across the network and can take a different route to get from point A to B. Also, a datagram is not checked at every node along the route; only the header (or address) information is checked for the location of the destination. As each packet is received, it is scanned for the destination address and then immediately routed along to the next node. No CRCs are done on the data. It is not the network's responsibility to guarantee the integrity of the data; that will be left to the receiving device using a higher-level protocol. With this concept of a different path for each packet, the idea of sequentially numbering the packets is lost. The network treats every packet as its own entity (or as packet number one). If a packet gets lost or something else happens, the network does not care. The sequencing and reassembly of the data back into its original form is handled by the receiving end's higher-layer protocols. An example of this is transmission control protocol with internet packeting (TCP/IP). IP packets are datagrams that are segments of the original data stream. It works at the layer 3 of the OSI model, as covered in Chap. 15. As a network layer protocol, IP's only concern is to break the data down into packets and make its best effort to deliver the packets.

TCP, at the receiving end, is responsible for reassembling the data back into the serial data stream and ensuring the integrity and sequencing of the information. TCP is not a network layer protocol; it works at the higher layers of the OSI model (transport and above). TCP/IP is the mainstream of the Internet and has been widely adopted as the protocol stack of choice in many LAN and WAN arenas. Because of its robustness and ability to deal with the packetization of the data, more organizations are using TCP/IP as their WAN protocol. TCP/IP is not without its problems, but the advantages far outweigh the disadvantages. The industry as a whole and the vendor community in particular has recognized the importance of TCP/IP in the mainstream of products. Just about every LAN network operating systems (NOS) vendor now supports the use of TCP/IP. From a physical network and data link layer, TCP/IP will run on most any topology. It is supported by the hardware suppliers of network interface cards (NICs).

19
Local Area
Networks (LANs)

What Are LANs?

The digital data communications technologies covered up to this point can function over essentially arbitrary distances. Whether at 1 mile or 10,000, they will work about the same. Nevertheless, certain data communications functions would be much more useful (e.g., file sharing and transmission) if higher speeds could be used than are generally available on a wide area basis. What if some slick optimization and engineering techniques were used to improve communications performance radically, at least for short distances? The productized answers to these questions are local area networks (LANs).

In the early 1980s, a new device made its way to the desktop. IBM introduced the PC, a single-diskette system that was functionally a terminal device that just happened to have its own intelligence. The single floppy diskette was sufficient to run the entire system and still allow the user plenty of storage space. The second version of IBM's desktop computing system (PC) was a dual floppy disk system, where one diskette was used to run the operating system for the machine and the other floppy was used for the data storage. Not much exciting happened during this era, but there were rumblings of the single- and dual diskette systems being too limited and awkward to use. The industry had no problems. Then, it happened!

The third evolution of the PC introduced by IBM had a hard disk. The disk operating system was stored on the hard disk so that the machine could operate without the user playing around with those floppies. This in itself was a major improvement, but this version of the PC was a bit pricey ($5000 to $6000). Furthermore, management and information systems personnel had to reckon with a different situation. The hard disk had a total capacity of 10 megabytes. These professionals knew back then, just as you and I would know, that no one would ever need 10 megabytes of hard disk space all to him- or herself. Therefore, a new sharing of resources was needed. The first LANs were resource sharing capabilities, so that five users could share that 10-megabyte disk. Hence the LAN was born. Because the LAN was hard disk sharing, it was typically set up in one department and was very localized. The actual LANs started out as serial port connections where users were daisy-chained together to access the disk.

A little later on, the introduction of the laser printer (HP's wizardry) led to a new problem. One specific user in the department might have a laser, while others had a dot matrix printer. In the early 1980s, this dot matrix output was nothing to write (or print) home about. Therefore, everyone wanted a personal laser printer rather than a dot matrix. Clearly, management did not want to spend $5000 for every employee to have his or her own laser printer. The LAN was the resource sharing service, so the printer moved onto the network.

What is a LAN? Originally, the industry defined *local area networks* as data communications facilities with the following key elements:

- High communications speed
- Very low error rate
- Geographic boundaries
- A single cable system or medium for multiple attached devices
- A sharing of resources, such as printers, modems, files, disks, and applications

Let's take these one at a time.

High Communications Speed

At the time (the 1970s), use of communications speeds above 1 Mbps on either a wide or local area basis was exceedingly rare, other than on terminals locally connected to a mainframe. Therefore, "high-speed communications" meant transfer rates in the millions of bits (or *megabits*) per second. Even then, it was obvious that submegabit speeds would become a significant speed limit unless they could be vastly increased.

Nevertheless, why not simply use the existing WAN protocols for LANs? They certainly can be so used. The reason, in a word, is *overhead*. One major source of overhead in WAN protocols is their routing capabilities. These range from relatively simple (e.g., IBM's SNA, which traditionally requires much human intervention to establish or change routes for network traffic) to extremely complex (e.g., Digital's DECNet, that can generally figure out where to send something without any assistance at all). But regardless of their sophistication, they add overhead not required in a local area network. Thus, LAN protocols eliminate routing overhead by functioning as though all nodes are physically adjacent. Another source of overhead is error checking, which is covered in the following paragraphs.

Most local area network technologies currently in use provide maximum communications speeds well above 1 Mbps (Table 19.1). This is a summary of the typical speeds available from the various approaches. We look at the more common of these transport systems individually in later chapters. For the moment, the products that produce these speeds are what we are interested in.

TABLE 19.1

Summary of Major
Architecture Speeds

LAN Name	Typical Speeds
Appletalk	256 Kbps
Arcnet	2.5 Mbps
Token Ring (original)	4 Mbps
Ethernet	10 Mbps
Token Ring (updated)	16 Mbps

AppleTalk is the only true LAN in the table that falls below 1 Mbps. It is nonetheless very popular due to its excellent ease of configuration (not to mention that it comes free with every Macintosh!).

Very Low Error Rate

An *error rate* is the typical percentage of bits that can be corrupted on a regular basis during routine transmissions. As mentioned in the discussion of protocols, error checking ensures that such damaged bits are retransmitted so that applications never see the errors. But the data link protocols in use for wide area networks must contend with much higher error rates than would normally be experienced using LAN technologies.

Wide area analog communications typically experience one error in somewhere between 10^5 and 10^6 bits transmitted. A *very low error rate* is at least two orders of magnitude better than this or one error in 10^8 transmitted bits. Several technologies meet this challenge.

In a typical WAN protocol stack, error checking takes place at multiple stack levels. At each level, that error checking requires both additional bits in the transmitted frames and additional processing time at both ends (at the sending end, to create the error-checking bits; at the receiving end, to determine whether any errors occurred). If some of that error-checking overhead can be eliminated because of fewer errors to catch, higher speeds can more easily be achieved—which, not coincidentally, is another of the prime characteristics of a LAN.

The physical elements of a LAN do not depend on the vagaries of the telephone companies' facilities. Those elements are digital and are designed specifically to deliver high-speed communications with very low error rates. Thus the higher levels of the protocols can be relied on to handle the few errors that do creep in. Most LAN protocols do little or no error checking, and this is one of the design elements that permits their higher speeds.

Geographic Boundaries

A local area network (LAN) is bounded by some geographic limitation. (Otherwise, by definition it would not be local.) Engineering considerations and physics impose this limitation; vendors did not impose it by fiat. The technologies employed to meet the other requirements (especially the requirement for high speed at a low error rate) tend to preclude more wide area transmission capabilities.

The geographic limitations of the various LAN technologies are expressed in different ways. There is usually a maximum distance from one node (or connected, network-addressable computer) to the next, expressed in tens or perhaps hundreds of meters. There is usually a maximum distance from one node to any other network node, typically hundreds or a few thousands of meters.

Network technologies that only reach up to the bounds of a single *campus,* or group of colocated buildings, satisfy these specific limitations. However, as we will see, a number of new technologies and connectivity options have worked together to mitigate the distance limitations significantly, which applies to LANs as they were originally defined and designed.

Single Cable System or Medium for Multiple Attached Devices

One of the less obvious but nonetheless essential characteristics of a LAN is that all devices on it are viewed and, to all intents and purposes, function as being adjacent. This implies that no special routing capabilities are required for communications traffic from one device to another; just put the information on the LAN, and it will reach its destination. To put this into the ISO context, no routing functions should be required for a LAN to operate.

To accomplish this feat, all the devices must reside or appear to reside (from the point of view of an attached device) on a single cable system. Thus, each end-node-to-end-node link is also functionally a point-to-point link. No routing choices need be made if every machine you can talk to is right next door. All signals from all devices propagate throughout that cable system. In reality, the LAN is a broadcast multipoint circuit on which all devices have equal access.

Different types of LANs accomplish this function in very different ways. Some of those ways are covered in this chapter; later chapters focus specifically on Ethernet and token ring signaling.

A LAN by Another Name

The previous definition of a LAN focused on the components used to provide connectivity among communicating devices. In fact, initially LANs tended to be configured to provide computer-to-computer communications on a more or less egalitarian, or many-to-many, basis. However, the rapid development of network operating systems (NOSs) gave rise to a different, although related, definition of a local area network.

What Do Users of LANs See When They Use the Network?

Configurations vary widely. However, when a user turns on his or her workstation, it normally runs a number of programs automatically until the user is prompted to enter identification (i.e., "user ID") and a password. This is called *logging in.* If you ask the user (with, of course, precise grammar), "Into what are you logging?" he or she will answer, "I'm logging in to the network, of course!"

Logging in to a bunch of wires? Well, not exactly. Because security and access to services is typically (although certainly not always) provided for a given user from a single file server, users often perceive and refer to that server as "the network" or "the LAN." Managers of those servers often go along with this usage in order to avoid confusion—or because they believe it to be the only definition anyway. For single-server shops, this causes no problems. In fact, not surprisingly, many of the NOS vendors also use the term LAN in this way. It is what they sell, so one cannot fault them for focusing attention on what to them is the important component of the environment.

However, referring to a server as a LAN can cause quite a bit of confusion in certain cases. Consider the case of having multiple servers on a single wiring facility. This is certainly a reasonable thing to do, particularly if the servers provide different types of services. Is each a LAN? Are they all, collectively, "the LAN?" If someone logs in to one of these servers from a different cable plant (connected using methods described later in this chapter), which LAN is he/she on?

Alternatively, consider the individual responsible for the design and implementation of the wiring plant, together with the connectivity devices described in the following paragraphs. That person will most likely refer to the wiring plant as the LAN. In this usage, LANs provide connectivity among devices, including servers. This book generally refers to the wiring plant together with the active devices providing connectivity as the LAN. File servers are just one category of servers, albeit a very important one.

Why They Are Used

For many, the answer to "Why use LANs?" is "How could we work without them?" Nevertheless, business operated for centuries without LANs, and automated data processing functioned successfully for more than a decade without them. What changed?

Performance

One major change was the increasing popularity of minicomputers (not personal computers; the development of LANs predates that of personal computers). Mainframe communications architectures had been optimized for use of large numbers of terminals. In fact, a special category of communications hardware (front-end processors) was developed in order to off-load the additional work necessary to manage such terminals.

In the minicomputer environment, however, front-end processors could not be used; the minicomputers were not built to handle the high input/output speeds necessary to service such processors. Instead, terminals were directly connected to serial communications boards, printed circuit boards with multiple RS-232 connections that plugged directly into the backplanes or buses of those minicomputers.

Minicomputers are designed to respond immediately to certain requests for services called *interrupts*. Every keystroke on terminals connected in this way generates an interrupt on the associated minicomputer. And "interrupt" is exactly what happens; all processing ceases for an instant while the single keystroke is processed. You can easily see that, as additional terminals are added to a minicomputer, the amount of substantive processing accomplished (versus that required merely to receive characters from attached terminals) decreases rapidly!

If the communicated information from the terminals could be presented to the minicomputers more efficiently (i.e., in groups without generating one interrupt per keystroke), the minicomputers could serve far more terminal users. A new input/output method was needed.

Wiring

Another issue, again in the minicomputer environment, was the snarl of wiring that communications was generating in computer rooms throughout the world as well as in buildings containing those same terminal users. Before local area networks, almost every logical (non-wide area) connection from a terminal to a computer required a separate set of physical wires from the terminal all the way back to the computer. Even if not every terminal needed access to every computer, wiring such as this was common. Figure 19.1 shows the spiderweb of wires that would be required to link the various devices together. This is expensive to install, difficult to manage, and inefficient in the use of ports and facilities.

Data switches, or data PBXs, were alternatives, but they could only support lower speeds—and to configure them economically, they still had to be large and centrally located, much like the computers! Such devices helped, but they were not an optimal solution. But what if a network device could be used to consolidate the terminal wiring? Terminal servers are LAN devices that can perform such a function (Fig. 19.2).

This might not look all that much better than the previous figure. Nevertheless, terminal servers can be located near the terminals, with one long cable back to the main computer system instead of one per terminal. In a real business setting, this greatly simplifies terminal wiring. A terminal server can allow any given terminal to have more than one session active simultaneously, either with one or several destination computers. If a new computer is added to the environment, only the authorizations change; no wiring other than the new computer's connection to the LAN is required. Best of all, the way LANs are implemented on minicomputers eliminates the interrupt-per-keystroke problem. Naturally, the sessions need to be responded to; however, a much smaller percentage of the minicomputer's resources are devoted strictly to communications functions.

Not only did these terminals generate wiring snarls, but also sites with multiple minicomputers were common. Suppose you had six computers, each with a need to reach all of the others on occasion. You could easily end up with a network wiring diagram like the one shown in Fig. 19.3. The wiring nightmares were so bad that many organizations joked that if the wiring was pulled out, the building would probably fall to the ground.

Imagine what a mess this would be if there were 8, 10, or 20 computers to connect! But if you used a common cable system instead, you might get something as the setup depicted in Fig. 19.4. Clearly, this is an improvement, but the real question is: Is it easier to manage? You bet! In addition, the number of components required, and therefore the cost, decreases geometrically.

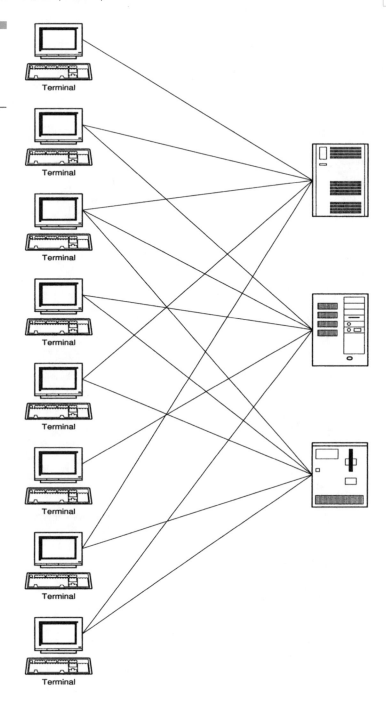

Figure 19.1
The typical wiring
scenario in a
multicomputing
environment.

Terminal

Terminal

Terminal

Terminal

Terminal

Terminal

Terminal

Terminal

Figure 19.2
The use of a terminal
server elimainates the
wiring mesh.

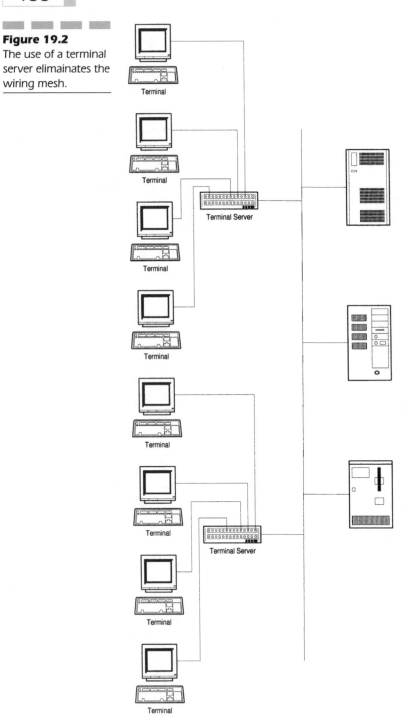

Figure 19.3
The minicomputer connectivity requires the same form of wiring mess.

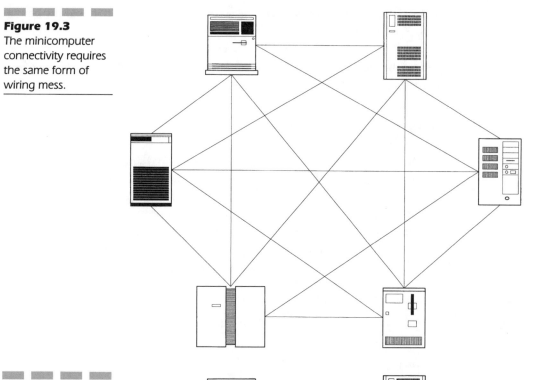

Figure 19.4
A single cabling structure might be a better approach.

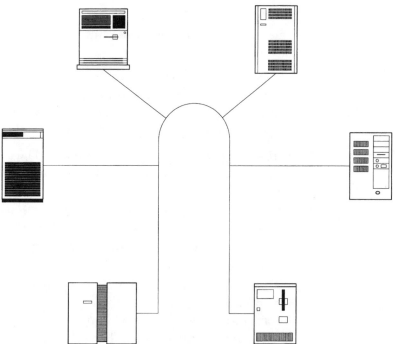

Initially, local area networks were developed to address these problems. Other issues, such as the sharing of printers, were initially not of concern—that issue came later; connection of terminals to the minicomputers provided access to the printers on those computers as well as WAN communications resources on those computers.

Shared Resources

Much of this progress occurred before personal computers were generally available. With the advent of personal computers, additional requirements surfaced. A general need appeared for resource sharing where the resources were not connected to a central computer. One of the two primary resource categories that were first shared was a printer. High-quality printers were initially quite expensive; management was understandably reluctant to provide one laser printer for each user of a personal computer.

Printers could be connected in three general ways:

- To a central server, perhaps providing file services as well as print services

- To a user's computer, allowing others use of that computer at the same time the user was using it

- Directly to the LAN by themselves, if the specific printer had LAN capability

The other major resource first shared was file services. File services provide a PC user the ability to reference files on another machine (the server) as though those files were actually located on the user's PC. Why not have them on the user's PC? There are several reasons.

- *Initially, large disks were also very expensive.* The cost per megabyte was lower for larger disk drivers, making sharing of such drives economically attractive.

- *The larger drives were also much faster.* Network managers in some cases could actually provide better disk input/output performance across the network than the user would experience were the information on a local drive.

- *Users do not back up data.* Yes, there are exceptions, but a manager would be foolish to assume that his or her users are among them. If information is stored centrally, then it can be backed up under management control rather than relying on users to protect their data.

The observant reader might ask at this point, "Why not just use a mainframe or minicomputer as before?" This is a very good question. Perhaps the answer is that the applications becoming available on PCs were so much more functional and attractive (remember "user-friendly?") than those applications users had access to on larger computers. Moreover, PC spreadsheets easily implemented many functions that hitherto could not be automated because of central IS departments' backlogs. Users were going to use their PCs anyway. Consequently, ways had to be found to support the users and their PCs properly and economically.

As network offerings matured, additional shared resources became common:

- Database servers

- Application servers

- Communications servers

- Backup servers

Database servers provide centralized processing capability optimized for efficient handling of databases. In practice, this typically means that they have large, very fast disk drives. Some mainframe environments also dedicate machines to handling databases. The difference here is that such a mainframe would typically cost well over $1 million; a minicomputer or PC in the same role will typically cost less than $500,000, sometimes less than $100,000.

One of the particularly useful capabilities of a LAN is to allow companies to provide individual processor configurations optimized for specific functions. The ability to provide such servers goes a long way to justify the extra management effort involved in managing LANs.

Application servers are machines dedicated to handling specific applications. Such a server might be marketed together with a specific application by a value-added reseller or VAR. Such a configuration can be used in either of two basic ways: via terminal access (as with mainframes, but for one application suite only) or a client/server approach.

Communications servers allow centralization of external communications facilities and hardware. Modern communications server capabilities include the ability to provide both in- and outbound modem pools (either separated or shared) along with routers and gateways (see below) that serve an entire user community. Without such servers, it is not uncommon to see a configuration involving one modem per PC user. Centralizing a modem pool allows acquisition of faster, better modems and more efficient use of them (if only a percentage of the user population will be using a modem at any one time).

Backup servers allow off-loading of backup functions, which would most likely otherwise reside on file servers. Because the failure of a file server is typically the single most disastrous possible occurrence in a LAN environment, it does not make sense to make the file server the primary backup device (i.e., the unit with an installed tape drive). If the server fails (e.g., because of a failed power supply), it becomes somewhat difficult to restore the data to another machine.

LANs are not limited to these categories of servers; these are just the most frequently implemented. Other possibilities include scanner servers and facsimile servers. Additional functions will most likely evolve over time.

Distributed Systems

The term *distributed processing* predates LANs. However, LANs make it much easier to implement. What is distributed processing? It is not simply spreading functions out among multiple computers, although many people believe that is all there is to it. Distributed processing has a few basic premises:

- Do the work close to where the results are required (minimizing communications costs, a key objective of distributed processing).
- Dedicate processor(s) configured appropriately to their specific functions.
- Use cheaper MIPS (a raw unit of processing power that is generally less expensive when purchased in smaller units).

For robustness—the ability to continue key functions in the presence of failures—provide a certain degree of redundancy. Do not have just one processor in any given function category.

Of course, distributed processing is also designed into wide area networks (WANs), but allowing computers to communicate at high speeds among themselves is much easier to accomplish in a LAN environment.

Client/Server Architecture

Client/server design is, depending on one's viewpoint, either an extension or a subset of distributed processing design. Whereas distributed processing typically begins with the premise that all computers are created equal—at least insofar as their communications patterns—client/server design uses a more asymmetric model. But there are more subtle differences.

The basic premise of client/server design is similar to that of distributed processing: dedicate processor(s) configured appropriately to their specific functions. But in this case, a certain amount of predefinition is applied. Specifically, it is usually assumed that, for a given application, precisely two processors will be involved: a desktop PC used by a human being (the client) and one other machine (the server). The division of labor between these two can be assigned in many ways. Many books have been written on the subject, so it is not pursued here. However, we mention the general approach because, in most cases, a critical characteristic of client/server design is high-speed communications between clients and servers, speed that can only be easily accomplished via LAN technologies.

Scalability

A key benefit often provided by both distributed systems and client/server architectures is scalability. Roughly speaking, scalability is the ability to increase the power and/or number of users in an environment smoothly without major redesigns or swapping out of equipment and software. In a mainframe environment, upgrade paths often looked like the chart shown in Fig. 19.5.

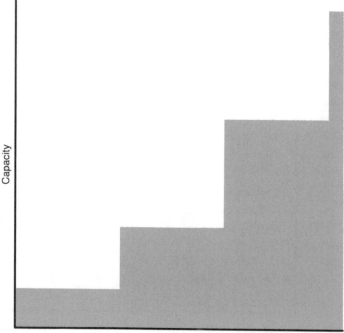

Figure 19.5
The function of capacity over time when dealing with the mainframe.

With smaller individual machines linked together with LANs, capacity upgrades tend to look more like the next chart (Fig. 19.6). In this chart, the capacity is added in smaller increments, as user needs dictate.

The key difference is that, with smaller machines, each upgrade is less expensive, and the added capacity is readily available. Each upgrade is less financially traumatic, so when the need arises, it is usually easier to make the appropriate acquisitions. With mainframes, each upgrade is a big bite, and it takes a while to exhaust the new capacity once you have it. Because each upgrade costs a lot, justification usually takes a while, often resulting in demand exceeding capacity until the approval comes through for an upgrade.

Scalability is more an argument in favor of many small processors rather than specifically for LANs. However, the easiest way to connect those processors is with a LAN.

How They Work

A LAN environment includes, at a minimum, nodes and wiring. We first cover what makes up a node.

Figure 19.6
Capacity for smaller-based machines over time.

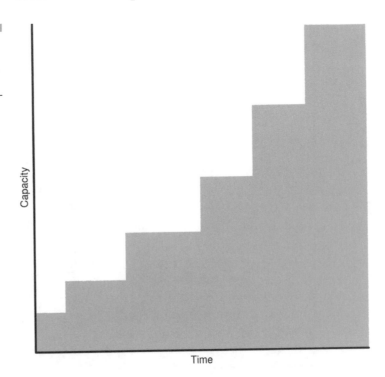

Capacity

Time

Node Configuration Elements

A LAN node is a computer that has a unique network address. That address consists of a sequence of letters and numbers (or possibly just numbers) that, just like your postal address, identifies the node so that messages sent to it can be delivered to the correct device. LAN nodes comprise certain basic elements (Fig. 19.7). The key components of the LAN include:

- Processor (computer, possibly but not necessarily including keyboard and monitor, etc.)
- Software
- LAN network interface card (NIC)

The processor can be just about any computer currently available. The software and NIC, however, deserve more discussion.

Figure 19.7
The basic components of a LAN.

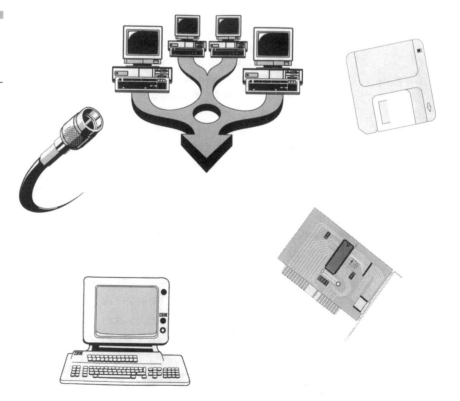

LAN Software for LAN Nodes

Every LAN node requires a complete, functional implementation of the International Standards Organization's (ISO's) Open Systems Interconnect (OSI) model. This does not mean that the specific software is the ISO standard software; rather, it simply means that all the functions described in the OSI model must be addressed, one way or another. With the exception of part of layer 1 (the physical layer), all of the OSI functions are accomplished in software, so it should be no surprise that most LAN software involves not one but rather several software modules that collectively allow a node to communicate.

At the bottom layer of the protocol stack (of these software modules) is certain software that is built into the NIC (in read-only memory, or ROM) as part of the physical layer. This is generally invisible to the user—or even to the installer—but it is critical. When installed, the onboard NIC software executes in the same "address space" used by other software on the node; thus certain address space might have to be set aside by the node in order for the software not to get stepped on by other programs. This might be handled directly by the node's operating system, or it might have to be addressed directly by the installer.

The next layer of software is the network driver. This software is the go-between or interface between the rest of the software on the node and the NIC. In the world of PCs, there are two prevalent standards for this software: Network Distributed Interface Standard (NDIS, from Microsoft) and Open Data Interface (ODI, from Novell). These standards define not how the software drivers operate, but rather the standard approach for the other software on the computer to interact with the NDIS or ODI drivers.

It is the responsibility of the NIC manufacturers to provide such drivers with their interface cards. Providing such standard software is much more difficult than you might assume. As with all of the other software on a node, the NIC interface software runs under the operating system supporting that node. Computers, especially personal computers, can run many different operating systems. A different driver implementing a given interface standard is required for each operating system. If a vendor supports both ODI and NDIS (and most do), that doubles the number of modules to be provided. Nowadays, the PC-based systems include all the necessary protocols to support either network card driver. An example is Windows 9X, which has all the necessary drivers for both NDIS and ODI.

Above the level of the driver software is the province of Network Operating Systems (NOS), such as Novell's NetWare, IBM's LAN Server, and Banyan's Vines. Because the driver standards insulate these environments from the hardware, all of these NOSs can support several different types of LANs. The NOS software on a node provides two major functions:

■ It provides all of the communications functions specified in the OSI model above level 2.

■ It intercepts and satisfies resource requests on the node for resources that would normally appear on the node but instead are to be provided via the LAN (e.g., file access, communications access, etc.).

These resource requests can originate directly from the keyboard (e.g., a user types a file from a server disk to the screen) or from a program (e.g., a word processor program retrieves a document from a remote server). Such requests are normally satisfied so transparently that the user or requesting program is completely oblivious to the fact that a LAN function occurred.

LAN NOS software is typically priced on a per-active-user basis. At this writing, such prices vary widely depending on the NOS as well as the volume of such licenses. They can range from less than $100 per license to as high as close to $300 each. Fortunately, when this pricing structure is used, there is normally not any additional charge for the server software.

LAN Network Interface Cards

Network interface cards (NICs) typically fit into an expansion slot on a computer. In some cases, however, the network interface is built into the computer in the factory. The latter approach usually results in a lower overall cost but can reduce or eliminate the opportunity to change or upgrade the network interface later.

Functions performed by a NIC include the following:

■ Providing a unique LAN address for the node that is built in by the manufacturer.

■ Performing the link level functions with other nodes appropriate to the physical connection (e.g., fiber optics) and protocol (e.g., Ethernet) used on the particular LAN.

■ Accepting protocol frames of information (see below) at full LAN speed, buffering (storing) them on the card until the computer is ready to process them.

■ Recognizing, examining the address information of, and ignoring received protocol frames that are not addressed to the node in which the NIC is installed.

■ Responding to certain management signals on the network. These might be inquiries as to the status of the card or recent communications loads imposed on it, or commands to stop communication on the network. (The latter might arrive if the card fails in such a way as to begin placing spurious traffic on the network.)

Like NOS software, NICs vary widely in price. Factors affecting price include:

- How many NICs are purchased (volume discounts)
- Type of LAN to be supported (Ethernet, token ring, etc.)
- Management capabilities, if any
- Glitzy add-on features provided by the NIC manufacturer
- Buffer size
- Speed
- Bus type

These last three—buffer size, speed, and bus type—bear more discussion.

All NOS node software functions in parallel with other software on a node. For example, the computer does not stop doing word processing when LAN communications occur. Thus, the node can on occasion get quite busy, perhaps so busy that frames arriving across the network might not always be dealt with promptly. However, those frames continue to arrive at the NIC at the rated speed of the LAN (e.g., if Ethernet, that speed is always 10 Mbps). What happens if the bits pile up in the NIC? If the buffer on the NIC fills and more bits continue to arrive, some bits will be lost ("dropped on the floor" in the vernacular). Obviously, the more buffer on the NIC, the less likely this is to happen.

There is an interaction between needed buffer space and the speed of the node. If the node's speed is well-matched to the LAN, then minimal buffers are needed. If the node is much slower than the LAN's rated speed, then larger buffers might be desirable. Smaller buffers will result in retransmissions of the data from the sender, reducing the overall throughput of the LAN.

The speed of the NIC is also a factor. However, you ask, does not the NIC communicate at the defined speed of the LAN? It does, on the LAN side. However, the speed with which it communicates to its node (the local speed) is independent of the LAN speed. The part of the NIC doing local communications might or might not be as fast as the node receiving the information. If it is not that fast, then the NIC is a communications bottleneck. If overall performance is not satisfactory, this is one possible cause.

The type of bus in the node can also have a major effect on the local NIC speed. We are not speaking here of the LAN "bus," described below, but rather the internal architecture of the node. In the case of IBM-compatible personal computers, there are currently at least four different types of buses:

- Industry standard architecture (ISA)
- Microchannel (MC) (literally obsolete now)
- Extended industry standard architecture (EISA)
- Peripheral component interconnect (PCI)

ISA is generally slower than all but the slowest LANs, no matter how fast the node processor or the NIC is. All of the other types can easily keep up with most LAN speeds, provided that the NIC internal speed is also fast enough.

Most other (non-IBM-compatible) nodes can accept data more quickly than most LANs can deliver it. Therefore, if there is a bottleneck at the node, it usually will be either the LAN itself or the NIC.

Topologies

The word *topology* is used in at least two different ways when discussing local area networks. Unfortunately, speakers rarely identify which of the two meanings they are using. Even worse, they often muddle the two meanings together. We try to distinguish the two meanings by qualifying the term into physical topology and logical topology.

Physical topology constitutes the way one lays out the wires in a building. The major physical topologies include:

- Bus
- Ring
- Star
- Combinations of the above (e.g., tree, double ring)

Logical topology describes the way signals travel on the wires. Unfortunately, logical topologies share much the same terminology as physical topologies. Signals can travel in the following fashions:

- Bus
- Ring
- Star (or switched)

So, is there any difference between physical and logical topologies? Emphatically, yes! As we will see, most of the logical topologies can be used on most of the physical topologies. Thus you can use bus communications techniques on a bus or star; ring communications techniques on a bus, ring, or tree; and star communications techniques on a bus or star. Confused? Read on—we sort the main variations out one at a time.

Physical Topologies

The bus physical topology has the following key characteristics:

■ All locations on the communications medium are directly electronically accessible to all other points.

■ Any signal placed on a bus becomes immediately available to all other nodes on the bus without requiring any form of retransmission.

Physical bus topologies are usually drawn as a line with nodes attached, as shown in Fig. 19.8. Note that the bus is a single high-speed cable with all devices connected to it. But as you will see when we cover media, some buses do not require any wires at all.

The ring physical topology has the following key characteristics:

Figure 19.8
The layout of a bus topology.

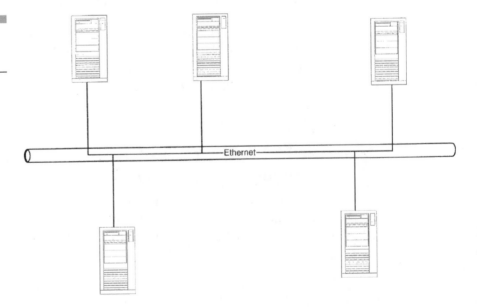

■ Each node is connected to only two other nodes.

■ Transmissions from one node to another pass via all intervening nodes in the ring.

Complete failure of a node, together with its NIC, can (in theory) cause the entire network to fail. We say "in theory" because in practice ring NICs are designed to fail "open"; traffic continues to pass.

Rings are typically drawn as shown in Fig. 19.9. There are variations to this ring that are covered in greater detail in Chap. 21.

Some types of rings (e.g., FDDI) have additional active devices to construct the ring itself (Fig. 19.10). This architecture is not a perfect circle as we think of it, but it is still a ring.

The star physical topology has the following key characteristics:

■ Each node connects to a central device with its own wire or set of wires.

■ The central device is responsible for ensuring that traffic for a given node reaches it.

Figure 19.9
The typical physical ring as it evolved.

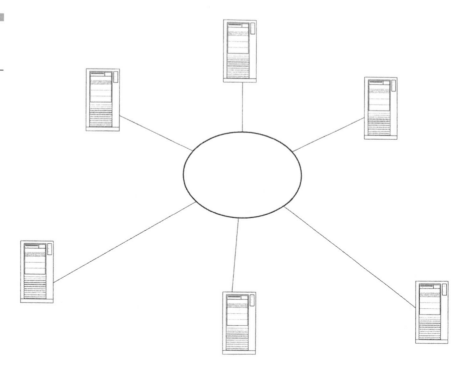

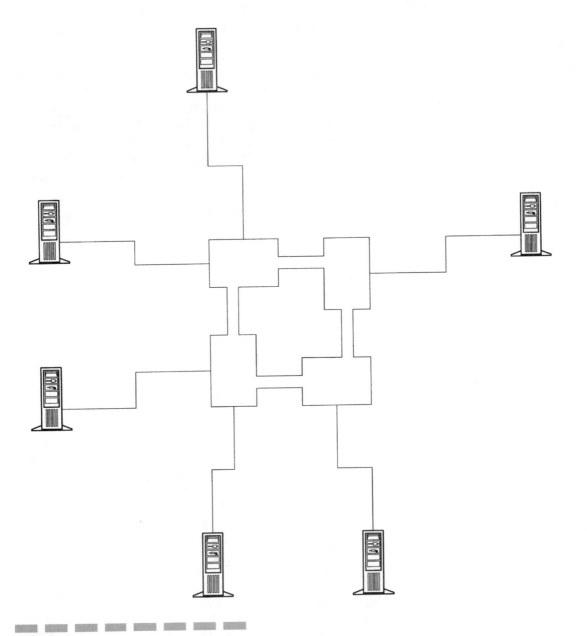

Figure 19.10
The FDDI uses a different type of ring. Not all rings are laid out in perfect circles.

■ Data might or might not pass through multiple nodes on the way to its destination, depending on the specific capabilities of the central device.

Star topologies often are drawn something like the one shown in Fig. 19.11.

Note that, although there is always some type of device at the center, star diagrams often do not show it. It depends on whether the focus of the diagram is the nodes or the way the network is built. In the latter case, the diagram would look like this variation, shown in Fig. 19.12.

Logical Topologies

As mentioned earlier, logical topologies include bus, ring, and star (or switched). A logical bus operates with each transmission visible to every other node on the LAN. The easiest way to do this, of course, is with a physical bus; all devices connect to the same wire.

Figure 19.11
The typical star
network layout.

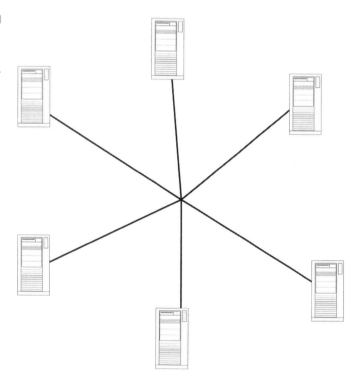

Figure 19.12
The star has some form of electronic hub that all devices connect to.

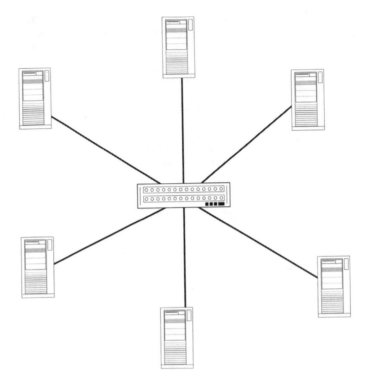

But consider the user of a star physical topology. If the central device instantly retransmits every frame it receives from one wire out to every other wire, how can the individual devices detect whether they are on a physical bus or physical star? Do they care? The answer to both questions is a qualified no. A certain amount of engineering goes into building such devices. However, Ethernet, for example, was originally designed to function on a physical bus. Nowadays, almost all Ethernet installations consist of physical stars.

Bus topologies require a method of handling collisions—times when two or more nodes attempt to transmit at the same or overlapping times. Different methods are used with different protocols.

A logical ring consists of a network where any given transmission passes from one device to another in a fixed sequence until it reaches its destination. It is naturally most obvious on a physical ring. However, it also can be implemented on a physical bus or a physical star. In the case of a physical bus, although all devices see all transmissions, each transmits on the common medium in a fixed sequence, sending packets only "downstream." An advantage of this tightly disciplined line protocol is that there can never be a "collision"; all devices know their turn and follow it. On the other hand, there can be quite a wait until one's turn arrives.

A logical star cannot be built on a physical ring or bus; it requires a central device to make intelligent choices as to the disposition of each frame, sending the frame only to the correct destination. Ethernet switches, a relatively recent development, operate in this fashion. Collisions cannot happen in this environment, although buffer overruns in the central device can occur if it is not quick enough to keep ahead of the frame arrival rate.

Mixed Topologies—The Real World

As we will see in later chapters, ideal designs as described above have never stopped engineers from building what will work. Token rings are really stars, Ethernet is also usually a star, and the fiber distributed data interface (FDDI) is the first real ring that was widely implemented—and it uses the variation shown with devices building the ring. Here is a brief list of some of the more popular LAN protocols, with notes as to their topologies:

Ethernet and token ring.   A later chapter discusses these protocols.

LocalTalk.   This Apple protocol uses a bus physical and logical architecture. Its contention management approach is called carrier sense with multiple access and collision avoidance (CSMA/CA). All devices listen continuously to media (carrier sense). Multiple devices connect to the same media (multiple access). When a device wishes to send a packet, it sends out a very small precursor packet requesting permission. Because the packet is very small (and is only sent out in the first place if the medium is idle), it is unlikely that it will collide with another packet. Presuming that it doesn't collide, all other devices are now on notice that the real packet is coming, so that they do not step on it with their own transmissions. This is LocalTalk's approach to collision avoidance.

Manufacturing automation protocol (MAP).   MAP uses a token-passing bus. The token-passing bus is designed to have the benefit of the bus (high bandwidth) and the predictability of the ring (token passing). Each device's transmissions are heard by all devices. However, every device transmits to the next in a fixed sequence, thus completely avoiding collisions—but lengthening the average time until any given device receives its turn to transmit.

Asynchronous Transfer Mode (ATM). ATM is both a late-model LAN and a WAN technology. It is designed to complement the LAN by performing noncontention LAN emulation. This means that the ATM will look like a star and provide dedicated bandwidth to each device on the network. In reality, the older LANs (called legacy LANs now) will not be sharing a single cable system and competing for the bandwidth as they do today.

Peer-to-peer versus server-based. This is an area of network design that sounds as though it is a topology-related subject, but really it is not. Peer-to-peer and server-based are two different approaches to designing network operating systems and applications. Whole volumes have been written on both approaches; we briefly summarize and distinguish between the two approaches in the next few paragraphs.

Peer-to-peer networks afford users great flexibility, because every node can potentially function as a server. In practice, this means that relatively small shops can set up many or all of their computers to provide file and other services to all other computers. In principle, this is an egalitarian and flexible approach. In reality, it can quickly turn into a nightmare if not managed carefully.

The problem is that managing any server properly requires that a certain amount of time and expertise be focused on that task. Users, not unreasonably, tend to focus their efforts on their own jobs to the exclusion of network management functions. If a user's machine is configured as a server and the user does not manage it, who does?

Another problem with this approach is that server functions consume computer and input/output resources (i.e., time). If you as a user are also providing services to other users, your local performance will be directly affected by the load others put on your machine. Because a major benefit of personal computers is that the power on your desktop is yours alone, peer-to-peer configurations can quickly cause frustration because of degraded performance—unless each machine has more power than is needed solely to satisfy its assigned user.

Having given some of the more serious drawbacks of peer-to-peer environments, it should be stated that literally thousands of small businesses function very well using this technology. The key is that they are small enough that the disadvantages described above never become serious enough to outweigh the advantages. Some of the advantages of peer-to-peer NOSs include:

■ Any system can be configured as a server, allowing very effective group access to any files desired, no matter where those files reside.

- Typically lower cost per node, partly because of lower software costs and partly because of the elimination of the need to purchase high-powered dedicated servers.

- Reduced management costs, partly because some management functions typically do not get performed, but also because these systems are designed to be more easily managed than the dedicated server systems.

Server-based networks predominate. Their advantages and disadvantages are partly the flip side of those listed in the peer-to-peer discussion. But two should be emphasized:

1. Server-based networks can be and typically are tuned so as to support larger numbers of users (anywhere from 20 to 300) on a single server. The management effort involved to support such numbers is much less on a per-user basis than would be the case were a peer-to-peer approach to be used for such a large user base.

2. Although it seems a self-fulfilling prophecy, the very large numbers of installations of server-based systems (especially NetWare) have resulted in an appropriately sized base of people expert in the configuration and management of such systems. Also, that same installed base has attracted many creators of after-market software products that complement and extend such environments, a far larger selection than is available for the peer-to-peer products.

Internetworking

Internetworking refers to the connecting together of two or more networks, either LANs, WANs, or a mixture of the two.

The creators of LANs designed them to be truly "local." This was not an arbitrarily imposed design restriction; by not requiring a LAN to be capable of wide area communications, the engineers could take advantage of a much more controlled environment. That control allowed use of techniques that, while not practical over a wide area, could deliver far higher communications performance than WAN communications of the day as well as reliability orders of magnitude faster than those same WAN communications.

But, people being what they are, a demand soon materialized for the ability to provide LAN-type communications (that is, fast, having a lower error rate, and using typical LAN protocols) over areas wider than those originally envisioned as being supportable by LANs. And engineers, being what they are, came up with not one but several ways of accomplishing this. All involve the use of additional devices that, in varying ways, take a LAN signal and send it further than the original LAN specification allows.

Terminology is important here. We cover four types of devices that provide such extended LAN connectivity. The first type (repeaters) is considered to extend the reach of a given LAN. As such, using repeaters is not technically "internetworking," although we describe it in this section because it nonetheless involves extending the reach of LAN technology in a given environment. All three of the other technologies (bridges, routers, and gateways) are considered to provide connections among or between different LANs, thus providing true internetworking capabilities. The reason for this division between repeaters and the rest should become apparent as we cover the devices in more detail. We begin with the simplest of the devices, and go on from there to more sophisticated products.

Repeaters

Repeaters are relatively unintelligent connections between two LAN segments of the same type (Ethernet, token ring, etc.). A repeater is sensitive to the traffic's content only at the ISO level 1 (physical) layer. An Ethernet repeater will transmit any traffic that can be sent on Ethernet, a token ring repeater will send on all token ring traffic, and so on. Repeaters satisfy only three key functions:

- Distance
- Electrical isolation
- Media conversion

Bridges

Bridges are intelligent connections between two LANs of the same type. They operate at layer 2 of the OSI reference model (data link). The bridge is responsible for linking the two segments or LANs together and transparently passing frames across the link to each other. No form of choices are used, and there are no format or protocol conversions or any other services other than providing the physical and logical paths between the two networks or segments.

The bridge can be intelligent, providing filtering and forwarding services across the link. A filter is used if an addressable device is on the network that is sending the frame. This is so a frame of information does not get sent across the wires to another segment or network that has no need to see the frame. Forwarding takes place when the bridge sees a frame that is addressed to a node on the other side of the link.

Routers

Routers are switching devices that connect two LANs where multiple paths exist (or a single path exists on a dial/leased line basis). Working at layer 3 of the OSI model (routing, or network), the router will be responsible for ensuring that the data gets to its destination. There are several considerations about this decision process, as the router is more sophisticated; therefore, it is more complex to manage. However, this added complexity brings significant gains to the users who are looking to internetwork their LANs. Many large organizations now use routers in their backbone network to connect segments (departments, floors on a building, etc.) together with alternative paths to get from one LAN segment to another. The feeling is that the resiliency is worth the added complexity.

Gateways

Gateways connect two networks of different types (i.e., a LAN to an X.25 network). These operate at higher levels of the OSI model, ISO levels 4–7 (transport, session, presentation, and application). In effect, the gateway handles everything above the network layer in the OSI model. The controlling portion of a session, the format and protocol conversion to make all things common, and the actual application interface reside in the gateway's domain for interconnectivity.

Benefits of using gateways include an initial cost savings, high performance, and high flexibility and scalability.

LAN Switching versus Nonswitched

When discussing driving habits, some people say, "Speed kills." That may be true, but in the case of networks, lack of speed can kill. There are many ways to design networks, but one of the most popular and straightforward is to design a backbone whereby there is a high-speed path for high-speed requirements at a central point (topologically speaking) and all subsidiary networks connect to that backbone. See Fig. 19.13 There are several assumptions in this model. One is that there is a need for connections between either the subnetworks or from each of the subnetworks to that central backbone, possibly for server or application access. These assumptions are not necessarily valid. For example, if the business primarily uses local area networks to provide office automation capabilities, there may be a whole series of islands of automation where it doesn't matter that they are islands. We certainly don't want to be loading applications across the wide area; and if all you are doing is loading applications and storing such data as word processing or spreadsheet information, then you really don't need a backbone to connect up with disparate networks.

There are many cases, however, where backbones are required, so speed is important. The technologies suitable for high-speed backbones are described later in this chapter. Here we are talking about a way to deliver somewhat higher speed to desktops or for small backbones.

The typical office automation network designer has to make a choice about where various components reside. The major components are data and applications. Ideally, we believe that data ought to be stored centrally, whereas applications ought to be executed from the desktop. Why? It's a matter of performance. The internal bus speeds of even medium-speed desktop computers far exceed the capabilities of cost-effective local area networks. Loading applications takes a long time, relatively speaking. When you have a lot of people doing this at the same time, it can bury a

Figure 19.13
A backbone
architecture.

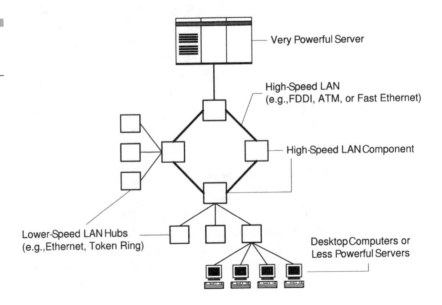

network. You can pay the money to upgrade in the way we're going to describe here, but an easier approach is simply to distribute the applications to the desktop. But, you say, that is a maintenance nightmare! In the past, it was. There are applications available today that will distribute software to the desktop in a very cost-effective and efficient manner. It is not necessary to touch every desktop in order to install or update applications if one has LAN connectivity to those desktops from a distribution server.

We realize, however, that not every company has installed such software distribution capabilities or is convinced that this is the way to go. So what do they do when the performance penalty of loading applications from servers to all the desktops begins to bury their networks? Wouldn't it be nice if they could simply change the central equipment in the network and get a speed kicker? In fact, they can do exactly that.

With older technology such as coaxial Ethernet, this wouldn't work. But the technology we're going to discuss, switching, does allow for a speed boost, a *real* capacity boost, without necessarily touching the desktops in any way.

The key fact to remember here is that most local area networks employ shared media. Generally, when we refer to *shared media,* we mean Ethernet hub-based LANs (as shown in Fig. 19.14) and token rings. Even though both topologies in modern networks are wired with radial or star-type wiring, the equipment at the center (as discussed elsewhere in this book) causes every node in a ring or Ethernet segment to see all of the traffic generated by all the other nodes. They are designed this way, and it is a good way to design a network for many reasons. However, when the traffic starts increasing, everybody sees all of the increases. That's a problem. How do we address it? We replace the central piece of equipment, an Ethernet hub or a token ring MAU, with a corresponding switch.

The job of a switch is twofold:

■ It has to make the traffic on the wires to the devices to which the switch is connected look exactly the way it was before, from the point

Figure 19.14
Shared LAN—packets go everywhere.

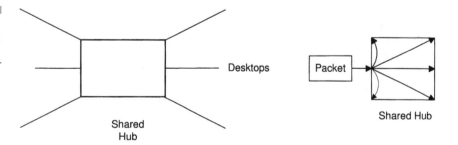

Desktops

Packet

Shared
Hub

Shared Hub

of view of the connected devices, because we are not planning on changing the electronics in those devices.

■ It examines every packet coming in on every wire and, on a packet-by-packet basis, directs those packets outbound to only the port or circuit where the packet has to go.

That may sound like routing, but it happens at a much lower (and therefore faster) level of the architecture than that at which routing takes place. Whereas routing is done at the middle levels on such protocols as IPX or TCP/IP, switching is done down at layer 2, the media access control (MAC) layer. It's done on the basis of the hardware addresses of the connected network devices. Therefore it's very fast. But the speed alone is not what creates the major benefit here. The major benefit is that the only traffic that is put on a wire going to a device is traffic destined for that device (plus broadcast traffic, unfortunately—see Fig. 19.15). So, that wire only sees traffic to and from that device; what this means is that, in effect, the wire now has the full rated bandwidth of the network dedicated to just that device and the connections it's talking to; that bandwidth is no longer shared. Whereas before we might have had an Ethernet running at 10 Mbps and 30 devices sharing that bandwidth, now each device sees 10 Mbps coming to its doorstep and no other traffic to degrade that. This is a major performance upgrade.

There are some limitations, however. For one thing, this is not necessarily a good way to upgrade server connectivity. If you think about it, you will realize that the connection going into the server is still 10 Mbps, or 16 Mbps in the case of token ring. Traffic on wires going to workstations without a switch would have lots of traffic that really wasn't designated for those workstations when using either Ethernet or token ring. But in a typical small network, all of the traffic is either going to or from the server if there is only one. So, giving it a switched connection doesn't really help very much because all of the traffic is still going to have to appear on that circuit (Fig. 19.16). A better way to go is to have a switched hub for either

Figure 19.15
Switched LAN—
packets go only to
correct destination.

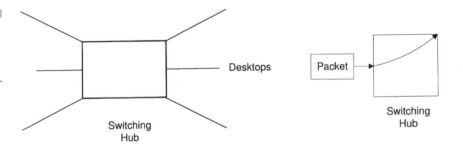

Switching
Hub

Desktops

Packet

Switching
Hub

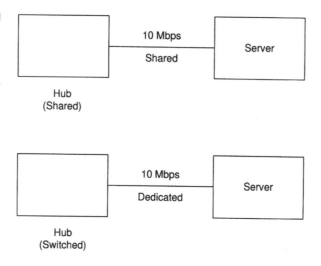

Figure 19.16

Switched connections
don't do servers
much good.

Hub
(Shared)

10 Mbps
Shared

Server

10 Mbps
Dedicated

Server

Hub
(Switched)

token ring or Ethernet that has a higher backbone speed connection into something like FDDI or ATM or any of the other technologies we will talk about later (Fig. 19.17). Now you've upgraded both the workstations and the server connection, and you have an overall increase that means you're less likely to have bottlenecks.

The beauty of this approach (switching to improve connectivity capacity to workstations) is that you only have to change the central hub. The electronics in the workstations remain the same, as does the wiring. Even if the hubs cost two or three times per port what a nonswitching hub would cost, it is still very cost-effective when you consider the alternatives.

In addition to a simple lowering of the traffic levels, collisions are usually avoided in this approach, since there aren't other devices on any par-

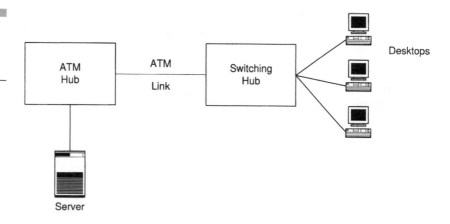

Figure 19.17

Upgrading the
server's connection
to ATM.

ATM
Hub

ATM
Link

Switching
Hub

Desktops

Server

ticular segment that are trying to get in at the same time. Put another way, the segment becomes a single wire connecting, or virtually connecting, two devices talking back and forth through the hub without any interfering traffic to get in the way.

There are a couple of interesting variations on this that are being marketed. One is full duplex Ethernet. If you recall, Ethernet is in fact a half duplex medium. Only one packet can be on the wire at a time, and it, of course, is going only in one direction, or at least it is only destined for a device that is in one direction along the wire. On the other hand, full duplex Ethernet, if it is switched, changes this in such a way that the devices can communicate both ways simultaneously over what is still a 10Base-T-type wiring cable run (Fig. 19.18). Unfortunately, this approach does require replacement of the network interface card in all machines participating in the network. It therefore does not result in a cheap upgrade, but it does give you potentially double the capacity on the connections to the workstations and/or servers compared to even what switched Ethernet can provide. But beware! In many cases, the amount of traffic on the network going from the workstations to the central site is truly minimal when compared to the traffic going toward the workstations. Putting in the technology that provides equal transmission capacity in both directions sounds great, but if you really don't need much more than a tiny fractional increase on the overall capacity on one of the directions, then in effect you are creating capacity that will never be used—and paying a pretty penny for it. One place where full duplex technology could be very valuable is video conferencing on the local area network. There you are sending video signals bidirectionally, generating a great deal of bandwidth requirements. But in general, we don't feel that full duplex local area network technology has much of a future, at least to desktops.

The next variation is more of a topological variant than a technological one. Switching technology was described in the preceding paragraphs

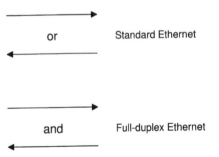

Figure 19.18
Standard Ethernet is half duplex; full duplex Ethernet works two ways at once.

in the context of delivering full bandwidth to individual desktops. But in many cases, workgroups need greater connection speed to a backbone. Moreover, many companies cascade Ethernet hubs in order to gain connectivity benefits while minimizing wiring costs (Fig. 19.19). The problem, of course, is that the connections become much more heavily loaded as one moves up the upside-down tree hierarchy that Ethernet uses for its physical configuration. Ethernet could be considered a poor man's backbone, where, at the lower end of the tree, we insert whole nonswitched Ethernet hubs for workgroups rather than individual workstations (Fig. 19.20). It's a much less expensive approach than using the much faster technologies described later. Of course, one doesn't get the major speed increases that these other technologies can deliver—but this solution may suffice for a while.

Switching can be used with other protocols, especially token ring, FDDI, and any of the fast Ethernet described below. Of course, some vendor needs to build the hardware, but vendors are amazingly ambitious, and we expect that you will be able to buy almost anything you want in a switching variant. Those ports will cost more, but adding switching to some of the following faster technologies in some cases could make the difference between those technologies being acceptably fast for a given backbone or not being quite fast enough.

There is at least one "gotcha" associated with implementing switching technology. In a shared (nonswitched) local area network technology such

Figure 19.19
Cascaded, shared hubs.

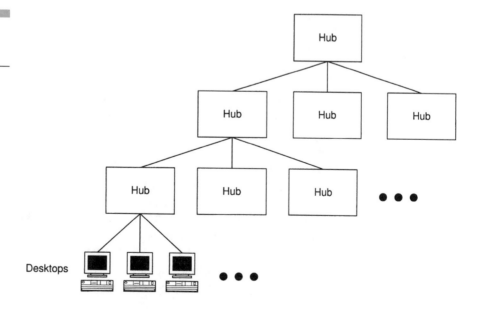

Desktops

Figure 19.20
Switched hubs—
a lightweight
backbone.

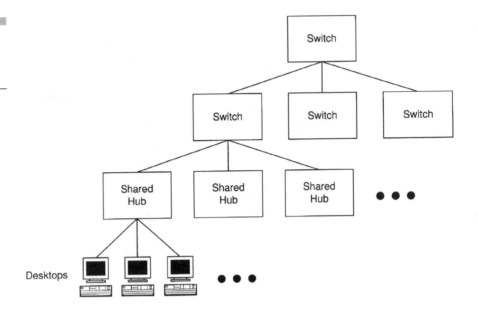

as token ring or Ethernet, it is possible to monitor a network by connecting up at any point and setting your equipment to *promiscuous mode.* This requires special software, but most modern NICs can do this. Although some consider this a security risk, it is extremely beneficial for network management purposes. If one wishes to monitor the level of traffic on a network or examine packet contents to figure out why things aren't working properly, this is a very valuable characteristic in an architecture. Switching technology defeats this capability. Since the only packets that go onto a wire are those destined for that wire, a PC connected up in promiscuous mode will still only receive packets addressed to that PC; packets going elsewhere only go to those other locations. Some network devices have special monitor ports that actually provide all the traffic seen on the backbone of those devices. But not all hubs and routers have this capability. It is a recognized problem, and we do expect hardware vendors to address it over time.

As mentioned above, switching technologies are sometimes used in order to provide a moderate upgrade in speed in an environment that is running out of gas. They are an alternative to changing technologies and installing something like a fast Ethernet or one of the other higher-speed technologies. So the question becomes: does it make sense to switch or use something else like a fast Ethernet? We believe that a mix of technologies will, in most cases, provide the biggest bang for your buck. Because of the elimination of collisions in a switched Ethernet environment vis-à-vis a shared Ethernet environment, switched Ethernet provides a significant improvement in to-the-desktop bandwidth. In fact, there are very few applications that would demand more than clear-channel 10-Mbps bandwidth. An example might be full-motion, uncompressed video teleconferencing. However, the vast majority, including downloads of application programs, can easily be satisfied with clear-channel 10-Mbps Ethernet. This cannot be said of shared Ethernet, which runs out of gas at a much lower level of utilization—and hits that utilization level much more easily. Now, the 100-Mbps Ethernet in an auto-detect mode can use either 10 Mbps when attached to a 10-Mbps switch or 100 Mbps when attached to a 100-Mbps switch. As the industry evolved, the introduction of the 1-Gbps Ethernet changed the way we perceive the use of the bandwidth.

However, we do not believe that switched Ethernet is a major backbone technology except in the smallest environments. Regarding the use of switched Ethernet in order to provide server connectivity is a case in point. A better approach is to use switched Ethernet as the lowest-level connectivity solution in an architecture and connect those switched Ethernet workgroups or switched token ring workgroups to a higher-speed backbone technology, perhaps one of those discussed elsewhere in this book.

20
Ethernet

Ethernet was the first commercial approach to using a LAN on a bus topology. Ethernet is the most commonly used LAN technology today, accounting for more than 80 percent of the installations worldwide. Although certain recently developed technologies are faster (including ATM, two varieties of fast Ethernet, FDDI, 20-Mbps ARCnet, 16-Mbps Token Ring, and the new 25-Mbps Token Ring/ATM, coming from IBM), Ethernet still enjoys continued popularity and growth for a number of reasons:

- For its speed, it is by far the least expensive networking approach.
- Its various standards supporting a wide variety of media are sufficiently well defined that products from multiple vendors can be mixed and matched with a reasonable expectation that they will work well together.
- Ethernet equipment exists to allow virtually any intelligent device to connect to an Ethernet.
- A variety of interfaces still exist for dumb terminal interconnection through terminal servers.
- It is fast enough for the vast majority of applications in use today, despite what we hear in the press or from the vendor community.

The Ethernet standard(s) define an approach to building a contention-based LAN on a bus topology. As originally defined and implemented by Xerox, Intel, and Digital Equipment Corporation (the DIX design), to connected devices the bus appears to be a single straight cable system. The cable is available to multiple users at the same time, even though only one user at a time can send data if a shared bus is installed. We present these concepts in more detail in the rest of this chapter.

A note on standards. Technically, the LAN protocol implemented most widely and still referred to as Ethernet is actually defined in the 802.3 standard, and is not precisely the same as either of the two "Ethernet" standards (I or II) defined by the Xerox-Intel-DEC triumvirate. However, most equipment available today can support both standards; also, in most cases the devices can interoperate on the same media. In any case, most people, when not attempting to be particularly precise, refer to 802.3 as *Ethernet*. We do also.

Concepts

Bus—CSMA/CD

The human ear is much better at picking out a single voice from among multiple simultaneous speakers than is any electronic receiver at selecting one signal from among many. Thus every local area network using a bus (i.e., a shared medium) design must incorporate a contention management or arbitration mechanism that results in only one signal being present at a time. No matter how many devices share the medium, they have to wait their turn. Otherwise, signals on the cable would conflict like voices on the floor of a commodities exchange—but with less profit. Ethernet uses a contention management approach called carrier sense, multiple access with collision detection (CSMA/CD) to ensure that devices "speak" politely, one at a time. CSMA/CD is designed to provide equal use of the network for all attached devices. CSMA/CD actually breaks down into three meaningful two-word terms:

CARRIER SENSE All stations listen to the cable continuously. Many non-Ethernet signaling approaches send a single unvarying signal (a carrier) and impose variations on it based on the desired information to be transmitted, whether data, voice, or music. Examples include AM and FM radio as well as television. Ethernet is different in that Ethernet nodes do not send out any signal at all unless they have something to say—kind of like Mr. Ed, the talking horse. Therefore, when stations monitor the cable, they are listening for the presence of a carrier signal, whether modulated or not. If a device that wants to transmit detects a carrier on the cable, then it knows that another device either is preparing to send, is sending, or has just finished sending information onto the cable. In any case, the waiting device will hold back until the carrier signal vanishes.

MULTIPLE ACCESS If no one is using the cable, any of the network devices attached to the Ethernet can transmit data onto the network at will; there is no central control, nor is there a need for any. This is a more powerful design element than might be apparent at first blush. Networks *degrade* (or experience reduced performance) in different ways, depending on their design. One factor that does not in and of itself degrade an Ethernet is the attachment of literally hundreds of devices (the theoretical limit of the number of addressable stations on the network is 1024, although most networks only have 30 to 35 nodes attached). Because the wire is truly idle, unless a device is transmitting, only active devices impose any load on the network. Thus a myriad of devices can be connected, so long as most of them are not transmitting at any one time.

Of course, there is a potential fly in this ointment: If all those attached devices need to transmit lots of data, great congestion will ensue. However, that might be acceptable for some networks—not the congestion, but the threat of it. Some networks are built to support primarily word processing, an occasional file transfer, and perhaps some electronic mail. With such networks, the light load per machine allows many devices to be present without degrading the network. Networks that will experience much traffic at all per device need to be engineered more carefully, perhaps by being divided up into multiple bridged segments.

COLLISION DETECTION In the event that two devices attempt to transmit at the same time, or even during overlapping time intervals, a collision will occur.

Think of two people simultaneously beginning to speak at a cocktail party. If they both continue to speak, neither will be understood. This might be fine for a New Yorker, who can talk and listen at the same time, but for the type of network under discussion, it can be devastating. On an Ethernet, if a collision occurs, the data is lost and each system must retransmit.

At the cocktail party (if the attendees are reasonably well behaved), both speakers will stop as soon as they become aware of the problem. On an Ethernet, the first device that detects the problem will stop the regular transmission and send out a special jamming signal, which tells all attached devices that a collision has occurred. The jam signal will be heard by every device because all devices continue to listen to the network even if they are transmitting.

At the cocktail party, the speakers will each hesitate a moment, then try again. The one that hesitates the shortest time gets the word in edgewise, while the other will wait for the next opportunity. On an Ethernet, any device that was attempting to transmit at the time the jam signal arrives will assume that a collision occurred. It will therefore calculate and wait a semirandom amount of time. What is semirandom? The calculation is an attempt to generate a random number, but the (guaranteed unique) address of each device is a part of the calculation. Thus, the amount of time waited in such circumstances will always be different for each involved device, ensuring that the next attempt by the device that waits the shortest time will succeed. Well, almost.

The problem is that the only devices that apply the waiting algorithm are those that experienced the collision. Any other device that has occasion to transmit will only follow the standard Ethernet approach (listen, then transmit). If a new participant "decides" that it is time to transmit, it could do so just as the "fastest waiter" in the previous group finishes its wait and goes ahead with its second try. If this happens, another collision will occur.

You might well ask, "how does anything ever make it through an Ethernet?" We must emphasize that all of the shilly-shallying described in the previous paragraphs happens in a small number of microseconds. And remember, unless devices attempt to transmit at almost exactly the same time, collisions will not happen; the second device waits its turn, then transmits.

Nonetheless, if you understand how the collision detection and avoidance mechanism works, it should be no surprise that as the number of communicating devices attached to an Ethernet increases, the number of collisions also increases. In practice, close management is indicated when steady traffic on an Ethernet exceeds about 30 to 40 percent of the 10-Mbps capacity of the network. This really means that an Ethernet with an average number of users on it will yield an effective throughput of approximately 3.3 to 4 Mbps. This assumes a mix of 30 to 1000 terminals, or 30 to 35 PCs, or 6 computer-aided design (CAD) workstations, or 2 minicomputers. In all of these examples, we are using averages; however, the authors know of organizations that have supported 1000+ terminals on large Ethernets. The difference was the type of work being performed. If all the work is word processing, as you might find in a legal business environment, then the demands on the network will be minimal. Yet, put five to six CAD terminals on the network and things will definitely be different. These five to six terminals will saturate the network and be prone to collision.

Half Duplex

If you have ever seen a fast typist using a terminal connected to a minicomputer via Ethernet, you might have concluded that Ethernet is a full duplex protocol. However, Ethernet is actually a half-duplex protocol. It is so fast, though, that it is functionally full duplex. Although no more than one frame can occupy the wire at a time—and in the case of a typist, it is possible that each 64-byte frame might literally include only 1 byte of data in each direction—the frames are moved back and forth so quickly that it looks to a human being as if the typing and the receiving of characters are simultaneous.

If two files were copied, one from node A to node B and the other from node B to node A, at exactly the same time, the flow of frames in both directions would so intermingle that again it would appear that the transmission was going in both ways simultaneously. Nevertheless, because at any one instant only one frame can occupy the wire at a time, the medium is actually only half duplex.

Bandwidth

Ethernet has always been a 10-Mbps facility (except one no-longer-marketed variant—AT&T's StarLAN, which ran at 1 to 2 Mbps). There is an advantage to making all implementations run at the same speed: no significant speed matching is required when connecting multiple Ethernet segments, even those built using different media.

However, although all Ethernets run at 10 Mbps, not all devices on an Ethernet actually communicate at 10 Mbps. This concept is difficult to grasp, but it is important. Every node connected to an Ethernet incorporates an interface capable of receiving from and transmitting onto the Ethernet at 10 Mbps—exactly. Otherwise, it does not comply with the Ethernet standard. But that interface card might not communicate to its own node at 10 Mbps. After all, a large percentage of existing desktop PCs cannot even accept data at 10 Mbps through their internal buses! Sometimes the interface card is not built to interface with the node at a high enough speed; in other cases, the card could go faster but the node cannot. This inability to keep up with the nominal speed of the Ethernet itself is one major source of performance problems, especially if the slow interface is found on a server.

As with all protocols, not all of the bandwidth on Ethernet is actually available for carrying application-related data. Some bits are required for overhead functions. Ethernet supports a variable-length "frame"; the minimum frame length is 64 bytes, the maximum 1518 bytes. The variation results from changing the amount of data transmitted. In the Ethernet (actually, 802.3) protocol, 18 bytes in each frame are dedicated to overhead, as shown in Table 20.1.

TABLE 20.1

Composition of an
802.3 Frame

Function	Bytes	Description
Destination address	6	Ethernet address of the node to which the frame is addressed
Source address	6	Ethernet address of the sending node
Length	2	Indicates the number of bytes of data in the next (data) field
Data	46–1500	The "payload"—the actual data being transmitted
CRC	4	Cyclic redundancy error detection bytes

Destination Address

Both the destination and source addresses are 6-byte fields, usually represented in hexadecimal format; for example, BB-BB-BB-BB-BB-BB.* Every Ethernet interface on a LAN receives and processes all frames transmitted on the LAN. The main job of the receiving interface is to compare the destination address in each received packet with its own address. If the addresses match, the frame is delivered to higher protocol layers at the node for further processing and eventual handoff to an application. If the addresses do not match, the frame is "dropped on the floor"; that is, it is discarded.

Source Address

As mentioned elsewhere, every Ethernet node contains an interface device, sometimes built in, other times provided as part of an add-on printed circuit board that implements the interface. Every single one of those devices worldwide contains (or should contain) a unique 48-bit Ethernet address, usually stored in programmable read-only memory (PROM). Every company that wishes to manufacture such Ethernet devices applies to the

---

*In our normal numbering system, decimal or *base 10*, the digits 0 through 9 are used to count. Note that even though we call it base 10, and there are 10 digits, there is no single digit that represents the quantity 10. Each position (ones, tens, hundreds, etc.) in a number contains a decimal digit (that is, 0 through 9) indicating how many of a power of 10 (100 = 1, 101 = 10, 102 = 100, and so on) goes into that number. Thus the number 264 is four 1s plus six 10s plus two 100s.

In the hexadecimal or *base 16* numbering system, we also use the digits 0 through 9 to count. But we need six additional symbols to represent the values 10 through 15 (there is no need for a single symbol for 16). Hexadecimal uses the capital letters A through F to represent the values 10 through 15, respectively. Each position in a hexadecimal number contains a hexadecimal digit (0 through F) indicating how many of a power of sixteen (160 = 1, 161 = 16, 162 = 256, and so on) goes into that number. Thus the hexadecimal number 264 is four 1s plus six 16s plus two 256s or, in total as represented in the decimal numbering system, 612. The hexadecimal number BB is eleven 1s plus eleven 16s, or a total of 187 decimal.

The "natural" internal numbering system used in most data processing and communications equipment is binary, or *base 2*, using only 1s and 0s. Network addresses (and many other items in data communications and data processing) are usually represented in hexadecimal notation because of the ease with which it can represent or be converted into 4-bit binary (base 2) numbers, as follows:

Hex digit	0	1	2	3	4	5	6	7	8	9	A	B	C	D	E	F
Binary value	0000	0001	0010	0011	0100	0101	0110	0111	1000	1001	1010	1011	1100	1101	1110	1111
Decimal value	0	1	2	3	4	5	6	7	8	9	10	11	12	13	14	15

Because it only takes one hexadecimal digit to represent any possible single 4-bit binary value (as illustrated here), and 1 byte equals 8 bits, it takes exactly 2 hexadecimal digits to represent the contents of 1 byte.

Institute of Electrical and Electronics Engineers (IEEE) (in the past, Xerox performed this function) for a unique 3-byte code. As the company manufactures the interfaces, it "burns in" its 3-byte code into the first 3 bytes of the 6-byte address of that device. The other 3 bytes of the address are normally assigned sequentially, 1 per device, but can in fact be assigned in whatever way the manufacturer wishes. The only requirement is that every device has a different 6-byte address. If properly followed, this system guarantees that every manufactured Ethernet device will have its own unique 6-byte physical address.

Apparently, at least one Far East manufacturer has "broken" this approach to guaranteeing address uniqueness. It acquired chip designs for Ethernet interfaces—and then proceeded to manufacture them identically, one after another, with identical addresses. If more than one of those interfaces were to be installed on the same local area network, they would be unable to communicate, either with each other or with other devices.

Other variants to this rule occurred when users were allowed to assign their own 3-byte extensions on the addressing mechanism. This allowed a manager of a network to assign specific addresses to servers, sometimes in a sequence that made sense to the manager for the appropriate LAN. However, this led to some complications where the manager would assign two identical addresses to the devices on the network, which led to conflicts. Most installations today have the standard addressing that comes on the card because of the past problems with this process.

Length

Valid lengths for Ethernet frames are from 64 to 1518 bytes inclusive. The length field in the frame indicates only the number of payload bytes, and that value can range from 46 to 1500 bytes inclusive. Notice that a frame cannot contain fewer than 46 bytes of data. What happens if the sending device provides fewer bytes to be transmitted? The controller will add *pad* bytes to bring the overall frame size up to the minimum length of 64 bytes. The undersized frames are called *runts*.

Why is there a minimum length? There is an interaction between minimum frame length and the ability to detect collisions. Each medium supported by the Ethernet standard has a defined maximum length for any one segment. One factor that goes into determining that length is the need to ensure that a device transmitting a frame onto the segment will still be listening if a collision occurs—and will recognize that the collision happened to its own transmitted frame. Devices stop listening for their own collisions a very short time after they finish transmitting.

Therefore, once a device begins transmitting, it must continue to do so long enough for the most distant device to receive at least the beginning of the frame, plus the time required to receive a jam signal from that distant device if such an event occurs.

If the minimum frame size were smaller, then the defined maximum segment sizes would also have to be shorter. To allow longer-than-specified segment lengths, the minimum frame size would have to be increased, requiring additional pad characters and reducing the efficiency of the Ethernet protocol. The 64-byte minimum cannot be claimed to be "ideal"; rather, it is an engineering judgment call by the original designers of the Ethernet protocol. If installers adhere to the rules regarding segment lengths, no undetected collisions can occur.

Data

This is the frame's payload, the reason the frame is being transmitted in the first place. The data can be up to 1500 bytes, although it does not have to be arranged in 8-bit octets (bytes); actually, it can be any pattern of up to 12,000 bits (that is, 8×1500). In most cases, the actual number of application-related bytes carried will be smaller than this, because some of the 1500 bytes will be used by higher-level protocols for their overhead bytes, sometimes many more per frame than are required by Ethernet itself. The average size Ethernet frame is judged to be between 300 and 350 bytes.

CRC

The cyclic redundancy check (CRC) digits allow the receiving node's Ethernet interface to determine whether a frame was received intact or not. If not, the frame is discarded. If the interface is one of the more expensive available, it might also keep track of the number of frames discarded in order to facilitate management and detection of network faults. Error checking was covered earlier in this book.

Components

Speaking very generally, the components used to implement a LAN tend to be similar across implementations; there must be an interface built into (or added to) each node, each node must have driver software installed that communicates between the nodes (network) operating system software and the interface, and the interface must be connected to the network medium. However, when one delves a little deeper, there are significant differences both among LAN architectures and even among variations in the same LAN architecture, such as Ethernet.

For example, Ethernet is probably supported over more different types of media (with truly standard implementations) than any other LAN protocol. And the detailed piece parts, as well as design considerations, vary greatly depending on the specific medium.

The rest of this chapter covers the various components that can be used to build an Ethernet. We also mix in some design information and rules. Before getting down into the details, though, we would like to indicate what happens to people who violate the rules. Actually, at most organizations there aren't network police. However, many network installers and users wish there were. Generally speaking, there are two kinds of consequences that can result when LAN design rules are violated. The network doesn't work. Obvious, right? What could be worse? What could be worse is the alternative: the network that does work...sort of.

The designers of the various LAN standards (not just Ethernet, of course) attempted to specify standards that would result in robust networks—networks that could continue to operate in the presence of certain kinds of problems. But, TANSTAAFL ("there ain't no such thing as a free lunch"). If a network does have some kind of problem and continues to operate anyway, that operation is usually somewhat degraded. In Ethernet's case, for example, intermittent short circuits appear to users as collisions. No data is lost, but overall performance can slow. Out-and-out failures are often much easier to diagnose and repair.

10BASE5

This would be a good time to explain the naming standard used for the various Ethernet substandards. Bearing in mind that these were defined by committees, here goes. Each term is made up of three parts. The first part (here, *10*) indicates that the network runs at 10 Mbps. The second part (*BASE*) indicates that the network is a baseband network (see below for 10Broad36). The last part (in this case, *5*) indicates some physical component of the cabling system. In the case of 10BASE5, it indicates that one segment can reach up to 500 meters (1640 feet) in length. We explain the variations in these terms as we cover each of Ethernet's supported media.

Ethernet was originally defined to operate over a particularly heavy type of coaxial cable, one thicker than most individuals' thumbs. People in the industry often referred to it as *orange hose*—a particularly apt name because of its usual color. A more conventional term for the 10BASE5 medium is thick wire.

10BASE5 PIECE PARTS   As with any LAN, every attached device must have a network interface card (NIC). In the case of thick wire, that NIC has a port called an attachment unit interface (AUI), a nine-pin interface designed to connect to a transceiver cable (sometimes also referred to as a drop cable).

The transceiver cable can extend up to 50 meters (164 feet), but usually is much shorter, somewhere between 2 and 10 meters. This cable is not Ethernet cable; it is not coaxial, but rather multipin, similar to RS232 but with fewer wires. The transceiver cable plugs into—naturally—a transceiver.

Transceivers include the componentry for a combined transmitter and receiver. Minimum spacing between transceivers on the cable is 2.5 meters (8.2 feet). When using the orange Ethernet cable, there are black marks spaced on the cable to indicate the appropriate spacing between the taps. If you do not adhere to the distance limits, other problems can occur. Ethernet transceivers are also sometimes referred to as *taps* or sometimes *vampire taps*.

Vampire tap is most descriptive. Thick-wire Ethernet taps can be attached to the orange hose without interrupting the network. This is done by inserting two sharp points into the cable, one deep enough to reach the core, the other stopping at the braid. (A special tool is usually used to predrill the holes.) The tap clamps onto the cable with the two points sunken into the cable. The design might seem somewhat baroque, but the ability to add or remove devices from an active network was a significant advance in network technology at the time Ethernet was first designed.

Not all networks—not even most other Ethernets—require separate transceivers, so it would be worthwhile to indicate their function here. Ethernet transceivers perform several important functions:

- Physical access to the main LAN cable, as described above
- Media conversion from coaxial to multistrand cable
- Electrical protection and isolation in both directions (LAN to NIC and vice versa)

But transceivers are essentially passive devices; the protocol smarts reside in the NICs.

Thick wire is the only Ethernet medium that must be configured as a physical bus. As mentioned above, thick wire can be configured in segments up to 500 meters long. Up to three segments can be connected in series using two Ethernet repeaters. If more segments are so connected, the (very small) delays unavoidably introduced by the repeaters might foul up the collision detection part of the protocol.

Up to 100 direct physical attachments can be made to one segment, although there are ways to attach far more network devices than that. Up to 1024 network addresses can reside on a segment or set of segments connected by repeaters. As with the other rules, violators of this configuration rule might be punished by undetected collisions.

Barrel connectors are small passive devices used to connect two pieces of coaxial cable to make a longer one. Barrel connectors are not defined parts of the Ethernet standard. They can be used, but the rules for segment lengths simply apply to the aggregate length built up of smaller wire pieces; that is, a 500-meter length can include one or more barrel connectors, but it still is limited to 500 meters.

Terminators are also small passive devices. But, unlike barrel connectors, terminators are defined Ethernet components and are essential to proper operation of a 10BASE5 Ethernet. Every 10BASE5 segment must have precisely two terminators installed, one on either end. One and only one of those terminators must be properly grounded. We used to tell students that this was required to ensure that the bits do not "fall out of the wire." But actually, in a matter of speaking, terminators are intended to ensure that the bits do "fall out of the wire"; improperly terminated coaxial cable ends can reflect transmissions back onto the wire, interfering with themselves and/or later transmissions and generating unnecessary collisions.

Configuration of 10BASE5 Parts

SETUP   Designing 10BASE5 networks is relatively simple. But installing them is less so. If only one segment is needed, typical configurations include those shown in Fig. 20.1.

Note a few characteristics of the 10BASE5 cable layouts:

■ No loop is created; the cable forms a line, however bendy.

■ The cable is strung in such a way that all parts of the building are within reasonable transceiver cable reach.

■ The cable is continuous and is most likely not configured to take advantage of any existing cable troughs. In fact, it is probably placed in the ceiling.

Figure 20.1
Typical 10BASE5
cable layouts.

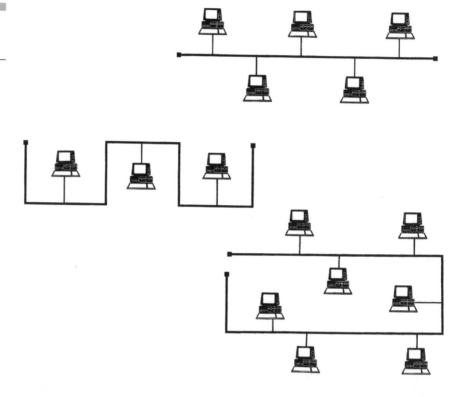

- The 500-meter segment length limitation applies to the length of the cable on the reel; how it is laid out does not affect the length limitation at all.

- Everything that attaches to a 10BASE5 cable attaches via a transceiver that is clamped to the cable or via something else that is attached via a transceiver. The cable ends never plug into anything but the appropriate terminator.

- If more than one segment is required, the segments connect as shown in Fig. 20.2.

Because of the complexity of the layout, the differences in building designs, and the difficulty in using this thick-wire coax, this has become a throwback in the installation business. Very few new 10BASE5 networks are being installed these days, so we will not go into great detail on their possible layouts. But it is a good idea to expand on the two-repeater rule here because it applies (sometimes in modified form) to other Ethernet topologies as well. The rule says, in effect, that "Thou shall not configure an Ethernet such that traffic must flow via more than two repeaters to get from any node to any other node." Seems straightforward, but there is

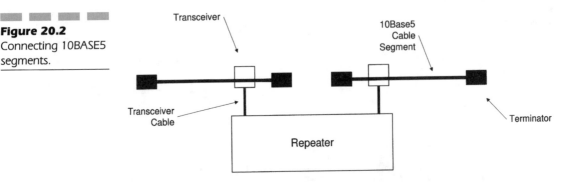

Figure 20.2
Connecting 10BASE5
segments.

at least one nonobvious subtlety. If two repeaters are configured so that the cable between them has no other drops (not even one), then for purposes of the two-repeater rule those two repeaters each count as one half of one repeater. Thus, the two add up to one. An Ethernet can thus have up to four repeaters in series without violating the two-repeater rule. Figure 20.3 illustrates a configuration with three repeaters.

The two-repeater rule is not as onerous as might be the case. It is possible to configure a 10BASE5 network with an almost arbitrarily large number of segments. For example, consider the "comb" configuration shown in Fig. 20.4. Here, no single packet must traverse more than two repeaters to reach anywhere on the network, although seven repeaters are in the configuration. Each segment can still be 500 meters; there is opportunity with such layout to connect a great many nodes, if necessary. Nodes can be placed on any segment, although in practice the top segment, which is connected to all seven repeaters, is often designated as a *backbone* and reserved for repeaters and perhaps network monitoring equipment.

Figure 20.3
Three repeaters
complying with the
two-repeater rule.

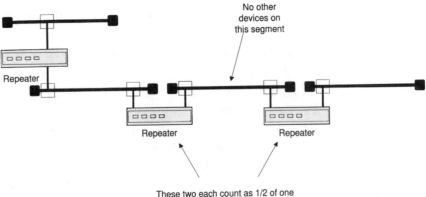

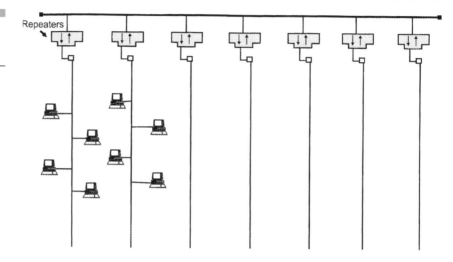

Figure 20.4

Comb 10BASE5 layout.

CHANGES Changing connections to a 10BASE5 network is simple; simply unplug or plug in components as needed, adhering to the configuration rules. No special software configuration is required for new nodes; removed nodes require no closeout procedures to remove them from any tables or anything else. (This simplicity might or might not extend to the upper layers of the protocols used on the Ethernet.)

However, physical access to the main cable itself is required. In an office environment, the main cable is typically strung in the ceiling, with transceiver cables leading to individual offices. So moves and changes can be quite an effort. For this reason and others, 10BASE5 is rarely installed anymore. See Fig. 20.5 for how accessing the cable and the need to add a drop cable from the ceiling to the device can be problematic.

10BASE2

10BASE5 wiring is extremely difficult to work with, and costly on a per-foot basis. With improving technology, companies determined that by applying some engineering smarts, they could design a new Ethernet standard (still 802.3) for use on a much lighter, flexible, and less expensive type of coaxial cable. 10BASE2, also often referred to as thin wire, or cheapernet, is specified to run on RG-58A coaxial cable. This cable is almost identical to cable that you probably have running into your home for cable television. However, there are quite a number of differences between the configuration of 10BASE5 and that of 10BASE2.

Figure 20.5
To move a station is very complex. It requires going above the ceiling. The drop behind the wall may not lend itself to removal.

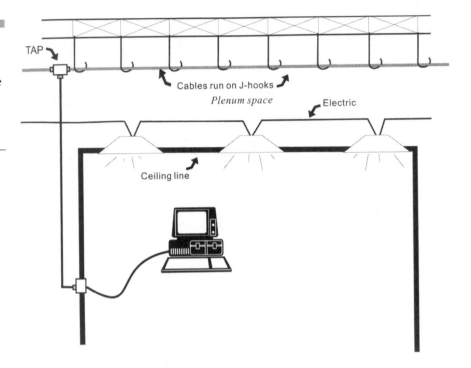

10BASE2 PIECE PARTS As with the other 802.3 media, 10BASE2 requires a NIC in each node. However, the NIC is much smarter and the "tap" much simpler (and less expensive) than in a 10BASE5 configuration. Also, there is no transceiver cable required or even allowed; instead, the tap—called a T connector—plugs right into the back of the NIC. The T connector splits the signal on the cable, allowing it to continue on and also sending a copy directly into the connector on the NIC—a coax connector of the same type as on the other two sides of the T. A T connector looks something like the one shown in Fig. 20.6.

T connectors have advantages and disadvantages compared to transceivers. They are far less expensive, about one-tenth the cost. They can be added or removed from a 10BASE2 daisy chain in seconds, much more quickly than a 10BASE5 transceiver. However, unlike with 10BASE5, T con-

Figure 20.6
The 10BASE2 T connector.

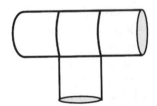

nectors are inserted in line in the coaxial cable; if a T connector is to be added or removed, the entire cable will be down for those few seconds during which the change is being made. This is true even if the T connector is being added to the end of the cable. Once a T connector is added to the cable, unplugging it from or plugging it into its attached node does not disrupt the cable—so long as the T connector itself remains in the cable. For this reason, many installations preconfigure many T connectors in each daisy chain to avoid later disruption.

The minimum distance between T connectors is 0.5 meters (1.6 feet). One 10BASE2 segment can be up to 185 meters (607 feet) long, much shorter than a 10BASE5 segment. As with 10BASE5, every segment must end with a terminator, in this case one designed for 10BASE2. As you can see in the next section, there are differences in how the coax is terminated and in how it is laid out when compared with 10BASE5.

Configuration of 10BASE2 Parts

SETUP 10BASE2 is supported in two different kinds of configurations and is also often used in a third not originally envisioned in the specification. The three configurations can briefly be summarized as follows:

- Star with daisy chains
- Pure star configuration
- Daisy chain bus

The original specification for 10BASE2 was a star configuration, with each radiating spoke a daisy chain bus. Such a configuration might look like Fig. 20.7.

A multiport repeater (MPR) takes every signal it receives on one daisy chain and sends it out over all of the other daisy chains to which it is connected. Up to 27 T connectors (some vendors support up to 30) can be installed on each daisy chain; the number of daisy chains is limited by the number of ports on the MPR. The last T connector on each daisy chain must have a terminator on one side. The MPR is an active device; it plugs into ac power. As such, it is grounded through its power cord and therefore provides an electrical ground for each attached segment. Thus no separate ground is required (or indeed allowed) for the individual segments. Note that each daisy chain is plugged directly into a port on the MPR; the cable end at the MPR does not require a T connector.

This MPR configuration of daisy chains is very cost effective compared to 10BASE5. However, wiring and maintaining it rather onerous. Until the

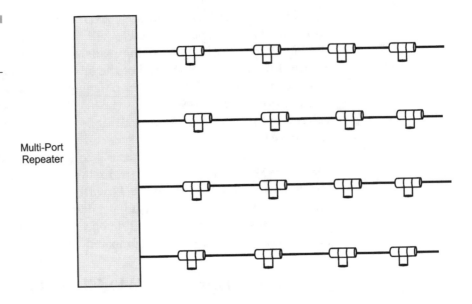

Figure 20.7
10BASE2 using a
multiport repeater.

last few years, using this approach for wall-plate installations (vs. just running the cable along the floor) was very difficult and often subject to frequent outages as individuals interrupted connections, disrupting the entire daisy chain. More recently, special wall plates have come onto the market that simplify such wiring; however, they are quite expensive. 10BASE2 installations can also use a pure star configuration, as shown in Fig. 20.8.

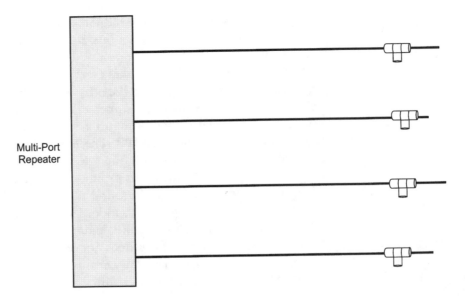

Figure 20.8
10BASE2 pure star.

With only one T connector (and therefore only one Ethernet node) per cable, the per-port cost of a 10BASE2 pure star is much higher than the daisy chain approach; the cost of the MPR itself is spread across far fewer devices than with the previous configuration. Nonetheless, the star has two major advantages over both the daisy chain and 10BASE5:

■ Because it is no longer a physical bus (although it is still a logical bus), it can use the physical communications paths installed in buildings for the purposes of telephone wiring because telephone wiring too is star configured.

■ Because each node is on its own cable, cable faults generally do not affect other nodes through the MPR.

Both of the two previous 10BASE2 configurations are officially supported by vendors. The following one is not. The pure daisy chain bus (Fig. 20.9), without use of any repeater, was not contemplated by the specifiers of the 10BASE2 standard (perhaps because the opportunity to sell equipment to implement it is minimal). The biggest problem with it is that, although every Ethernet segment is supposed to be grounded, daisy chains without repeaters usually are not in practice grounded.

Because such networks are by definition very small (not to exceed 30 T connectors), the problems brought on by not grounding them properly usually do not cause major interruptions. Certainly this is probably the least expensive type of Ethernet configuration on a per-port basis that supports more than two nodes. All of the cabling components together (except the NICs and software) can in most cases be assembled for about $10 to $25 per Ethernet node total.

CHANGES   Removing a node from any 10BASE2 network is simple; just unplug the T connector from the back of the NIC, leaving the T connector in the cable. Adding a node to either of the daisy chain configurations is more difficult. If dealing with preterminated coaxial cable lengths, adding a node requires at least one additional cable length and a T connector.

The problem is not so much acquiring the materials as getting at the portion of cable to which the new cable length is to be added. In addition, unless the new connectors are used (and these must be preinstalled), the

Figure 20.9
10BASE2 daisy chain
bus without repeater.

entire daisy chain is down until the new T connector is inserted and connected.

The pure star approach is typically precabled to every work location, just as is normal telephone wiring. With appropriate use of patch panels at a central location, moves and changes in such an environment are easier than moving telephones. (Normally, telephone moves involve telephone number changes; Ethernet moves require only connecting the new location up to any active MPR port.) Locations not requiring Ethernet connectivity simply do not have their coaxial cable runs connected to an active MPR port. If 10BASE-T wiring had not been invented, the 10BASE2 approach might have become much more popular.

10BASE-T

10BASE-T defines a standard for running Ethernet over unshielded twisted pair (UTP) wiring. The prospect of using existing, in-place (but unused) telephone wiring provided much of the impetus for the development of this standard. Many organizations had already been using some of the existing wiring in their buildings after the breakup of the Bell System in 1984. What actually happened was that Bell came along and told users that the in-house (station) wiring was theirs. Bell would no longer be in the in-house wiring business and would not maintain the wires as they had in the past. Realizing that the number of wires in the building just for the telephone systems was extraordinary,* and that installing new wiring was expensive, many organizations saw the opportunity to use the existing spare telephone wires to run their higher-speed LAN connections. However, in practice, little of the wiring that was in place (when the standard came out) could meet Ethernet performance requirements. Even if these wires could meet the standards, the vendors required that they be certified. This meant that the vendor would send in a team of wiring experts who would test and certify each cable run. Reality revealed that it was more expensive to certify the existing wiring than to install all new wires. This proved to be a catch-22 for the end user.

Nonetheless, the much lower wiring cost of 10BASE-T (primarily because of the lower cost of telephone wiring as opposed to coaxial cable), as well as the availability of people who know how to work with UTP wiring,

---

*Back then, Bell used to install a 25-pair cable to every desktop, even if only a single pair was needed. This enabled them to avoid having to rewire when a phone was changed or moved.

has made this cabling approach extremely popular. The vast majority of new installations today use 10BASE-T for horizontal (that is, to the desktop) wiring.

The possibility of running Ethernet over UTP excited many companies before the actual standard was developed. Several vendors, including Synoptics,* DEC, and others came out with Ethernet-over-UTP variants before the 10BASE-T standard was finalized. These variants worked, but naturally, it was impossible to intermix components from these vendors to build an Ethernet-over-UTP configuration. Most of the vendors did, however, have the integrity to guarantee free conversion, or at least plausible migration paths, from their proprietary hardware to hardware compliant with the 10BASE-T standard when the latter became available.

That promise does not mean, of course, that all customers bothered to take advantage of such upgrades when they became available. You might find yourself dealing with an existing UTP-based Ethernet installation with which you have had no previous involvement. If so, we suggest you verify that the hardware is indeed 10BASE-T compliant rather than an older proprietary variant before blithely assuming that added 10BASE-T hardware will integrate properly.

10BASE-T PIECE PARTS   10BASE-T NICs are functionally similar to 10BASE2 NICs. That is, there is no separate transceiver as is required with a 10BASE5 architecture. The connections are quite different, however. Ethernet UTP wire is terminated at each end with an eight-pin modular RJ-45 jack (only four pins are used). The modular jack plugs directly into the NIC; no T connector is required. Each cable supports only one Ethernet node, as with the pure star thin net approach. See Fig. 20.10 for a representation of the 10BASE-T environment. At least category 3 wire must be used (see the chapter on media for more on categories of UTP wire). Its length must not exceed 100 meters (328 feet).

---

*Synoptics later merged with Wellfleet (a router manufacturer), creating Bay Networks. Bay was acquired by Nortel in 1998.

Figure 20.10
The 10BASE-T
connection.

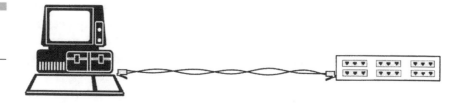

A 10BASE-T hub (sometimes also called a concentrator) is usually required. This is another version of the LAN in a box. The box or hub is the backbone network, whereas the horizontal run to the desktop is the drop. The only exception is if only two devices are to be connected; two 10BASE-T devices can be linked via a crossover 10BASE-T cable. 10BASE-T hubs are functionally identical to multiport repeaters, although some vendors provide sufficiently fast central hubs that the two-repeater rule can be slightly relaxed.

10BASE-T hubs (and, to a large extent, other hubs as well) come in two broad categories:

- Work group hubs, sometimes referred to as stackable hubs
- Backplane chassis, sometimes referred to as concentrators

Work group hubs are relatively inexpensive (as low as $200 to $300) devices that typically support 12 or 16 connections. If they incorporate SNMP management agents, the cost increases by several hundred dollars. Work group hubs, by definition, have very few options in their configurations. Nevertheless, their low cost and simple environmental requirements (they can be mounted in a rack or simply placed on a table or on the floor) make them extremely popular.

Chassis-based concentrators offer great modular flexibility. Such devices consist of a power supply, some control logic, and several slots into which interface boards can be inserted. Each interface board can support several 10BASE-T cables, often 12. Depending on the manufacturer, other boards might be available to support different types of networks out of the same chassis. Those other network types (for example, Token Ring) might be interconnected to the Ethernet board either via an additional interface board just for that purpose or via an external device not inserted into the chassis.

10BASE-F (Fiber)

Newer uses of various cabling systems have always been explored. The fiber-optic medium has been catching everyone's attention because of the declining costs of the glass and the electrical characteristics (that is: the freedom from electrical and radio frequency interference and the isolation from electrical components being nonissues) of the fiber. When users discovered that the fiber could be used inexpensively in the backbone network, they were thrilled. If they could use a high-speed medium that was impervious to the electrical, mechanical, and radio frequency interference, then the network could be stretched to areas that were previously unavailable. Furthermore, none of the grounding and bonding issues that come with copper cable were issues on the fiber network. When the Ethernet standard ramped up to higher speeds (100-Mbps Ethernet, for example), the fiber would support these higher speeds with ease. Thus, the backbone saw more implementations with the 10BASE-F installed.

The 10-Mbps standard was still used in the backbone network, but UTP wiring was still the least expensive proposition for wiring to the desktop. Therefore, a medium changer is required. In the closets where the LAN will run from floor to floor, or from closet to closet, fiber is used. At the hub, ports exist for the connection from hub to hub via the fiber backbone. Inside the hub, the electronics are present to convert the fiber backbone to a copper station drop, or, in more precise terms, from the light to the electrical pulses needed for the copper. A typical configuration of the 10BASE-F environment is shown in Fig. 20.11. This could be extended to the desktop, but, as already stated, the cost of putting fiber cards acting as transceivers in a PC or workstation is too steep for the average user. Therefore, this is done in a mixed environment.

10Broad36

Some networks have been built based on a broadband coaxial cable in the backbone. This might be more prevalent in a campus area (such as a college, corporate park, or hospital) where multiple buildings are interlinked. The use of the broadband cable was an expedient to support the needs of the organization's voice, data, and video needs because all of these components were analog transmissions that were connected between buildings. Therefore, the broadband backbone cable (such as a CATV cable) was used when the Ethernet was introduced. Customers who needed to link multiple buildings together were constrained by the 500-meter distance limitations of the Ethernet baseband coax. Even with the two-repeater rule, the most that the cable could be extended was 1500 meters, just under a mile. Although this might sound like a lot of cable, which should reach just about anywhere, that one mile of cable gets used very quickly when it needs to be snaked through ceilings, up and down from the ceiling heights, and so on.

Therefore, the existing broadband cable offered a significant distance increase to Ethernet users who had to exceed the distance allowances. Using a 10-Mbps transmission speed on a broadband cable, the distance

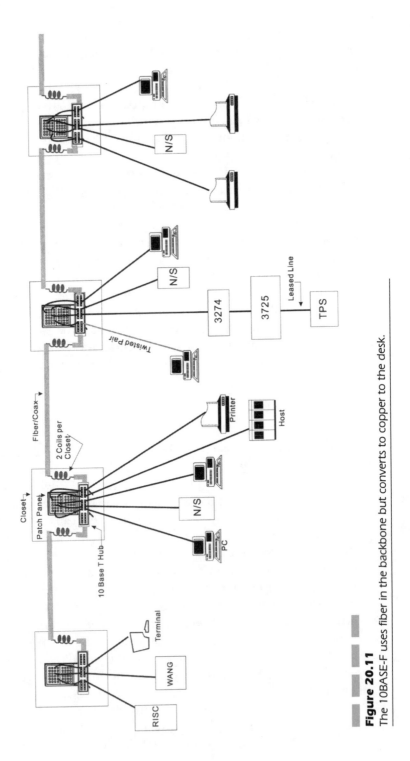

Figure 20.11

The 10BASE-F uses fiber in the backbone but converts to copper to the desk.

limitations were increased to 3600 meters. This is a sevenfold increase in the distances over which the cable could be run. So the vendors came up with a solution to the distance and the medium needs for the end user. Broadband cable inherently is an analog transmission system. Special devices called frequency agile modems (FAMs) were attached to the cable to modulate the electrical signal onto the carrier. The cable is normally broken down into various channels of 6 MHz each. Using a bridging arrangement, two channels could be connected together as a 12-MHz channel and used to transmit 10-Mbps Ethernet on this coax. This allowed more flexibility in overall networking ability because the Ethernet, as stated in the beginning of this chapter, has been implemented in a variety of media.

Digital Equipment Corporation (DEC) was a supporter of this form of connectivity in their DECNet on Ethernet from the onset. Others also came up with the necessary piece parts to use the broadband cable systems for the 10-Mbps Ethernet standards. This has also been accepted as a connection under the IEEE standards committee. See Fig. 20.12 for a cable using the broadband Ethernet connection.

Figure 20.12
A broadband (CATV) application using 10 Mbits/s on coax.

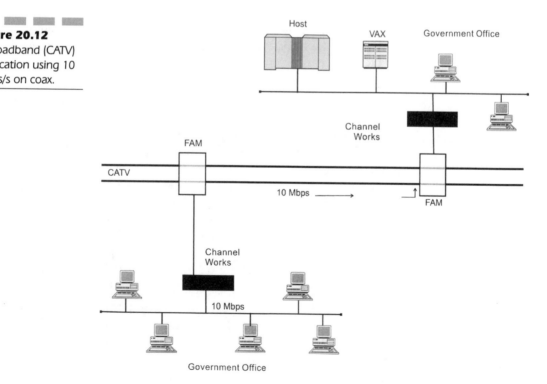

An innovation from a spin-off of DEC called LANCity, Inc. has produced a device or a concentrator called *Channel Works* that allows direct connection to a CATV operator's cable system to deliver 10-Mbps Ethernet connectivity across the cable. In this regard, the channel works product is supposed to be able to garner up to 83 separate Ethernets on a single cable. The primary application here has been to link multiple buildings in a metropolitan area network (MAN) for government, schools, universities, and hospitals that have multiple buildings all across town. This has been a pretty well-accepted innovation in many communities. LANCity was later acquired from DEC by Bay Networks, which in turn was acquired by Nortel in 1998. One of the most widely used CATV interfaces for the cable companies as they roll out Internet access is through the use of these channel works boxes.

Fast Ethernet

Why Do We Need Fast Ethernet?

As described elsewhere, the uses of local area networks have changed over the years. While in the past the primary use was for movement of files and access to data by applications, a frequent use of local area networks now is to load applications from servers. Also, the character of those applications has changed. The executables themselves have moved into the multiple megabyte range, and in some cases graphic files are also now in the millions of bits. What all this means is that faster networks are needed. Another factor driving the need for speed is larger, higher-resolution monitors. A modern 1024×768 monitor displays slightly over 1 million bits for one screen's worth of information. That would require, ideally, slightly more than $1/10$ of a second to traverse an Ethernet. In practice, it would take considerably longer. One-tenth of a second may not sound like a great deal, but on a small network of 10 users, a whole second might be eaten up just by the one screen to each user. In practice, it would be worse since contention would slow things down.

Some applications are considerably worse than this. Large, high-resolution graphics files make some environments entirely unsupportable by a standard Ethernet. For example, the size of a single digitized magazine cover can approach 70 megabytes. Bottom line: we need more speed.

The culture of the workforce has also changed considerably. The days of just typing ASCII text characters on a screen have gone. With Internet and Intranet access, the user now does more information processing. Moreover, that information is represented through a GUI, in which the graphics interfaces are far more dense. The amount of data traveling across our networks is rapidly becoming exponential. The 10-Mbps Ethernet is stepping aside for the 100-Mbps Ethernet. Many organizations have moved to a 10-Mbps switched Ethernet as a means of forestalling the change of all the NIC cards in the PCs. This made sense, but only begs for the actual upgrade later. Now newer switches and hubs are auto detecting where a mix of 10- and 100-Mbps Ethernet NIC cards can be used on the same network. This complicates the issue slightly but also gains time for the changes that are inevitable.

User organizations are now waiting for the PCs to be changed out so that they can buy the new NIC cards. In fact, many computer manufacturers are now installing 10/100-Mbps Ethernet cards directly in the PCs when they are made. This saves the LAN administrator from having to buy the cards separately. There is a minor issue associated with the manufacturers installing these cards. Not all NIC cards are created equal. Many are made with special features and functions not available to others. Therefore, there could be some conflict on the LAN when these mixed variations are all on the same network.

As the convergence of voice and data caught on in the industry during the late 1990s, the move is to incorporate voice, data, and streaming video at the desktop on the LAN. Given the limits of the 10-Mbps Ethernet with the specifications discussed above, the movement to 100 Mbps was escalated.

What Are the Alternatives?

If one is looking for alternatives to standard 10Base-T Ethernet, many are available. As described elsewhere, switched Ethernet is an excellent alternative. However, switched Ethernet doesn't have the headroom of some of the more capable technologies; what it really does is eliminate the adverse effects of contention. Token ring is an alternative to Ethernet but it is not all that much faster and it still costs more than Ethernet—about 10-fold. Switched token ring is still faster, but it isn't a mainstream technology at this point and probably never will be. It may make sense for organizations already using token ring, but not for anyone else. The other two major alternatives to Fast Ethernet are Asynchronous Transfer Mode (ATM, described in a separate chapter) and Fiber Distributed Data Interface (FDDI, also described elsewhere). Both ATM and FDDI are excellent alternatives. However, the cost to upgrade to either one is considerably more than that required to upgrade to any of the varieties of Fast Ethernet. In any case, please refer to the chapters on these technologies to help you decide whether they are right for you.

The term *Fast Ethernet* is primarily used today to refer to a technology technically described as 100Base-T. In 1996, when the technology was introduced, Fast Ethernet actually referred to any of several technologies. There is some confusion here. One of the technologies was switched Ethernet. Switched Ethernet is described in another chapter—but generally speaking, Fast Ethernet no longer refers to switched 10-Mbps Ethernet. Likewise, full duplex Ethernet, at least the full duplex variation of 10Base-T, is no longer referred to as Fast Ethernet. Just as with 1200-bps modems, which once were described as "high-speed," the advent of better technologies has resulted in changing the terminology used to describe the older technologies.

Currently there are two technologies that really deserve the name Fast Ethernet. Of the two, the vast majority of analysts have declared 100Base-T the winner. The other is 100VGAnyLAN. We'll discuss both technologies in this chapter, but make no mistake; 100VGAnyLAN is dead. We'll also briefly describe Gigabit Ethernet, a technology that was standardized in 1998.

100Base-T

When developers were trying to determine what the successor to Ethernet should be, they had two basic choices: try to stick to the basic technology as closely as possible, just increasing its speed, or try to make fundamental improvements to solve flaws that had become more obvious over time. Two camps developed, one taking each approach. The winning camp is the one that took the first approach. Their product: 100Base-T. The good news about 100Base-T is that the frame size and characteristics are identical to those in standard Ethernet. It uses CSMA/CD and in general can be understood and managed by network administrators and software in the same way that 100Base-T could be managed. The bad news is that there are still collisions, and the product still taps out at considerably less than the nominal 100 Mbps that it is supposed to be able to handle. One might describe it as *Ethernet on steroids*.

It is, however, the most popular replacement for standard Ethernet by far. In large part this is due to good marketing, but there is definitely something to be said for not having to do major retraining of your network administrators and technicians. In 1998, 50 percent of all new LANs installed were at the 100 Mbps rate. This varied between the shared bus concept in a hub and the switched version using a switch. Regardless, the costs, although higher than 10-Mbps NICs, are sufficiently attractive to merit the installation regardless of whether the speed is needed at the desktop today.

Like the original Ethernet, 100Base-T supports several media. Notably absent from this list is coaxial cable. Both standard copper wiring and fiber are supported. As with all high-speed technologies, the copper implementation has severe distance limitations. Most of the high-speed technologies do not support end devices more than 100 meters away from their central electronics locations; Fast Ethernet is no exception.

Fast Ethernet can be handled on three variations of copper wire. The first is category 3 UTP (unshielded twisted pair). If you are going to use category 3, however, you must use four pairs of twisted pair wiring. This is an extremely unpopular option. It is not difficult to see why. While many installations have some copper wiring available, it is extremely unusual to have four pairs that are unused. If one is going to have to put in new wiring, there is no reason to put in anything less capable than category 5. In addition, Fast Ethernet on category 3 only supports half-duplex operation. If Fast Ethernet is used in half duplex mode on four pairs of either category 3 or category 5 wiring, it is described as 100Base-T4.

The second copper variation supported is category 5 UTP. In the case of category 5, Fast Ethernet can operate in either half or full duplex on two pairs of wire rather than the four required for category 3. Operating on two pairs, Fast Ethernet is referred to as 100Base-TX. This is by far the most popular implementation of Fast Ethernet.

The third variation supported on copper for Fast Ethernet is shielded twisted pair (STP). Generally speaking, STP should only be installed in special cases where the potential interference either to or from the network is an issue. STP is much more difficult to install, with more demanding grounding requirements than UTP. It doesn't increase speed, nor does it improve usage of the pairs in the cable. As far as network performance is concerned, it works the same as on UTP.

In addition to copper wiring, 100Base-T is supported on fiber. As with category 5, fiber supports both half and full duplex operations. Why might you use fiber? It doesn't improve the speed. The reason is fiber's general superiority over copper, as described elsewhere in this book. In the case of Fast Ethernet, the main reason to use fiber would be to extend the available distances. Fast Ethernet over copper, as mentioned above, only reaches 100 meters to the desktop. A more severe limitation is that distances between copper-connected Fast Ethernet hubs must not exceed 5 meters (Fig. 20.13). The use of fiber considerably relaxes these limitations.

As with standard Ethernet, there are variations on the way the wiring (either copper or fiber) can be used. Buyers must select among shared, switched, and/or full duplex use of the media and obtain the right electronics to install at both ends of the cable. The default is shared half-duplex. One should be aware that just as in standard Ethernet, the contention experienced with a shared half-duplex configuration means that in no case except a two-user network will you be able to count on actually getting anywhere near the 100-Mbps nominal performance of a Fast Ethernet. However, that should not usually be a major problem. Just as with standard Ethernet, medium or even large work groups can be set up that still get a significant performance improvement over standard Ethernet by using shared half duplex Fast Ethernet.

If what you're doing is building a backbone, you probably want to use switched full duplex Fast Ethernet. This will deliver close to the rated 100-Mbit performance to each node, and it costs considerably less than the available alternatives (FDDI and ATM are the ones that come to mind). Note that for anything but shared half duplex, one must use either category 5 UTP or fiber cabling. Naturally, the cost of the electronics involved to deliver switched or full duplex is considerably more than that for shared half duplex—in fact, three to five times the cost of the less expensive approach.

Changing your network over to Fast Ethernet has been made much easier than it might have been by clever manufacturers that have addressed the problem. Since all Fast Ethernet implementations are based on either copper or fiber, the cabling is done in a star or radial configu-

Figure 20.13
Copper-connected
Fast Ethernet hubs
have to be close!

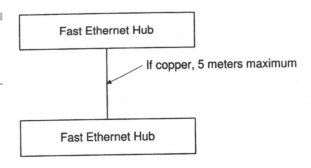

ration; no physical bus topologies need apply. This means that in theory and in practice, each node can be upgraded individually. The manufacturers build the central electronics (hubs, switches, etc.) with the capability, in most cases, to handle either 10- or 100-Mbps Ethernet on a port-by-port basis. This means that better electronics can be swapped at the central location, the network can be brought back up, and one-by-one individual users' machines can be upgraded (Fig. 20.14).

Better than this, however, is the advent of dual-speed NICs. Several manufacturers make NICs that operate at both 10 and 100 Mbps. These are described as 10/100 cards. But any given computer can run at only one speed at a time—why do we need the dual speed? Most environments upgrade individual computers in groups; certainly it is rare to replace all of the computers in an environment at one time. Often, they are replaced one at a time, as users' needs change. With 10/100 cards, a company can make a long-term decision to move toward 100Base-T without having to upgrade all the central hubs. The company can upgrade the desktops and then upgrade the hubs as it makes sense to do so. In most cases, the 10/100 cards automatically sense the speed of the connected wire (that is, of the central electronics port) at the time the PC is booted.

A changeover plan could be developed in one of two ways. One plan would be to replace all of the NICs in desktops over time with 10/100 cards. These cards are not significantly more expensive than standard Ethernet cards. Once all of the cards are replaced, then replace the central electronics with 100Base-T switches and hubs, reboot all the desktops, and

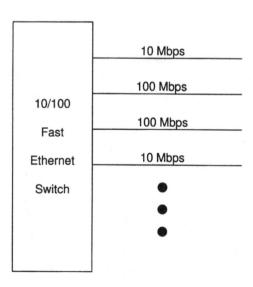

Figure 20.14
Implementing Fast Ethernet—a 10/100 hub.

you are done. The second approach would be to replace the central electronics with switches and hubs with units that have 10/100 ports. These are more costly, but give a degree of flexibility that isn't there in the first scenario; not all desktops have to be upgraded—ever. Once the central electronics are all upgraded, one can begin upgrading the desktops. In a pinch, some desktops can be upgraded in parallel, although this complicates the management efforts.

100Base-T Performance Expectations

The most obvious alternative to implementing 100Base-T is switched 10Base-T Ethernet. Surprisingly, the latter may, under some circumstances, deliver better performance than shared 100Base-T. One may think that this is an unfair comparison; however, the costs are similar, whereas the cost of switched 100Base-T is probably much too high for desktop use in most cases. So we will consider switched 10Base-T and shared 100Base-T as alternatives.

It all boils down to the numbers. In an environment with small workgroups (fewer than 20), shared 100Base-T will probably deliver much better performance to the desktop than shared Ethernet could. Why? With 100Base-T, each user will average at most 5 Mbits of capacity on the hub if all are active simultaneously. But in most real-world scenarios, not all users are truly simultaneously active. In the typical case, fewer than 10 percent of the users are doing anything on the network at any given instant. Ten percent of 20 people is 2 people. These two active users, in most cases, would share 100 Mbps of capacity, or 50 Mbps each—obviously better than switched Ethernet's delivery of 10 Mbps at best to each user. However, let's cook the numbers a little differently. If there are 100 users on a switched Ethernet segment, each will still get 10 Mbps delivered to the desktop all the time. Fast Ethernet at 100 Mbps, in the worst case, will deliver only 1 Mbps to the desktop. If 10 percent of users are active at a time, then each will get about 10 Mbps. In practice, contention on the 100-Mbps hub will reduce that by 20 to 30 percent. In this case, switched Ethernet might be a better choice than the Fast Ethernet.

A major decision factor is the type of work the users are doing. If each is doing on a continuing basis network-intensive activities that can be satisfied by a 10-Mbps pipe, switched Ethernet is probably the technology of choice. On the other hand, if the user population makes sporadic use of the network, but each use tends to benefit from a truly high bandwidth (editing large graphic files is an application that comes to mind), then 100

Mbps (or higher) would definitely be the better choice since each user would experience close to the maximum capacity when needed, if it's only lightly used.

100VGAnyLAN

Back before 100Base-T had established its market share, a group of technically oriented companies including Hewlett-Packard felt that the second approach to upgrading Ethernet—improving its flaws as opposed to just increasing its speed—was the better approach. The result of their efforts was 100VGAnyLAN. This technology, like the Betamax VCR, is clearly superior to its competitor (100Base-T). Unfortunately, as with the Betamax, it didn't win over the market. One can still buy 100VGAnyLAN products; however, the system is regarded by most as an also-ran.

100VGAnyLAN had much to recommend it. Its designers chose to eliminate the CSMA/CD method of allocating capacity. Instead, it uses what is called a demand-priority allocation scheme, where each NIC notifies its corresponding hub that it has something to send. The hub pays attention to these notifications and solicits the input once notified. It also is possible to designate high-priority traffic, such as video, enabling a better pattern of performance for such traffic in this technology.

100VGAnyLAN also had the capability to support either Ethernet or token ring traffic. This traffic cannot be on the same hub, but one can use the same kind of hubs for both, significantly reducing the complexity of a LAN implementation environment. The rules for cascading 100VG hubs are no more restrictive than those for standard Ethernet, and considerably more flexible than those for 100Base-T. As with 100Base-T, the cable running out to the workstations can be 100 meters long; however, the distance between hubs also can be 100 meters, 20 times the limitation of 100Base-T.

There are some disadvantages to 100VGAnyLAN, partially explaining its failure in the marketplace. For one thing, its initial specification required four pairs of unshielded twisted pair wiring, no matter which category is being used. Another disadvantage is more political than technological. Since the beginning, Ethernet has used CSMA/CD as its media allocation scheme; 100VGAnyLAN does not. It uses demand priority. In fact, this is an advantage; however, purists in the Ethernet camp have attacked 100VGAnyLAN on this basis. It is not a very good reason not to use the technology; however, enough advocates of Ethernet went the other way because of this that it seriously and adversely effected

100VGAnyLAN's marketshare. Very few vendors subscribed to the 100VGAnyLAN technological specification. At this point, it is really only enthusiastically supported by Hewlett-Packard—and even HP has moved on to support 100Base-T.

Gigabit Ethernet

While we do not know of anyone who alleges that full duplex switched 100Base-T is insufficient to handle any likely desktop requirement, if one puts together enough high-speed desktop LANs, the backbone to serve them needs to be a better technology than that desktop LAN technology, whatever it is. Moreover, if the use of voice, data, video, and multimedia finally becomes a reality on a single architecture, then the need for more than 100 Mbps to the desk may become a reality. A new technology, Gigabit Ethernet, was developed with the goal of providing these applications to the individual user in the future. In 1998, products were introduced before the standards had been adopted. As with many new technologies, some vendors came out with products early and promised to change or upgrade those products, at no cost to the initial implementers, to whatever the standard becomes. This allowed the vendors to get a leap on the market and assure their market share before others could get there. So the question is, when would you use Gigabit Ethernet and when would you not?

Like 100Base-T, Gigabit Ethernet comes in half and full duplex variations. In its half duplex implementation, it has some very severe limitations regarding delivery of actual capacity. There is an interaction in all Ethernets between minimum packet size and size in terms of radius of the actual maximum network configuration. The smaller the minimum frame size, the smaller the maximum radius of the network. This relationship is compounded in a bad way by increasing the speed of the network transmission. Since the goal in upgrading to a faster variety of Ethernet is to keep the same frame size limitations, the radius of the Gigabit Ethernet without using a special technique is limited to about 20 meters—obviously impractical. What the designers have done is implement a special technique that involves transmission of a special signal in any case where a frame smaller than 512 bytes is transmitted on a Gigabit Ethernet. This means that for large frames (e.g., those associated with large file transfers), Gigabit Ethernet might actually deliver close to its full speed. However, many networks being used by human beings as opposed to machines generate large numbers of small frames averaging only 300 to

350 bytes. In the worst case such a network, even if implemented on Gigabit Ethernet, would experience performance little better than that delivered by 100Base-T in its full duplex incarnation. The technique allows for a 200-meter network diameter. This limitation means that any network implementer needs to seriously analyze what kind of traffic is going to be carried by that network, and evaluate the likely bandwidth to actually be delivered by Gigabit Ethernet.

Unlike ATM, Gigabit Ethernet also provides no quality of service (QoS) guarantees, although the proponents of Gigabit Ethernet are working to provide QoS-like services. This means that even though its nominal speed is very high it still is not an ideal technology for satisfying isochronous type traffic (for example, video).

If one implements full duplex Gigabit Ethernet, the issue regarding distance pretty much disappears. Gigabit Ethernet requires the use of fiber. There are two kinds of fiber: multimode and single mode. Multimode should be able to reach up to about 550 meters, whereas single mode should be able to reach distances approaching 3 kilometers.

One other caution: Intel-based servers at the present time are generally incapable of accepting or transmitting information at speeds significantly greater than those that can be provided by 100Base-T. Thus hooking a server directly to a Gigabit "pipe" would not necessarily deliver improved performance over Fast Ethernet. Higher-end UNIX servers can deliver performance two to four times the above; but even they cannot yet receive or transmit information at true Gigabit speeds. Backbone components (routers, etc.) also are just now coming on the market that can handle speeds in this range. Therefore, simply buying Gigabit devices and dropping them into the network is not likely to be a successful strategy.

Gigabit Ethernet does promise to be a less expensive technology than ATM. In cases where backbones can be successfully implemented with Gigabit Ethernet, it may be more cost effective. However, ATM is available at speed ranges that are at least competitive with Gigabit Ethernet. With the additional capabilities of ATM and its large (carrier-based) installed base, ATM is probably the preferred technology for high-capacity backbones at least for the next couple of years.

Just when we thought it was safe to come out of the LAN closets and use the technologies, with the knowledge that Gigabit Ethernet is coming along to deal with bandwidth issues, the next in a series of evolutionary steps emerged. Now the talk of the new millennium is 10-Gigabit Ethernet. Nortel networks announced their strategy in 1999 with the use of 10-Gbps Ethernet over SONET OC-192 based networks. This strategy also points out that the end user may not need this speed yet, but the carriers

(Telcos and ISPs and new providers) will be looking for this throughput. If a carrier chooses to offer 1 Gbps to a consumer (large organization), then the 10 Gbps is necessary for the provider to guarantee the QoS and the effective data rates. This is particularly true when serving the ISPs and ITSPs.

What is next? Who knows where this is all leading? However, the talk of offering 1 Gbps to the desktop for the convergence of voice, data, video, and multimedia has become commonplace since the introduction of these two high-level steps.

CHAPTER 21

Token Ring LANs

Token Rings

Another access method that can be used in a local area network (LAN) is the token-passing concept. The token-passing concept is an access method that was initially developed to work on a physical and a logical ring. Remember, LANs are used in both physical and logical topologies. A topology is the layout of the physical wiring plan to provide a single shared cabling system. If you think about some of the constraints and limitations of the bus topology covered in the previous chapter, the issue is the collisions that can occur on a network. The issue that seems to arise in the discussion of the topology of the physical Ethernet is that the signal (information) has to propagate (travel) down to the ends of the cabling system. At the ends of the system are terminators (no, not Arnold Schwarzenegger) that absorb, or remove, the electrical pulses from the cable. This means that in order to send information and be assured that the information reaches its destination, the transmitting device can only use 50 percent of the time to generate the information onto the wire. The other 50 percent of the time allotted to this device is devoted to listening to the cable to make sure the signal makes it all the way to the ends without colliding with some other device's data. This would appear to be somewhat wasteful, because only a 50 percent duty cycle is allowed. To overcome this situation, the use of a ring topology was introduced. Along with the topology, a collision avoidance scheme was also introduced. If a device wants to transmit, it must have a "permission slip" to do so. The permission slip is in the form of an electrical signal, called a token, which travels around the network constantly. Only one user will be on the network at a time. Only one token is needed.

The IBM Token Concept

In 1984, IBM announced plans to produce a set of network interfaces that adhered to the IEEE 802.5 token-passing concept. This access method was IBM's defense against the use of Ethernets. As you will recall, Ethernet was produced by Digital Equipment Corporation, Intel, and Xerox as the network topology and access method of choice. With the proliferation of LANs in the industry, IBM had not endorsed the Ethernet, yet had no counter—other than Systems Network Architecture (SNA) for a networking environment. To arrive at an IBM-stamped LAN, then, the token-passing ring was announced. In 1985, the token ring LANs were beginning to appear. End users who had not changed over to the bus topology were waiting for the IBM LAN. They immediately went for token rings: after all, "no one ever got fired by buying IBM." So a new set of boundaries was established. Users were in one of two camps, either supporting or criticizing peer implementations of the LAN topologies. IBM created a new approach in defense against the Ethernet supporters. Many of the IBM systems would not work on an Ethernet, so immediately a new connectivity solution was introduced to support the big iron and the in-between hardware. The range of devices from the desktop, in either an industry standard architecture (ISA) or a microchannel architecture (MCA) to the mainframe computing platforms could all be introduced to the LAN.

IBM's version and the standardized IEEE version of tokens included the use of a deterministic system. By using a single token, the network allows only one device at a time to transmit. Therefore, the risks and problems associated with the CSMA/CD LANs are eliminated. Each device gets its fair share of use on the network. In order to do this, the devices are allowed a time to use the token as it arrives.

Initial Layout

In the very beginning days of the token ring, the cables were installed as a physical ring. In Fig. 21.1, the actual layout of the cabling is shown. In this initial layout, each device installed on the ring is connected to the physical and closed ring of wires. Wires run to an inbound slot on a device—actually to the network interface card (NIC), which is now called a token interface card (TIC)—then out to an outbound side of the device. What this actually means is that every device has a wire in/wire out connection. This is done on a four-wire cable. As the devices are attached to the network, they are wired from the upstream neighbor and to the downstream neighbor. This continues around the LAN until the circle is closed.

Problems Encountered

Nothing ever works the way we want it to all the time. Unfortunately, several problems arose with the use of this system of wiring devices together physically. These include, but are not limited to, the following:

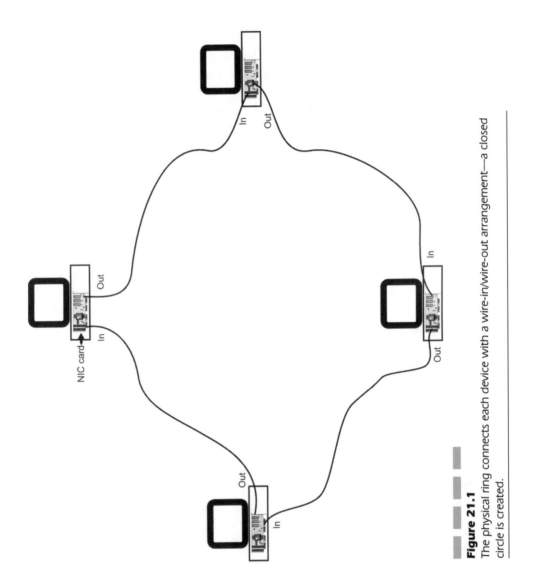

Figure 21.1

The physical ring connects each device with a wire-in/wire-out arrangement—a closed circle is created.

Lack of Power

When the devices are wired together, the electrical signal (or data) runs down the wires through one device to the next. This means that the electricity comes into the TIC card, is read, and then is regenerated to go out to the cable. However, this implies that the TIC card must be active at all times. When users on the ring finished their daily activities, or if they did not come into work, a problem was created. In both of these cases, the user's device (PC) was not powered on. The TIC card that is installed inside the PC in an expansion slot draws its power from the PC. However, if the PC isn't on, the TIC card is dead, which means that the LAN is down (Fig. 21.2). This left many a LAN administrator in a quandary because, it was not initially obvious. Many hours were spent as administrators ran around the building looking for the cause of the problem on the LAN.

Lack of Connection

After an end user has waited for connection to the LAN for what they perceive as forever, the LAN administrator finally gets the wires installed,

Figure 21.2
If a PC is powered off, the network doesn't work.

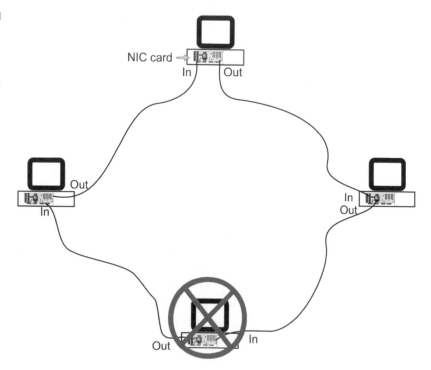

the PC configured, and the software loaded. This usually happens off-hours or on weekends. At any rate, on the following business day, the user comes into the office. Initially, the end user is ecstatic about finally being connected to the LAN. However, this wears off quickly and the user then decides that the device is not located in a convenient location. Therefore the user picks up the PC and moves it across the room to another location in the office or work area. Unfortunately, as shown in Fig. 21.3, this also pulls the wires out of the wall and the jack. The broken wires render the LAN unusable because electrical continuity is lost. The LAN must be wired as a closed circle so that the electrical current will flow around the wires consistently.

Constant Changes

The last scenario involves the dynamics of the networking environment. After a LAN administrator painstakingly sets up the network, the inevitable happens. A new user must be added to the network. Sometimes this can be quite simple, yet other cases require a whole new layout of the wiring system to reach the new user. In any case, however, the result is the same. The LAN administrator has to cut the connection and run the wires from the existing termination on a PC to the new one added to the network. This means that the LAN is out of service or *down* for the time it takes to get the new wires run and connected. Only when everything is put back into the physical ring will the network be back up again. This is shown in Fig. 21.4, where a new user must be added.

Obviously, the results of these scenarios left one conclusion: the LAN is down a lot! Something had to be done to preclude this from continually happening. IBM decided to create the "LAN in a box" concept.

The Solution to Physical Problems

To overcome the downtime and the lost credibility, the evolution to a centralized network began. IBM developed a modification to the physical structure of the ring. Using a version of a star network, the multistation access unit (MAU) was introduced. In Fig. 21.5, a multistation access unit was installed in a star configuration in a telephone closet. The ring now was a physical star and a logical ring. *Physical* defines the structure of the

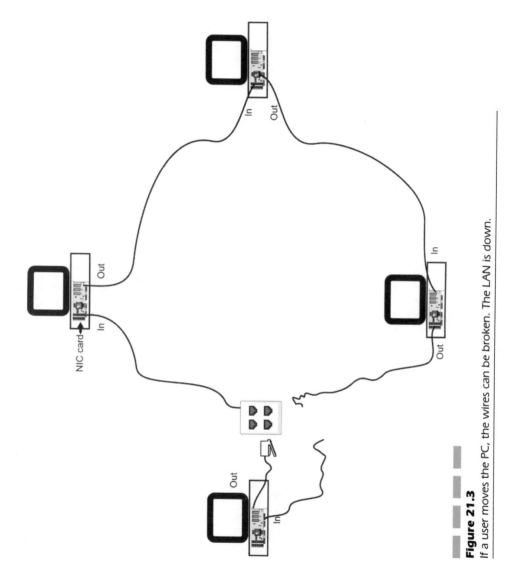

Figure 21.3

If a user moves the PC, the wires can be broken. The LAN is down.

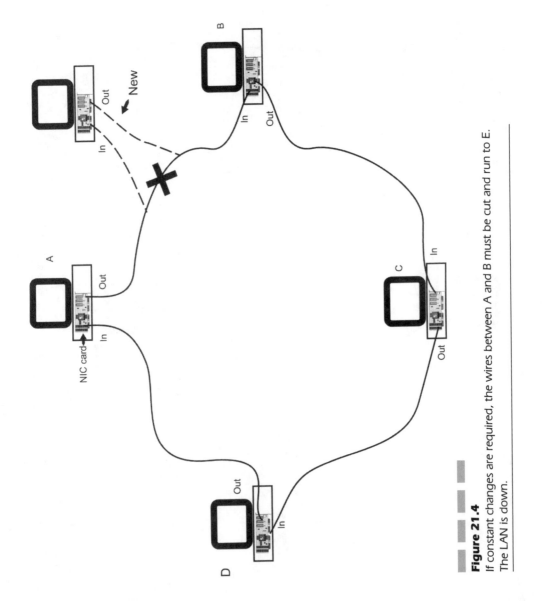

Figure 21.4

If constant changes are required, the wires between A and B must be cut and run to E. The LAN is down.

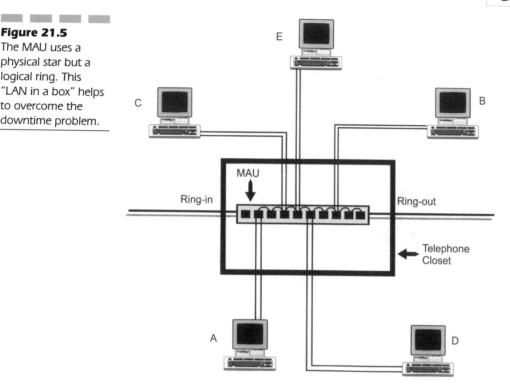

Figure 21.5

The MAU uses a physical star but a logical ring. This "LAN in a box" helps to overcome the downtime problem.

wiring, whereas *logical* defines the signal (energy) flow. Using a basic telephony concept, wires were no longer pulled from device A to B to C to D to A as they had been. Instead, they were all pulled from location X. A single cable from X to A, X to B, and so on was used. This allows for some other unique characteristics.

Additional capabilities of the newer token ring networks allow for the self-healing of a cable or device problem. Using the physical ring from MAU to MAU, if the cable gets broken (Fig. 21.6), the electronics running the ring is intelligent enough to close off the ring and reverse itself. This creates a far better approach to keeping the LAN up and running than what was used in the past. If the cable gets cut or disconnected, the LAN senses the open condition, drops a shorting bar, and loops the transmit and receive wires together. The resulting network is shown in Fig. 21.7. Although this is not a perfect solution, it does help to keep the uptime needs of the organization closer to the design and intent. Because the ring closes and reverses itself, designing the cable lengths is very important. What used to be a short connection between MAUs has now doubled in

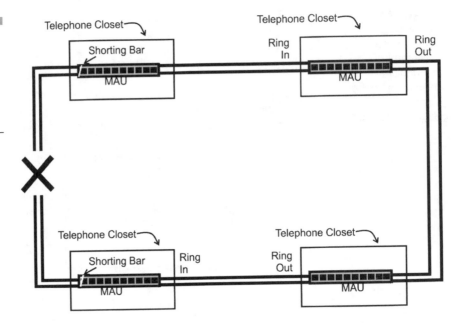

length because the signal must run on a longer cable to get from device to device. Distance limitations are still a factor insofar as the signal must run on the wires. By reversing the wires, the cable distances are doubled. This merely means that the LAN administrator must be cognizant of this limitation.

The original token ring topology was delivered at 4 Mbps. Distances of wires run from device to device could not exceed a total length of 1600 feet. Newer rings operate at 16 Mbps, but the distance was cut in half for a total cable run of 800 feet. Too many organizations realized that this distance limitation was in the way. To solve this problem, IBM included a repeater function in the MAU so that each port on the MAU (Fig. 21.8) would regenerate the signal and overcome the risk of signal loss. The MAU typically had only eight ports, as shown in the figure, plus a ring-in and a ring-out port. This allows multiple MAUs to be hooked together in a ring fashion and allows for more than eight devices to be clustered into a single logical ring.

Further, each of the ports in the MAU also includes electronics in the componentry that have shorting bars. A shorting bar allows for a single (or multiple) device to be unplugged from the network without causing the network to go down. The original, as you recall, required that all devices be physically operating and plugged together in order to work. Now users can unplug a device or power it off without risk to the network.

Figure 21.7
The resultant
network is a closed
loop around the
problem.

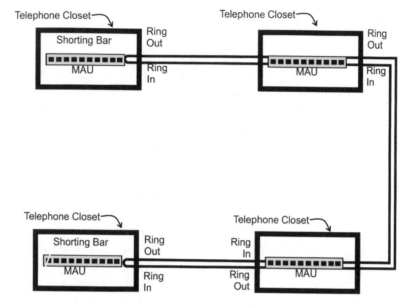

Figure 21.8
A repeater function is
included in the MAU
to overcome the
distance limitations.
Unplugged PCs are
bypassed with
shorting bars in the
MAU.

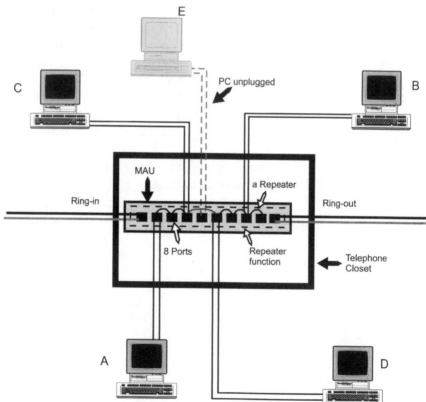

Cable Types Used

When IBM introduced the token ring, a second part of the network design was the introduction of the structured wiring plan. To ensure that the networks would operate in an office or factory environment, different types of cables were introduced. The cable types allowed for various environmental and operational conditions to exist. This was not an inexpensive cabling solution, but if one wanted the IBM seal of approval, the IBM cabling system had to be used. In Table 21.1, the variations of the cabling are shown. This wiring included the typical twisted pairs similar (but different) to telephone wiring. Telephone wiring existed in buildings for years and many LAN administrators thought that a simple installation would be to use the existing telephone wires. This led to devastating results. Cables reflect energy. They also act like antennae and draw in electrical energy. Unless the wires are protected from each other (bleed off of energy) and from other sources of disturbances in a building (such as electric motors, electrical welders, fluorescent light ballast, etc.), the results will be degradation of the network and corruption of the data. Shielded wires were introduced by IBM to preserve the integrity of the data and prevent the influx of noise onto the cable. Thicker wires were also used. Instead of the traditional telephone wire, which is very thin (26 American wire gauge, or AWG), IBM used a thicker wire (22 AWG). In some cases, a fiber solution was also recommended to carry the higher speed data. The cabling design included both stranded and solid conductor copper cables, shielded and unshielded, and varying gauges. What IBM was attempting to accomplish was a one-stop shopping arrangement for all of an organization's cabling needs. Many felt that this was overkill typical of IBM. Furthermore, the pricing for such a cabling system left many organizations with sticker

TABLE 21.1

Summary of the IBM Cable Types

IBM Type	Number of Wires and Type
1	Two pairs of shielded solid core 22 AWG
2	Two pairs of shielded solid core 22 AWG and four pairs of unshielded solid 22 AWG
3	Four pairs of unshielded solid core 22 AWG
5	Fiber-optic cable, multiple strand
6	Two pairs of shielded, stranded 26 AWG
8	Flat cable to go under carpets
9	Two pairs of shielded solid core 26 AWG

shock after being quoted installation prices in excess of $1 million. This cost was greater than anyone ever could have imagined—especially when everyone was accustomed to using the less expensive telephone wiring.

Although this was a very expensive and a very bulky wiring scheme, there are organizations that installed this system nearly 10 years ago and are still supporting the original and the upgraded capacities and speeds on the same wires. To prevent problems with radiation and cross talk, IBM also used a device called a media filter. The purpose of the media filter was to reduce or eliminate the interfering emanations from the wires that would cause corruption of the data. These, again, have worked for the duration.

Shielding of the cables when higher data rates are being used is important. However, the shielding is dependent on proper grounding. The ground wire should be connected to a proper building ground. The building ground should have a good solid earth ground. This is a possible problem for all types of buildings and subsequently can cause problems on the LAN. If the LAN wiring is not properly grounded, disruptions are likely to occur. If the building ground is suspect, it should be checked by a certified engineer and problems should be corrected. Old ground rods corrode or deteriorate, losing the proper depth to maintain a good ground. Installers like to use a cold-water pipe to tap the ground, but fail to recognize that many buildings now use plastic, not copper, pipes, eliminating the ground. One must be very specific in requesting a proper ground.

Speeds

As already mentioned, IBM's first version of the token ring operated at 4 Mbps. A lot of these rings are still in use today, nearly 10 years after the announced product. The industry and the Ethernet proponents scoffed at the slower-speed network when they heard of the limited speed. They perceived the situation this way: IBM was so desperate to get a network out into the marketplace that they shortchanged the user. What most people did not recognize was the significance of the deterministic rather than CSMA/CD network access control. A well-installed Ethernet operating at 10 Mbps will support up to 1024 nodes and yield approximately a 40 percent net throughput. This equates to about 4 Mbps of effective utilization. In actuality, the Ethernet yields about 3.3 Mbps. The main reason for this difference is the propagation time, the potential for collisions, and other timing delays that are prevalent on the bus. If collisions begin to occur, the backoff algorithm used by each of the NIC cards prevents either of the two conflicting devices from transmitting until a period of time passes. If continued collisions occur, the backoff algorithm used by the NIC cards begins to become exponentially longer and longer. This means that more devices than can use the cable at a given time are waiting to use the cable.

Conversely, the deterministic nature of the ring using a token-passing access control does not have the same problems. There will be no collisions, because only a single token exists and only one device can transmit when it controls the token. Consequently, the ring can be more effectively used. The effective throughput of a 4-Mbps token-passing ring is approximately 3.3 Mbps. This parallels the performance of the Ethernet even though the raw speeds are different. Further, IBM's second version of the token ring card introduced a higher rate of speed. Operating at 16 Mbps, the faster speed using a deterministic access control method yields approximately 12 Mbps. These speed comparisons are shown in Table 21.2 as a quick reference in the event this issue ever has to be reviewed. Because two camps always exist, pro or con on Ethernet or token ring, this one point is usually a neutralizing factor. Therefore we wanted it to be clearly understood and visible for future reference.

Media Access Control Layer

As with any LAN, the token-passing ring requires certain access control methods. To summarize from earlier discussions, the LAN really operates at the bottom two layers of the OSI model—the physical layer and the data link layer. The data link layer is subdivided into two parts: the media access control (MAC) and the logical link control functions. A token ring therefore can be shown (Fig. 21.9) as it stacks onto the model. This comparison shows that the following apply:

- The physical link deals with the physical structure of the cabling system and the access method used at the physical layer. This is labeled layer 1 in the figure.

TABLE 21.2

Comparing Effective Throughput on LANs

Access Method	Raw Speed	Effective Throughput
Ethernet CSMA/CD	10 Mbits/s	3.3 Mbits/s
Token-passing ring	4 Mbits/s	3.3 Mbits/s
Token-passing ring	16 Mbits/s	12.0 Mbits/s

Figure 21.9
A token ring as it
stacks onto the
model.

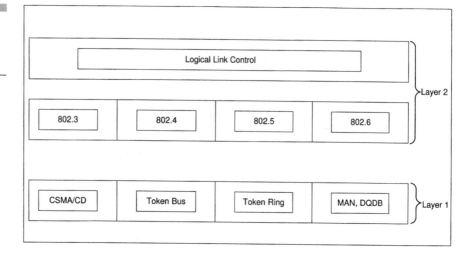

Figure 21.9
A token ring as it
stacks onto the
model.

- The data link sublayer, called the media access control layer, shows the access control in support of the IEEE and the International Telecommunication Union (ITU) model for access control.
- The logical link control (LLC) brings the various topologies together in a common format. The LLC represents the upper portion of layer 2 of the OSI model.

In each of the cases, the other forms of LAN access methods are shown in the same layers of the model. Layer one deals with the forms shown in Table 21.3.

The Frame

The token ring uses three different types of frames in its ongoing operation. These three frames are determined by the service that they provide on the network. They are:

TABLE 21.3

Comparing
Standards and
Topologies

Topology/Class	IEEE Standard
Ethernet	802.3
Token-passing bus	802.4
Token-passing ring	802.5
DQDB, CATV services	802.6

- The token
- The frame
- The abort message

We will look at each of these and their purpose so that the differences can be seen more clearly. In many ways, the token ring is considered the most popular network concept (whether this is true or not) by users who support this infrastructure. The ills of the past Ethernet concept were addressed on the ring so that less confusion resides on the ring, and troubleshooting, maintenance, and diagnostics can all be simpler. The word *simpler* might frustrate many a network administrator, but the intent is to simplify the detection of errors and the resolution of problems. We (Don and Bud) differ in our opinions about which of the two network strategies is better and why. Obviously, the intent of this book is not to steer any user into a specific product or service, but toward a better understanding of what each does. From this perspective, the control placed on the token ring yields a better management scheme. The network facilitates the use of the cabling system in the way it was designed to be used. Now back to the frame formats. Our intent is not to bog the discussion down with a lot of bits and bytes. But if you don't understand what the ring is doing and what the reasoning behind it is, then the whole situation becomes a moot point. Again, the differences are significant in the way they operate; the functionality that they serve is similar.

The Token

The token is made up of 24 bits of information (three octets). The three octets define the start and stop sequence within this frame. A token will traverse the network constantly, even if no traffic exists. Refer to Fig. 21.10 for the frame format of the token as we continue through this discussion. The first octet in the token is the start delimiter (SD). It is made up of a combination of good bits and violation bits (in the way we produce digital pulses—in this case, called Manchester Coding). The reason for the good and bad bits is to differentiate a token from real information. Therefore the start delimiter is a unique set of 1s and 0s that can be differentiated by all devices on the network as being the beginning of a token. These 8 bits are always formed in the same way.

Skipping to the third octet in the token is the end delimiter (ED), which contains a combination of good bits and bad bits. Much like the start

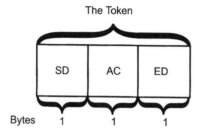

delimiter, the end delimiter is used as a specific sequence to signal the end of a token. Again, this is a unique combination, so there is no confusion about what is being presented to the network.

Now back to the middle octet in the token. This octet is called the access control (AC) byte, which really equates to the working portion of the token. The 8 bits are made up of a series of working bits that define what is happening inside the token (Fig. 21.11). Bits 1 through 3 in the access control byte are marked as priority (P) bits. Each device (or node, as they are called) on the network has an assigned priority by the system administrator. Because there are 3 bits, a total of eight priorities can be used on the network, working from 000 through 111. In order of succession, 000 will be the lowest-priority device and 111 will be the highest-priority device. In order for a node to transmit information on the network, it must have the token (of course, the token must be free). The device must have equal or greater priority than the token in order to use it. This will allow certain devices to have a higher priority to send infor-

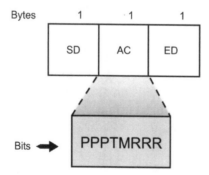

P= Priority (2^3=8)
T= Token or Frame
M= Monitor
R= Reservation

mation than others. An example of this might be that the server on the network would have a higher priority than a regular low-end user who only performs word processing functions. This is not a caste system, but a means of providing service and controls on the network so that one device cannot take over the network.

The fourth bit is called the token (T) bit; it signals what is present, a token (T = 0) or a frame of data (T = 1). Two possibilities exist as the token comes around: it is either a free token or a busy token containing user information. This token bit can establish the ability to let the server use the token more often than an end user, thereby allowing all others on the network to gain access to the server and receive their information. You can only imagine how the performance would be viewed if the server has to wait to send a file, save a file, and so on. Everyone on the network would think that the network is performing slowly, and in reality it would be. Regardless of the reason, if the user does not get an instantaneous response, then the network is slow. There are no other options.

Bit number 5 in this AC byte is used as a monitor (M) bit for network token control. Think of this bit as a traffic cop on the network. A device on the network is assigned the responsibility of monitoring what happens on the network. When a token is sent around the network, it would not be wise to let it circle around and around continuously. Therefore, the monitor station is responsible to see that this doesn't happen. The transmitting station (sender) sets the M bit in the token to a 0. As the token goes by the active monitor, it sets the M bit to a 1. If the monitor station receives an incoming priority token or a frame with the M = 1 set, it knows that the transmitting station did not take the information off the network after a round trip. The monitor station then takes the token off and cleans it up. Then it issues a new token to be used on the network. There is still only one token on the network at this point. To prevent any major problems on such a network, the monitor station is set up to handle the problems with tokens on the network. Thus, if something happened to the monitor, one or two neighbors are designated as the standby monitors to take over the responsibility (Fig. 21.12).

Bits 6, 7, and 8 of the AC byte are allocated as reservation (R) bits. Remember the priority levels? Well, if a server or any other device has information to send across the network and it has a high priority, it can reserve the token for its use on the next pass around the network. These 3 bits are assigned for use so that a single device cannot control the network and send frame after frame of information, denying access to all other stations on the network. Therefore, as the token goes by, the NIC inside the node will set the reservation bits and request the next available token.

Figure 21.12

One or two devices will be selected as standby monitor nodes. The monitor node is usually a high-priority node.

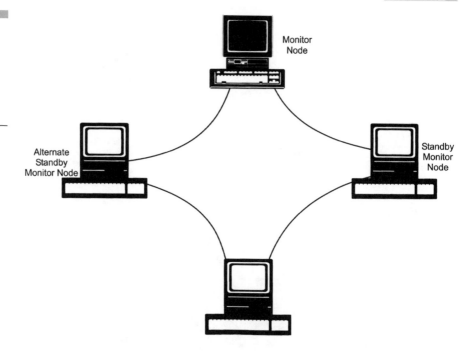

Monitor Node

Alternate Standby Monitor Node

Standby Monitor Node

Certain constraints can be allowed on the network where a single node might only be allowed one use of the token, after which it must release the token. Other higher-priority devices might be allowed to use the token two or three times before relinquishing control. It is through these bits in the access control field that this is implemented so that other devices have a fair share of network usage. This was a well-thought-out plan for the network.

The Abort Sequence

Figure 21.13 shows the abort sequence that is used in the token ring network. In the event that problems exist on the network, the nodes are designed to recognize the problem and discard the token. The abort sequence includes the ability to read the start delimiter and the end delimiter, place them together, and issue this to the network by the detecting node. This alerts all nodes on the network that the problem exists. Things that can go wrong will go wrong. This can include such problems as data corruption, lost tokens, time outs, address problems, and so on.

Figure 21.13
By placing the start
and end delimiters
together, a node
detecting a bad
frame will alert all
other nodes that a
problem exists.

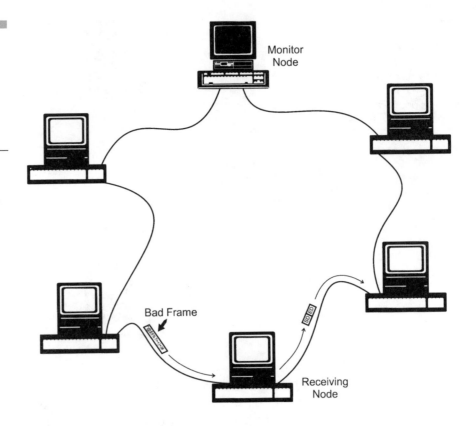

The Frame

Above and beyond the token format, the ring carries a frame of information known as the *data frame* (Fig. 21.14). This frame has a variable length based on the data inside. This is comprised of the following pieces:

■ The start delimiter (SD) in the token frame

■ Access control (AC)

These constitute the first 2 bytes of the data frame. Following these 2 bytes come:

Figure 21.14
The data frame layout
in a token ring
(802.5) network.

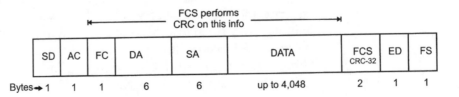

- The frame control (FC) byte, which indicates the type of data in the frame.

- The destination address (DA) of the token, a definition of where the frame is being sent. The destination address is 6 bytes (48 bits) long.

- The source address (SA), the indicator of where the frame came from or who sent it. This address is also 6 bytes long.

- The data. In this field, the actual information being sent is provided. It can be just information, or it can also contain such information as MAC or LLC information or routing information. This allows all physical and data link overhead to be contained inside the data packet transparently. The data field is a variable amount of information that can be up to 4048 bytes on a 4-Mbps network and up to 16,192 bytes on a 16-Mbps network.

- The frame check sequence (FCS), a CRC-32 error-detection pattern that checks the validity of the information in the FC, DA, SA, data, and FCS bytes. Using a 32-bit CRC, the error-detection capability is better than 99.99995 percent. If a single bit error occurs, in the transmission of the frame, the frame will be tossed out. This is fairly stringent, but must be used to ensure the integrity of the data on the network.

- The end delimiter (ED) from the original token. This is a single byte of information.

- The frame status (FS) byte, which indicates the status of the actual frame. In this byte control, mechanisms are used to determine what has happened since the data was initially sent by the transmitting station.

In Fig. 21.15, the various bits used in the FS byte are shown. There are reasons for this technique. First the figure shows that the byte is composed of a sequence that mimics *ACxxACxx* where:

- The *A* bits are used by the receiving device to indicate that the destination address was recognized.

Figure 21.15
The frame status byte includes the *A* and *C* bits. *A* is used to designate a recognized address (to a receiver), and *C* is used to designate the data has been copied.

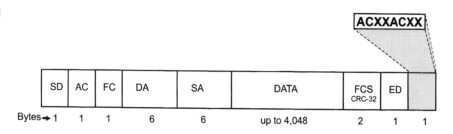

SD	AC	FC	DA	SA	DATA	FCS CRC-32	ED	
Bytes→ 1	1	1	6	6	up to 4,048	2	1	1

ACXXACXX

- The *C* bits are used by the receiving device to indicate that the information (data) was copied.

- The *x* bits are not used and therefore ignored.

- The *A* and *C* bits inside the FS byte are important for the overall control of the network.

Because the transmitter is using a one-way cable transmission system, the information is sent out onto the wire. The sender must know that the data arrived at the desired location. However, because this is a one-way system, the only indication to the sender occurs if the receiver sends back a "return receipt requested" notification. That is where these bits come in to play. If the sender does not receive the frame back (in the full circle) with the *A* and *C* bits set properly, then the message must be retransmitted. If, however, the *A* and *C* bits are properly set, the transmitter knows that everything worked according to plan. Therefore the transmitter will remove the information from the network, free up the token, and send it back onto the network for another device's use. Note that there are two *A* and *C* bits in this byte. The frame status byte is not checked by a CRC; therefore, as a check and balance on this arrangement, the two *A* and two *C* bits must be identical. In Table 21.4, the combinations of the *A* and *C* bits inside the frame status are shown. If an *A* or *C* bit is set to 0 it is not recognized or used, whereas if set to a 1, it means that everything is okay.

TABLE 21.4

Summary of the Frame Status Bits Being Used

Pattern of Frame Status Byte	Explanation
00*xx*00*xx*	Nothing was recognized or copied. The station may be out of order or turned off or the frame discarded.
10*xx*10*xx*	The receiver recognized its address and responded, indicating that it is on the network. However, the receiver did not copy the information. The data CRC may have been corrupted, or some other problem may have occurred.
01*xx*01*xx*	If this happens, you have a major problem. The addressee did not respond to the frame of data but somehow it copied the frame. This is theoretically not possible, or there could be a promiscuous device that set the *C* bits but received someone else's data.
11*xx*11*xx*	The addressee recognized that the information was for it, and the data was in fact copied. Everything worked the way it was intended to.

The following scenario will hopefully summarize the use of a token ring network simply. We recognize that many readers are seeing this information for the first time and that many of the framing formats and control sequences are complicated and confusing. Sometimes the information needs to be read and reread. Therefore we attempt to use the network as best we can through an analogy. The process is fairly straightforward, but the concept can get quite complicated if you do not have a basic understanding of just what is taking place. So, fasten your seat belts and enjoy the ride around the ring.

Case Example

We wish to set up a token ring, so we establish the number of users who need to be connected. As we begin the process, we use an older format of physical and logical ring. Therefore we find that we have five users that need to share the resources on the network. In Fig. 21.16, we run the cables from device to device. Here we have a wire connected to the in port of the NIC card at station A. The second wire is connected to the out port of the

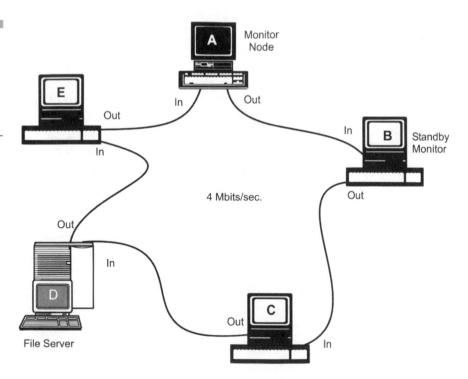

Figure 21.16
The physical and logical ring in place. A 4-Mbits/s ring is created, and devices/ nodes are tagged as functions require.

NIC in device A, but the out wire runs to the in port of device B. At device B, we run the out wire from B to the in port at C. From C, we run the out port wire to the in port on device D. At D we run the out port wire to the in port wire at device E, and at E we connect the out port to device A. Now we have a physical ring. The circle is closed and continuity exists between and among all devices.

Next, we install the LAN drivers in each of the PCs to basically let each PC know how to use the NIC card that has been installed in the expansion slot. The card has several components, including the electrical connection to the backplane on the PC, a chip set that operates at 4 Mbps, buffers to hold the data, and a framer that will take the necessary data to be transmitted around the network in chunks. (Remember that in a 4-Mbps token ring, the frame can be as large as 4048 bytes.) Next, we select device A as the monitor station and B as the backup (or standby monitor). We begin by assigning the connected device D as a file server. Now we are just about ready to begin the process.

As each device is powered up, the 6-byte address is recognized as being the card's location. So the card's CPU recognizes its own address. At the physical layer, we now have full connectivity.

When we load up the network operating system (NOS), we also build a database of the network devices attached to this ring. This will also be the time that we grant certain rights and privileges to the individual PC, including the services that can be accessed, the priority of the individual PC, its address, and the upstream and downstream neighboring information.

Next we begin to operate the network. User A wants to send information to the file server. Therefore user A begins by using a *save* command on a file in word processing. The PC and its operating components (whether in DOS, Windows, or other) send the information to the NIC card for delivery to the file server. The NIC card stores the data in a buffer. Next the NIC central processing unit (CPU) formats the data into a frame of information. In Fig. 21.17 the data is being prepared into a frame (envelope). Into this envelope the NIC stuffs 4048 bytes of information. For all intents and purposes, this is two pages of information. So the two pages of the information have been inserted into the empty envelope.

In Fig. 21.18, the addressing information is placed on the frame. We now have a frame just about ready to go. There is more information that must be attached to this packet—but that is the permission slip referenced earlier in this chapter. The permission slip comes from the network.

The NIC monitors the incoming port to see whether anything is coming down the wires. It waits for the token to come down the path to col-

Figure 21.17

The process of preparing a frame for transmission. The frame is formatted by the NIC and stored in a buffer until the node can send to the server.

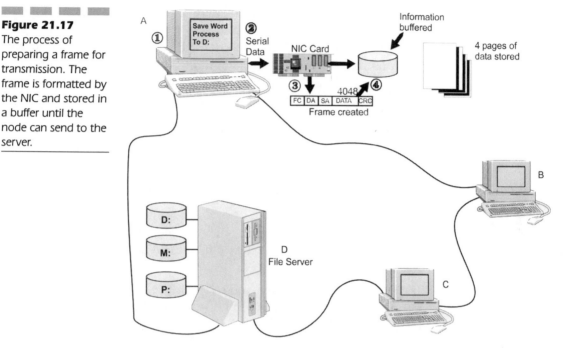

Figure 21.18

The NIC inserts the addressing information and performs a CRC on the information in the frame.

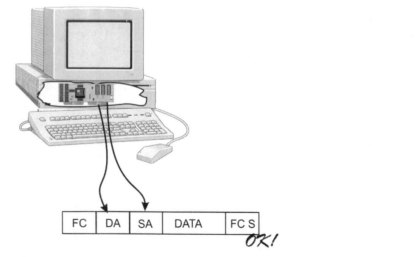

lect the data. Now the NIC sees the token coming in on the in port and grabs the information. In a very quick analysis, the NIC determines that the token has other information attached to it. Unfortunately, the token comes by, but is busy. This is a service where only one token at a time will be available to use. Therefore, the NIC just passes the busy token along through the out port back onto the network so that the information can get to its designated address. The NIC now waits for the token's next pass.

Here it comes again. This time the NIC grabs the information and determines that the token is free (Fig. 21.19). The NIC does several things:

- Grabs the free token and makes it busy by changing a bit field in the AC byte

- Appends its data to the token, starting with the frame control, the destination address, the source address, the information field (4048 bytes), and the frame check sequence

Behind this information are the end delimiter and the frame status bytes.

- Now the NIC sends the frame out to the network. As the frame is moving down the wire, the next NIC located at device B sees the token coming and grabs it.

- NIC B then reads the token and sees that this is a frame, so it checks the destination address. Realizing that the frame is not for B, it then

Figure 21.19
The NIC grabs the free token and appends the data frame. Note the sequence.

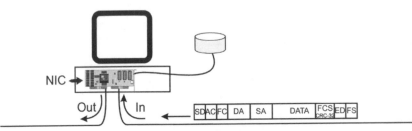

① The NIC sees the inbound frame, examines it and sees it's busy, then sends the entire frame back out to the network.

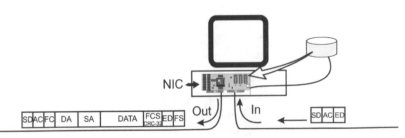

② The NIC sees the free token on the next pass so it inserts the buffered frame into the token and sends it out to the network destined for the file server.

immediately sends the frame directly back on the network to its downstream neighbor. The process continues through C.

- As C sends the frame out to the network to D, things change. The addressee is the file server at D. So when D gets the frame, it reads the information and sees that the destination address is D.

- D now makes a copy of the frame and stores it in memory on the card. Now D sets the *A* (address recognized) and *C* (frame copied) bits in the frame status byte to 1.

- Leaving the entire frame intact, and having modified the *A* and *C* bits, D sends the frame back out to the network, where it will go through device E then on to device A (the originator).

- When A receives the frame back, it analyzes the frame and sees that the *A* and *C* bits have been set to 1 (return receipt requested). Therefore A strips off the frame field, sets the token as idle, and sends the free token out to the network for the next station waiting to transmit to use. A node typically only gets one use of the token and then must wait for the next pass before using the token again.

- The process restarts all over again, depending on the source and destination addresses used.

As you see, this can be an efficient use of a network service. The larger the network gets, however, the more the devices will be waiting to transmit rather than transmitting. Growth and latency on these networks will slow the network down by 1 bit time per node attached. Therefore, when IBM introduced its souped-up token ring network at 16 Mbps, the frame size increased by fourfold and two tokens were allowed to exist on the network. This should be helpful in keeping things moving quickly.

Not much changed over time since the second version of the token ring. When a user needed more than 16 Mbps, then the migration on a ring was to FDDI (discussed later). These NIC cards are significantly more expensive than the Ethernet cards (regardless of the speed). An example of this is shown in Table 21.5. These numbers are representative of the costs for the cards today.

One can see where the industries got a little confused about why they should buy IBM's version of LAN. In fact, at the end of 1999 the token ring amounts to approximately 12 percent of all LANs installed worldwide (compared to 80+ percent Ethernet). The largest corporations in the world use the token ring, and based on their data needs, price is not the issue.

Therefore, the token ring presence was diminished by the pricing model for smaller LANs and for smaller organizations. In 1998 and 1999 a

TABLE 21.5

Comparing Cost
of NIC Cards

NIC Card Type	Speed	Cost per Card in $
IBM token ring	4 Mbps	$295
IBM token ring	16 Mbps	$495–695
Ethernet	10 Mbps	$19
Ethernet	10/100 Mbps	$25
Ethernet	100 Mbps	$45–50
Ethernet	1000 Mbps	$450
ATM	50 Mbps (full duplex)	$400

task force got together with the express intent of developing the 100- and 1000-Mbps token ring to keep pace with the evolution of the Ethernet. Some progress has been made, but more work needs to be done before a standardized fast token arrives on the market.

Moreover, the emphasis that IBM took was on the ATM NIC cards at the desktop. ATM was initially developed to deliver 50 Mbps full duplex service to the end user. IBM conducted separate studies and determined that the end user did not need 50 Mbps full duplex. Instead, the end user only needed 25.9 Mbps full duplex. IBM recommended this change to the ATM Forum and the ITU. This recommendation was met with much resistance, and finally the ATM Forum and ITU accepted a modified version at 12.8 and 25.6 Mbps throughput for ATM to the desktop.

While this activity was under way, Bay Networks (who acquired Centillion Switch Company) slipped into the lead of suppliers of the IBM token concept. However, as stated, not much else took place until the end of the 1990s.

22

Baseband versus Broadband

When LANs were first emerging in the late 1970s, they were called local nets. These local networking services and techniques would be the foundation of the LANs of the future. As with any other emergence, two opinions were formed as to how these services might be provided. In our earlier discussions in this book, we said that there were really two factions in the industry: the voice and the data communications factions.

Whenever a voice person got involved in the picture, all data needs were satisfied by the traditional data communications techniques. This meant that the voice analog dial-up or leased line services were installed. After that, a modem was used to convert the digital signals of a computer into the analog signals of the voice world for transmission. This was the same for local and long-haul communications needs. Most voice people felt comfortable with the use of modulation techniques on the link.

However, whenever a data communications person got involved, the picture changed. Most of what was installed in the data world evolved around the single largest player in the industry, IBM. IBM's strategy was to use a coaxial cable to allow high-speed communications for computing platforms. This included the hierarchical structure of the SNA world. We've already covered most of these concepts in past chapters, so by now the acronyms should be familiar. At any rate, the coaxial world was introduced to overcome the limitations of the noisy twisted pair wiring that the voice folks always used. This coax was traditionally used in a baseband arena. A single cable was extended to the end user device (3270 terminal, 3274 controller or other). Nothing else needed to be done, because IBM would directly apply the electrical voltages on the coax for propagation down the wire to the next hierarchical device. This was a baseband application of the cable. So a nonmodulated approach was used. Electricity was directly applied to the cable and it propagated (ran) down to the ends unchanged.

Another faction that cropped up in the industry came from the radio frequency proponents. The use of radio-based systems was also prevalent in the industry prior to this era. Therefore, the radio folks looked at the need to send simultaneous signals down a medium to an end device. They began to use radio frequency (RF) techniques to carry multiple data signals. Where localized communications was the only requirement, the RF folks decided to apply some of the characteristics of the CATV business in the local data communications business. Local nets were set up with the use of a broadband communications capacity, the equivalent of CATV, to support the need for multiple, simultaneous data streams on a single medium. The voice half of the business favored this approach because a modulation technique (i.e., modems) was used.

Baseband and Broadband

Baseband and *broadband* are two different types of signaling that can be used on a wire. We are speaking here of the basic electrical waveforms that are injected onto the medium. Nevertheless, do not tune out at this point just because you are not an electrical or radio engineer (or even if you are)! We cover these methods in the context of benefits and choices rather than instructing the reader on the design and construction of the associated components. The emergence from the local nets into a formalized LAN arena introduced a new battleground between these factions in the industry: those that felt comfortable with the use of a baseband service (the data communications groups) and those who felt more comfortable with the use of a higher-speed communications channel capacity (the voice communications, broadband RF, and electrical engineers). The stage was set for the next set of arguments. In the early 1980s, this was a hot battle, with each side claiming that it had the best solution and that the other side did not know about what it was talking.

Many an organization had to take this decision to a very high level of management for approval. When you think about that concept, why was management bothered with these complex decisions? All management wanted was a communications solution that met the organizational needs without wasting money. Here, management was asked to decide on a technical decision that it was not qualified to make. Hence outside consultants were brought in to assist with the decision process. Unfortunately, the external consultants were equally split in their recommendations. We have seen cases where management brought in a consultant at the recommendation of the baseband side of the business. Then a management consultant was brought in from the broadband side of the business. Management now had two recommendations, each different, so a third (independent) consultant was brought in to be the tiebreaker. What a waste of time and money!

Let's look at some of the differences between these two concepts. First let's review the fundamental motivations that were used in determining whether to use a LAN or other connectivity arrangement.

Motivations Driving the LAN Decision

If you go back to when these decisions were being wrestled with, there is some historical perspective that will help you to understand the argument in its basic form. When local nets and the ensuing LANs were being researched, the fundamental needs were clear.

The numbers of terminal users in the late 1970s and early 1980s were limited. Many very large organizations had a total number of terminals based on a 50:1 or 100:1 ratio. For every 50 or 100 total employees, there would only be one terminal device. This might be a shared device among users or a common function (data entry) whereby all employees brought their data to a keypunch operator for input/output.

The industry was claiming that by the mid-1980s the terminal ratio would be closer to 10:1 and closing in on 1:1. Office automation would mandate that the deployment of such terminal devices would be required for every "knowledge worker."

The alternative to massive wiring to support this increasing population of users was to use a form of the backbone system with easier connectivity to a central computer, rather than "home runs" to the host from the user terminal. However, the use of a LAN to replace or simplify the cabling nightmare and the expanding population goes far beyond this. Following is a group of other reasons for installing a LAN on a single-shared medium.

Data Rates

The original data rates for terminals were normally handled by local cable runs or dial-up communications. The newer terminal devices that use bit map overlays, graphical user interfaces, and the density of the data require much higher data speeds than were previously available. The LAN is designed to support much higher throughput speeds than traditional dial-up communications and local cable attachments.

Interconnection

Older terminals were hard-wired back to the host, or controller, or were connected by a dial-up line. The use of a LAN moves away from the concept of a central computer control into a distributed control architecture. This allows a server to support users locally and only connect to the host when necessary. This interconnectivity is a function of the LAN. Figure 22.1 shows the other problem with interconnectivity. A user might have a need to access three different computers; therefore, a terminal attached to each was required. The lack of desk space was a problem, and the cost of such a solution is exorbitant.

Integrated Resources

Voice, data, text, and video were all on separate networks and wiring schemes (Fig. 22.2). The LAN was touted as the first solution to integrate all these into a single infrastructure. Although this is still not a reality, except in limited tests, the concept of a single access to any service was a motivation to consider LANs (Fig. 22.3).

Compatibility Issues

Various access and protocol stacks were used. The LAN was to provide a common interface for access to any connected service. The operative

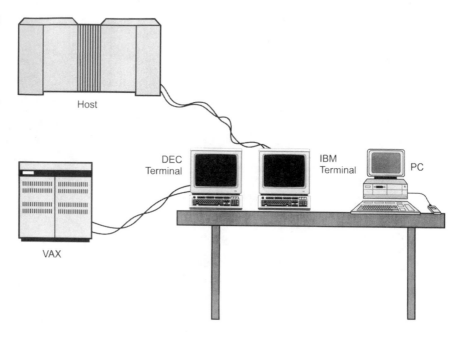

Figure 22.1
The cost for a separate terminal connection to each computer was high.

Figure 22.2
Four separate wiring systems were used to provide appropriate services. This is an expensive, nonintegrated solution.

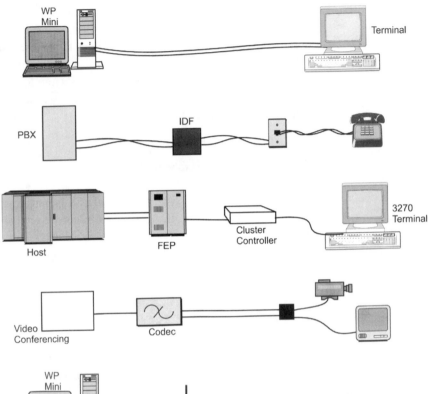

Figure 22.3
An integrated solution with a single cable was the ideal goal of a LAN.

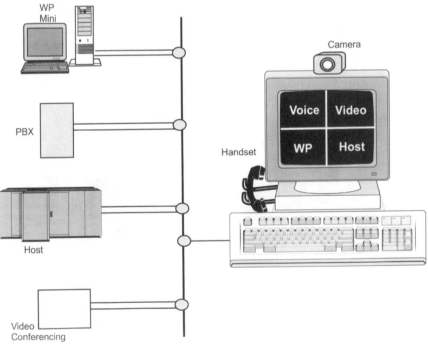

phrase is *common communications among various vendors.* LAN interfaces were to be highly intelligent, so that it did not matter what the need was as long as the common path could be established.

Cost Implications

As already mentioned, the cost of separate systems and wiring solutions was exorbitant. The LAN single connection was to provide this at a much more reasonable rate (Table 22.1).

TABLE 22.1

Primary
Motivations for
Considering the
LAN

Reason	Discussion
Data rates	The original data rates for terminals were normally handled by local cable runs or dial-up connections. The newer terminal devices that use bit map overlays, graphical user interfaces, and the density of the data require much higher data speeds than what was available. The LAN is designed to support much higher throughput speeds than traditional dial-up communications and local cable attachments.
Interconnection	Most terminals were hard-wired back to the host, or controller, or were connected by a dial-up line. The use of a LAN moves away from the concept of a central computer control into a distributed control architecture. This allows a server to support users locally and only connect to the host when necessary. This interconnectivity is a function of the LAN. In Fig. 22.1 we see the other problem with interconnectivity. A user may have a need to access three different computers; therefore a terminal attached to each was required. The lack of desk space created problems, and the cost of such a solution is exorbitant.
Integrated resources	Voice, data, text, and video were all on separate networks and wiring schemes as shown in Fig. 22.2. The LAN was touted as the first solution to integrate all these into a single infrastructure. Although this is not still a reality, except in limited tests, the concept of a single access to any service was a motivation to consider LANs as shown in Fig. 22.3.
Compatibility issues	Various access and protocol stacks were used. The LAN was to provide a common interface for access to any connected service. The operative phrase is *common communications among various vendors.* LAN interfaces were to be highly intelligent, so that it did not matter what the need was as long as the common path could be established.
Cost implications	As already mentioned above, the cost of separate systems and wiring solutions was exorbitant. The LAN single connection was to provide this at a much more reasonable rate to the organization.

Table 22.2 represents the cost comparisons of the alternatives. In this table, we use example prices only because the actual connections and services go beyond what we can address for every organization. The numbers are generic, but roughly approximate what was found in an organization back then. We looked at the cost of wiring a voice connection to the PBX, a data terminal via coax to the host, a text-based terminal to a word processing system (specialized or minicomputer), and a video connection. Then we compared this to the LAN connection costs that were offered back then.

This was an appropriate comparison, but a problem emerged with this philosophy. The various systems being offered in the LAN arena would not support all of these services on a single cable system. The reasons for this argument are as follows:

- Voice, although a baseband system, is a two-way simultaneous conversation, full duplex. Although we saw that voice operates primarily as a half-duplex service, the need to be fully duplexed is always there. Conversations are built around the need for two people (or more) to interact as they would in a face-to-face conversation. The LANs would not specifically allow this in the Ethernet and token-passing ring installations. They are baseband.

- Both voice and video require a constant bit rate of transmission. They are not data communications transport. Voice and video are not easily packetized or placed in frames like data. Data and LANs deal

TABLE 22.2

Comparison of Costs for a Single versus Multiple Cabling System Connection

Needed Connection	Separate Connection	LAN Connection
Voice, twisted pair to the PBX	$ 500	
Data terminal, coax cable 327X device	$1000	
Word processing connection to a specialized system or minicomputer on a coax or twisted pair	$ 450	
Video connection on a coax CATV cable	$ 500	
Voice, data, video, and word processing integrated on a LAN		$ 800
Total	$2450	$ 800
Differences		($1650)

with more variability in their need to transmit, but can suffice for the packetization or framed format of transport. If an integrated cable is used, various streams will be required on the cable simultaneously in a full and half-duplex method. This was formerly not an available service because baseband LANs are primarily one-way transmission systems.

■ As this was a shared cable system, the use of the cable by a single user at a time on the entire bandwidth was not appropriate. Voice does not need 4, 10, or 16 Mbps. Video can use this much, but we have already compressed this into much smaller bandwidth requirements. Hence, the baseband LAN was only perceived as a single transport for data, not voice and video. This underpinned the benefits of a fully integrated solution.

Three ways were available to provide the integrated solution; these included the following.

The PBX

Integrated PBX solutions were being touted for their robustness in the business environment. Because most organizations already had a telephone system, the PBX manufacturers were pushing their wares as the solution for voice, data, and other needs. However, the PBX was limited in its ability to handle the higher speeds integral to the LAN desires and data transport. Digital PBXs were still relegated to the 64-Kbits/s digital or 9.6-Kbits/s analog dial-up data communications connection. The PBX players were quick to state that the average LAN transmits at 4 or 10 Mbps, one way. However, the digital PBX can handle much higher aggregated speeds than a LAN. Voice is two way, so the PBX has a two-way simultaneous connection. Table 22.3 shows how the PBX suppliers were touting their bandwidth capacities.

Although you might argue the numbers, the PBX makers also conducted studies and found that most computer systems were limited to 64-Kbits/s data transfer rates, and that the average user needed no more than that. We have to give them credit for trying! PBXs introduced one added risk that everyone noticed right away. The system is a centralized architecture, which meant that all of the associated risks of a single box on a network are associated here. Further, this was a digression from the distributed architecture that the LAN players were trying to accomplish.

Chapter Twenty-Two

TABLE 22.3

PBX Manufacturers'
Comeback to Their
LAN Service

Item	PBX	LAN Bus	LAN Ring
Number of users that can be simultaneously wired	10,000	1000 in Ethernet	260 in token ring
Maximum speed per connection	64 Kbits/s	10 Mbits/s	16 Mbits/s
Two-way benefits	64 Kbits/s	Not available	Not available
Total number of users communicating at once	5000*	1	1
Number of users times aggregated bandwidth	640 Mbits/s[†]	10 Mbits/s	16 Mbits/s
Ability to handle video	Yes	No	No

*This assumes that 5000 users can simultaneously call the other 5000 users in the system, meaning that 50 percent of the people can call the other 50 percent.

[†]This number is derived from the following: 5000 users transmitting and receiving at 64 Kbits/s is 128 Kbits/s times 5000 connections or 640 Mbits/s. This is arguable, but the companies were there in the running and hyped what they had to offer.

Baseband Cable Systems

As mentioned in earlier chapters, an Ethernet and token ring normally use baseband signaling. Baseband signaling is actually rather simple in concept. Devices using this technique place a digital (square waveforms—that is, 1s and 0s) signal directly on the attached medium (the cable). Only one device can communicate at a time on a baseband system. There is no means for allowing two devices to originate a signal on the medium during overlapping time intervals. If two devices did try to transmit simultaneously, both digital signals would instantly be garbled beyond recognition. In the case of a bus network, several elaborate methods are used to ensure that transmitters have the medium to themselves (see the Ethernet discussion). In token ring and star networks, any given cable has only two connected devices following rigid protocols; no conflict arises (see the token ring discussion).

As with any digital signal, the square waveforms degrade over distance somewhat more quickly than would be the case using analog transmission techniques. However, so long as baseband networks are engineered within their designed distance limitations, they experience a very low level of errors indeed.

In Fig. 22.4, a representation of the degradation of the square wave is shown. Distances are limited in a coax cable system such as the Ethernet.

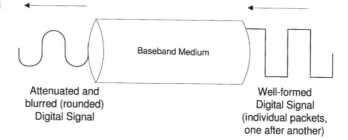

Figure 22.4
The square wave degrades over distances, limiting the overall LAN length.

Baseband Medium

Attenuated and
blurred (rounded)
Digital Signal

Well-formed
Digital Signal
(individual packets,
one after another)

Specification for a thick-wire Ethernet is 500 meters. Yes, repeaters can be used to extend the distances, as seen in the chapter on Ethernet. Two repeaters can extend the linear distance to approximately 1500 meters (just under a mile of cable). However, this is still a single-user access at a time on the entire bandwidth available. The degradation of the signal is only one part of the equation.

A key characteristic of actual baseband network standards is that in general they have been designed to allow a "by-the-numbers" design approach. If one uses standardized components and stays within the very specific guidelines, the networks should work fine (barring failed components, of course). Also, the equipment, because it only deals with that one signal at a time, is less costly than digital broadband components. Proponents of the baseband systems were diehards in their assessment that this was the least expensive and the most robust operation to meet the needs of the LAN for the future. Realizing that these LANs were being built for transport of data only, these people felt that there was no need to do anything else. Xerox, of Ethernet acclaim, was one of the single largest proponents of the baseband approach. This was natural, because Xerox was in that market. But several comments made by Xerox as an organization led people to believe that this was the technology that they would live and die on.

Several issues must be taken with the idea of the single bus topology or the token ring topology that only allows one user at a time to send in one direction. There has already been much said in the chapters on Ethernet and token ring about how the network works. Therefore, we concentrate here on the differences between baseband and broadband. We are not trying to suggest that a "choose-or-not-choose" approach is covered here. We merely try to cover all of the issues in an unbiased fashion. Following are some counters to the arguments that always surface.

Baseband Is Cheaper Than Broadband and PBX Solutions

Although this is a valid argument, the jury never sits for a global statement in this industry. Whereas the Ethernet cards are very inexpensive, and the ancillary equipment pieces such as hubs and twisted-pair wiring are inexpensive today, the PBX prices for 64-Kbits/s transport have also dropped through the floor. Second, CATV has been in use for longer than we can remember. The parts are also available and mass produced. Thus a per-tap arrangement makes this a somewhat moot point, because the cost per drop is about the same. If we look at the token ring services, the MAUs and the token ring cards are more expensive, as seen in the previous chapter. A 16-Mbps token ring card can cost upward of $695, and the MAU can be as expensive as $600 to $700 for an eight-port tap into the baseband LAN.

Baseband Is a Very Scalable LAN Service

In fact, it is not as scalable as one might be led to believe. More than 20 years have passed since the initial baseband LANs were first contrived. Only recently has the ability existed to take the baseband from the 10- or 16-Mbps speeds to the 100/1000-Mbps range. There is a fundamental need that is addressed with a baseband service. The speed constraints are not suitable for the application of trying to merge voice and video on the LAN. Therefore, a full duplex LAN in either Ethernet or token ring is now an option. The addition of voice and video will demand duplex and higher speeds, or else the network will quickly come to a grinding halt.

Baseband Is Highly Efficient

Any protocol that we add onto any topology and medium can be efficient. The use of baseband, Ethernet, and token ring in particular has a certain amount of overhead. These protocols tend to make the baseband decision different than you might think. As we saw in the Ethernet discussion, there are minimum sizes for frames. This is so the systems attached to the baseband cable can see the timing and listen to the information during specific time slots. All data transferred must abide by the rules, or the system doesn't work. The overhead and the "wait until it's your turn to send" approach make the 10-Mbps throughput drop to approximately 3.3 to 4 Mbps. That happens to be 60 to 67 percent inefficient.

Therefore, the argument for the baseband system is not as straight-forward as you might believe. The heavier the load on a baseband system, the greater the efficiency loss. For example, a 10-Mbps Ethernet typically yields a net throughput of only 3 to 4 Mbps. The larger the population of users or the density of the data, the greater the risk of collisions. See Fig. 22.5 for the collision. If I send "peanut butter" and you send "jelly," halfway down the cable we collide. Now you have jelly in my peanut butter! This isn't what I sent, so it must be discarded and resent. In a baseband token ring, things are more deterministic, but the greater the population on the network, the greater the waiting time to finally transmit. See Fig. 22.6 for a small-scale look at the performances of CSMA/CD and token-passing rings.

Broadband Cable Systems

The term *broadband* is used in several different ways in the communications industry. Descriptions of two of these uses follow.

Figure 22.5
The denser the data or the greater the number of users, the greater the likelihood of collisions.

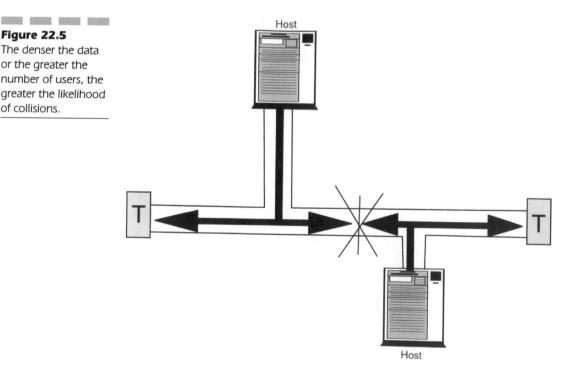

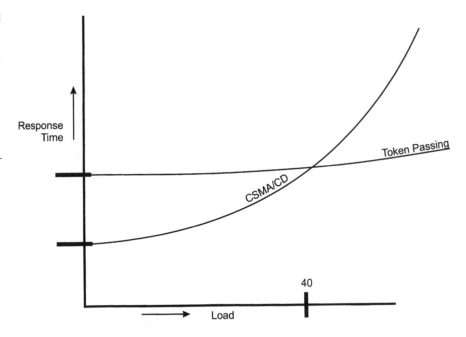

Figure 22.6
A comparison of load and response time for baseband networks. Token passing is a smoother degradation, whereas CSMA/CD works efficiently at lower loads.

- Any wide-area communications channel of significant bandwidth. *Significant* in this case used to mean any circuit whose capacity equaled or exceeded 56 Kbps. Some analog facilities fit this description. Now, if a circuit is not at least a T1 (1.544 Mbps), it is rarely referred to as a broadband facility. In fact, many of the T1 services and below are referred to as *narrowband,* T1 up to T3 are called *wideband,* and everything above the T3 level is referred to as *broadband.* Like anything else in this industry, all of this is according to your own interpretation. Analog wide area facilities, of whatever capacity, are rarely referred to as being broadband.

- Any of a specific set of service offerings from common carriers (e.g., SMDS, T1 and T3, ATM, frame relay, SONET, etc.). The specific technique of using coaxial cable to carry multiple channels over LAN distances.

This last meaning is the one discussed in this chapter. We have spent a lot of time throughout this book in defining the applications, uses, and definitions of baseband, yet very little time discussing broadband communications. Therefore, we finally delve into the subject here. Although most people do not think in terms of baseband and broadband, there are significant differences and reasons that they exist. To fully understand the broadband communications, we use a CATV system as a comparison.

The LAN arena has always argued that baseband was the simplest and least expensive solution to installing a network. Fundamental differences exist with such a global statement. Further, the broadband communications on a cable system is perceived as an inferior product offering.

Broadband signaling on coaxial cable is one method of designing a local area network. It is unique in several respects:

- Multiple signals coexist simultaneously (really at the same time, not just looking as though they are happening at the same time because of speed).

- Broadband signaling is inherently analog—but digital signals can nonetheless be sent using it, just as digital signals can be sent over the public telephone network.

- It uses many components originally designed for a completely unrelated industry—cable television.

- It is not "plug-and-play," unlike, for example, Ethernet and token ring.

- It is targeted primarily at industrial environments, because it runs only on coaxial cable, a medium more resistant than most (not counting fiber) to interference by industrial processes (e.g., arc welders).

Cables that use broadband signaling function like a superhighway with a few hundred lanes, all divided by solid double yellow lines (with no breaks). Each lane carries traffic more or less independently of traffic in other lanes. If traffic drifts across lanes, there will be accidents that disrupt traffic. However, this doesn't normally happen, because the lanes are separated by buffers (the double yellow lines provide some physical spacing, although not so much as to be wasteful).

Broadband LANs are based on the same underlying technology as cable television. To understand this, you only need to consider a typical CATV system in a hotel or large apartment complex. The building is served from some transmission system located outside the confines of the physical plant. This serving end of the cable is called a *head end*. From the head end, the television signals are distributed along the cable and the cable forks as necessary to cover all locations. On the cable there are many TV signals at the same time. The user of the cable interface unit (the TV in this case) selects a particular channel from the whole list of available channels on the cable. This is done by tuning into a specific frequency, what you do when you use the infrared remote control changer (Fig. 22.7). The set is then tuned to the channel designator selected on the remote controller. The TV does this by tuning in on a specific frequency range or a *frequency band*. Because television channels need to carry a lot of information in the

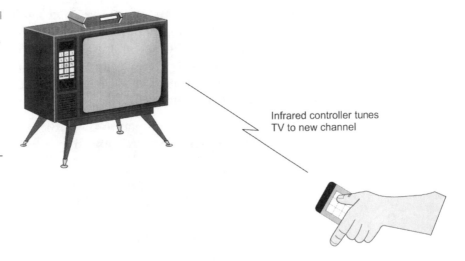

Figure 22.7
The remote controller causes the cable box and TV to tune to a new channel. All channels are on the cable at one time, but only one can be viewed.

Infrared controller tunes
TV to new channel

form of motion video (albeit one-way broadcast video), the bandwidth of the channel has been allocated a 6-MHz capacity (6 MHz or 6 million cycles per second). The interval from the lowest to the highest frequencies passed on this channel is 6 MHz. The channels are run next to each other similarly to the way the telephone network operates. Channel 2, for example, on a CATV might operate on a frequency band of 48 to 54 MHz. Channel 3 will operate at 54 to 60 MHz. Standard cable systems have an immense amount of bandwidth available to them.

As the TV signals propagate down the cable, they will lose some of their strength. This loss of power is similar to that occurring on the telephone network and is called *attenuation*. If the signal drops below a certain threshold, it becomes very difficult to separate the signal from the electrical noise always present on the cable. Of course, when this happens, the signal quality, or the output, to the TV drops and a subquality TV picture is produced. To overcome this problem, an amplifier is placed at certain intervals along the cable. The amplifier boosts the signal strength so that it can continue to move along the wire in a usable manner.

Although there is only a single physical cable, a number of channels can simultaneously be present on the cable. This is accomplished by sharing the frequency spectrum of the cable and assigning different frequency ranges to different channels. This was covered earlier in the chapter on data communications and is called *frequency division multiplexing* (FDM). A broadband cable communications system operates exactly like this CATV system, and in fact uses the same types of cable and connectors. A counter to this FDM technique is division of the cable not divided into frequency bands but into time slots. This is called *time-division multiplexing* (TDM); here each station can grab the entire bandwidth of the cable, but only for a very short period. The cable is currently capable of carrying up to 750–950 MHz, depending on the vintage in use. Early cable systems using the FDM techniques divided the channels into a number of dedicated logical channels as strictly an alternative to twisted pair wiring. This FDM technique did not address the interconnectivity solutions required by a LAN. Past innovations have allowed a time-division multiple access (TDMA) within an FDM channel. A single cable can service a large amount of terminal communications. On this same cable, but on different FDM channels, data circuits, voice channel, and video signals can all coexist.

Modern systems allow multiplexing on two separate dimensions. The channels are separated by FDM, and in a specific frequency band the channel can then be shared via TDMA among multiple users. The most common of the TDMA access methods on a broadband cable has been the CSMA/CD developed by Xerox Corporation for Ethernet.

The considerable raw bandwidth of the cable is divided into many individual channels, each wide enough to carry one digital signal. However, the underlying structure of the network is different. On a broadband network, there is a single head end device that receives every broadcast signal on one of two major sets of channels, then retransmits the same signal on the corresponding outbound channel. Broadband bandwidth allocations are illustrated in Fig. 22.8.

The actual cable runs might look something like those in Fig. 22.9, a sample broadband cable layout. This diagram provides a clue as to why small proportions of Ethernets are implemented on broadband cable plants; they are too complicated!

Figure 22.8
Broadband bandwidth allocation.

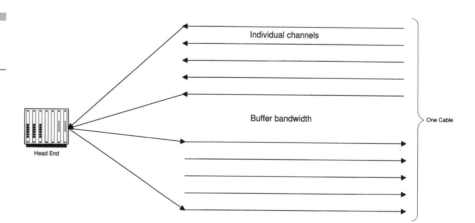

Being a truly multichannel medium, broadband cable permits a surprising variety of applications to reside on a single cable plant. On one cable, a company can carry security video signals, terminal channels for mainframe access, one- or two-way video signals for classroom use, and Ethernet. Ethernet? But Ethernet is baseband!

Well, yes, Ethernet is baseband. Baseband signaling, as stated, requires a dedicated channel on which one frame at a time is placed. But the Ethernet can use multiple media. Why not carve out part of a broadband cable's bandwidth for an Ethernet (or even several Ethernets—up to three at once are supported)?

Broadband cables in the CATV world have an inherent difference; they are designed around delivering the TV programming on a single cable in a unidirectional flow. All signals come from the CATV head end. A couple of variations are allowed to turn this cable into a bidirectional transmission. These are as follows.

Using a single cable, a midsplit arrangement can accommodate two-way simultaneous transmission. A midsplit divides the cable into separate pieces. A portion of the cable, called a *reverse direction,* is used to send signals to the head end from the transmitting devices. The rest of the cable, called the *forward direction,* is used to send from the head end to the receiving devices. Figure 22.10 shows this arrangement. A portion of the cable is allocated for a guard band to prevent the overlap between the channels. The head end's job is to receive the signals from the devices on the reverse direction, change the frequency, and send the signal back out to the cable on the forward direction.

Figure 22.9
Sample broadband
cable layout.

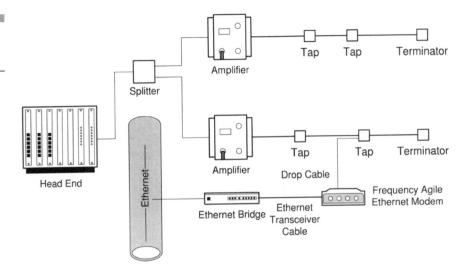

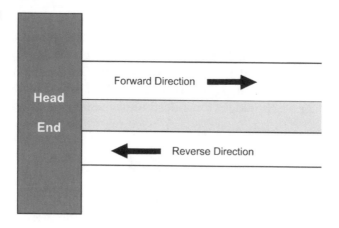

Figure 22.10
The midsplit cable.

The second choice is to use a dual cable system. In this case, the entire spectrum of a single cable is used in the reverse direction to the head end. The second cable is used for the forward direction from the head end to the devices attached to this cable. Figure 22.11 shows a summary of how this will look. For those of us who remember the Wang net approach, Wang used a dual cable system. Each of the frequencies was split up to provide services and utility bands for connecting other devices such as Ethernets to their broadband cable system.

Broadband is inherently an analog signaling method. Because video cameras, for example, are also analog devices, a signal from a video camera (or video recorder) can be directly transmitted onto a broadband cable channel in RGB (red, green, blue) format. No conversions are required, as shown in

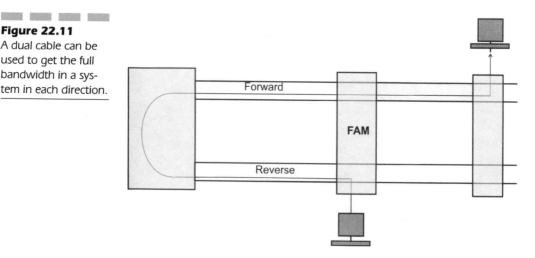

Figure 22.11
A dual cable can be used to get the full bandwidth in a system in each direction.

Fig. 22.12. Voice signals and conversations can also be modulated onto the broadband cable easily. In the case of a PBX on a site or campus, voice tie lines can be used (Fig. 22.13). Dedicated high-speed data links can also be set up on a broadband cabling system. Further, telemetry and paging circuits can be modulated directly on and off the cable. Finally, Ethernet and other digital technologies are indeed implemented on broadband cables.

Figure 22.12
Video can be modulated directly onto the broadband cable without any conversion.

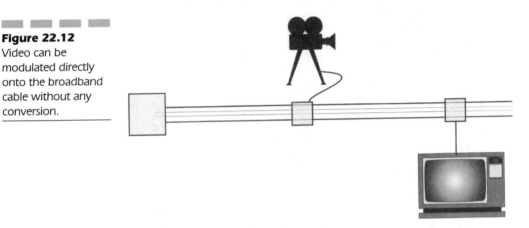

Figure 22.13
Voice-grade tie lines can be sent directly across the broadband cable.

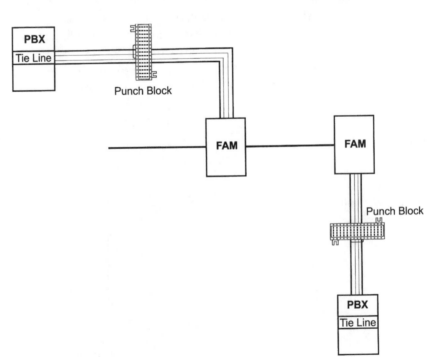

But additional hardware is needed. From earlier chapters, you might remember another situation where digital signals were required to travel over an analog network, in that case the public telephone network. As in that case, to transmit digital signals over a broadband network we must use special modems in addition to all the normal components required to implement a cable television system. Broadband modems are called frequency agile modems (FAMs). Because frames are sent on one frequency and received on another frequency, they must be agile indeed.

However, all of the typical voice, data, video, and LAN services can be applied in some form onto the broadband cable. This was one of the motivating factors in making the decision to go to a LAN, wasn't it? The ability to create an integrated cabling solution for all of our communications needs has quite a bit of merit. Yet, unfortunately or fortunately, baseband cabling persevered and became the norm. Some of the reasons for this are the complexities of managing and maintaining the system. Table 22.4 summarizes the capabilities of using a broadband cabling system.

What more can we ask a single cabling system to do? Everything that we need to accomplish on a single site can be performed and modulated onto a single or dual cable system. Yet the opponents of this cabling scheme have a tendency to disparage it and claim that it is foolish to even consider this as an alternative.

One significant disadvantage to broadband (as opposed to baseband) is that it must be installed by technicians qualified in radio technology. Moreover, as with all radio technology, it must:

- Be periodically checked to ensure compliance with FCC regulations.

- Be returned when significant changes are made to the configuration, particularly when additional taps, which would cause added loss, are placed on the cable. The use of more amplifiers might also be required.

TABLE 22.4

Summary of What Can Be Run on the Broadband Cable

Voice tie lines such as four-wire E&M

Logical connections of 3270 terminals to the cable and to the controller

CAD/CAM

Video conferencing

Closed-circuit security systems

Building management systems such as energy management, security, alarms, etc.

LAN a la Ethernet

In short, it is more of a hassle than baseband technologies. Thus the baseband war was won. There have been some very strong arguments for both sides of this equation. In a large campus environment, broadband might carry a lot of benefits, whereas on a department-by-department basis, installation and maintenance of baseband was less of a problem, so this was the preferred route to follow.

Baseband versus Broadband

Why use one over the other? Baseband should be your default choice because of its simplicity and lower cost. It is far more widespread; parts and service are much more readily available.

As with any default, there are a number of factors that might push one in the direction of other solutions. Those needs that might mandate selection of a broadband rather than a baseband solution include:

- A cable television plant (with available bandwidth) is already in place in the facility to be wired (thus helping to limit costs).

- Video signaling is also required for a significant portion of the locations to be wired.

- Several separated Ethernets (but no more than three) are required, and which Ethernet a drop is to be on might change. (This reason is weaker today than in the past; such a configuration can now easily be accommodated with Ethernet hubs and appropriate 10BASE-T wiring.)

- The distances to be covered are longer than those that other Ethernet cabling can support, but within those supported by broadband.

- The network is to be installed in a highly electrically noisy environment—for example, on a factory floor. (Fiber is an alternative solution to broadband that should also be considered in such cases.)

23

Fiber Distributed Data Interface

When we discussed the speed of the LANs as we looked at the Ethernet and token ring topologies, we were specific in terms of what can be expected. The older Ethernet operates at 10 Mbps, whereas the initial version of the token ring operates at 4 Mbps, with an updated version operating at 16 Mbps. These speeds were fine when the LANs were first introduced, but as we approach the end of the century LAN throughput is becoming a potential bottleneck. The speeds were initially introduced and shared among 20 to 30 users to perform word processing, small spreadsheets, and the like. However, as the LANs ballooned in the industry, with devices appearing on nearly every desk, and the applications became more intense, the 4-, 10-, and 16-Mbps capacities were clearly signaling that there was a problem looming.

More users are sharing more data on a single cable. We could use a subnetworking technique to buy time, but this introduces smaller networks competing for a shared bandwidth. Furthermore, the more we subnetwork, the more pieces on the network can fail. Consequently, the industry set out to develop a much faster network arrangement. The result was a technique known as fiber distributed data interface (FDDI). FDDI was designed under the auspices of ANSI as a high-speed, highly reliable data transport. The significance of the FDDI standard is that it is a national standard supported by the entire data communications industry. Because it is a national standard, FDDI workstations, computers, and other peripheral devices work together regardless of the manufacturer. Moving away from the limitations of a 10-Mbps bus and the high-end 16-Mbps ring, FDDI was designed to have an effective throughput of 100 Mbps. Using a fiber-optic backbone and a deterministic token-passing approach, the technology overcomes the constraints of the older systems. This is not a panacea for all LANs, but a stepping-stone into higher-speed digital communications. There are certain strengths and limitations when considering FDDI. It is not for everyone, yet for those organizations that have exhausted the throughput capabilities of the LAN, FDDI might offer some breathing room.

FDDI Design

When ANSI first started this long and arduous task back in 1988, the goal was to overcome more than just the throughput problems. Surely, a higher speed capability was an earnest desire, and at 100 Mbps, no one would argue that a 6- to 10-fold increase in throughput was not warranted. However, there were other constraints that were also concentrated on; these are summarized in Table 23.1, which looks at the limitations of the bus and

TABLE 23.1

Summary of
Existing Topology
Problems and
Limitations

Topology/Item	Limitation
Bus speed	10 Mbits/s with effective throughput of 3.3–4 Mbits/s
Ring speed	4 Mbits/s with effective throughput of 3.3 Mbits/s 16 Mbits/s with effective throughput of 12 Mbits/s
Bus attachments	Up to 1024 addressable nodes, but limited by cable lengths of 1500 meters
Ring attachments	72 addressable nodes on unshielded twisted pairs and 260 addressable nodes on shielded twisted pairs depending on distances
Unshielded twisted pair	Error-prone due to wire characteristics and electrical, and mechanical interference
Shielded twisted pair	More expensive, bulky, and subject to ground loops
Bus and ring topologies	Subject to disruptions from cable breaks and other cable-related problems
Bus topology	Collisions when the network gets 40% busy, resulting in less throughput

the rings of old combined. These were the targets for the standards committees and should be presented right up front.

To overcome these and many other limitations, the FDDI standard was written and technology was developed to address each of these areas. Therefore, FDDI is designed to:

- Support higher data rates at up to 100 Mbps
- Work in a deterministic token-passing environment, preventing collisions
- Support attachment of up to 1000 addressable nodes
- Extend the distance of the ring up to 200 km (approximately 124 miles) for a single ring and 100 km (approximately 62 miles) for a dual ring
- Provide robustness, in that a dual ring can be used as a backup or to wrap around after equipment and cable failures
- Use optical fiber as the preferred medium to overcome the physical limitations of electrical and mechanical interference
- Allow spacing of up to 2 km (approximately 1.25 miles) between repeaters

Using the parameters listed, FDDI was born. You might expect that the industry and LAN administrators jumped for joy. Unfortunately, this did

Chapter Twenty-Three

TABLE 23.2

Summary of Costs
for a PC Equipped
for the LAN
Connection

Item	Cost
Change to PC, Intel-based 486D × 4/100 or Pentium 90-MHz w/8 mbytes RAM, 1.0-Gbyte hard drive, single floppy, and SVGA monitor	$2500
LAN interface card Ethernet	$ 99*
LAN interface card ring (4/16)	$ 295
Cables and connectors	$ 25
Total	$2624–2820

*This is an either-or decision with the Ethernet or the token ring card. Thus the range of costs for the PC is included with either option.

not happen. FDDI received a cool reception. The single largest reason (one we have stated over and over) was money! FDDI is just too expensive for most LAN administrators to implement. As FDDI was introduced, a card operating in a high-end server could be purchased for $15,000 to $23,000. The individual cards that were designed to work in a PC or PC-based server sold for $1500 to $2000. Can you imagine what the cost for a 100-user LAN might be? For the sake of keeping it simple, we have summarized the costs in Table 23.2. At the time FDDI came to the desktop, the PC hardware costs were rapidly dropping. The table summarizes the cost of a typical PC.

From a cost of a $2600 to $2800—for the PC, reduce the incremental charge of the NIC card ($99 to $295) but add the cost of an FDDI card ($1500 to $2000) and you would double the cost per device on the network very quickly. Now, using this example, a price comparison is shown for different-size networks, with the resultant differences as shown in Table 23.3. This is so that a true picture of the financial implications can be achieved. The table brings the reluctance factor home really nicely.

Can you imagine the way management would feel if we were to request a LAN using FDDI and show the cost of a 100-user network? The PC-based solution with a standard NIC card is $200,000 less than the FDDI solution. What would be the benefit achieved for an added $200,000 in investment? But wait! There's more. The prices listed are examples only, so we should make sure that all the costs are associated with this comparison. We didn't add the differences for the server connection, the wiring differential, and the concentrators (MAUs) that work in the closet. So, taking this one step further, these costs must be considered. To do so, we summarize the cost per connection as follows in Table 23.4 (by now you see we are getting cost crazy).

TABLE 23.3

Summary of Cost Difference in a LAN

Number of Users	Cost/PC Std NIC 2600	Cost/PC FDDI 4600	Difference
10	26000	46000	20000
20	52000	92000	40000
30	78000	138000	60000
50	130000	230000	100000
100	260000	460000	200000
250	650000	1150000	500000
500	1300000	2300000	1000000
1000	2600000	4600000	2000000

We are not trying to dissuade the reader from looking at the FDDI solution, but are trying to make sure that a true picture is achieved. Thus, if we take that 100-user LAN we dealt with in the previous example, and add to the cost of that LAN the variation (Table 23.4), a comparison of the LAN to the FDDI now totals up differently (Table 23.5). This is for comparative purposes only, so that a set of numbers can be used. The actual prices will be completely different when you look at the building and work group specifics in your own network. Having said that, we continue our comparison.

This is a shocking result, because the actual cost to use the FDDI is double that of a regular twisted-pair LAN. So why use it? Well, let's back up a little. We have been using a 100-user LAN with Ethernet operating at 10 Mbps. For only twice the investment, we get a 10-fold increase in the throughput! Regardless of what management has to say, there are some benefits to be had with this type of effective throughput. This is espe-

TABLE 23.4

Additional Costs of Installing FDDI to the Desktop Above Regular LAN Connections

Item	Cost Per Port/User FDDI	Totals
Servers for 100 users, 2 used	2 FDDI cards @ $15,000	$30,000
Wiring for fiber to desk, with connectors, etc.	$50,000 for station wiring, $14,000 for backbone wiring, $4,000 for connectors	$68,000
MAUs (called concentrators) for FDDI backbone	4 @ $500	$2,000
Total difference		$100,000

TABLE 23.5

Truer Comparison
for the 100-User
LAN with FDDI

Original Cost of Connectivity	Added Costs for FDDI	Cumulative	Total
100-user LAN, hardware, etc., at $2600 per station	0	$260,000	
Cost of wiring for normal LAN at $350 per port/station	0	$ 35,000	$295,000
100-user LAN, hardware, etc., with FDDI to the desk	$200,000	$260,000 plus $200,000	
Cost of wiring for fiber to the desk and fiber backbone, connectors, etc., from Table 23.4	$100,000	$100,000	$560,000
Difference for 100 users			$265,000

cially true when mission-critical applications reside on the LAN, when CAD/computer-assisted manufacturing (CAM) or graphics applications are running, or when heavy database and spreadsheet activity is resident. A bit more on the graphics applications: When you are running a network operating system from any manufacturer and you combine a graphical user interface (such as Windows), you have added to the density of the data through the GUI. To further compound the issue, when users are using Windows on the network and they use a screen saver, they have all just added an immense amount of graphics on the LAN. This is known by many administrators who saw their 4-, 10-, and 16-Mbps LANs grind to a snail's pace, only to find that the GUI and screen savers were the culprits. Meanwhile, back to FDDI: this would help to significantly reduce this problem.

FDDI Configuration

As we mentioned, FDDI is designed to run on a token-passing ring operating at 100 Mbps. The fiber used is a 62.5-μm center core with a 125-μm outer cladding. This constitutes a multimode fiber system. Other multimode fiber sizes can be used, but the specification states 62.5/125 as the transmission medium. The fiber is configured as a single ring (Fig. 23.1). The single ring uses two strands of fiber to provide the resiliency of the ring to heal itself in the event of a cable problem. An optional second ring (Fig. 23.2) can be installed. The second ring will be used as a backup to the first in the event of a major failure. When two

Figure 23.1
FDDI on a single-ring topology.

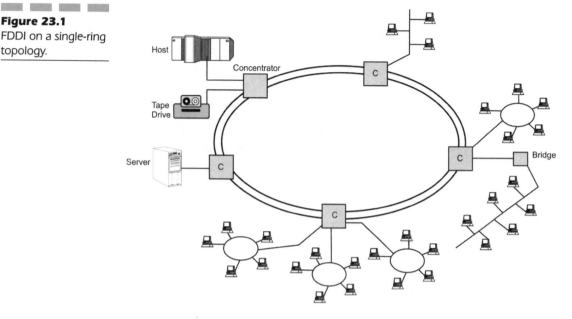

Figure 23.2
FDDI dual ring is operational.

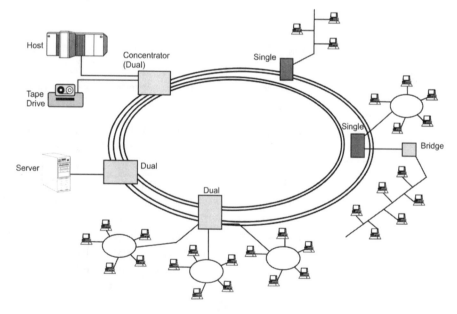

rings are used, they operate in a dual counterrotating manner. This means that they transmit the data in different directions. If dual rings are used, then four strands (or fibers) are run to each station. An alternative to a full dual ring architecture can be used for critical applications and for critical user groups. This alternative is shown in Fig. 23.3,

Figure 23.3
A mix of single ring
and dual ring can be
linked together.

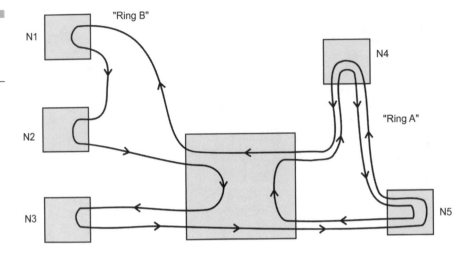

where a single ring and a dual ring are intertwined. The single ring users are noncritical, whereas the dual ring users are the more critical and downtime cannot be tolerated.

Stations on the fiber can be connected differently because of these options. They can be:

- *Singly attached stations.* These devices are noncritical and are attached only to a single ring.

- *Dual attached stations.* These devices have the circuitry and componentry to be attached to dual rings.

A definition is in order here because we have a tendency to intermingle words in this industry. We talk about critical and noncritical users. Then we introduce the concept of a *station.* Other times we talk about *nodes.* Are they the same? The answer is no! There are some subtle and logical differences in defining these terms. So let's try it this way.

A *user* is any workstation, PC, server, or other device used to perform work. A *station* is an addressable device on the network. This is the closest proximity to a user there is. The user is human, the station is the mechanical device. However, being an addressable device, the station can either input data into or extract data from the FDDI.

A *node* is an active device on the network. The node can be a repeater or a device that completes the passing of information around the FDDI network. A node is nonaddressable. All stations are also nodes, but all nodes are not stations.

Isn't that crystal clear? A node on the network can be a concentrator, the equivalent of a MAU or a multiport repeater. When we refer to attaching users either singly or dually to the FDDI, we are referring to the station. However, the physical attachment is likely through a node (a concentrator).

FDDI on the OSI Model

You knew this was coming! Where does all of this fit in our model? When we looked at the LAN architecture, we stated that the LAN typically works at the bottom two layers of the OSI model (physical and data link). Therefore, because we are really dealing with a LAN that is "on steroids" to get it up to a faster 100 Mbps, it should be obvious that the FDDI standards deal with the lower layers of the OSI also. The reference here is shown in Fig. 23.4.

At the physical layer, the FDDI is broken into two separate sublayers, as follows:

- Physical-media-dependent (PMD) sublayer at the bottom. This PMD is responsible for such things as the specifications of sending and receiving the signals; making sure that the proper power levels for the light are applied; and specifying the physical cables and connectors used.

- Physical layer protocol (PHY) is designed to be media independent. This means that this upper portion of layer 1 doesn't care what the cables are. It does define the coding of the information to be sent, the decoding of the signals that arrive, the timing on the network, the status of the lines (wires or fibers), and the framing conventions.

Figure 23.4
The FDDI stacks up against the bottom two layers of the OSI model.

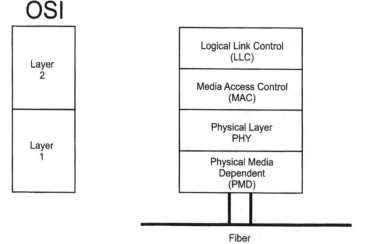

The data link layer is again a sublayered architecture. The two sublayers and their responsibilities are as follows:

■ Media access control (MAC) sublayer protocols define the rules for formatting the frame of information to be sent/received, error checking, handling the token on the FDDI, and managing the data link addressing. Some of this is contained in what is called *station management* (SMT). SMT is responsible for managing the station attached to the FDDI. It provides the rules for node configuration, statistics gathering on errors, recovery in the event of an error, and managing the connections.

■ Logical link control (LLC) is, as always, the common control function of the LAN. It defines the protocols to be used in the upper layers of the model, because most LANs these days use multiple protocols. An example of this is a LAN running Novell Netware 3.X or 4.X. On this LAN, Novell's SPX/IPX protocols are prevalent, but depending on the network, TCP/IP might also be running on the network. LLC manages which of these two stacks is used in delivering the information to the upper-layer devices.

In this chapter, we stated that the PHY layer is responsible for the coding of the information, the timing on the network, and the framing conventions to be used. This is a fairly sophisticated set of responsibilities. When you think about it, this is heavy stuff. The coding is a classic example. FDDI uses symbols in the transmission of data onto the cable. To send the pulses down to the fiber, the PHY is responsible for performing a 4B/5B coding conversion. Four bits (4B) of information are delivered to the PHY. The PHY converts these 4 bits into a 5-bit symbol (5B). This allows for some synchronization (timing) on the FDDI and makes use of the extra bits inserted into the symbol by ensuring that enough pulses (1s) are used to keep the power of the light on the fiber. If there are too many 0s, then the equipment on the fiber will get amnesia and lose its place. If that happens the equipment will have to resynchronize (time) between ends. No data will be transmitted while the systems are trying to retime, so a disruption will occur. Along with this creation of symbols, the FDDI uses electrical characteristics that are called nonreturn to zero inverted (NRZI). The use of 1s and 0s will occur with a transition from the 1s. The provision of the symbols allows the transition to occur on a regular basis. This is all taken care of through the standards and specifications. Figure 23.5 shows the process as the actual data goes through the 4B/5B and the encoding being used in the NRZI format. For the actual data throughput, the 4B/5B encoding allows for a transmission rate of 100 Mbps on a 125-Mbaud link. A binary 1 is equal to a pulse of light; the light is on. A binary 0 is equal to no light; the light is off.

Figure 23.5
The encoding and decoding process at the physical layer.

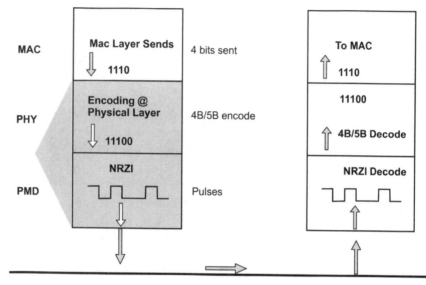

Building on this discussion, the MAC layer is responsible for the construction of the frames at the sending side and the interpretation of the frames at the receiving side of the link. It is also responsible for sending frames to and receiving frames from the LLC, repeating frames similarly to the token ring process. A number of other functions occur at the MAC layer, but a crucial one to be aware of is that it provides access to the ring using a timed token protocol. A station on an FDDI network can send information after it detects an idle token. To do this the station must:

- Capture the token
- Remove the token from the ring
- Send the data (this can be a single frame or multiple frames)
- On completion of transmission, put the token back onto the network so another station can send

After transmitting information onto the network, it is the responsibility of the sending station to perform the cleanup operation by removing its own information from the ring. As the data makes a full circle and comes by the sender, the sending station will remove the data from the ring. If something is corrupted or changed, the MAC layer will not try to solve the problem. The LLC or higher layers are responsible for error recovery, not the MAC layer.

The timing of the token passing on FDDI is different from a typical ring topology. A timed token-passing concept is used. What this means is that as a station is ready to transmit its information, synchronous slots are made available to the station. When the station sends all of its data frames, if it does not have any more reserved synchronous traffic to send (the synchronous traffic is the guaranteed bandwidth necessary for the particular station), it can use the extra time left over to send asynchronous traffic. Asynchronous traffic is referred to as random arrival or a bandwidth of demand capability for excess services, not the start/stop communications process commonly used in data communications. Each station has a small clocking mechanism (a stopwatch) to determine how long it will take the token to go around the ring (rotation) and pass by again. On the basis of this timing sequence, the station can determine how long it can hold the token and transmit its information. Once the station determines the timing, it sends all of its synchronous data on the network. If any spare time is still left over, the station can send its asynchronous traffic.

The FDDI Frame

You knew that we would discuss this piece, so here it is. The FDDI uses a frame size of 4500 bytes, which is different than the size of the other LAN frames we have already discussed. A comparison of the frame sizes based on the LAN type used is shown in Table 23.6. This is for comparative purposes, so that the size of the frames can be seen to work for the FDDI and other LANs.

The actual frame consists of similar size and structures of the existing LANs, but there are some subtle differences. This is a mechanism that transmits the data around the network. The frame will be broken down into several components (Fig. 23.6). The token for FDDI prior to data is as shown in Fig. 23.7. This is not unlike the token used in a token-passing

TABLE 23.6

Comparison of
Frame Sizes Based
on LAN Type Used

LAN	Frame Size
Ethernet 10 Mbits/s	1,518 bytes
Token ring 4 Mbits/s	4,048 bytes
Token ring 16 Mbits/s	16,192 bytes
FDDI 100 Mbits/s	4,500 bytes

Figure 23.6
The FDDI frame.

P	SD	FC	DA	SA	DATA	FCS	ED	FS

	P	SD	FC	DA	SA	DATA	FCS	ED	FS
Bytes	6	1	1	6	6	0≤ 4500	4	½	2+
4B Symb.	12	2	2	4 or 12*	4 or 12*	~	8	1	3+

* Some variation here based on whether 16 or 48 bit address

ring. After all, FDDI is a token-passing concept operating at higher speed. In Table 23.7 the actual frame codes are listed for reference. It would be wise to refer to this table when reviewing the figure so that the abbreviations will be easier to understand.

The frame definitions resemble those of a standard token-passing ring frame, with the exception of the preamble. The preamble is designed to set the timing and synchronization of the frame for each of the devices.

Synchronization

Because there are timing arrangements used on an FDDI, it is critical that all devices keep the appropriate clocking mechanism. The PHY defines the clocking arrangement for the network. Every FDDI station on the network has its own independent clock. The receiver is designed to synchronize its clock to the incoming data and use its clock to decode the data taken off the network. It then retransmits the data with a new timing (clocking) that is generated by the local station. Sounds complex, but it is really quite simple. This keeps the network operating efficiently. The preamble in the beginning of the frame is designed to have pulses to create the clock synchronization. This again keeps everything in good working order. It should be noted that the bulk of the networks using the FDDI

Figure 23.7
The FDDI token.

P	SD	FC	ED

	P	SD	FC	ED
Symbols	≥ 12	2	2	1

	Abbreviation	Description

TABLE 23.7

Summary of Abbreviations Used in the FDDI Frame

Abbreviation	Description
P	*Preamble,* a presynchronization pattern that is made up of at least 6 bytes consisting of all 1s. The 6-byte (48-bit) minimum is broken down into twelve 4-bit patterns to create the 5-bit symbols (4B/5B).
SD	*Starting delimiter,* made up of 1 byte (8 bits) to indicate the start of a frame. When using the 4B/5B symbols, the starting delimiter is divided into two 4-bit patterns and converted into two 5-bit symbols.
FC	*Frame control,* a 1-byte sequence broken down into two 4-bit sequences and encoded into two 5-bit symbols.
DA	*Destination address,* a 6-byte address that becomes twelve 4-bit patterns and is encoded into twelve 5-bit symbols.
SA	*Source address,* a 6-byte address that is broken down into twelve 4-bit patterns and then encoded into twelve 5-bit symbols.
DATA	The data that is being carried; amount varies from 0 up to 4500 bytes. This is be broken down into the 4-bit patterns and encoded into the 5-bit symbols.
FCS	*Frame check sequence,* a CRC-32 computed from the FC through the FCS. This is a 4-byte sequence (32 bits) or 8 symbols.
ED	*Ending delimiter,* the signal to indicate the end of the frame. This is a 1-byte pattern that goes through the same 4B/5B sequence.
FS	*Frame status,* at least 12 bits or three 4-bit patterns consisting of 3 pieces of information. The first is the *E* bit that will be used to denote whether an error occurred in the frame. Next the *A* bit signifies whether the address is recognized or not. Third is the *C* bit to indicate whether the information was copied. The EAC will be repeated at least twice as a check for errors since there is no CRD performed on the frame status.

systems operate with almost no errors and that the performance of the components is exceptional. This is a highly reliable networking service.

FDDI Applications

The primary application of FDDI is in the backbone for many LANs. This high-speed backbone is used to connect a group of lower-speed LANs, such as token rings and Ethernets. Although there are other ways of using this FDDI network, users saw it as a means of extending the LAN into a campus area network (CAN). Using a high-rise as one example, the backbone service is achieved. Figure 23.8 shows a group of individual LANs, operating with various speeds and topologies, connected to the FDDI backbone. Here is where the FDDI standards shine. FDDI is topology and protocol independent when connecting these other services. The high-rise idea works nicely in many older buildings where the cables might be in elevator shafts, running close to electrical closets, or in proximity to other

Figure 23.8
Various speeds on topologies can connect to the FDDI ring.

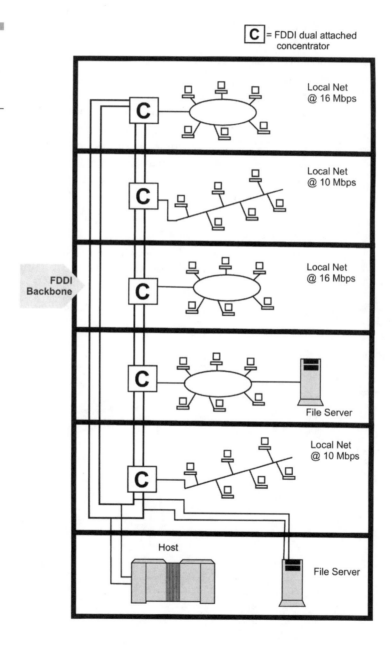

$\boxed{C}$ = FDDI dual attached concentrator

forms of interference (electrical, mechanical, or radio) that would corrupt the data on a copper-based backbone. Further, where riser space is at a premium, fiber is less bulky and offers a very reliable connection between floors without taking up much cable or wall space. Moving out from the backbone in the high-rise, the FDDI can also be used in the CAN. In this particular case, the fiber is immune from electrical hazards such as lightning spikes and the like. This means that the typical protection normally used on copper cabling between buildings and the grounding and bonding requirements required by local fire and electric codes are nonissues. FDDI stands to provide the higher throughput between these buildings on a campus because the extended distances allow greater flexibility. Working on the 62/125-µm fiber and allowing for up to 2 km between repeaters extends the distances over those of conventional LANs.

Figure 23.9 is a representation of the campus connection. The cables require far less conduit space, so the positive is that added ducts will likely not be required. Another positive is that if the outer conduit runs are stuffed with existing copper wiring, the fiber can be laid in electrical conduits, which might be far less packed. If need be, fiber can be run through the same right-of-way with water pipes, something we could never do with copper. The water pipes can act as grounds and attract electrical spikes. Too many applications have been hindered because of the

Figure 23.9

A campus environment with FDDI as the host connector.

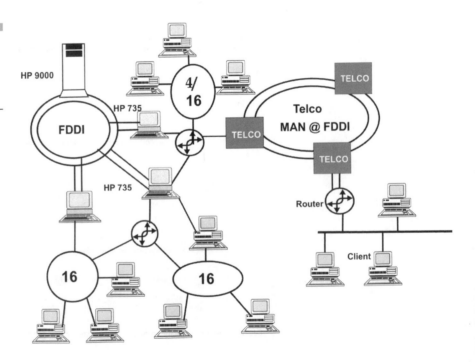

lack of bandwidth and connectivity in the past. FDDI offers the opportunity to overcome these limitations.

Figure 23.10 shows another variation of the application for FDDI. This case uses the FDDI as a collapsed backbone arrangement, where multiple closets in a high-rise or a campus can be linked with the fiber and an FDDI interconnection can support the higher-speed throughputs. The

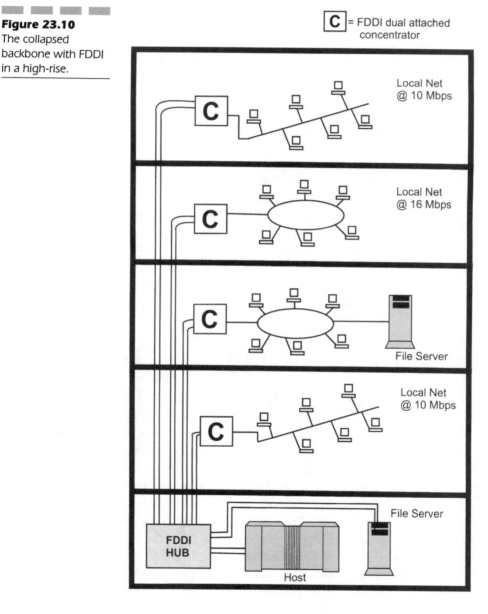

Figure 23.10
The collapsed backbone with FDDI in a high-rise.

ability to bridge and route traffic across the FDDI is another positive. Because FDDI is protocol independent, one or two different brands of equipment can be attached to the fiber ring. Types of bridges and routers are summarized in the following.

Encapsulating Bridges

The data from the bridge (frame) is inserted directly into an FDDI frame and remains transparent to the network. As long as the addressing mechanisms are addressable, there is little concern about the use of an encapsulation technique. Although the FDDI nodes will not be concerned with the frame format, the stations attached to individual LANs will be. If encapsulation on FDDI is performed, the frame must be delivered to a station that can read the frame format.

Translating Bridges

The data from a bridge (a frame) is stripped of the overhead that is associated with the original network topology and reformatted with new overhead to make the frame 100 percent FDDI compatible. This allows various equipment and topologies to coexist on the network harmoniously. Translation is more expensive than encapsulation, but the differences are becoming minuscule.

FDDI in the MAN

Recent developments in the local exchange carriers' areas have indicated that FDDI has the robustness and reliability to be delivered in the metropolitan area. The LECs are quickly moving ahead with plans to offer 100-Mbps throughput across their fiber rings in the major metropolitan areas. A particular scenario includes what is going on in the nation's capitol. Bell is providing an FDDI ring throughout Capitol Hill to interconnect LANs in various government agencies and buildings across the town. This form of a MAN is shown in Fig. 23.11 with the MAN in place. The LECs have been using some conservative approaches in their deployment so that they do not overstate their offerings or create a bottleneck. Table 23.8 summarizes the differences of the FDDI standard and the way many of the LECs will offer this service.

Figure 23.11
FDDI in the MAN.

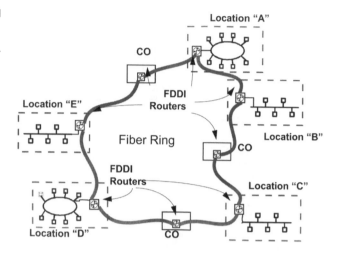

FDDI Recovery

Clearly, the LECs and users alike will be careful to overcome risk where possible. Yet when placing an FDDI into service, the risk is exponential; all of the traffic on the network becomes supercritical. The FDDI provisions call for the ability to select how the network will be used and recovered. FDDI allows for a bypass option so that a failed component on the network will be bypassed, preventing down time on the network. Figure 23.12 shows the ability to perform the bypass of a failed component. Using the aforementioned dual counterrotating ring topology allows an optical bypass capability in the node electronics. If a station is not powered on, as was the case in the token ring problem lists, then the optical bypass allows the network to continue to operate. This is one solid way of guaranteeing less problems on such a network.

TABLE 23.8

Comparison of FDDI Standards and LEC Implementations

LEC	Standards
Multimode 62/125-µm cabling, optional single-mode fiber 9/125 µm	Multimode fiber 62/125 µm Specs are completed for single-mode fiber
500 stations	1000 stations
100 Mbits/s using single or dual ring	100 Mbits/s with a single ring, optioned at 200 Mbits/s with dual ring
100 km maximum distance either ring	200 km maximum distance with a single ring, 100 km with a dual ring
Repeaters spaced up to 2 km apart	Repeaters spaced up to 2 km apart

Figure 23.12
A dual counter-
rotating ring in
healing wrap
bypasses problems.

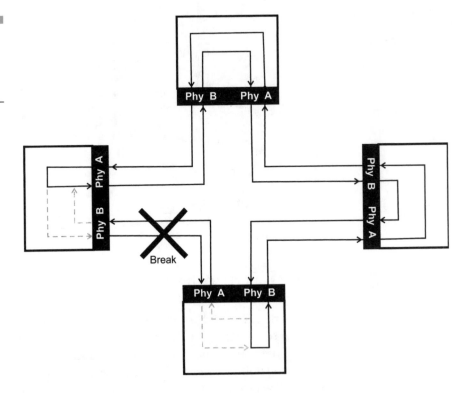

Also, stations attached to the ring can be single attached stations or dual attached stations. The single attached stations are not perceived as mission critical. If a problem occurs with a cable break, the ring can close itself off and go around a problem. However, if the ring has multiple complications or failures, disaster will strike. With deeper pockets in the FDDI world, a user can take advantage of dual attached stations that are connected to both the primary and secondary FDDI cable.

FDDI-II

Where FDDI is designed for the bursty nature of our data needs, there was no other alternative for other packet forms of transfers. Voice and video were considered to have a constant bit rate, whereas data transmission is considered to be a variable-bit-rate service. Therefore you could not attempt to put these other telecommunications services on the FDDI as originally specified. Newer specifications have been written to accommodate the voice and video needs of organizations. This is designed to give native throughput on a LAN-to-LAN connection, as well as pack-etized isochronous services such as video and voice.

Clearly, FDDI-II was destined for greatness. Or so we thought! The evo-lution of FDDI-II has been slow in coming, partly because of the relative slowness of the standards bodies. However, the issue now is: will it ever come about? With the evolution of switched multimegabit data services (SMDS) and asynchronous transfer mode (ATM), there are other alterna-tives to the FDDI-II standard. Many customers are looking at ATM as a better way to handle the integration of voice, video, data, and LAN traffic. Where FDDI only gives the user a total of 100 Mbps, ATM starts at 50 and 155 Mbps and aggregates the bandwidth up into the 2.4+-Gbits/s rates of today. When you need added bandwidth, ATM allows you to add another card that increases cumulative bandwidth up to the new rate added. FDDI will remain a shared resource for the future, so the cumulative bandwidth without any form of subnetworking will remain at 100 Mbps. You need only wait a few short years to see whether these two transport systems are complementary or competing service offerings.

Fast Ethernet

Another risk in FDDI acceptance is the migration to the 100-Mbps Ether-net standards. In this particular case, these two separate approaches rival the FDDI standard. The 100BASE-T standard will ramp up the Ethernet over twisted-pair wiring in place (at level 5 wire) to a shared resource at 100 Mbps. This will also allow for the integration of the existing 10-Mbps Ethernet standard frame format into the new fast Ethernet standard.

Another approach is the 100VGAnyLAN. It is a different concept, but it will offer 100 Mbps of throughput regardless of the topology and access method used. This technique will use the ring or bus capacities and allow high-speed throughput at up to 100 Mbps. Sponsored and endorsed by Hewlett-Packard, the 100VGAnyLAN will be an inexpensive alternative to FDDI.

Part of the acceptance will depend on:

■ The price of the interface cards that will work in a standard AT bus PC. Currently the price of the 100VGAnyLAN cards is rapidly drop-ping. This might be a sign of acceptance and production cost decreases, or it might be an attempt by HP to drive the demand on the basis of the cost of the cards.

■ The fact that the cabling in place is already suitable on a twisted-pair environment and adapted to the higher-speed LAN capability.

■ The ability to support any bus or a ring topology without a major change. Too many organizations want to provide some stability and avoid mass migration to newer, high-priced technologies.

The jury is still out on this acceptance and rollout. However, as we mentioned in the beginning of this chapter, FDDI is still expensive. The cards for the individual workstation have not dropped to the reasonable level yet. The least expensive card seen on the market was priced at $1295, whereas the fast Ethernet and VGAnyLAN cards are competing in the $695 range. The concept of ATM to the desktop is now appearing, with cards priced at the $895 range, but recent developments offer the price of an ATM card in a PC at $400. This will put significant pressure on the FDDI manufacturers to come up with lower-cost chip sets if they plan to be competitive and stay in the market.

Ethernet Switching

Just when it seemed safe to think about the options for faster LAN service with a myriad of capacities, such as FDDI, FDDI-II, ATM, fast Ethernet, and the 100VGAnyLAN technologies and transport systems, a new twist is occurring. Vendors are now introducing Ethernet switching systems that use a high-end hub in the office environment that will have an aggregated bandwidth of multiple Gbits/s in the backplane. Under this arrangement, users will have multiple subnets that are all connected to the high-end hub. The high-end hub will support multiple ports (8-12-16), each operating at 10-Mbps speeds today. Future boxes like this will more than likely double the number of inputs. Using a cross-connect switching matrix inside this hub, users can switch from port to port inside the matrix. With the 16-port model, an organization can have eight simultaneous LAN-to-LAN connections, each operating at a true 10 Mbps across the connection. In effect, this gives the entire 10 Mbps to each network (departmental) of a smaller group of users without the risk of congestion that goes with a bridge or router.

This 10 Mbps of untouched bandwidth is a switched resource, and so the constraints of meshing the networks together go away. Further, congestion management states that if a problem occurs, a new port can be added and a new subnetwork can be incorporated into the LAN. This is unproved at the time of this writing because it is still in the definition and development stages. However, the opportunity to let user organizations stick with the installed base of Ethernet cards at 10 Mbps, which are also reasonably priced (under $20), and to provide the aggregated bandwidth in a collapsed backbone hub is very attractive to some. No one really knows just how much bandwidth is required at the desktop. But the industry is pointing to this massive traffic jam looming. How the individual LAN administrator handles this problem varies exponentially with the price, technology, and comfort levels applicable to any specific vendor product.

Again, this is all in a nether zone and the winners and losers are not yet defined. It is possible that any one or all of these services will be around for years to come. Still, it is possible that only one or two will survive.

24

Switched Multimegabit Data Services

The following information deals with a service rather than a technology, but we will probably bounce between the two fields in such a definition. The use of high-speed data transfer has been around in various forms, and the last few chapters discussed ways of moving data across a network faster and more efficiently. These are goals in the industry that never get totally satisfied. As soon as an enhancement is developed, the industry says, "That was good, but I want it faster, better, cheaper—now!" We have mentioned catch-22 environments throughout this book, and this is one of them. We ask for services to be faster, and when this is delivered, we want them to be faster yet! Switched multimegabit data services (SMDS) is a similar situation, in that it is a service that delivers the data throughput at a very high rate of speed by today's standards. Yet everyone has looked at it and said, "too little."

So, back to the drawing board. One quick word of caution for the reader at this point: SMDS and some of the other high-speed communications services and technologies we are discussing go beyond the scope of a voice/data primer. We need to discuss these because the techniques are already here and many of the industry gurus toss the acronyms around. Therefore it is important for a user to have a generic overview of the SMDS concept. Unfortunately, this can get quite detailed and technical. There is not a lot we can do about it, so bear with it and someday it will all make sense.

What Is SMDS?

As already mentioned, SMDS is a service provided on a high-speed communications channel to support what is called a metropolitan area network (MAN). A MAN is defined under the ANSI and IEEE specifications for an 802.6 network arrangement. In many cases the user needs to provide connections between LANs in a major metropolitan area. This had been done in the past with leased lines, such as 56-Kbits/s and T1 lines. The use of a leased line was limited in a couple of ways.

First, the leased line might not have enough speed to connect various LAN-to-LAN services. If a user is running a 4-, 10-, or 16-Mbits/s LAN, the T1 will limit the throughput to 1.536 Mbits/s, leaving the possibility of severe congestion at the network interface. This has been the single largest concern in the industry. The 56/64-Kbits/s channels were even worse as far as throughput. A 4-, 10-, or 16-Mbits/s LAN trying to send data frames from LAN to LAN would certainly become congested and produce significant delays in the transfer of the data.

Second, the costs of the leased lines are not insignificant. If the network shown in Fig. 24.1 is connected with a full-meshed arrangement, and (as in this case) four locations require a total of 12 leased access lines, the cost of a T1 at a representative cost of $700 per link per month would be $8400 monthly, or slightly over $100,000 per year. Other ways of providing this connection could be used, but in each case, if the network is configured with fewer links, the congestion would become exponentially more severe. Using a 56- or 64-Kbits/s link to each of the locations does not achieve the desired end result, so we will not go into the math on this type of linkage.

What if congestion is really bad? Here the user was faced with a dilemma. T1 was not all that attractive pricewise, but the only other choice to get more bandwidth between the sites was to step into a T3. This would be an astronomical fee to pay to provide the connectivity. In many parts of the country, the cost of a T3 access to a central office is more than $7000 per month. The cost now to connect these four locations in a meshed network (Fig. 24.2) would approximate $84,000 per month, or $1,000,000 annually. Few organizations can afford this form of connectivity.

Adding to the problems of too much or too little bandwidth, the excess capacity of a T3 would be wasted in most cases because this is a localized

Figure 24.1
A meshed T1 network requires too many access lines, which creates a very expensive network.

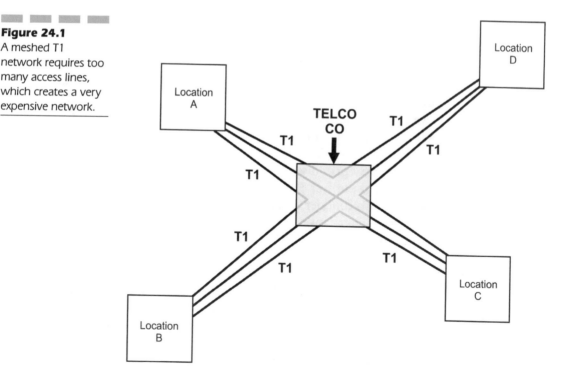

Figure 24.2
T3 installations to
get native LAN
throughput are an
order of magnitude
more expensive.

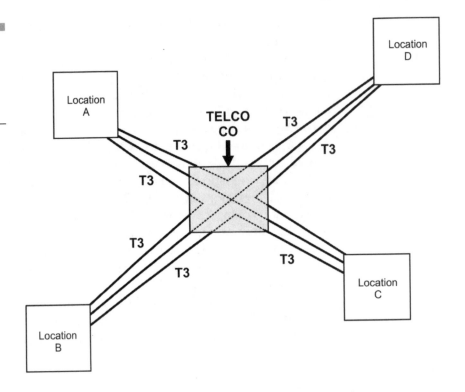

Figure 24.2
T3 installations to get native LAN throughput are an order of magnitude more expensive.

form of communications. Although you could multiplex voice and other data transfers onto this network (at the T3 level), the cost of the multiplexers would likely be exorbitant. Additionally, the cost to place a call or to transfer data to these local sites would be either free or very inexpensive. It would not be prudent to try to force this traffic onto the T3 when a less expensive technique already exists.

These results left users and LAN administrators in a bind. There was no reasonably priced service that would provide native LAN connectivity and throughput, yet the demands of the networks and resources were escalating. To conquer this problem, several of the standards bodies began to look at alternatives. This includes ANSI, Bellcore, and IEEE in the United States.

SMDS is another form of high-speed public connectionless packet switching. As such, the SMDS services will handle the transfer of data across a MAN at rates of 4, 10, 16, or 34 Mbits/s. Users can commit to a burst rate that guarantees them the throughput they require. Since SMDS is a service and not a technology, the data delivery is at whatever rate the subscriber signs on for. This service is designed to assist users in building network connections that require high-speed linkages without the added overhead and costs associated with fixed capacities. It is, therefore, a shared resource between and among various organizations in a metropolitan area. The connectivity can be intracompany or intercompany. As a LAN-to-LAN service, SMDS uses a connectionless orientation. The users do not have to dial up or be connected directly to each other. Similarly to the LAN, the frames or packets are sent out across the network with a destination address. All units on the network read the frame, but only those devices that are addressed will copy the data.

The Local Environment

The LECs are noticeably absent from the discussions of the higher-speed services in this book. Most of the higher-speed connections have been through the links provided by the IECs. Even in the case of leased lines, where the IEC must use LEC facilities (Fig. 24.3), the IEC is still in control of the service. Most of our networks are wide-area-network (WAN)-based. Therefore, as a result of the divestiture in 1984, the LECs have been restrained from participating in this arena. Where the LEC did provide the local loop to the customer premises for the IEC, the LEC prices were considered too expensive. The tariffs did not allow for the negotiation of

Figure 24.3
Even though the WAN physically passes through the LEC office, the connection is logically transparent to the user/IEC.

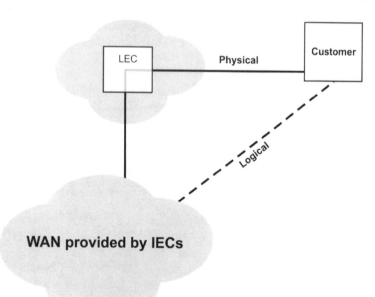

better prices because LEC services are supposed to be universally applied to any customer. The IECs in turn began to suggest ways to bypass the LEC connection, such as through the CAPs and CATV companies. These competitors are covered in earlier chapters. Consequently, the local environment for MAN services was limited and options were not readily available.

Bellcore developed the specifications of SMDS in the local arena for the Regional Bell Operating Companies (RBOCs) that fund the Bellcore research and development efforts. Just about all of the RBOCs have endorsed the use of SMDS (Table 24.1).

Using SMDS as a shared service among many users, the LECs now have an alternative to the competitive arrangements. Furthermore, as SMDS is being developed, the LECs are offering these services as value-added services rather than having the costs rolled into the embedded base pricing structure. This means that the services are not tariffed the same, and that competitive pricing and special deals can be had. SMDS is designed for applications that are bursty in nature, requiring a burst of bandwidth for short periods of time. These applications do not need committed capacities and bandwidth all the time, but when they need it, they grab it and use it. Unfortunately, SMDS is designed for the data traffic carried on LANs, not for the higher-density, quality graphics and interactive video applications that are emerging. These applications will be addressed in the future. SMDS does support interactive sessions between users, so all is not lost under the restrictions and constraints of an SMDS network. Thus the interactive applications that can be supported in a localized environment include:

- Computer-assisted design (CAD)
- Computer-assisted manufacturing (CAM)

TABLE 24.1

Status of RBOCs Supporting SMDS

RBOC	Status
Ameritech	Adopted and available
Bell Atlantic	Adopted and available
Bell South	Adopted and available
NYNEX	Not adopted, no plans
PacTel	Adopted and available
SouthWestern Bell	Adopted, available in future
US West	Adopted and available

- Source code transfer from host to host
- LAN-to-LAN native transfers
- Imaging services (magnetic resonance imaging, document imaging, etc.)
- Publishing services (compound documents)

Technology Used in SMDS

We have stated that SMDS is a service and not a technology, so we should also say that SMDS uses a technology called distributed queue on a dual bus (DQDB) to handle the transfer of the information. As a technology, DQDB is straightforward; it distributes the ability of users on the network into queues, and it uses dual bus architectures that are unidirectional (one-way) each. In LANs, most of what we discussed is a unidirectional data transfer. This architecture uses a controlled sequencing on a fiber-optic cable. However, the primary implementations have been single-direction transfers (one cable transmits one way, while the second cable transmits the opposite direction). Using a distributed queue, a group of time slots is allocated on the network. Each device that wishes to transmit data across the cable must wait its turn for a queue slot. When the queue slot becomes available, the transmitter can send data.

The SMDS Goal

The goal of SMDS, then, is to provide the high-speed data transfer for customer systems and yet preserve the customer equipment investments. Too many systems and services in the past have required the use of *forklift technology*. When you needed more, you brought in the forklift and removed the old equipment. Then the lift brought in and plunked down the new equipment. This did not meet the user needs, nor did it encourage the migration to the higher-speed communications services until the equipment had been fully depreciated under IRS and financial rules. A second goal of SMDS is to enter the workplace as quickly as possible. If there is no incentive, then everyone will take a wait-and-see attitude. If users are not encouraged to implement SMDS, they will seek other avenues of providing the services—meaning competitors. The LECs don't want this to happen; they want in as fast as possible. If they are the first providers and meet the needs quickly, users will be less likely to go out and seek other alternatives. Further, if the SMDS proliferation can be rolled out quickly, users can switch to higher rates of speed in the future as they are developed. In the future SMDS will carry up to 150 Mbits/s in the MAN.

Access Rates to SMDS

As mentioned, this is a primary service offering from the LECs. As such, the access rates delivered today deal with the transport services of the past. This means that there are two basic platforms for accessing an SMDS network: T1 (1.544 Mbits/s) or T3 (44.736 Mbits/s). Unfortunately there is little in between. However, the LECs have been offering this access at very attractive rates. Many of the LECs around the United States, and the PTTs around the world (18 countries have already adopted the SMDS standards), are rolling out the service and the products. In the European marketplace, the PTTs are delivering speeds of 2 Mbits/s (called E1), 34 Mbits/s (E3), and 140 Mbits/s (E4), or 155 and 622 Mbits/s on what is termed synchronous digital hierarchy (SDH). These higher speeds are still in beta, but will be the thrust of the future.

The LECs that have been aggressively pushing this service have come up with some very attractive rates to encourage its use. In the East, the LECs offer unlimited usage of the network with T1 access for around $500 per month/per site. This eliminates the need for the meshed network and allows a single connection into the network for on-demand bandwidth up to 1.5 Mbits/s (Fig. 24.4).

Accessing the MAN

Next, look at the access method shown in Fig. 24.5. The LEC will install a twisted-pair (four-wire circuit) T1 access for the lower-end user, or a fiber loop for T3 access. Variations are shown in which the LEC can run a fiber to the door and will deliver any rate on that fiber. Still another variation is the use of a fiber ring into the customer premises where a T1 or T3 access can be provided. If fiber is not available to the door, the LECs might bring the fiber to the curb and deliver a metallic link (coax) into the customer location at T3 speeds. These are all variations of the same service

Figure 24.4

The revised network allows access up to T1 speeds on demand. This is much less expensive, but it can be very robust.

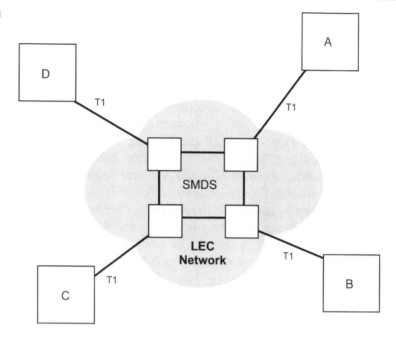

Figure 24.5

Various options exist for the LEC to deliver access into the SMDS network.

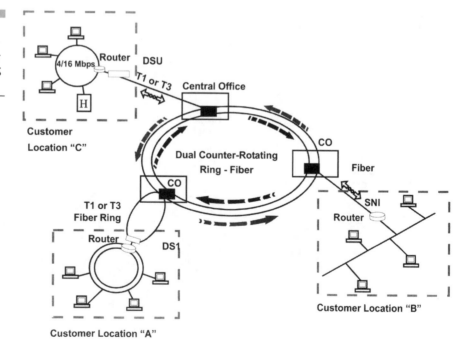

provisioning dependent on what the LEC has available to use. In the future SMDS might well wind up on microwave radio if that alternative is already in place. However, the primary method will be deployment across a fiber-based network.

Once the local loop is installed, the customer will use a high-speed router at T1 or T3 speeds. An interface called the SMDS network interface (SNI) will be used as the connection at the customer location. Just a quick note here—many of the LECs have been delivering SMDS service to the customer premises and providing the router. It seems that they have no means of controlling the data that is sent across the fiber and that if the customer provides the router, there is no means of stopping the customer from adding cards in the router, thereby grabbing additional bandwidth for free. Because this is a value-added service and not covered under the normal tariffs, this can be accomplished. The demarcation point on the fiber has now moved beyond the entrance of the customer's building into the equipment. The DEMARC will now be at the high-speed interface on the router card. With the SNI, the customer now attaches to the router (Fig. 24.6). The link might be on a single thread or a ring using a dual thread; it doesn't matter. The customer equipment, such as a LAN or a host computer, is then connected to this network. Access to the network is achieved using a DQDB technique, as shown. The access using DQDB through the SNI then proceeds out to a switching system in the network. Protocols are written to provide the interswitching systems services to make this all transparent to the end user. MCI has announced plans to provide connections on a long-distance basis with Bell Atlantic and

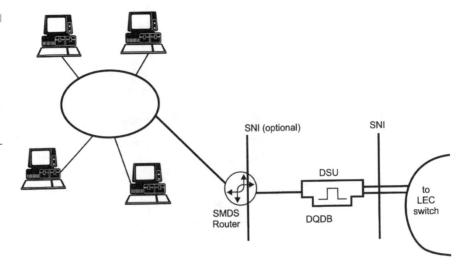

Figure 24.6
Customer connection at the DEMARC may change depending on who provides the router. The SNI is the interface between customer and LEC.

Pacific Bell, creating a true cross-country SMDS network. This is shown in Fig. 24.7 with the intercarrier interfaces in place. This is a pilot service at present, but we can expect to see more in the future. In the event the LECs choose to connect to another service, called ATM (covered in Chap. 27), the overlay of SMDS will be directly onto an ATM network. Once again, this will be transparent to the end user.

The IEEE 802.6 Architecture

As with this entire book, we attempt to cover all the bases from a brief overview of the techniques, systems, and services to more detailed discussion. It is fitting, therefore, to explain just how the SMDS and DQDB work together in an architecture of protocols and services and how they stack up against the IEEE 802.6 standards and protocols. The architecture is shown in Fig. 24.8 with the dual bus at the physical layer, and DQDB at the link layer, as they compare to the OSI model. Note that sitting on top of the DQDB (data link layer) are four different services. The first is a

Figure 24.7
IEC-to-LEC arrangements are now falling into place, taking SMDS across the WAN. The intercarrier interfaces have been defined.

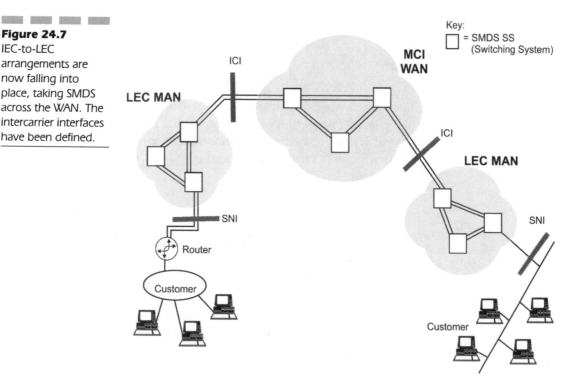

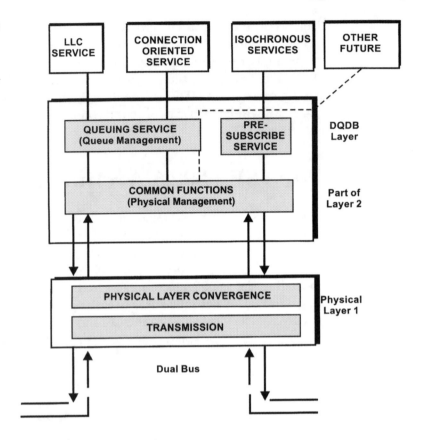

medium access control to logical link control service, still working at the data link layer. The second of these services is a connection-oriented service. Third is a provision for isochronous service (such as voice, video, etc.). The fourth service is termed *other,* and is under study for future use. The MAC layer contains several services.

The DQDB Architecture

Logically, the DQDB is structured as a dual bus, although in many cases it might also be in the form of a ring. In Fig. 24.9 the dual bus architecture is laid out. There are two separate unidirectional buses. Each transmits in one direction, but combined, they allow for a full duplex operation. The buses are labeled *A* and *B* to provide the necessary full duplex operation between any two nodes on the network. An LEC lays out the cables in one

Figure 24.9
The dual bus
topology as defined.

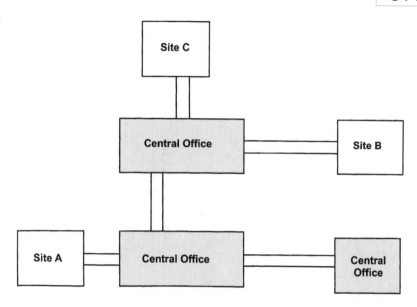

of two ways. The cables can be laid out with a head end in two different nodes (Fig. 24.10). In this case, the head end of each of the buses appears at different locations, signifying the direction of the bus. In Fig. 24.11, the two head ends of the dual buses are located in the same end point, establishing a looped arrangement. This resembles a ring, but in reality, it does not create the ring because both ends of the fiber are open ended, meaning they are not closed off into a physical ring. Using either physical installation, the provision exists to recover the network in the event of an equipment failure or a problem on the fiber, such as "backhoe fade" where the cable gets cut. A break in the cable on the open bus architecture is healed as shown in Fig. 24.12. The network will automatically reconfigure itself as two separate networks. The system in Fig. 24.13 uses a looped architecture; if a failure occurs, the network will bypass the failed component or cable and reconfigure it into what constitutes an open bus. These robust arrangements will ensure more availability and more user satisfaction.

Figure 24.10
The open bus is one
way to lay out the
MAN.

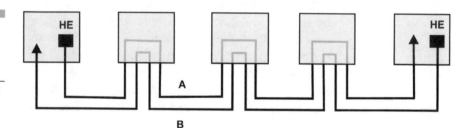

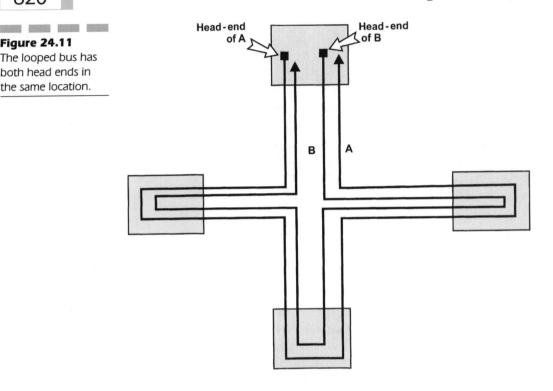

Figure 24.11
The looped bus has
both head ends in
the same location.

The MAN Access Unit

Similar to a multistation access unit (MAU) on a LAN, each node on a MAN has an access unit called a MAN access unit. At the risk of confusing the user, we will refer to this as an access unit instead of a MAU. Each access unit is attached to both buses. The access unit can support a single node or act as a concentrator handling multiple inputs from a cluster. The access unit is used to read and write into the DQDB slots at the rate of 8000 samples per second. The buses have a slotted queue arrangement. As the access unit reads or writes into these slots, it can deliver information, read information, or request time on the network by reservation. The 8000 samples per second means that the network is timed to all of the synchronous time-division multiplexing systems in use for voice and data transmission, as well as the synchronous optical network (SONET) standards. Figure 24.14 is a graphic representation of the functionality of the MAN access unit. The nodes read and copy data from the slots created in the queue. They gain access by writing to the slots.

Figure 24.12
In an open bus, the result of a cable break is subnetworks formed into two networks.

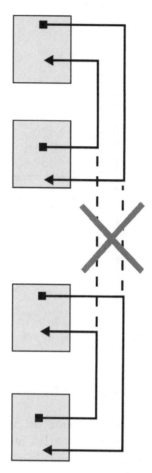

The DQDB protocol is straightforward, based on a counter system. The counter system is used on each bus to determine whether a slot is available to be reserved or written into. What happens is that the DQDB node reserves slots on one bus in order to use the slots on the second bus. Every time a slot comes by with a reservation tagged to it, the node increments a counter. It also watches the other bus and decrements its counter with every slot that goes by empty. Essentially what happens is that the node reserves slots on one bus to be used on the other. It sends a reservation request so that all of its upstream and downstream neighbors know that it wishes to send information. Then it begins to count the number of empty and reserved slots as they go by. When the counter equals zero, the node knows that it can now send its data. This is a form of carrier sense multiple access with collision avoidance (CSMA/CA) that prevents a problem from occurring when nodes try to send data at the

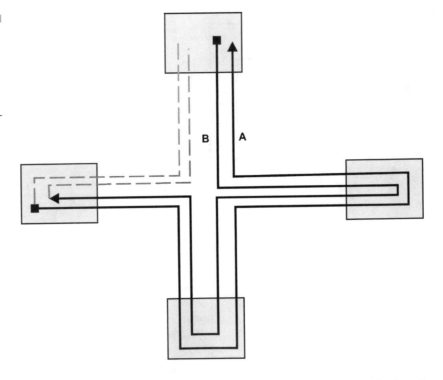

same time. Figure 24.15 shows a sample of what happens on the bus from the access unit.

Because this is called distributed queue on a dual bus, it should stand to reason that each node maintains a list of queues for each bus. Although these two queues operate identically, they are kept separate

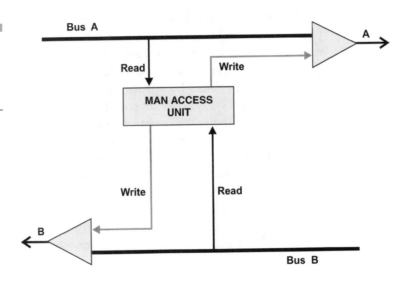

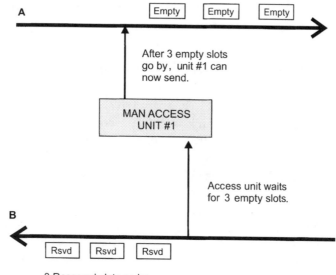

Figure 24.15
When unit 1 wants to send, it counts the reserved slots on bus B, and then it decrements the counter on bus A until a slot is available.

from each other. This creates a deterministic approach to the dual bus very similar to a token-passing arrangement. To prevent any one node from saturating the network, provisions are in place to periodically send empty queue slots onto the network so that this deterministic MAN can be achieved. No one node is allowed to reserve all the slots on the network; slots are negotiated on the basis of quality of service (QOS), which equates somewhat to a throughput rate. You buy the throughput you need, and you pay for the QOS that you want.

The Data Handling

A fair way to get into data transfer on a MAN is to understand the formulation of the data units. The 802.6 can accept data up to 9188 bytes (octets) long. Then a three-step process is used to break the data down into more manageable portions. The end result is that the data will be transferred across the network in the slots (queues) at 53 bytes of information at a time. We will take the description of this process slowly to try to keep it going smoothly. It is analogous to ATM, which is covered later, but it is not the same. To prepare the data for transfer, the equipment is used to define the protocol data units of up to 9188 bytes. Several convergence layers (processes) act on the data to make it more manageable. These convergence layers are defined by the layer they are handled in.

Think of this as slicing and dicing the data into smaller chunks, similar to the X.25 transport.

Segmentation and Reassembly

Within the DQDB layer, the media access control convergence produces a segmentation and reassembly process (SAR). This process takes the very large frame of information and breaks it down into smaller pieces that can be put back together at the receiving end. If we start at the top (Fig. 24.16), a large frame (MAC layer) of information is delivered from the logical link control unit. In this case, it might be a bridge, router, or LAN interface. At this stage the header and trailer information is added to the MAC layer information. This is called an initial MAC protocol data unit (IMPDU). The IMPDU is then chopped up into fixed-length segments called derived MAC protocol data units (DMPDUs), and header and trailer information is attached to it. The numbering of these DMPDUs is handled by the relative position in the slicing and dicing process. The DMPDU can be a beginning, a continuing, or an ending protocol data unit. The numbering is just that—beginning, continuing, or ending unit. As you can see from Fig. 24.16, the data units prepared for the slots in the DQDB format will be 53-byte segments. This will include 44 bytes of payload (the data) plus 7 bytes of header and 2 bytes of trailer information. This fixed-length segment is then sent across the DQDB in a time slot.

Figure 24.16
As the MAC layer frame (9188 bytes) is delivered, a header and trailer are added to create the IMPDU. Then it is sliced and diced into segments, and the header info is added, creating a slot (53 bytes).

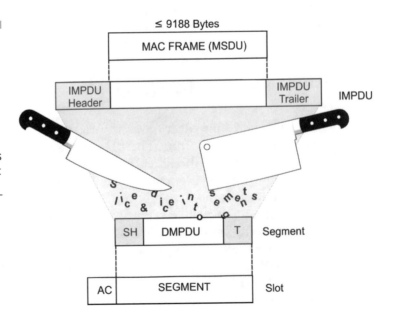

The SMDS Data Unit

The encapsulation of an SMDS data unit is very similar in the way the data is transferred across the network. First, up to 9188 bytes of data is delivered to the interface equipment. Then the process begins of slicing and dicing the data into smaller pieces that will wind up in the 53-byte slots using an SMDS interface protocol (SIP). Figure 24.17 shows the process from the very beginning. The 9188-byte data unit, plus the associated overhead, is added to the data unit. This is a layer 3 SIP-PDU. Table 24.2 describes the abbreviations of the information contained in the layer 3 SIP-PDU. You might want to keep this available as you review the diagram.

The layer 3 SIP-PDU then gets broken down into smaller pieces by going through another process that breaks the data into layer 2 SIP-PDUs. The layer 2 SIP-PDUs will be broken down into 44 bytes of data and an additional 7 bytes of header and 2 bytes of trailer information. This is shown in Fig. 24.18. Table 24.3 is a summary of the abbreviated definitions for this figure. Again, you might wish to keep this available as you review the framed format in the graphic.

This slot or segment is then delivered to the physical medium for transport across the network (layer 1 SIP-PDU). The physical layer is divided into two pieces or sublayers called the *physical layer convergence process* and the *transmission system*.

SMDS in a LAN Environment

Enough of the bits and bytes for now. When using a LAN-to-LAN in a metropolitan area, the concern is how to transmit the information transparently. Most LANs use some protocols that provide for the reliable delivery of the actual data. In many LANs today, the protocol of choice is TCP/IP. If we stack this up on an architecture, we need to see where the pieces all fit together. Figure 24.19 shows a LAN architecture running TCP/IP. This stack uses the TCP portion of the stack to guarantee reliable data transfer at layer 4 of the OSI model. The IP is a packet-switching protocol working at layer 3 of the OSI model. IP breaks the data down into smaller chunks for delivery across a network, then hands the packets down to layer 2 of the OSI, which is the data link layer. Here is where we can enter the SMDS world by using the equipment to transfer the data across the network. At the data link layer, the stack deals with the three SMDS interface protocols (SIPs). The data will be segmented into slots, or 53-byte segments, and will be prepared to be sent across the network. At the center of all this is the stack. This is where the convergence takes place, using the

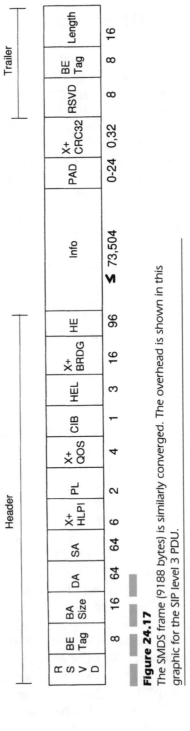

				Header						Info		Trailer		

R S V D	BE Tag	BA Size	DA	SA	X+ HLPI	PL	X+ QOS	CIB	HEL	X+ BRDG	HE	Info	PAD	X+ CRC32	RSVD	BE Tag	Length
8	8	16	64	64	6	2	4	1	3	16	96	≤ 73,504	0-24	0,32	8	8	16

Figure 24.17

The SMDS frame (9188 bytes) is similarly converged. The overhead is shown in this graphic for the SIP level 3 PDU.

TABLE 24.2

Summary of Fields
in the Layer 3
SIP-PDU

Abbreviation	Length	Description
RSVD	8 bits	Reserved
BEtag	8 bits	Beginning and ending tag
BAsize	16 bits	Buffer allocation size (length)
DA	64 bits	Destination address (to)
SA	64 bits	Source address (from)
X+		Ignored by the network
X+HLPI	6 bits	Higher-level protocol ID
PL	2 bits	PAD length (if used)
X+QOS	4 bits	Quality of service
CIB	1 bit	CRC presence indicator bit (indicates the presence or absence of CRC)
HEL	3 bits	Header extension length (describes the header extension)
X+BRDG	16 bits	Bridging
HE	96 bits	Header information
PAD	0—24 bits	Padding or filler
X+CRC32	0—32 bits	Cyclic redundancy check (if used)
L	16 bits	Length (must agree with the BAsize)

SOURCE: © 1991, Bell Communications Research, Inc.

original TCP/IP data and sending it across the network in SMDS format. At the receiving end, the data is reassembled and sent back up the protocol stack to be delivered in a usable and understandable format.

Information Throughput

As mentioned earlier, the user will subscribe to an element of throughput. The phrase used in the SMDS world is *sustained information rate* (SIR). This is very similar to the frame relay term called the *committed information rate* (CIR). The SIR is based on a rate of service called classes. There are five classes of service defined for use depending on the application being served. The classes of service are shown in Table 24.4. These SIRs very closely match the LAN throughput speeds that we are familiar with today.

Figure 24.18
The layer 3 PDU is segmented into layer 2 PDUs at 53 bytes.

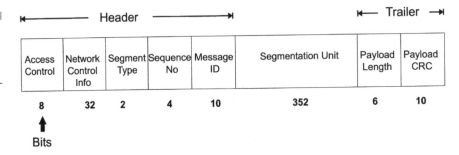

In several cases we have discussed the use of 4, 10, or 16 Mbits/s. Two other speeds exist. One is 25 Mbits/s, which is a LAN speed several vendors will be offering in the future, supporting IBM's efforts to get LANs at ATM speeds. Although ATM is defined as a 50- or 155-Mbits/s rate today, IBM has conjured up the notion that a full duplex LAN operating at 25 Mbits/s will yield an effective LAN speed of 50 Mbits/s: that's ATM. This is not yet a supported speed by the standards bodies and the ATM Forum, but IBM and 32 other companies are planning on the introduction of this service and the chip sets necessary to deliver this speed. It will likely become a de facto standard in the future. The last SMDS class of service supports 34 Mbits/s throughput. The intent of classes of service is to provide control over how much information any node can place on the network as well as the ability to prevent or control congestion on the network.

TABLE 24.3

Summary of Abbreviations for the Layer 2 SIP-PD

Abbreviation	Length	Description
AC	8 bits	Access control
VCI	20 bits	Virtual channel identifier
PT	2 bits	Payload type
PP	2 bits	Payload priority
HCS	8 bits	Header check sequence (a CRC on the header)
ST	2 bits	Segment type (beginning, continuing, or ending)
SN	4 bits	Sequence number
MID	10 bits	Message identification
PL	6 bits	Payload length
PCRC	10 bits	Payload CRC

SOURCE: © 1991, Bell Communications Research, Inc.

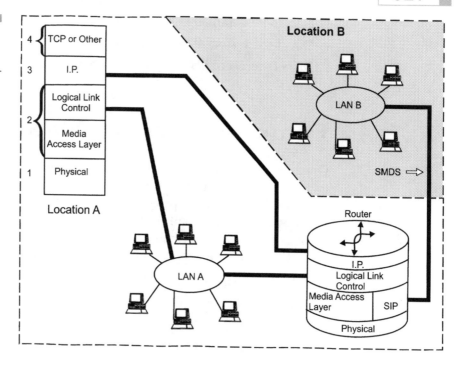

Figure 24.19
Typical LAN running
TCP/IP into SMDS.

The use of SMDS and the DQDB formats and protocols can deliver a myriad of services across the MAN. In the future, new capacities will be added. Whereas today the MAN is defined to provide LAN extension services, the ability to add other services will be introduced in the future. The current use of SMDS in the industry includes the following:

- LAN-to-LAN in a metropolitan area
- Intra- and intercompany document transfer and sharing
- Collaborative processing and development

TABLE 24.4

Summary of SMDS
Classes of Service

Access Class of Service	SIR
1	4 Mbits/s
2	10 Mbits/s
3	16 Mbits/s
4	25 Mbits/s
5	34 Mbits/s

- Host-to-host transfers
- Direct access storage device (DASD) mirroring
- Disaster recovery planning for remote hot site

Future Services on SMDS

Once the added speeds and service levels are provided, work will be done to provide several other services on the MAN, as well as a connection to the WAN. MCI is already positioned to offer nationwide SMDS services at up to 34 Mbits/s, but will be looking for the 100- and 155-Mbits/s and higher throughput speeds. These are definitely under study for future use at this time. New services will include the following:

- Video conferencing
- Interactive TV services
- Multimedia
- Broadband ISDN (B-ISDN) services and interfaces
- Voice
- Medical imaging
- Teleradiology
- Telelearning

We can only guess when all of these services and applications will be ready to roll out across the MAN and WAN. However, it would be a safe bet to suggest that many will be deployed by the LECs by 1997–1998. The LECs are moving in this direction and will want to be the first in the game. As added services are deployed, the LECs will try to preserve their customer base and offer these services as reasonably as possible. The pilots that have been introduced and the rollout that should ensue indicate that customers will be looking for SMDS quickly. One publisher in Canada has already deployed a 14-site SMDS connection to Canadian and U.S.-based locations. A major University in the East has already integrated over 30 LANs into a MAN and plans to roll out more connectivity. The case histories of these developments are all positive. Delays across the network are insignificant, pricing is reasonable, and access is fairly straightforward. The protocols and rules are well defined. Look for a lot more on SMDS in the future as you journey into the communications world.

25

Frame Relay

One of the newer forms of data transport in the industry is a higher-speed packet switching technique called *frame relay.* Frames of information are generated by most all of the data communications processes today. Although we call them by different names, such as *packets, frames,* or *cells,* they all are really just a means of transmitting a specific amount of information across a network in some logical order that is understandable to the device at the other end. The use of frame relay is new, as stated, and is just now becoming an acceptable means of transporting information across a designated network. Although frame relay has been specified as a transport system for several years, its introduction and acceptance has been met with only mild enthusiasm. This is not to say that it is not an efficient means of transferring the data, merely that the industry was confused about what the intended goals would be.

What Is Frame Relay?

Frame relay is a high-performance, cost-efficient means of connecting an organization's multiple LANs and Systems Network Architecture (SNA) services with various techniques. Like the older X.25 packet switching services, frame relay uses the transmission links only when they are needed. Essentially, the virtual circuit concept applies here as much as in any other network service.

Virtual means *almost* or *not quite.* A virtual circuit, then, is almost but not quite a circuit. There is a physical connection (Fig. 25.1) where the customer rents or leases a circuit into the cloud. This circuit into the cloud is then terminated onto a port in a computer system. The computer system is trained to recognize where the connecting ends of the wires are located and to grab the use of the full set of wires when the customer has traffic to pass to the end location. Thus, a virtual network or virtual circuit connection is established into the cloud for future use as the customer's needs dictate. Because the connection is not always "nailed up," other customers connected to the same network supplier can also generate and transmit traffic across the same physical pairs of wires within the cloud. Additionally, like private line connections, frame relay transports data very quickly, with only a limited amount of delay for network processing to take place.

When comparing frame relay to the X.25 services, much less processing is required inside the switches (network processors) and therefore the reduced amount of processing allows the transport of data much more

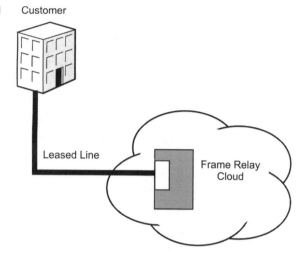

Customer

Figure 25.1
The customer leases/
rents a dedicated
circuit into the cloud
and terminates it on
a port.

quickly. By eliminating the overhead in each of the processors, the network merely looks at an address in the frame and passes the frame along to the next node in the network.

Why Was Frame Relay Developed?

Major trends in the industry led to the development of frame relay services. These can be categorized into four major trends as follows:

- *The increased need for speed across the network platforms within the end user and the carrier networks.* The need for higher speeds is driven by the move away from the original text-based services to the current graphics-oriented services and the bursty, time-sensitive data needs of the user through new applications. The proliferation of LANs and now the client/server architectures that are being deployed have shifted the paradigm of computing platforms. The demands of these services will exceed by hundreds to thousands of times the data transport needs of the older text-based services. Users demand more readily available connectivity and the speed to assure quick and reliable communications between systems or services. Fortunately, the bursty nature of the way we conduct our business allows the sharing of resources among many users who thereby share the bandwidth available. To accommodate this connectivity in a quick manner, some changes had to be made and the protocol dependency and processing of the networks had to be minimized. One way to accommodate the reduced overhead associated with the network is to eliminate some of the processing, mainly in the error detection and correction schemes.

■ *Increasing intelligence of the devices attached to the network.* The use of data transfer between and among devices on the network has moved many of the processing functions to the desk top. Because the processing is now being conducted at the local device, as opposed to using dumb terminals and a single host processor, the capacity to move the information around the network must meet the demands of each attached device. Increased functionality must be met with increases in the bandwidth allocation for these devices.

■ *Improved transmission facilities.* The days of "dirty" or poor-quality transmission lines required the use of overcorrecting protocols such as X.25 and SNA. Because the network now performs better, a newer transmission capability is needed.

■ *The need to connect LANs to WANs and the internetworking capabilities.* Today's users want to connect LANs across the boundaries of the wide area, unshackling themselves from the bounds of the LAN. The users demand and expect the same speed and accuracy across the WAN that they get on the local networks. Therefore, a newer transport system to support the higher-speed connections across a wider area was needed. The LAN-to-WAN internetworking works fine in a simple point-to-point arrangement, except that the network is dynamic and the ability to connect to multiple sites concurrently must be robust enough to meet this newer need.

The Significance of Frame Relay

The network was originally brought up through the older analog transmission techniques that have been addressed several times throughout this book. As an analog transmission system, the network was extremely noisy and produced a significant amount of network errors and data corruption. This element was most frustrating to the data processing departments. When data errors were introduced, a retransmission was required. The more retransmissions were necessary, the less effective was throughput on the network. In fact, several years ago, the use of a 4800-bps-transmission service on the analog dial-up network might well have produced an effective throughput of only 400 bps after all the errors and retransmissions. This was intolerable and had to be corrected. To solve this problem, the network introduced the X.25 services, also called packet switching (addressed fully in Chap. 18). Whereas the X.25 was originally designed to handle the customer's asynchronous traffic, frame relay was designed to take advantage of the network's ability to transport data on a low-error, high-performance digital network and to meet the needs of the intelligent, synchronous use of the newer and more sophisticated user applications.

When compared to private leased lines, frame relay makes the design of a network much simpler. The private line network shown in Fig. 25.2 requires a detailed analysis to set all the right connections in place; this further accentuates the traffic-sensitive needs of the user network. The meshed network uses a series of connections that are N1 points with links running from site to site. Therefore, if 10 sites exist in the network, 9 links will run from each site to every other site. This allows for speed of connectivity, but the network costs are much higher. Further, depending on the nature of the data traffic, the bursty data needs of a LAN-to-LAN or LAN-to-WAN connection are not required full time. We therefore spend significantly more of the organization's money to support the meshed leased line network. In Fig. 25.3, a frame relay access from each site is provided into the network cloud, requiring only a single connection point rather than the nine of the earlier network. Data

Figure 25.2
A leased line network requires more detailed analysis depending on the communities of interest.

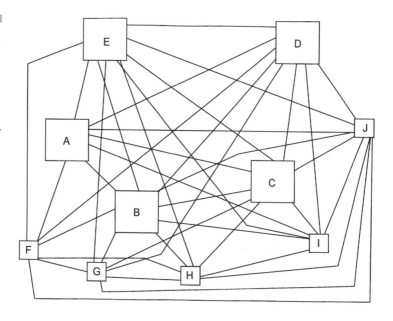

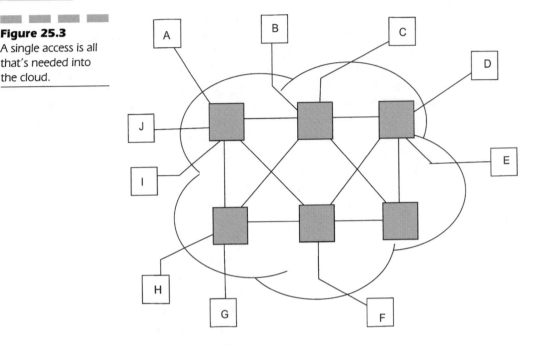

Figure 25.3
A single access is all
that's needed into
the cloud.

transported across the network will be interleaved on a frame-by-frame basis. Multiple sessions can run on the same link concurrently. Communications from a single site to any of the other sites can be easily accommodated using the predefined network connections of the virtual circuits. In frame relay, these connections use *permanent logical links* (PLLs), more commonly referred to as *permanent virtual circuits* (PVCs). Each of the PVCs connects two sites just as a private line would, but in this case the bandwidth is shared among multiple users rather than being dedicated to the one site for access to a single site. Using this multiple-site connectivity on a single link reduces the costs associated with customer premises equipment, such as CPU ports, router ports, or other connectivity arrangements. Because fewer ports are required, fewer connection devices are required; therefore, the customer saves money.

Because the PVCs are predefined for each in a pair of end-to-end connections, a network path is always available for the customer's application to run and transport data across the network. This eliminates the call setup time associated with the dial-up lines and the X.25 packet arrangements. The connection is always ready for the devices to ship data in a framed format as the need arises. This takes away the need for the constant fine-tuning of a private line or a dial-up network link arrangement.

Comparing Frame Relay to Other Services

When the network suppliers and the standards bodies were attempting to define the benefits of frame relay services, they looked at a comparison of the time-division multiplexed (TDM) switched circuit and the packet switched network services.

TDM Circuit Switching

TDM circuit switching creates a full-time connection or a dedicated circuit between any two attached devices for the duration of the connection. TDM divides the bandwidth down into fixed time slots in which there can be multiple time slots, each with its own fixed capacity, available. This is shown in Fig. 25.4, where each attached device on the network is assigned a fixed portion of the bandwidth using one or more time slots depending on the need for speed. When the device is in transmit mode, the data is merely placed in this time slot without any extra overhead such as processing or translations. Therefore TDM is protocol transparent

Figure 25.4
TDM switching forces the transmission into a fixed-capacity time slot, and the channels are virtually wasted.

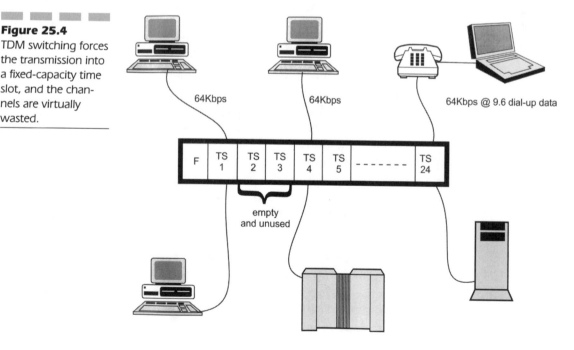

64Kbps   64Kbps   64Kbps @ 9.6 dial-up data

| F | TS 1 | TS 2 | TS 3 | TS 4 | TS 5 | ------- | TS 24 |

empty
and unused

to the traffic being carried. Unfortunately, when this attached device is not sending data, the time slots remain empty, thereby wasting the use of the bandwidth. A higher-speed device on the network can be slowed down or bottled up waiting to transmit data, but the capacity that sits idle cannot be allocated to this higher device for the duration of the transmission. TDM is not well suited for the bursts of data that are becoming the norm for the data needs in today's organization.

X.25 Packet Switching

Because the TDM world had its limitations of fixed bandwidth alloca-tion, a newer service was created that allowed the bandwidth to be allo-cated on the fly. Instead of simply putting the data into a fixed time slot, the user data is broken down into smaller pieces called packets, each con-taining both the source and destination addressing information as well as other control functional information. When a user sends data in a burst, multiple packets are generated and routed across the network based on the addresses contained in the packets. The network creates a virtual cir-cuit from each source to each destination to keep track of the packets on each connection. This is shown in Fig. 25.5, where multiple virtual circuits can be active on a line at the same time. This is a different form of multi-plexing, more of a statistical time-division-multiplexing scheme. STDM uses the analyses of the past users to allocate more interleaved packet slots to the heavier users and fewer interleaved slots to the lighter users. The major drawback to this scheme is the penalty paid in speed of delivery. Guaranteed data delivery and integrity was a prerequisite for the devel-opment of the X.25 networks; the delay in processing the data across the network was the price we paid. Clearly, something had to be done to over-come these limitations.

What the standards bodies attempted to accomplish was to use the best from all systems. Taking the features of the switched network and the packet arrangements, the network arrived at a frame relay service that should meet the needs of the user. This comparison took into account the following service connections:

- *Speed.* The speeds available to the user are important. As the capabili-ties of the dial-up digital network are now supporting the high-speed connections of up to 2.048 Mbps, the X.25 network only provides speeds of up to 64 Kbps. Thus the decision to use a digital connection of up to 2.048 Mbps.

■■■ ■■■ ■■■ ■■■

Figure 25.5

Multiple connections
on virtual circuits are
active on the line
simultaneously.

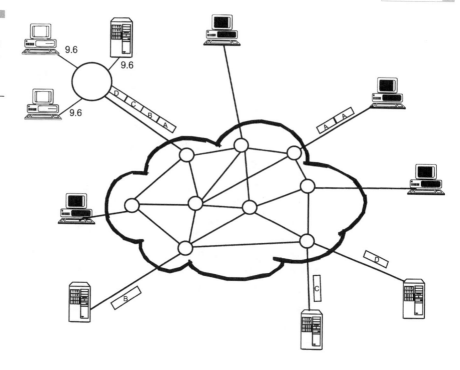

- *Network setup delay.* Call setup time on a digital switched network is relatively low. This is also true for the X.25 network, in general terms. However, as network congestion builds up the switched network circuits of the X.25, the network can introduce extensive delays. Thus, the network suppliers were looking to implement a low-delay call setup. The PVC allows for this because the call setup time is eliminated.

- *Routing.* The routing of the switched network is static. When one uses a switched call setup, the routing is used once to establish the link. From the perspective of a network delay or a failed link, the user will have to reestablish the call through a new dial sequence. This is not as robust as expected. The X.25 network is robust in that if a problem occurs on the link, the packet nodes in the network will immediately reestablish the connection with the next packet that runs across the network. Therefore, the suppliers and standards bodies opted for the robust and dynamic call setup in which delay would be very low.

- *Signaling.* In the switched network design, the signaling is also static in that it is used only in the initial setup or teardown of the call. Therefore, if the call is interrupted, the signaling disappears. In the packet switched networks, the signaling is dynamic—it is contained in every packet and thus is easily reestablished. The standards bodies opted for the dynamic signaling arrangements of the packet switched networks.

- *Bandwidth.* Again, the bandwidth consideration is important. In the circuit switched networks the bandwidth is fixed. If you dial up a 56-Kbps circuit, you get a full 56 Kbps whether you need it or not. However, the variable length of a packet allows for use of much less than the maximum of 56 or 64 Kbps. The disadvantage here is that the speed cannot exceed the range of 56 or 64 Kbps. The robustness rests in the packet switched service as opposed to the circuit switched world. Therefore, the standards bodies opted to arrive at a dynamically allocated bandwidth-on-demand concept.

- *Costs.* Circuit switched services are relatively low priced. The costs per minute for 56, 64, 128, 384, or up to 2.048 Mbps are very reasonable. Pricing approximates $0.50 per minute for a switched service, and a T1 switched service is dropping well below the costs that you might expect, putting pressure on the costs of a dial-up analog service. In the packet switched services, the expense is very low, priced at a cost per kilopacket. Some numbers approximate $0.23 per kilopacket. This is extremely low in terms of the dial-up network. The only issue here is what the best method of call processing will be if frames are routed across a PVC.

Unfortunately, this one issue was the obstacle to wide acceptance and use of the frame relay networks. Pricing was introduced in various scales that went from low to very high. The standards bodies do not get involved with pricing issues. The network suppliers have changed their pricing plans and are getting very aggressive. In fact, because frame relay falls under the auspices of a value-added service, it is not governed as a tariffed service. The network suppliers are now offering frame relay access and use on an individual case basis. They might require a nondisclosure agreement before they will submit a proposal for pricing a frame relay network. This was initially a turnoff for customers who were interested in pilot programs introducing frame relay for the WAN connections. The problem was solved through the creative pricing arrangements, and now frame relay is becoming more readily accepted.

In Table 25.1 the issues and comparisons from this information are summarized. This is how the standards bodies arrived at the service offerings of frame relay.

Service levels with frame relay are also a concern before critical data is placed on the network. The level of service is something that must be seriously considered. Response time across the network should equal that

TABLE 25.1

Comparing Frame
Relay Services to
Switched and
Packet Services

Offering	Circuit Switched	Packet Switched	Frame Relay or Fast Packet
Speed	64 Kbps 2.048 Mbps	64 Kbps maximum	0—2.048 Mbps
Delay	Low	Low to high	Low
Routing	Static (call by call)	Dynamic (packet by packet)	Dynamic (frame by frame)
Signaling	Static (session)	Dynamic (packet)	Dynamic (frame)
Bandwidth	Fixed	Variable, dynamic	Dynamic
Costs	Low	Very low	Variable (low to very high)

of a dedicated line. However, the use of a frame-based service must be carefully designed with the rates that will deliver the desired response time. Table 25.2 shows a comparison of what can be expected from the use of a frame relay network. This assumes that the appropriate rates are used and that the number of hops allowed in the network is kept at a minimum. The rates are defined when the network is set up and the ends are established. The number of hops is the number of nodes the frame must pass through within the frame relay network to get to the end. This table shows what can be typically expected from the use of this service.

TABLE 25.2

Expected Service
Delivery with
Frame Relay

U.S. Domestic Network Service Levels	Values
Percentage of availability	Node to node: 99.99% End to end: 99.85%
Bit error rates expected	Approximately 10^{-8}
Block error rates expected	Approximately 10^{-5}
Burst rates	To access channel rates or CIR
Burst timing	≤ 2 s if available
Delay on network	≤ 20 ms average
Average round trip network delay	≤ 400 ms maximum depending on CIR
Data rates	0—2.048 Mbps, varies by vendor
Maximum number of hops	4—5 depending on network

Frame Relay Speeds

Frame relay was designed initially to start from 64 Kbps up to 1.544 Mbps in North America. Speeds of 2.048 Mbps were approved in the rest of the world. This speed is based on the use of T1 or E1 for the access link. A small company came along and introduced speeds of 50 Mbps. This company, Cascade Communications, broke all the barriers.[*] Cascade wanted the network to be robust, not limited to old data rates.

When designing a frame relay service, the speed of access is important both prior to and after installation. The customer must be aware of the need for and select a specified delivery rate. There are various ways of assigning the speed from both an access and a pricing perspective. For small locations, such as branch offices with little predictable traffic, the customer might consider the lowest possible access speed. The frame relay suppliers offer speeds that are flat rate, usage sensitive, and flat/usage sensitive combined. The flat-rate service offers the speed of service at a fixed rate of speed, whereas the usage-based service might include no flat-rate service, but a pay-as-you-go rate for all usage. The combined service is a mix of both offerings. The customer selects a certain committed information rate (CIR). The committed information rate is a guaranteed rate of throughput when using frame relay. The CIR is assigned to each of the permanent virtual circuits selected by the user. Each PVC is assigned a CIR consistent with the average expected volume of traffic to the destination port. Because frame relay is a duplex service (data can be transmitted in each direction simultaneously), a different CIR can be assigned in each direction. This produces an asymmetrical throughput based on demand. For example, a customer in Boston might use a 64-Kbps service between Boston and San Francisco for this connection, yet for the San Francisco—to—Boston PVC a rate of 192 Kbps can be used. This allows added flexibility to meet the customer's needs for transport. However, because the nature of LANs is that of bursty traffic, the CIR can be burst over and above the fixed rate for 2 seconds at a time in some carriers' networks. This burst rate (*Br*) is up to the access channel rate, but many of the carriers limit the burst rate to twice the speed of the CIR. When the network is not very busy, the customer could still burst data onto the network at an even higher rate. The burst excess rate (*Be*) can be an additional speed of up to the channel capacity, or in some carrier's networks it can be 50 percent

---

[*]Cascade was later acquired by Ascend Communications. Ascend was acquired by Lucent Technologies in 1999.

above the burst rate. Combining these rates, an example can be drawn as follows:

$$CIR + Br + Be = \text{Total throughput}$$

$$128 \text{ Kbps} + 128 \text{ Kbps} + 64 \text{ Kbps} = 320 \text{ Kbps total}$$

Remember that the burst and the burst excess rates are for 2 seconds or less, depending on the carrier used. Some carriers that will not allow any bursting across the network. Rather, they require that the maximum throughput be limited to the committed information rate. What we are emphasizing here is that no standard offerings exist. Remember to ask, rather than assuming that all frame relay networks are the same.

Another variation of this arrangement is the flat-rate service offering, where a fixed price arrangement is provided regardless of the amount of traffic generated. This arrangement also allows the customer to select various CIRs at the home site or any site.

The variable PVC arrangement with differing CIRs is a flexible arrangement. However, care must be taken in regards to this arrangement because there is a fee associated with the port on the switching node within the network that can be substantial. This might be shown by comparing a quick rate structure that was once a tariff. The vendor in question used a published rate of $275 per month per site for a 64-Kbps CIR and a charge of $5895 per month for a CIR at 1.544 Mbps. This is exclusive of the leased line charges from the local telephone company for the access into the closest POP that handled frame relay services. Still other costs were included in the tariff. In some cases, the cost per account was charged at an added $10.00 per month plus $0.50 per kiloframe. This might suggest that the costs are variable—and they are.

In the initial example used above with the small branch office, a 0-bps CIR was selected; therefore, every single frame generated on the network is considered a burst rate. The burst is a little more expensive than the CIR. When dealing with the unknown it sometimes makes sense to hedge your bets until a few months' traffic is handled and the adjustment to a committed rate can be made. The CIR rates can vary in 4-, 8-, or 16-Kbps increments, so there is a lot of flexibility in the selection process. Again, the variable rates are based on a compilation of the various vendor products. Some vendors might have these and others might not. The adoption of frame relay was significantly delayed when customers attempted to figure out the pricing arrangements and make commitments to the vendor. The wrong choice can be costly in either direction. Too much capacity is expensive; too little might delay traffic.

Guaranteed Delivery

Another issue that comes into play with the use of frame relay is guarantee of delivery. When a committed information rate is used, the guarantee is that the network will make all effort to deliver traffic (frames) at the CIR, but bursts are another situation. In reality, the way the network works is that you have a connection-oriented protocol. As you send data frames into the network they will follow the same logical connection in sequence. Therefore, there should be no out-of-sequencing, and there should be no loss of frames. Unfortunately, there is no real guarantee! Frames can be lost, discarded, or delayed while en route. The best effort to deliver is more the norm. When using the burst rate or the burst excess rate, the network will make its best attempt to deliver the frames; but no guarantees are made. As each frame bursts out through the network, it is marked within its overhead with a *discard eligibility bit*. This means that as the network nodes attempt to serve higher rates of throughput, the frames are given a designator by the end user equipment. This designator lets the other hops on the network know that if the network suddenly begins to get congested, the frames riding the network beyond the CIR can be discarded and other customer frames within the CIR will have priority. In essence, the network provides some breathing room for users by expanding and contracting based on how busy it is. The less busy the network, the higher a customer can burst without risk. However, because bursts are for 2 seconds or less, the customer must be aware that the network can bog down quickly.

Because the network will only make its best attempt to deliver the data, the end user equipment must be intelligent enough to recognize that frames have been discarded. In Fig. 25.6, the frame format is shown with the setup of the bytes in the header information, creating the discard eligibility setting. In this framed overhead, other pieces of information are also contained. Designators in each frame that are set by the network nodes along the path alert the nodes and customer equipment when con-

Figure 25.6
The frame format used in frame relay.

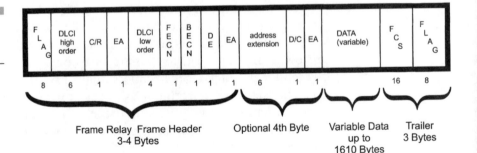

gestion is occurring. The forward explicit congestion notification (FECN) and a backward explicit congestion notification (BECN) are used for this purpose. These are shown in the framing overhead.

Another piece of the frame layout is the data link control identification (DLCI) that marks the PVC addressing scheme. The FECN and BECN, along with the DLCI, are parts of the 2-byte sequence. Encapsulated in the frame after the header is the user data, a variable-length frame of user information that can carry a LAN packet or frame of up to 1610 bytes. Cascade Communications also changed the rules for the initial payload. When Cascade introduced a frame relay switch, it could handle payloads of up to 4096 bytes. The 4096 bytes was sufficient to satisfy the IBM data needs. The initial 1610-byte payload can carry a full Ethernet frame with some overhead. The IBM token ring uses a frame of up to 4048 bytes. Other manufacturers therefore had to adjust their frame size upward to meet the Cascade size. Cascade literally stole the market with their aggressiveness and innovation. All other vendors were caught off-guard and had to react to Cascade's moves.

The bytes are not important at this point, but the variable length is. Because this is a variable, there are some inherent delays in the processing of the data across the network. For this reason, the packet switching networks have received a negative image. The X.25 and the frame relay networks have timers and delimiters that the nodes must use to process the data across the network. These are set to certain parameters to allow frames or packets to be transmitted at differing sizes. Along with the variable-length data frames, the trailer in Fig. 25.6 represents the overhead associated with the error checking (frame check sequence), which is a form of CRC. However, when transmitting data across the network, buffers must be allocated to receive the frames. Because the data are variable, the full size buffer must be allocated to receive a frame. This happens even if the frame is only half-full. The switch cannot process the data until the beginning flag and the ending flag are received and the CRC calculated on what is in between. Therefore, added buffering is required for this system to work. Variable works in some cases, but in others it introduces extra overhead and latency.

However, the frame relay nodes across the network only perform the error checking of the addressing; they don't check the user data contained inside. Actually, this is how some of the speed increases were achieved. Each node on the network only checks the address and passes the frame along the predefined path (the PVC) to the next node. Nowhere is the data verified. This is no longer the responsibility of the network, as it was in the X.25 world. It is the responsibility of the end devices (customer equipment) to do the error checking and retransmission requests. The data integrity is assumed an end user responsibility. This is a major digression from the packet switching world. Each node along the way was responsible for performing a full CRC on every packet and accept or reject the packet on the basis of the result of the CRC. Delivery and integrity of the packets were guaranteed; not so for the frames. How can this be? When the packet switching techniques were first established, data integrity was all-important because of the nature of the analog network. The data errors were significant in that older environment.

To overcome the problem, the sequencing and the integrity checks were necessary at each step along the way to ensure that the customer received the correct data. This was at a price—the time and the delays inherent in this delivery mechanism. A trade-off had to be made: either reliable but delayed data or faster but unreliable data. When the frame relay system emerged in the network, the digital world was already well in place. Therefore, with digital transport and the fiber-optic networks, errors on the network are much less likely to occur. Not that they do not occur, but there are less of them because of the characteristics of the fiber-based systems. As we have already seen, the fiber-optic networks are not prone to the same kind of errors that were specific to the radio and copper-based networks. These radio- and copper-based networks are prone to disruptions and errors from noise that will not be a problem for the fiber-based frame relay services. In addition, the copper-based networks are prone to errors from electromagnetic interference (EMI); again, the fiber networks are impervious to this type of disruption. Thus, the frame relay services can afford the customer a choice, as mentioned. Because you can expect fewer errors from the transmission medium, then the intense error checking at each step along the network can be eliminated and save valuable transport time.

Advantages of Frame Relay Services

The benefits and advantages of frame relay services are many. At a minimum, these will include the following:

Increased Utilization and Efficiency

The network uses a combination of frame relay and cell relay services to allow the user network to "breathe" when necessary and dynamically allocate the bandwidth to a logical pool of capacity in real time. The support of multiple connections simultaneously allows multiple sessions to be on-line concurrently. The nature of the bursty data transmissions on our applications does not require the full-time allocation of bandwidth. This allows shared resources on corporate LANs in remote sites.

Savings Through Network Consolidations

Data from various sources can be applied to the network in various ways, such as SNA traffic from the host, LAN bursty traffic from the desktop devices, and traffic from other application-specific devices. The use of a single connection to support all of these connections at various speeds and at variable times allows the end user to consolidate services and there-fore save money. The frame relay service is typically used to connect three or more sites; therefore, it provides cost-effective full logical mesh connec-tivity. Frame relay enables the user to save on transmission and switching costs.

Improved Network Up Time

Network down time is a phrase that will make even the most staunch data processing person shudder. The risk of down time was always a major con-sideration in the design efforts of the past. Redundant links were required to prevent a single event from bringing the entire network to a halt. This redundancy consumed a significant portion of the data processing bud-gets. The use of a single link to any point across the network risked "back-hoe fade" or other network disruption. Automatic rerouting of the links was not easily accommodated. The use of a frame relay service into the virtual cloud, however, allows the reestablishment of the network connec-tions within the cloud automatically. The only single point of failure that really remains is in the local loop or last mile. This can be redundantly protected, if necessary, at a greatly reduced rate. Because the network sup-plier is responsible for the robustness of the recovery and delivery mecha-nism, the customer can save a considerable amount of funds that can be better spent on other areas. Up time can achieve 99.9 percent availability on a network of this sort. Furthermore, only a single connection is required. The customer can reduce the equipment costs for the interface equipment (i.e., DSU/CSU, routers, mux ports, etc.) because the number and type of interfaces can be limited to one.

Improvements in Response Time

With direct logical connectivity with the PVC to the multiple locations on the network, a single interface improves the response times to which we have been accustomed. Limiting the number of hops that the frames will traverse through, along with eliminating call setup time, vastly improves response times. Allocation of bandwidth to support the bursty data needs also improves the response times, especially when a burst rate can be accommodated. Allowing the link to support higher throughputs than originally allocated, the user can achieve greater throughput in a shorter period by 2-second bursts across the network. Allowing the network to give breathing room to support the initial bursts of data, as opposed to the fixed bandwidth of the leased line concept, users will achieve higher throughput. Frame relay is the best transport protocol for high-performance and fast-response interconnection of intelligent end user devices, providing the highest ubiquitous speed, lowest-overhead protocol standard available for the WAN.

Easily Modifiable and Fast Growth

The logically connected PVCs can be adjusted easily through the network administration group of the carrier. New PVCs can be assigned on a single link quickly, or existing PVCs can be increased in capacity up to the access channel capacity needed to support the dynamics of the organization's transport needs. With the older private-line meshed networks, when a new site needed to be included into the network, a whole new configuration was required. Special time frames were required to rework the entire network or a regional portion of an existing network. With the PVC concept, a virtual connection is created quickly and without major modifications to the network. One orders the physical link from the local exchange carrier into the frame relay supplier's point of presence (POP). Then the frame relay supplier configures the logical connections using PVC arrangements. The speed of service can be cost-effective and much less tedious than was the case in the networks of the past. The frame relay service supports interconnection of LANs running multiple protocols and transports the data over the WAN protocol transparent. Some of the LAN protocols supported across the frame relay networks include Appletalk, TCP/IP, IPX/SPX, XNS, OSI, DECNet,

SNA, and NetBIOS.* This is a fairly robust set of protocols in any environment today.

Standards Based

The use of frame relay falls into the CCITT (now the ITU-TSS) and the ANSI standards infrastructure. The interoperability between various switching platforms has been proven, and it is a logical progression toward the future switching services for broader bandwidth services, such as B-ISDN and ATM, allowing a smooth transition to these newer, emerging services. Future services that may possibly be interfaced with the system of frame relay might include video conferencing (using compressed motion at 384 Kbps and packetized transfer under the H.261 standard), facsimile services (CCITT groups III or IV), X.25 services, and integration or gateway functions onto a packet switching network.

Services Available

The following services and protocols are usually readily available from the carriers, allowing the transparent flow of data across the network. Those that are not will be added as the network services proliferate and the interfaces are developed to support the various service offerings.

TCP/IP and Novell IPX/SPX

Using the packet-level routing capabilities of the frame relay network through a certified or type-accepted router (the word internationally is

---

*Appletalk is the Apple Computer protocol.
TCP/IP is the Internet protocol (transmission control protocol with Internet protocol).
IPX/SPX is the Novell Netware protocol (Internet Packet Exchange/Sequenced Packet Exchange).
XNS is the Xerox networking protocol (Xerox Networking System).
OSI is the open systems interface reference model (Open Systems Interconnect Protocol).
DecNet is the Digital Equipment Corporation protocol (Digital Equipment Corporation Networking Protocol).
SNA is IBM's standard protocol (Systems Network Architecture).
NetBIOS is the IBM standard protocol (Network Basic Input/Output Systems Protocol).

homologated), any type of data can be transported—although each respective router must be checked for compliance. Some routers might not support all of the protocols needed within an enterprise's network. The particular router to support the specific protocols is normally determined by the carrier in its recommendations for the network connections. In the event a customer does not have a router capable of internetworking with the suggested network or other pieces, the carriers are in a position to rent, lease, or sell the specific routers that will work on their networks. The routers support a variety of LAN topologies such as Ethernet, token ring (at 4 or 16 Mbps), and FDDI. These routers will also support a wide range of media interfaces from a coaxial twisted-pair wiring (shielded or unshielded) and fiber optics. Newer versions are currently being explored to support the wireless connections also.

CCITT X.25 Protocol

Convenient interfaces to the X.25 networks are being introduced to allow a customer to connect to and transport data to sites that might not have the frame relay services available (an example is sites in Mexico and South America, where the only packet switching service available is X.25) or that cannot justify this connection financially. The X.25 interface can connect the host-computing environment to the network interface across a public packet switching service, then into an X.25 gateway on the public network (Fig. 25.7). Or packets of data can be rerouted across the frame relay network through the router at the customer location (Fig. 25.8). In either case, the connection, not the location of the connection, is the important issue.

Facsimile (CCITT Group III or IV) Traffic

The use of an X.400 protocol electronic messaging service for domestic and international messaging services can be accommodated by the carrier or through the interfaces at the customer location. Facsimile traffic is extremely time-sensitive, and the packetized fax has not been truly developed. X.400 is a "store-and-forward service" to accommodate this form of communications. Future developments might well change this into a frame relay capability.

Figure 25.7
Access to frame relay
can be through the
public packet
switching service to
an X.25 gateway.

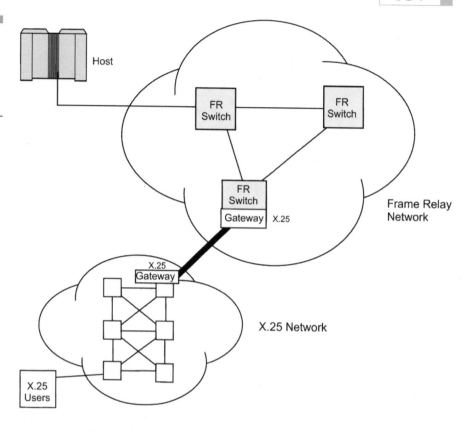

Frame Relay Standards

Clearly, then, the standard form of frame relay has some merit. The
standards bodies have completed a significant amount of work, but
much still remains to be done because of the varied implementations
by the carriers, each with some proprietary or modified standard in use.
Many of the network suppliers (although not all by any stretch of the
imagination) use a standard frame relay switching system. The system
of choice is the StrataCom* IPX switching system. The vendors had to
arrive at this choice to support the high-speed architecture of the frame
relay networks. This system uses a dual bus 32-Mbps midplane, using
redundancy at every module, bus, power, and so on. The IPX system was
chosen because of the flexibility of the services offered in circuit mode,

---

*StrataCom was acquired by Cisco.

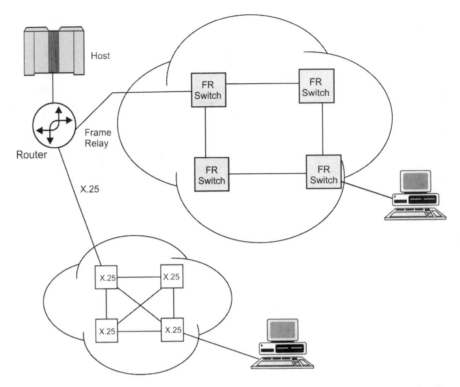

Figure 25.8
An alternative is to use X.25 to a customer's router that redirects the packet onto frame relay.

packet mode, or frame mode capabilities. The IPX system is a typically proprietary hardware/software implementation of the standards, yet it works efficiently and was the first frame relay switch used. All other switches were based on the de facto standard. The benefit of the IPX system is that it offers the single platform that most vendors have accepted and therefore the greatest chance of compatibility with systems in the network. Many of the major North American providers (including AT&T and MCI/WorldCom) use the IPX platform in their backbone networks.[*]

The Major Players

The two major standards players in the specification and support of frame relay services are the CCITT and ANSI. Together, they have defined a standard data (packet mode) interface for ISDN networks and frame relay services. The basics of frame relay were actually developed in CCITT

---

[*]Who can forget the outage in 1998, when the entire AT&T network went down, and again in 1999, when the same failure occurred in MCI's network.

blue book I.122 (*Packet Mode Bearer Services*) in 1988. Because this specification has been around for a while, you can see that the implementation was slow to start. The speed at which frame relay was developed from that point on has been directly attributable to the demand for a simple, easy-to-use, high-performance service for LAN-to-LAN connectivity. Additional standards are covered in the following paragraphs.

Others

In 1989, StrataCom (now a Cisco Company), Digital Equipment Corporation (now Compaq), Northern Telecom, Inc. (Nortel), and Cisco Systems began a joint development effort to specify added frame relay services, under the auspices of the Frame Relay Forum.* Their efforts were specifically geared toward addressing the needs of LAN-to-LAN communications, including the need for a local management interface (LMI). The collaboration of these groups led to the publication of a joint specification for the first implementations of frame relay by numerous vendors in 1990. Portions of that specification were adopted by CCITT and ANSI for inclusion in the standards. The standards that are associated with the frame relay services are shown in Table 25.3.

*StrataCom was one of the founding members of the Frame Relay Forum, a vendor interest group that accelerates the acceptance of frame relay in the vendor community. After the joint publication, StrataCom wanted to form a vendor interest group to accelerate implementation of frame relay within the structure of the standards. StrataCom was recently acquired by Cisco.

TABLE 25.3

CCITT and ANSI Standards Involved with Frame Relay Services

ANSI Standards	
T1.606	Frame Relay Architectural Specification
T1.618	Data Transfer Protocol
T1.617	Signaling Specifications (control protocol)

CITT Standards	
I.122	Framework for Providing Additional Packet Mode Bearer Services
Q922	ISDN Data Layer Specification for Frame Mode Bearer Services (LAP-F)
Q931	ISDN User-Network Interface Layer 3 Specification (blue book)
Q933	ISDN DSS1 Signaling Specification for Frame Mode Bearer Services

In 1988, CCITT approved Recommendation I.122, *Framework for Additional Packet Mode Bearer Services,* which is part of a series of ISDN-related specifications. ISDN developers had been using a protocol called link access protocol with D channel (LAP-D) to carry signaling information on the D channel of ISDN (LAPD) as defined by CCITT Recommendation Q921. LAPD has characteristics that could be useful in other applications, such as provision for multiplexing of virtual circuits at level 2 in the frame level (as opposed to level 3 in the X.25 networks). Therefore, I.122 was written to provide a framework outlining how such a protocol might be used in applications other than ISDN signaling.

LMI Specification

In September 1990, the joint group of DEC, StrataCom, Nortel, and Cisco Systems wrote the added specification known as *link management interface* (LMI) or *interim link management interface* (ILMI). For its basic operation within frame relay, it simply refers to the ANSI standard. LMI defines an added interface function, the local management interface, as a proposed method of exchanging status information between the user device (e.g., router or switch) and the network. LMI also defines a number of other useful specifications and options to simplify the implementation of frame relay and enhance its operation. More than 50 vendors accepted support of the functions of LMI, so ANSI included portions of the LMI specification (ANSI standard T1.617) with some minor modifications.

What the Standards State

To fully understand how frame relay works, the following three areas must be incorporated in the networking technology:

- The basic data flow of frame relay
- The interface signaling used at the network/user interface
- The internal operation of a frame relay network

The frame relay standards define an interface and the flow of the data, not the whole network. The internal operation of the network depends on the vendor's implementation of goods and support services.

The Basic Data Flow

In most popular synchronous protocols, data is carried across a communications line based on very similar structures. The standard HDLC frame format is used in a myriad of these protocols and services. Frame relay makes a very slight change to the basic frame structure, redefining the header at the beginning of the frame (2 bytes long). The basic frame structure for other synchronous protocols is shown in Fig. 25.9. In this case, the HDLC frame header consists of an address and control information. For frame relay, the header is changed as shown in Fig. 25.10, which uses the 2 bytes (octets) to define the following pieces:

- The data link connection identifier (DLCI). This identifier uses up to 1024 LCNs.

- The command/response bit (C/R). This is not used in frame relay.

- The extension bit (EA). When set to 0, it extends the DLCI address.

- Forward explicit congestion notification (FECN). This is set in the frames going out into the network toward the destination address.

- Backward explicit congestion notification (BECN). This is set in frames returning from the network to the source address.

- Discard eligibility bit (DE). This bit is used by the source equipment to denote whether the frame is eligible to be discarded by the net-

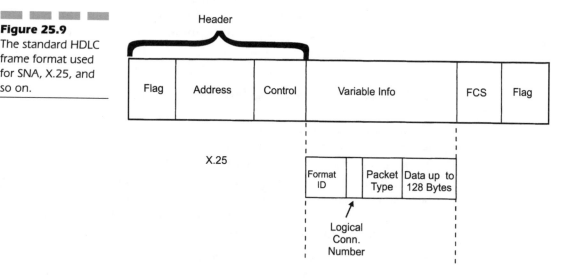

Figure 25.9
The standard HDLC frame format used for SNA, X.25, and so on.

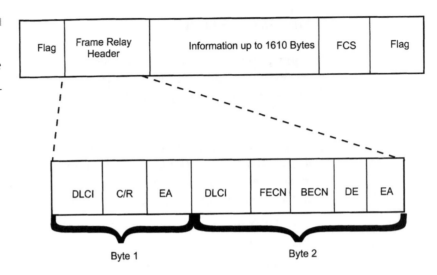

Figure 25.10
The change is in the HDLC frame for frame relay.

work if the network gets congested. When set to 1, it indicates that the frame is eligible to be discarded during congestion period.

■ Extension bit (EA). When set to 1, it is used to end the DLCI.

The FECN, BECN, and DE parts of the addresses are designed to alert the end user equipment on the status of the network's ability to process frames. Higher-level protocols use a window size control procedure or other form of control to handle traffic flow. These flow control mechanisms are designed to alert the network components and the end user equipment to slow the delivery of frames to the network. If discarding is taking place while a router or other piece of equipment is in burst mode, the router or other equipment will begin to buffer the frames until the congestion is resolved.

Interface Signaling for Control

When frame relay was first proposed to the standards bodies, the concept was fairly straightforward: keep the network protocols simple and let the higher-layer protocols at the other end of the link deal with any problems. To this end, the proposal was very practical. However, it soon became apparent that to bring about the deployment of frame relay in the industry where real-world situations will arise, the networks would need to have a signaling mechanism to address several areas:

■ Allowing the network to notify devices when congestion exists and becoming aware of the status of the permanent virtual circuits (PVCs)

- Guaranteeing throughput and equality for all users on the network, so that a single user would not occupy the network and deny others their just access

- Creating the opportunity to expand the services and features of the network in the future

Therefore, the standards players had to incorporate the use of a signaling system to solve these issues. Some of them make use of certain bits within the 2-byte header. Others use certain DLCIs as the interface control mechanism. Unfortunately, these mechanisms add to the complexity of the frame relay network, so the actual use of the signaling mechanisms is optional. If a vendor decides to implement frame relay services, but chooses not to implement the signaling standards, the result will be a compliant network, so the data will still flow. However, if a vendor chooses not to implement the signaling, other complications can arise. The effective throughput of the network, the expected vs. actual response times on the network, and the efficiencies gained from frame relay might all be lost. You must be aware of how the carrier or vendor has implemented the frame relay service. This could be a critical portion of an evaluation of the use of frame relay services.

Internal Networking

All of the standards and modifications define what is supposed to happen at the frame relay interface, normally called the user-to-network interface (UNI). How well the network performs is also determined by what happens inside the cloud; unfortunately, the standards do not attempt to specify how to determine this. Some of the functions that occur within the cloud include:

- Determining the path for the PVCs in use

- Estimating when congestion is building up, and deciding what to do to stop it

- Responding quickly to mitigate the congestion when it does occur

- Deciding which frames to discard when this becomes necessary

- Providing guaranteed rates or quality of service levels to users

- Establishing or permitting multiple levels of priority traffic

- Providing the correct throughput levels of service

- Effectively monitoring performance and producing reports on statistics, routes, and so on

Voice over Frame Relay

Because frame relay is easy to set up and conceptually a fast packet relay concept, it allows users to build virtual connections from just about anywhere to anywhere. The industry has moved away from point-to-point leased line connections and has rolled out a frame relay virtual network in lieu of the leased lines. Much of this is because the virtual network provides for a robust connectivity solution at a less expensive cost. It is from this robust network that a data network emerged. Frame relay initially was developed around a packet switching data communications need, especially for LAN-to-LAN traffic; however, more users began to look at the cost benefits of using frame relay as a WAN technology. What resulted was the question, "If data is cheaper, why not voice or fax?"

Consequently, the industry started to look for a means of carrying voice communications on a frame relay network. Some of the considerations used in looking at voice over frame relay include the following:

■ Frame relay travels across the public switched telephone network, so the provider does not have to dedicate specific loops across the long-distance portion of the network and maintain specific copper connections. Therefore, it is less expensive, because the overlay to the public switched telephone network can be used for multiple customers.

■ Frame relay virtual circuits afford the variability in bandwidth with the bursting capabilities already discussed. A typical frame relay capability specifies both the committed information rate (CIR) and the committed burst rate.

■ Many organizations have excess capacity available; therefore, carrying packets of voice inside the frames is straightforward.

In view of the capabilities of the committed information and the committed burst rates, vendors are more able to flexibly and dynamically allocate the capacities without having excess service remaining idle. Consequently, vendors have been beefing up the backbone networks using high-speed ATM (ATM is discussed in a later chapter) so that the capacity is readily available to provide the services necessary. Therefore, when discussion about using frame relay to carry voice communications occurred, the carriers were not overly distraught. However, their caveats were acknowledged and acted on. Frame relay is good for bursty traffic, as opposed to the sustained information rates of voice. Frame relay guarantees only on the committed information rate, not on the bursting rates. If voice is to be used on frame, the risk of frames not arriving at their final destination still exists. If voice frames are discarded, lost, or even delayed, there is no mechanism to go back for retransmission. The retransmission requirement would be activated when something arrives out of cycle, the receiver on the far end (the human) would have to use a self-correcting method (*"What?"*). It is through the request for retransmission that a new frame would then have to be retransmitted, but it would not be one small frame of information, it could very well be a complete sentence or discussion. This of course does require some fine-tuning on a network before much voice is run across frame.

Actually, the question really is, "Why bother?" Voice is already becoming a commodity, with prices now (in the late 1990s) ranging from 4 to 6¢ a minute for long distance. Even on international traffic, the cost per minute is dropping down to the 10 to 15¢-per-minute range for most foreign countries. Consequently, if one looks at using frame relay, which is a nonguaranteed service, the question is, "Is it worth it?" Using commodity-based pricing for voice communications, organizations have still looked at a free ride for voice on a data network. This of course means that if an organization chooses to run frame relay, the likelihood is that it will use compressed voice. Several standards have been adopted to provide voice over frame relay, not to mention the G.723A and G.729 ITU standard for voice compression techniques. The G.729 standard offers what most people consider the best compromise in both speed and compression to preserve voice quality. This is typically done at 16 Kbps, and is the direction the majority of the users are heading. Several vendors have produced the products, and many test beds are already out and running. However, one of the predominant suppliers in the market has indicated that frame relay is exceptional when committed rates can be assigned for voice communications. This supplier's recommendation is that frame be used on private line frame relay services to carry voice. This allows the actual bandwidth allocation to sustain the data rates for digitized voice and still meets the expectations of the lag and delay across the network. That same vendor, however, does not recommend frame on the public switched frame relay networks. What this means is that users would have to be very cautious and carefully approach how they would run voice over frame relay. Clearly, the idea of getting a free ride for voice on a data network is a novel one. For years, the data has always traversed the network as a free ride behind voice communications. Therefore, this change in concept and ideology brings a whole new perspective on the use of voice communications. One need only look at the amount of traffic that is carried from site to site on a voice communications network for employees to chitchat with each other. This amounts to a significant amount of money, and the idea of putting it onto a backbone network already in place to carry data offers some attractive possibilities. If voice can ride for free and the data network can still sustain the amount of information processing that must take place, then perhaps this is worth considering. As the network is put in, users must be somewhat careful to ensure that enough communications links are available. Frustrating users, as in the old days of our queued trunking, would only serve to push people away from the use of frame relay and force them back on the public switched telephone network. Therefore enough circuits and lines would have to be installed to support the site-to-site traffic that would be carried on frame relay.

Equipment

Several of the FRADs (frame relay access devices) now have the capability of interfacing directly through a PC-based card mounted in a server that would allow the voice-over-frame capability. These cards have built-in DSPs to receive the analog information from a PBX trunk, digitize it, and then compress it for access onto the frame relay backbone. Through this technique, the compression portion allows for the 16 Kbps of voice throughput. Many of the DSP cards that would be mounted in a server-based platform are delivered with either two or four ports built in. If only two to four ports are installed between two high-end sites, obviously there may not be enough. Therefore the equipment suppliers should be challenged to determine just how many ports can be installed on their particular cards. The equipment manufacturers are producing products that cost approximately $1000 to $1500 per voice port installed, which compares favorably with the cost of a circuit card in a PBX architecture. Using 16 Kbps of voice is usually fairly safe. It is when organizations compress even further, down to 4 to 8 Kbps, that the voice quality can start to really break up. However, at 16 Kbps, high-compression voice multiplexing has been done very successfully over the years. The equipment also offers connectivity through RS 232, V.35, or 10Base-T, so that it can be plugged into an existing voice multiplexer, a router, or a PBX. Some of the standard interfaces that would be supported on these communications boxes include analog ports, digital ports, four-wire E&M, or DID circuits. In many cases the equipment vendors would require a separate access code such as the digit 7 or 8, as opposed to the normal 9 for outbound trunking. The equipment can also be integrated into the network management platforms that work within the LAN environment, such as simple network management protocol (SNMP).

International

It is now becoming increasingly possible and common for small, start-up ventures to enter into and define a segment of the marketplace. Due in part to equal access and deregulation in the United States and the forthcoming international agreements to standardize, deregulate, and in some cases privatize much of the world's telecommunications infrastructure, open communications will continue to create opportunities for growth in almost all segments of the telecommunications marketplace. Traditionally protected countries such as Japan and Germany have started to deregulate the industry and to encourage more efficient markets through increased competition. Taking advantage of the business trends of telephony and communications, organizations are now starting to build worldwide networks capable of transmitting voice, data, fax, and possibly video for a fraction of the cost traditionally incurred on the dial-up switched network. These organizations, working on an international network, have begun to deploy voice over frame relay services. In many cases there are two differing approaches on how communications capabilities will be offered: either usage-sensitive or flat-rate pricing. Under the usage-sensitive approach, the organization will dial into a public switch using the local seven-digit telephone number, and from there pass the traffic of the internationally placed call through the switch to a FRAD. This FRAD would then use an incorporation of an IP translation so that the frame relay packets could be routed to a far-end FRAD or router for retranslation at the far end. Because the local user is dialing into a local PBX-type port, the traffic can be captured for billing purposes on a cost-per-minute basis. One can expect that this cost-per-minute pricing scheme will be significantly less expensive than running a circuit switched dial-up telephone call to the same international locations.

On the other hand, organizations (third parties) are now considering the use of flat-rate pricing. Using a flat-rate schedule for a single monthly access fee, the clients will have dial-up voice, fax, and data access to and from a number of the world's business centers without incurring any additional long-distance cost regardless of the time of day or the length of call. Any calls made to locations outside those cities are at substantially lower rates because the backbone will carry the frame relay call to an end point where it then might be ported out onto the public switched telephone network for local calling capability. This again offers some unique opportunities but still runs the risk of congestion across the network.

As international marketplaces emerge, the cost of frame relay services will continue to drop. As this drop in price occurs, the use of frame relay will become more attractive to carry voice communications. If today's current technology allows for 50¢ per minute between the United States and Japan, for example, the frame relay services might well be offered at 20–25¢ per minute. This in turn would provide a public frame relay switched access method for half the price. Only the larger organizations, which have sustained traffic needs between the same two end points, would benefit from going to their own private frame relay services to carry their own voice. Instead, the small to midsize companies might see benefits from the use of the public switched frame relay access for carrying voice and fax in particular. But then, later, data communications will be the normal evolution, so that organizations will try to converge and integrate all of their technologies onto a single dial-up communications medium. *Sounds like the voice network, doesn't it?*

So the jury is still out as to whether frame relay will really emerge to be the voice-carrying network of the future, as we wait for the higher-speed throughput capabilities of ATM. Again, there is always doubt as to whether a data network is going to successfully carry voice and how contention will be handled when data and voice try to coexist on the same network in the future. Many organizations still have separation between the voice and MIS departments. Consequently, they will not be very excited about merging the voice and data on the same communications network. There are advantages and disadvantages to both sides of this coin. Frame relay was originally designed to carry data transmission for computer networks. It is a shared bandwidth system and operates most efficiently at speeds of up to T1 or E1. Eliminating the need for many dedicated circuits, it is very efficient and is used extensively for Internet transmission applications. However, like every system, it has both advantages and disadvantages.

Advantages

The chief advantage of this application revolves around the way it is priced. Comparatively inexpensive, frame is priced by bandwidth and by permanent virtual circuit (PVC) access point. It is not distance or usage based as a norm. This is of critical significance when evaluating the potential for the usage-based voice that would be carried.

It is also estimated that approximately 80 percent of Internet servers run on frame relay. This is partly because of its low cost, particularly among Internet access suppliers who provide service on a monthly access fee with unlimited usage. This has been incorporated in the general strategies of the frame relay deployment.

A second key advantage is that the frame relay services are carried over existing infrastructure by the communications network suppliers. Much of the upgrade, maintenance, restoration, and repair and other concerns about managing a network are the responsibility of the frame relay carrier as opposed to the end user. This removes most of the networking burden from the end user. Just about everything between the sites is domed, operated, managed, and maintained by the communications provider, removing all of that responsibility from the end-user organization.

Disadvantages

The prime limitation to frame relay has been the ability to transmit time-sensitive applications (such as voice, fax, and video) due to the burst transmission characteristics of the system. The original design was for a data communications network where these time sensitivities were not as critical. Digital conversion, compression, and multiplexing technologies have not advanced sufficiently to satisfactorily handle conversion of the common 56–64-Kbps analog telephone network capabilities. Furthermore, outside the United States, the acceptance and availability of frame relay has been somewhat sporadic. A number of countries still do not offer frame relay access or service capabilities, limiting where frame relay can be used.

However, organizations continue to plod forward, attempting to use the voice and frame services together. Simplified and limited use of voice over frame relay has not caused any major potential degradation on the frame relay networks, so many organizations will continue to use it. It is only after large deployment or significant investments have been made that we will determine whether or not frame relay will be used as a viable alternative to the circuit switched dial-up telephone network for voice communications. These approaches are but deviations and variations from the original voice telephone network as built by the Bell System over 120 years ago. As already stated, the judge and the jury still remain out.

CHAPTER 26

Integrated Services Digital Network

What Is ISDN?

Integrated services digital network (ISDN) is one of the carrier product offerings designed to support what the carriers perceive as the customers' need for the transport of information. Simply stated, ISDN is a mechanism that allows the on-demand transport of voice and nonvoice services on a call-by-call basis. ISDN was conceived in the 1970s, then produced in the early 1980s by the world's telephone companies as the next-generation network. The existing voice networks didn't deal well with data. One had to use modems to transmit data, the data rates were only around 4800 to 9600 bps, and the connections (worldwide) were unreliable. ISDN is based on all-digital transmission, using state-of-the-art technology to allow access to the raw bandwidth that the carrier believes is needed by the customer. In Fig. 26.1, the basis of ISDN is shown as the combination of present and future services.

The industry as a whole is somewhat in a state of turmoil. Many users feel that their needs for ISDN have not yet materialized. Some feel that it is too little too late. Still others feel that ISDN is part of the future, allowing the flexibility and the capability to address present and future needs for telecommunications services.

ISDN Defined

ISDN, as defined by the CCITT (or ITU-TSS, as it is now called), is a network service, evolving from the telephony networks, that provides end-to-end digital connectivity to support a wide range of services, including voice and nonvoice services, that users will have access to through a limited set of standard multipurpose interfaces.

Proponents of ISDN state that the services include the ability to:

■ Transmit the name and number of the calling party to your telephone before you answer the call. This would be of some interest to the residential user as a means of screening calls and allowing that user to refrain from answering a call from an unknown caller. Now these features can be provided without ISDN because of enhancements made to the central office switching systems. Features and functions that were earmarked for ISDN are now delivered on the plain old telephone service (POTS) networks. Think of the benefits to any telephone-intensive business as an incoming call delivers the customer's information (phone number, contact person, account status,

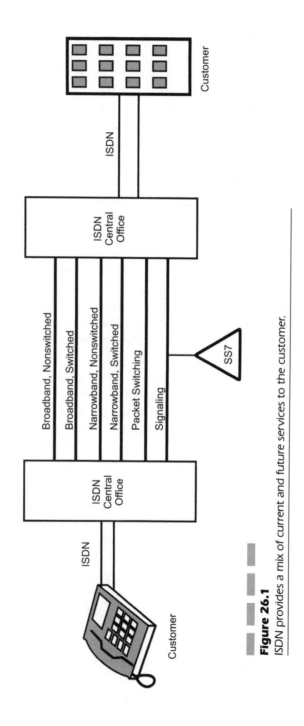

Figure 26.1

ISDN provides a mix of current and future services to the customer.

etc.) as the agent answers the call. The discussion here revolves around what is called automatic number identification (ANI). This is being done on some ACD systems, but it requires special arrangements. Estimates of the use of this caller ID are that 20 to 30 seconds could be shaved off each call. This, of course, means that improved service and lower labor costs are the resultant benefits of this service.

■ Allow access to multipoint and dial-up voice and data conference calls from the desktop. The use of voice and data calls through digital technology will be at a rate of 64 Kbps. You can get a customer on the phone, access a database, and display your marketing information on his terminal, PC, or other device all at the same time. Furthermore, as the conversation rolls along, a conference call to a major supplier or distributor can be accomplished without disrupting the initial call.

■ Pick and choose the services to be provided on a line-by-line basis, with access to various services as needed. The use of voice, data, and video conferencing can be allowed on a call-by-call basis. Time and billing information for the current call can be transmitted to a telephone or PBX system for accountability purposes, eliminating separate time slips.

■ Use the same line to transmit telemetry information without interfering with the ongoing voice and data calls in progress. The utility companies see the benefit of using this to remotely read meters in a residential application. These techniques would also be invaluable to large corporations by providing dial access to remote sensing systems or industrial control systems, or, in the case of medical applications, remote monitoring of patients' vital signs.

You can imagine the impact these features would have on the industry as a whole if the need for the labor-intensive resources that are becoming scarce at an alarming rate were to be reduced further.

Therefore, ISDN is a series of products and services designed by the carriers, for the users, to increase the productivity and contain the costs associated with doing business. Wrong!

No doubt, these enhancements offer a variety of applications that will only be limited by the user's imagination. However, the truest concept of ISDN is that it is a transport vehicle that delivers the digital data (voice and data services). It is not a product and it is not a package of hardware offerings from the suppliers.

ISDN is also an evolutionary grouping of transport bandwidths. Therefore, users have come up with a series of comical statements regarding the use of ISDN and what the acronym means, such as:

- I still don't need
- I still don't know
- I smell dollars now
- Innovations subscribers don't need

The list is endless, but the message is the same. How do users perceive the usefulness of a network topology and the ensuing benefits that can be derived? For many, the offerings of ISDN today match what can be done today without ISDN. For example, the caller ID display, which equates to automatic number identification (ANI), can be accomplished today using the central office digital switching technology that's already there. Another example would be the use of ISDN for telemetry purposes; by using a dial-up modem and some mechanical interfaces, the same function can be accomplished.

So why all the hype? If everything can be done using existing technology, why do we even need ISDN? The answer is obvious. Once you look at the possibilities of using the same lines and interfaces that you have today, integrated across the network on call-by-call basis, you can eliminate the need for special lines dedicated to specific applications. Moreover, because the total functionality is evolving, the additional features to be derived from the network in the future will far surpass the technologies we know today. Think of the possibilities—let your imagination be your only limiting factor.

Who Is Making the Rules?

The Consultative Committee for International Telephony and Telegraphy (CCITT)* is the responsible body for the development of ISDN standards on a global basis. CCITT's emphasis will be to ensure that ISDN will be just as ubiquitous as the present-day public switched telephone network (PSTN) when fully deployed. Because ISDN is an enhancement to the PSTN, this should pose no great threat. However, the CCITT has also expressed the need for portability in the deployment of ISDN in its universal state. This means that an ISDN terminal should be capable of being plugged into an ISDN anywhere in the world and still providing full features and functions.

---

*CCITT is called the ITU now, but at the time of specification it was still called the CCITT. Therefore, the references are to the standards groups of the time.

This creates a dilemma for a CCITT because the PSTN really hasn't achieved the same portability on a global basis. The interfaces are different, the line terminating impedances aren't the same, and the signaling is different. So how can the CCITT expect the evolution of universal access to a network from anywhere in the world? In fact, this hasn't been accomplished in our old analog telephony network, which has been evolving for over 100 years. The rest of the world typically has one or two networks at the most.

The ITU-T (formerly CCITT) is a UN treaty organization, and as such, each country is entitled to send representatives to any committee meeting. The representative typically came from the government-run post office, telephone, and telegraph (PTT) monopoly. The world's telcos are becoming privatized and competition is being permitted. This creates an interesting struggle within each country to determine who will represent that country's interest at the ITU-T. The name *ITU-T* came about due to the privatization trend separating telephone service from the post office and the general elimination of telegraph service. Because its members were no longer PTTs the organization couldn't properly be called the CCITT. The CCITT was always a consulting committee to the International Telecommunications Union. So, the ITU-TSS is the Telecommunications Standard subsection of the ITU. The CCITT was comprised of study groups (SGs). Each SG has its own area of expertise. Some of the better known ones related to ISDN are:

SG VII	Public data network (X.25) X. series standards
SG VIII	Terminal equipment for telematic services
SG XI	ISDN and telephone network switching and signaling
SG XII	Transmission performance of telephone networks and terminals
SG XV	Transmission systems
SG XVII	Data transmission over public telephone networks
SG XVIII	Digital networks including ISDN

Although they are called standards, the CCITT and ITU-T technically publish *recommendations*. The philosophy of the ISDN committee (essentially the telcos) was to specify the customer interface first and then figure out how to support it in the network—the theory being that if they could get all the peripheral or end equipment makers to make equipment based on their specifications, then when they got ready to roll out the service, the store shelves would be stocked with inexpensive ISDN interface equipment.

The CCITT has created an ISDN reference model or a picture of what an ISDN should look like. This model identifies a series of functional groupings of equipment and reference points where the critical functions

Figure 26.2
The CCITT ISDN reference model.

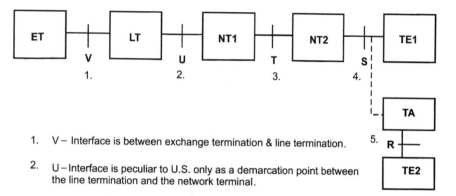

1. V – Interface is between exchange termination & line termination.

2. U – Interface is peculiar to U.S. only as a demarcation point between the line termination and the network terminal.

3. T – Interface is between network termination 1 (i.e., CSU, MUX) and network termination 2 (PBX, LAN, etc.)

4. S – Interface is between the network termination device (1 or 2) and the subscriber equipment, called terminal equipment **1**. S is also used between the NT and a terminal adapter (TA), which allows non-ISDN equipment to be used on the network.

5. R – Interface is between the non-ISDN equipment and **TA**.

must be performed to ensure communication between the groupings. Don't confuse the groupings and functions as different pieces of equipment; rather, they are functions that must be performed. These recommendations were contained in the findings, known as the red book, that came out in 1984. Subsequent work has been done on the definition and implementation of ISDN services. The additional information is contained in the blue book, dated 1988. Figure 26.2 is a representation of the ISDN reference model.

The implementation of these recommendations is the hot topic in the industry. Many vendors and carriers purport to have a fully ISDN-compatible terminal or network, when in fact the recommendations still aren't completed. Thus these implementations are at some risk of failing to gain acceptance when the final standards are set. Changes might be required when finalization is complete.

Why Do We Need ISDN?

The carriers and vendors would have us believe that we need ISDN more now than ever, and that the only way we can successfully use the all-digital end-to-end capabilities of the network is through the implementation of

ISDN. These providers might be correct, but they also have a tendency to oversell the capabilities and possibilities. The most overused and abused term these days is the concept of ISDN. Everyone wants us to believe that the system and network are there today for our unmitigated use of all of the features and capabilities. Unfortunately, nothing could be further from the truth. Sure, pieces exist, and trials have been under way for some time to evaluate the worthiness of the ISDN. However, full implementation of the entire network will take quite some time. The main reason is money. In order to implement a digital network at the local loop level, substantial investment by the local exchange carriers is required.

So who will pay for all these upgrades? You and I will! Unfortunately, the way this works is that the LEC builds the prices into its rate base, which must be approved by the regulatory bodies governing the setting of prices (such as the Public Utilities Commission). Many of the LECs have tariffs in place to offer the service. Unfortunately, the tariffs are as varied as the service offerings that they provide, even if it is not truly ISDN.

Meanwhile, back to the question of why we need ISDN. As the need for higher-speed transport increases, the limits of the available analog interface to the LEC and the IEC will be quickly surpassed. More and more users are demanding added call-carrying capacity to move information around the world. The primary areas that will consume bandwidth are additional speed for data communications, transfer of files from host to host, host to PC, PC to host, LAN to LAN, LAN to WAN, and so on. Video conferencing will also strain our telecommunications systems to the maximum. The bulk of the ISDN services being installed are data applications to surf the Web on the Internet.

Another need is the demand to transfer graphic information. As the desktop applications communicating complex documents from workstation to workstation are enhanced, even greater carrying capacities of bandwidth are necessary to move these files. The typical multipage, text-based document will be fine, but the complex document that incorporates the use of graphics, annotation with voice, data, text, video, and any other format will require the transmission of files at the multimegabit rates.

Thus, as more complex information is entered into the systems and the demand for greater bandwidth increases, the use of ISDN clear-channel capacity (which was covered in Chap. 14) will increase productivity and reduce costs. This doesn't portend that you must have ISDN to send documents at higher speeds. Specific needs will dictate the use of higher throughput and the technology to be used.

The main thrust is that ISDN is supposed to reduce call lengths because of the higher throughput on a dial-up basis. The shorter the call, the less expensive to support it. From the standpoint of the carrier (both the LEC and the IEC), the shorter call duration will aid in providing services to a greater number of users without adding circuit capacity and overbuilding the network. Therefore the carriers also stand to improve efficiency in the use of their networks, which should reduce their operating costs. This sounds too good to be true: both the supplier and the user will have lower costs! In theory, it should work; the reality remains to be seen. In fact, ISDN usage has increased call durations as users surf the Internet. This causes the LECS a lot of concern.

Now you can probably appreciate why the push is on for the implementation, acceptance, and standardization of the ground rules for this capability. The other obvious benefits of ISDN include:

- More flexibility for the user on a call-by-call basis.
- Better control over the network by the carriers.
- All-digital transmission is more reliable and less expensive than analog.
- Quicker setup time on calls.
- Additional services from the carriers, which equates to more revenue from new sources.
- Bypass defense by the LECs in providing service to the customer.
- Fewer new cable facilities needed into the office and the residence, because multiple simultaneous services can be used over a single two-wire connection at the local loop.
- Analog to digital/digital to analog conversions become the customer's responsibility.
- Signaling out of band becomes the customer's responsibility.
- Carriers can provide improved network management and maintenance control.
- High-speed facsimile, dial-up video, slow-scan video, Internet access, and packet switching will become more cost advantageous, thereby encouraging more use of these services.

The Overall ISDN Concept

The objective of ISDN was originally introduced in 1979: the evolution of telephony applications that provide digital end-to-end connectivity to support a wide range of services, including voice and nonvoice, through a set of limited multipurpose user interfaces. The reason for this concept was the obsolescence and cost-prohibitive nature of the older analog technology. Since the days when a dial-up voice connection was riddled with crackling and popping on the line, users have been clamoring for the local and long-haul carriers to improve service. The evolution of analog-to-digital transmission techniques significantly improved the transmission of voice and facilitated the ultimate introduction of higher-speed data communications. It would be appropriate to summarize some of the telecommunications requirements placed on the networks and carriers. The following are the candidates for integrated digital communications through the evolved networking techniques.

Telegraph

The telegraph was the first electrical communications system to be introduced. It used direct current pulse signals on single-wire earth-return lines, and later on two-wire lines. It was possible to detect signals over long distances by the use of sensitized galvanometers. Transatlantic cables were in use in the mid-1800s. Pulse codes such as Morse code were devised, and operators who were highly skilled could achieve speeds of 30 words per minute. The use of the telegraph and the ensuing delays over the transmission path distorted the actual human side of communications. Unnecessary words were eliminated as an expedient in getting the transmission across the cable. The telegraph introduced the first forms of digital coding and used a protocol to ensure accuracy and the reception of messages.

Telephone

The invention of the telephone by Alexander Graham Bell in the late 1800s, and the subsequent proliferation of telephones in the United States, transformed telecommunications. Users could now communicate directly with each other without the need for a protocol or the intervention of a skilled operator to interpret the message. After the operators were removed from the picture, telephone communications needed improvements to allow a novice caller to use the system comfortably over great distances. Automatic switching, an invention necessitated in the network and invented by an undertaker (Almond Strowger), increased the ease of use and the ability of unskilled callers to access other telephones without human intervention.

Telex

The development and manufacture of complex machines prepared the way for the production of delicate mechanisms for telephone switching. At about the same time Alexander Graham Bell was promoting the telephone, the typewriter (invented in 1867 by Christopher Sholes) was finding its way into the office. It took only a small adaptation to use an electric typewriter to send telegraph code for each letter and number on the keyboard. It wasn't until the 1930s, however, that interconnection using the Telex actually expanded without the need for the skilled telegraph operator. One of the reasons this process was slow to be accepted was the protocol needed for the machines to call and answer back with a discrete identity. Telex machines that used a punched tape constituted the first attempt to introduce direct machine-to-machine communications without the intervention of the operator. Many people in the industry thought this was a dead technology, especially because of the proliferation of facsimile and view data terminals—however, Telex is still alive and strong on an international spectrum.

Data Communications

Telex and telegraph are actually the original data networks introduced into the industry. The term *data communications* became more commonplace after the introduction of digital computers, which encouraged the concept of time-sharing tasks requested by many terminals and devices. From the original dedicated star network in the office or factory, data networks have become true worldwide transport networks interconnecting mainframe computers with compatible terminals and other compatible devices. Today, the use of data communications over the public switched telephone network is continually on the rise, using a binary code converted into tones (or analog signals) operating within the speech bandwidth spectrum (300–3300 Hz). Speech bandwidth as provided by the local exchange carriers has recently come to limit the speed of information flow between communications devices (up to 33.6 Kbps on dial-up analog lines). Therefore, the use of dedicated data lines and data networks has expanded to allow more data throughput in the communications world. The need to provide private line service impedes the user's overall flexibility in connecting to any device during peak loading periods. Many circuits have been justified on the basis of part-time use, while the line sits idle the rest of the time. This does not constitute effective use of our communications expenditures.

Packet Switching

Using a dedicated or a circuit switched facility to communicate information has been one of our primary means of communicating data. Public packet switching techniques were introduced as an evolution of these facilities. Public packet switched networks were modeled after the bursty nature of data communications and the concept that all data calls do not necessarily need simultaneous two-way communications. Several users' packets might be traversing the network on the same virtual link without any degradation of the response or loss of data integrity. This method allows the carriers to gain greater use of their facilities and allows users to communicate at decent rates of speed without the use of dedicated or private lines. Refer back to Chap. 18 on X.25 packet switching as a refresher.

Integration

The use and existence of separate networks to provide the above services has in some ways been an advantage in gaining access to the services. However, using separate lines and access methods to these services has also been a disadvantage. To gain access to the various services, the user had to subscribe to that network. The user was presented with an overhead and management problem in administering the effective use of each of the facilities provided. Moreover, in periods of peak loading, many facilities were overburdened while others sat idle. To eliminate this problem, users and suppliers alike began to explore and demand an integrated plan. This plan calls for access, through gateways or other methods, to any service by any user through a common set of lines. The existence of integrated networks will allow for user management of the entire telecommunications function through the single thread to the outside world. This is what the overall objective of the ISDN really is: to provide the user with easy access to multiple services over a single connection to the network. The least important element of the title *ISDN* is the word *digital*, even though it is the digital network that allows for this universal and simple access to the wide range of services to be provided. To meet this objective, the comparisons shown in Table 26.1 must be implemented.

These are no simple tasks. The concept is complex to define and even more difficult to perform. This is particularly true in the regulatory environment here in the United States, where clear lines of demarcation have

TABLE 26.1

Comparing the Objectives of ISDN

Locally	Publicly
Duplex operation of simultaneous voice, telemetry, and signaling on a single link must exist.	Capacity must exist to independently data, switch voice, data, telemetry, and signaling information from a subscriber location over a single line.
Transparency must exist on the message content used in the various services.	The ability must be put in place to extract the message address information, check information for validity, and transmit the message information to the desired address.
Operation must take place over the two-wire connection (twisted pair) within the distance limitations and other characteristics of the copper wire.	The ability to access specialized networks (packet, for example) and pass message information to and from these networks in a protocol-transparent way must be provided.
Extended areas must be serviced through the use of digital regenerators. The old analog amplification equipment must be removed.	Charges and billing information must be identified and billed across the various services between the providers.
Rapid access and setup times must be available. Power consumption across the line (span power) must be kept at a minimum during idle stages. Setup times must be provided end to end within 2 s, replacing the older methods of call setup requiring 6 to 30 s.	The capability to address or subaddress the customer premises equipment and ensure proper operation when addressing and connection takes place must be provided.
Transmission and timing systems to detect framing and bit timing must be provided in both directions.	Some form of local power to the customer premises equipment from the network switching exchange must be provided. This must be able to switch on local power to "power up/power on" the customer's equipment.
Good error-free performance (a typical error rate of 10^{-7} is expected) must exist.	
Compatibility in interfaces to other forms of digital transmission media such , as radio, fiber optics, cellular, and coax must exist.	
Protection must be provided against power cross-connect faults, lightning, etc.	
Remote equipment must be suitable for mounting independently on premises for integration with customer-provided terminal equipment.	

been established with the divestiture of AT&T from the telephone company subsidiaries.

The ISDN Architecture

ISDN is based on the open systems interconnection (OSI) layered communications model adopted by the CCITT. The architecture generalizes the OSI model and applies it to information and signaling transfer and system management. The inclusion of the transmission media in the model highlights the need to evolve ISDN from today's metallic environment in the local loop plant to a fiber-optic environment and on-premises wiring to higher grades of twisted pairs or fiber optics, while the system management reflects the market's need for the customer-controllable and easily maintainable systems and services.

Physical View

Two CCITT ISDN user network interfaces are used for connection to end user devices. They are the basic rate interface (BRI), 2B+D, and the primary rate interface (PRI), 23B+D/30B+D.* These are covered later. However, see Fig. 26.3 for the basic rate concept and Fig. 26.4 for the primary rate concept. While the most common PRI configuration is a single D channel per physical facility, multiple links (up to 20) can share a single D channel. This will lead to increased information-carrying capacity for the interface (the AT&T method of sharing a single D channel for multiple PRIs). The D channel on one 23B+D can be used to control a number of 24B PRI connections.

The PRI offers an economic alternative for connecting digital PBXs, LANs, host computers, and other devices to the network. The BRI brings Centrex and PBX customers integrated voice and data as well as advanced voice features. As opposed to many of the proprietary digital interfaces available in many of the PBXs on the market, the BRI is a standard interface and offers the user the benefit of multivendor compatibility. This statement assumes that the PBX makers will implement a standard, which is questionable today. An example of this point lies in the realization that every PBX manufacturer advertises its system as ISDN compatible. If this

---

*30B+D is the international version of ISDN PRI; it uses an E1 with the framing and signaling channels.

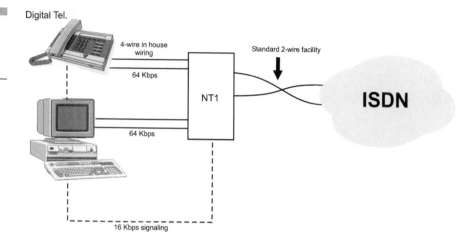

Figure 26.3
The basic rate interface (BRI) supports 2B+D.

is a true statement, then a user should be able to buy one manufacturer's system (Lucent Technologies, for example), plug in a different manufacturer's (Rolm/Siemens, for example) digital ISDN telephone set, and have the two work transparently. This is not the case today, because every manufacturer builds a proprietary means of using its equipment. The future offers some hope, but the standards must be cast in concrete first.

To maximize terminal portability and ensure easy change from one network to another, the same BRI and PRI interfaces are used for inter-

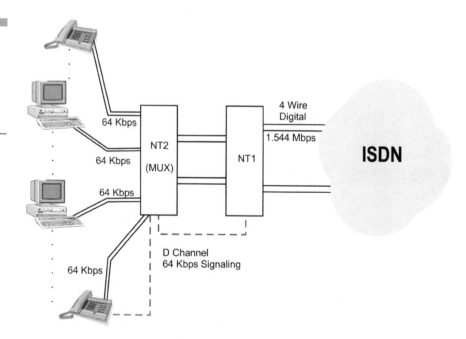

Figure 26.4
The primary rate interface (PRI) supports 23B+D in the United States and 30B+D internationally.

connecting terminals, hosts, digital PBX, and other devices in the premises network. The ISDN architecture includes premises applications processors that provide information movement and management and end customer control features. Examples of some of these features are:

- Voice messaging
- Text messaging
- Message center attendants
- Message detail recording
- Electronic directory
- Traffic data
- Customer station rearrangement

At the network ISDN node, the integrated channelized access is separated into components (either physically or logically, depending on the implementation) and diverted to the appropriate functions. The network ISDN node also provides interconnections to four types of networks:

- Channel-switched networks
- Circuit-switched networks
- Packet-switched networks
- Common channel signaling networks

Another important ISDN interface unique to the United States is the internetwork interface between local exchange carriers (LECs) and interexchange carriers (IECs). The court system and the FCC require that equal access be provided between the networks of the LECs and the IECs. Today, network application nodes are working hand-in-hand with network switches to provide sophisticated services. With progress in distributed processing, the use of application nodes will expand.

Logical View

Circuit Mode Services

The 1984 CCITT red book defines, in Q931 (I.451), messages and procedures for basic call setup and disconnect, and includes a potential framework for supplementary voice services. Carriers use LAPD as the link layer protocol and the extended Q931 as the network layer protocol for the D channel. The extension furnishes signaling for implementing all of today's Centrex and advanced PBX features in ISDN. New functional messages are added, as well as messages conveying activation and deactivation of feature buttons and dial feature/access codes. The extended version of Q931 protocol is being expanded to include functional messages for advanced signaling in the premises network and for features unique to ISDN. For an overview of the message format, see Fig. 26.5. This is the HDLC frame format for LAP-D messages that will traverse the data channel.

ISDN provides transparent circuit mode services for the transfer of information. Protocols for layers 2 to 7 are supplied by end user devices.

The out-of-band signaling infrastructure for the premises network is provided by the D channel in the BRI and PRI. It can also be provided by the public common-channel signaling (CCS) network that transports the user-to-user signaling transparently from one premises switch to another. This extends the benefit of today's CCS network to premises network users and facilitates the interconnection of geographically scattered premises switches.

In the public network, the ISDN user part (ISUP) and the signaling connection control part (SCCP) of the CCITT signaling system 7 (SS7) support end-to-end ISDN features, while the transaction capability (TCAP) supports communications between the network switches and network application nodes. The interworking between Q931 and ISUP is being

Figure 26.5

The HDLC frame format for LAP-D signaling.

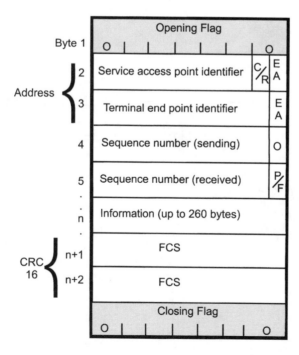

defined. In particular, mapping of messages and procedures for supporting network-wide features, such as call waiting, originating, and automatic callback are addressed. The automatic callback feature should work from any user to any user, no matter where the user initiates a call or what carrier service he or she uses.

Packet Mode Services

The ISDN architecture supports packet switched data services on the 16-Kbps D channel and the 64-Kbps B channel, using X.25, X.31 (an enhanced ISDN version of X.25), and the LAPD-based packet mode protocol.

X.31 extends X.25 to provide dial-up packet switched services in an ISDN. It uses Q931 signaling to establish the physical channel connection to the packet handler, after which the conventional X.25 in-band call control is used to establish the logical connection (Fig. 26.6).

The LAPD-based packet mode protocol is the next step toward integrated signaling. It uses Q931 signaling to establish the physical and logical connections. The LAPD-based packet mode protocol also decomposes

Figure 26.6
The Q.931 signaling establishes a physical connection to a packet switching (X.25) handler.

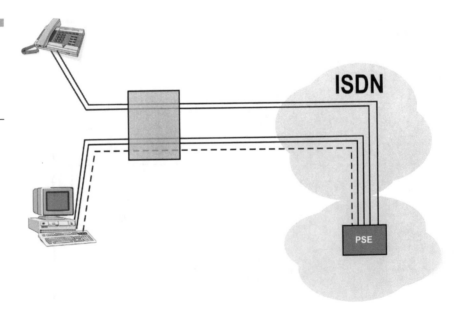

bearer capabilities into building blocks to allow the freedom to distribute functions between the network and the premises system. This allows tailoring of individual services to application needs.

LAPD layer 2 supports multiple logical links on a single physical link. The minimum-functionality packet switched service uses a portion of LAPD consisting of:

- Multiplexing
- Error detection
- Frame delimiting

Other LAPD functions like error retransmission and packet sequencing and acknowledgment can be performed on an end-to-end basis by end user devices. The LAPD-based packet mode protocol initially supports X.25-like services. Unlike an X.25 service, full X.25 is terminated only by the end user devices; the intermediate switch provides the minimum-functionality packet switching. Consequently, the X.25 services supported by the LAPD-based packet mode protocol have lower transnetwork delay and higher throughput. The LAPD-based packet mode protocol, together with its Q931 signaling procedure, provides a mechanism for the customer to request services on the basis of application needs on a call-by-call basis. This means that a user can use the signaling channel to establish a voice call on the link at 64 Kbps as an independent action (Fig. 26.7). The call can proceed as normal until the two parties agree to con-

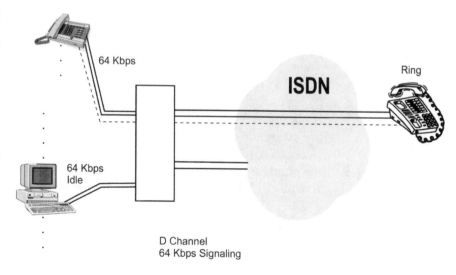

Figure 26.7
The user can establish a 64-Kbits/s voice connection on ISDN in subsecond times. The call will be physically connected until the parties hang up.

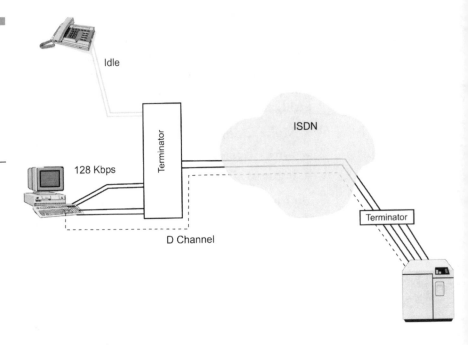

Figure 26.8
Immediately after a
voice call, the user
can signal the
network to establish
a 128-Kbits/s
connection to a
remote host
computer.

clude the conversation and mutually hang up, ending the connection
between them.

Immediately following this call, the user can now use the signaling system to link the two B channels together and form a 128-Kbps data connection to a remote host (Fig. 26.8) to conduct a file transfer to the host system.

Application Services

The marriage of the terminal and telephone expands the horizon of services that the network must provide. More and more required functions
go beyond information movement into information management and its
applications. Voice messages are left on recording systems that can be controlled by the end user for playback and editing. Electronic directory services can request that the switch dial a number directly when a button on
the phone is pushed. These applications require the higher-layer protocols
available in ISDN. CCITT's recommendations for the various layers are:

- X.224 as the transport layer protocol
- X.225 as the session layer protocol
- X.226 as the presentation layer protocol

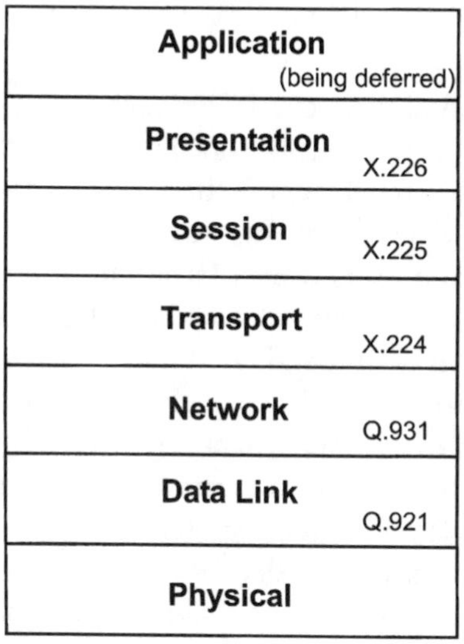

Figure 26.9
The ISDN protocols
as they compare to
the OSI reference
model.

Application
(being deferred)

Presentation
X.226

Session
X.225

Transport
X.224

Network
Q.931

Data Link
Q.921

Physical

These are shown in the context of a comparison with the OSI model in Fig. 26.9. At the application layer, CCITT and ISO are still working on defining the protocols.

Architecture Reality

The rollout of products and services to implement the ISDN concept has been under way for some time now. In the network, the new comers have approached the implementation in three categories:

- Exchange carrier networks (Bell and PTT networks)
- Interexchange networks (interexchange carriers)
- Premises networks (customer premises equipment)

The flagship vehicles for these services are the 5E-ESS and the DMS-100, which offer the features and capabilities in ISDN. Lucent's switches have had ISDN-like interfaces for years at the BRI and PRI levels. From that base, all vendors have continued development of interfaces, beginning with the PRI, to the ISDN standards to take full advantage of ISDN services in the public network.

The Exchange Carrier Networks

The key element in realizing the ISDN exchange carrier networks is the local switch. As long as the carrier (LEC) has a switch that offers the basic interfaces to ISDN, the switch acts as the ISDN node. The switch (typically the end office) will provide simultaneous voice and data services to Centrex and local dial-tone users over a basic rate interface. This switch will support the full range of Centrex voice features using the extended version of Q931, as well as circuit switched and packet switched data. In addition, the switch will more than likely support modem pooling for internetworking with non-ISDN data terminals through the public network. Figure 26.10 shows this in a modem pooling arrangement. This moves the services (a la data circuit switched) onto the network. Users will access these services on an as-needed basis. The network now becomes the service provider instead of just the dial-tone provider.

The switch will also support the end user applications processor environment. This application processor will provide message services, electronic directory, message detail recording, and traffic data statistics. Using SS7 transaction capability (TCAP) and switching services point (SSP), the switch will interact with network databases to process calls (Fig. 26.11). This will allow the exchange carrier to implement an intelligent network architecture and offer such services as virtual private networks, enhanced services, and centralized network control as part of the ISDN capabilities. The use of the primary rate interface connected to the switch will allow the transparent end-to-end transport of user-to-user information for applications such as caller name, security check, and feature transparency.

Additional flexibility can be derived through the use of remote switching modules (RSMs) and subscriber loop carrier (SLC) carrier systems, shown in Fig. 26.12. Using these devices, ISDN services can be provided to remote locations. On the network side (IEC), using signaling system 7 (SS7) in conjunction with the ISDN user part, access can be provided. The combination of channel, circuit switching, X.25 packet switching, and common-channel signaling is derived through a series of ISDN gateways for the inter-/intra-LATA services.

The Interexchange Network

The interexchange carrier network services can be accessed through a major ISDN node either by switched services through the local exchange

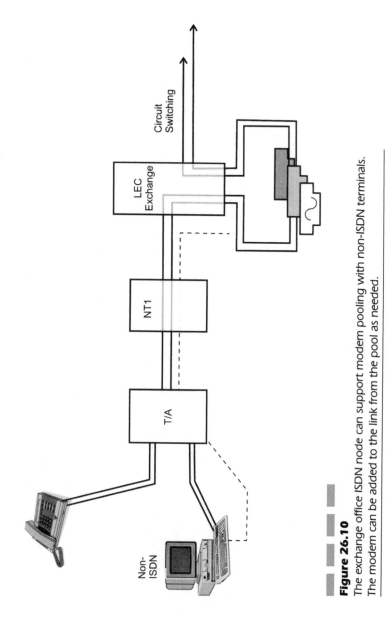

Figure 26.10

The exchange office ISDN node can support modem pooling with non-ISDN terminals. The modem can be added to the link from the pool as needed.

Figure 26.11
The ISDN carrier office allows access to a full range of services on the network. This allows call processing and features in an intelligent network architecture.

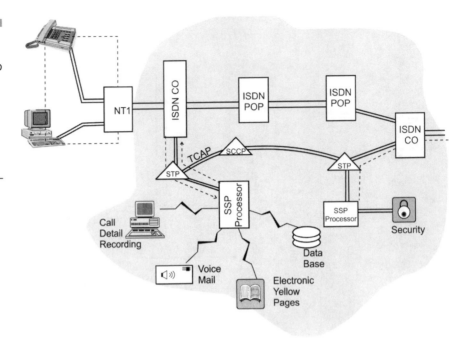

Figure 26.12
The LEC-to-IEC handoff will function through gateways to combine long-distance circuit/ packet/signal switch. Remote areas will be served from the LEC via remote partitions.

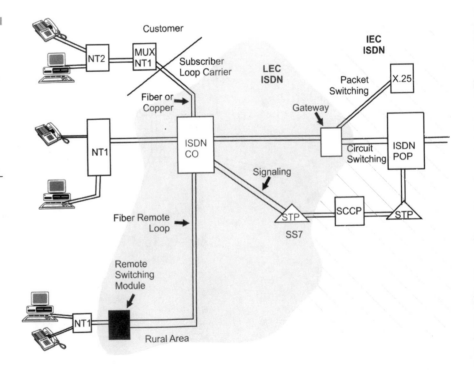

Figure 26.13
Access to the interexchange carrier network can be made through the LEC or direct to the IEC by bypassing the local carrier. At the IEC network, 384–1536 Kbits/s can be accessed through a digital cross-connect system (DCS).

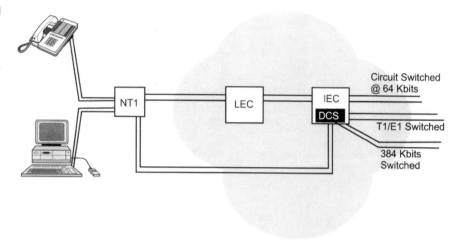

carrier or directly using private line services. The same four supporting transit components are accessible through the IXC network services. Service functions can be attained through a digital cross-connect system (DCS) supporting circuit mode service speeds of 64, 384, and 1536 Kbps as shown in Fig. 26.13. In the AT&T network environment, the use of a 4 ESS with digital access cross-connect systems (DACS) or a 5 ESS supports these services. The use of a packet switching system (IPSS) supports the X.25 packet data network. DACS-based private line services and customer-controlled reconfiguration (CCR) are also provided. AT&T ISDN services also add capabilities through network applications processors that operate off the common-channel signaling network to provide advanced 800/888 services, software-defined networks, and so on. Furthermore, the use of CCR provides the customer dynamic reconfiguration of private line networks.

The Premises Network

The premises line of services and products is designed around the full integration of voice and nonvoice applications, as shown in Fig. 26.14. Intelligence is also provided through a network gateway or node in an ISDN environment. ISDN-capable PBXs acting as the gateways to the public network, as provided by the exchange carriers, are accessed through the PRI. Using this arrangement, a variety of transport services and facilities can be accessed, taking full advantage of the call-by-call service selection capability of Q931. Add-on capabilities such as the following will be available through these ISDN-capable PBXs:

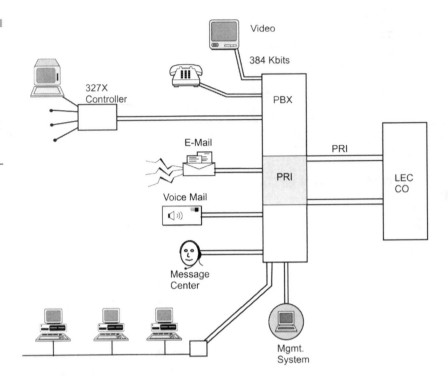

Figure 26.14
A full range of services will be available through the PBX for voice and nonvoice applications. The PBX will use the primary rate interface.

- Message center services
- Voice messaging (voice mail)
- Office telecommunications services (OTS)
- Electronic mail (E-mail)
- Centralized management systems
- Least-cost routing (LCR)
- Address translations
- Security
- Feature transparency
- Station message detail recording (SMDR)

Locally, the PBX will interface to a wide range of equipment for connectivity on the premises. This will include:

- Analog telephones
- Digital telephones
- Switched data services for asynchronous and synchronous terminals
- PCs and integrated workstations

- Cluster controllers (327X) and terminals that are typically hard-wired
- Integration of LANs
- Integration to WANs
- Integration of services of (Lucent) Electronic Tandem Network (ETN)/(NORTEL) Electronic Switched Networks (ESNs)

Basic Operating Characteristics

The basic structure for ISDN operating characteristics is digitally encoded data using a standard of 64 Kbps as the basic rate and 1536 Kbps as the primary rate. This is a derived channel using pulse code modulation techniques, the North American standard for digital transmission.

However, the concept of ISDN is based on a global perspective. Therefore, the use of North American standards for the concept of network transport has to accommodate interfaces to an international standard. In order to fully appreciate the magnitude of the digital hierarchy, perhaps a quick review of digital technology is in order.

First, the analog signal is converted to digital signals using pulse code modulation (PCM) techniques. For a refresher on the pulse code modulation technique, refer back to Chap. 14 in the discussion of T1.

North American T1 differs from the international or European standard of 30 channels at 64 Kbps, plus overhead of one channel at 64 Kbps, yielding a digital stream of 2.048 Mbps—the E1. Other differences exist in the encoding and companding techniques that require translation from North American to European standards and vice versa.

When subrate channels are needed or desired, a different encoding and modulation technique is used. The most common is adaptive differential pulse code modulation (ADPCM), which produces 44 channels at 32 Kbps each. The need for extra overhead for signaling and control dictates that only 44 instead of 48 channels are derived. ADPCM is a standard rate accepted by CCITT for use in ISDN.

Other rates of bandwidth are used in the ISDN concept. They use fractional portions of the DS1/E1 for speeds higher than the 64-Kbps rate. This is also referred to as multirate ISDN. The use of fractional rates works under a nonchannelized format today, but will be defined as a standard in ISDN. See Fig. 26.15 for a representation of this.

Although the channel speeds are defined at 24 or 30 channels at 64 Kbps, the requirement of providing signaling for status (dial tone, ringing, busy, off hook, on hook) leaves an effective data throughput or

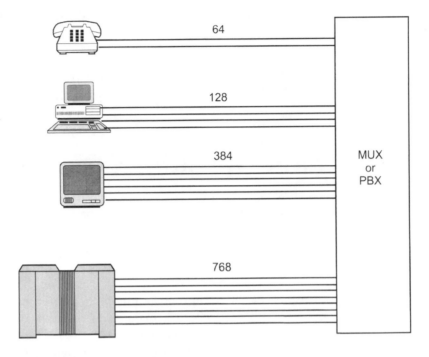

usable bandwidth of 56 Kbps. Again, a refresher on T1 (Chap. 14) will help to bring this back in focus. Bandwidths of 64 Kbps are available using B825.

CCITT and the ISDN study groups are agreed on the use of 8 Kbps of overhead for framing channels on a T1, and the format to be used. North American standards call for ESF under both the AT&T guidelines and the ANSI recommendations.

Bearer Services

ISDN works on the principle of transport services known as *bearer services*. The basic operation of the bearer service is the 64-Kbps channel capacity. The bearer service offers the ability to transport digital voice or nonvoice services using this standard. Packet services can also be transported across the bearer service at higher rates of speed (up to 64 Kbps). Bearer services, as mentioned before, are divided into two categories. The first is the basic rate interface (BRI), which uses two bearer services at 64 Kbps and one data channel operating at 16 Kbps, with an additional 48 Kbps of overhead, yielding a 192-Kbps data stream over the typical two

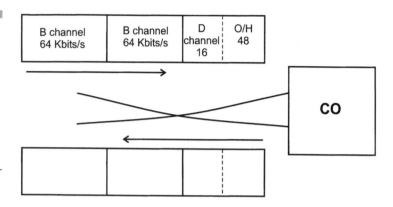

twisted wires (Fig. 26.16) from the business or residence to the local exchange. The second category is the primary rate interface (PRI), which uses 23 bearer services and one data channel, all at the 64-Kbps per channel speed, yielding a 1.536-Mbps data stream from the business to the local exchange. As mentioned before, this is the North American standard (used in the United States, Canada, and Japan), whereas the international standard uses 30 bearer services plus one data service all at the 64-Kbps rate, yielding a 2.048-Mbps data stream. These framing formats are shown in Fig. 26.17.

The primary rate under either the North American or international standard will use existing T1/E1 technology and be the main emphasis for the business world. Applications will be based on the use of WATS, In WATS (800/888/877), FX, tie lines, and data lines all combined into a single primary rate. Because T1/E1 technology has proliferated so much in the United States and the rest of the world over the past few years due to the economics involved, the evolution to a PRI should be simple. The connection through a digital PBX, multiplexer, network node, DCS, or other T1/E1-compatible equipment will require little change in the network and simple card changes in the customer premises network.

Initially, the basic rate interface posed some additional problems. These have since been resolved. For example, the two-wire connection to the small business, small branch office of a larger company, or residential service is typically an analog service. The problem to be addressed entailed how to digitize the customer information at the customer premises and carry the digital information across the two-wire circuit. Digital circuitry has a four-wire, full duplex capability. The North American approach to this is to use a technique known as 2B1Q (Fig. 26.18).

T1

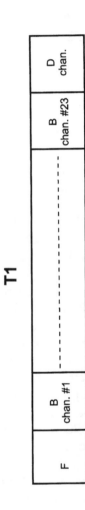

F	B chan. #1	-----	B chan. #23	D chan.

E1

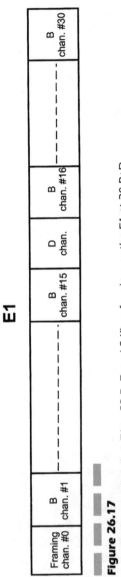

Framing chan. #0	B chan. #1	-----	B chan. #15	D chan.	B chan. #16	-----	B chan. #30

Figure 26.17

The PRI format will deliver T1 at 23 B+D and 8 Kbps framing, or the E1 at 30 B+D.

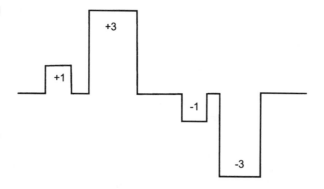

Figure 26.18
The 2BIQ technique allows the two-wire facility to provide ISDN, reduce noise, and cancel echo, and it allows full duplex operation.

In all line coding techniques there is a need to provide for:

- Continuously alternating signals with low bandwidth and good facilities for timing recovery.

- Low attenuation and therefore long line lengths (>2.5 miles).

- Protection from noise interference, in particular near-end cross talk (the transmitted signal interfering with the received signal at the near-end terminal from which it was transmitted).

- Low complexity to the echo cancellation function. CCITT has specified the use of echo cancellation in the transmission mode of ISDN.

- Fast convergence of echo cancellation. At switch-on, there is a perceptible time delay before the echo canceling routine gets into sync. On the subscribers' systems, which discontinue transmission during idle time, the delay must be kept to a minimum.

- High signal-to-noise ratios. This is affected by the number of decision levels in the signal. Pure binary signals with just two decisions will, all things being equal, have a higher signal-to-noise ratio than will multilevel signals.

Line coding methods can be divided into two basic categories: those that involve watching and converting data as it passes (called linear coding methods) and those that require recoding the bit stream through a lookup table (called block coding methods). Alternate mark inversion (AMI) and high-density binary modulo 3 (HDB3) are examples of linear coding methods. Four binary bits to 3 ternary (4B3T) is an example of block coding. Early line coding systems used biphase techniques, while the two main contenders under CCITT guidelines have been AMI and 4B3T. North American standards have introduced the newer contender, 2 binary bits to 1 quaternary (2B1Q).

The linear codes with two decision levels tend to send information to the line at the same rate as it is received in binary form (160 Kbps for the CCITT basic access). The multilevel codes tend to reduce the sending rate because more information redundancy is available in the code. Therefore such codes are more attractive, even though they require more complexity to combat cross talk and intersymbol interference. 4B3T, for example, uses a 120-Kbps line rate and 2B1Q uses 80 Kbps, but at the expense of lower signal-to-noise ratios and more intersystem interference (ISI) (lower baud rate). Many of the LECs have chosen 2B1Q as the method of delivering ISDN to the business and household, for explicit cost reasons.

The original concept of ISDN in the users and the exchange carriers' perceptions was that fiber optics or four-wire circuits would be required to introduce ISDN to the basic rate access. This included business or the residence service. However, to bring fiber to the door, or to reinforce the copper facilities to every door in North America, would have been cost prohibitive. 2B1Q allows the introduction of BRI services over the standard two-wire interface with no appreciable plant facilities costs. This will permit the introduction and proliferation of ISDN services. One estimate of the cost to bring fiber to every door was $482 billion, whereas the installed plant was only valued at $275 billion. Thus an investment of this magnitude would have impeded, if not killed, the ISDN concept of services here in the United States.

Clear-Channel Capability

The use of bandwidth in an ISDN world requires clear-channel capability. *Clear-channel* means that the full 64 Kbps is available for the data stream. As mentioned earlier in Chap. 14, there is a problem with clear-channel capability in the PCM techniques used in the North American standards. Many of the LECs have held back in deploying the use of clear-channel services because of the financial burden and the indecision of the standards committees here in the United States. As the LECs were evaluating clear-channel capacities, the costs of upgrading central office technology to support this were quite high.

In North America, the technique used is B8ZS. Recall the discussion of B8ZS from our discussion in Chap. 14. Data is now transparently transmitted through the network without conforming to the imposition of the 1s density rule. A good deal of activity has taken place to occasion a migration to B8ZS to provide the customer a clear-channel 64-Kbps transport.

Applications for ISDN

A myriad of potential applications exist for ISDN in the marketplace. Many of the vendors have announced products and services that support attachment of other devices to an ISDN network. The name of the game is transparent, digital, switched connections, from any service to any feature or service. In itself, this definition covers a broad range of applications that might have many users skeptical of the real ability of ISDN to perform all things for all people.

Skepticism is probably a good way to approach any new technology. But is ISDN really a new technology? The carriers who have begun to roll out the network services believe that this is nothing new, only a modification of the evolutionary network services. The newness really comes in the form of the reference points and the interface standardization, which users have been after for years. The carrier network has been all-digital for quite some time now, with the analog network remaining at the local loop level. Even the local loop has been evolving to a digital access point over the past decade, with the introduction of T1/T3 as a customer available service.

The only limitations to the applications for voice and nonvoice service are the ability to introduce the technology and the user's creativity. Some of the applications have been demonstrated in this list:

- Internet access
- Desktop conferencing
- Asynchronous protocol conversion to SNA/Synchronous Data Link Control (SDLC)
- Virtual telecommunications access method—internetwork node (VTAM INN) link replacements
- Asynchronous protocol conversion to SNA/SDLC with muxing
- 327X emulation
- Asynchronous access to private packet switched networks
- Asynchronous access to UNIX hosts
- Asynchronous access to UNIX hosts via PAD and X.25 muxing
- 327X coax elimination
- Extended LAN
- Asynchronous access to Ethernet terminal servers
- Replacement of dedicated lines (64 Kbps)
- Asynchronous modem pooling (in or out)
- ISDN wide-area networking (WAN)

- Access to Apple Talk Network and Gateway
- Ethernet LAN bridging and routing
- Customer service call screening/handling
- Incoming call management

Obviously, more applications and demonstrated capabilities exist, but it would be too difficult to generate an all-inclusive list in this format.

ISDN Centrex versus PBX Service

ISDN isn't really a product or a service but a transport mechanism to carry raw bandwidth of user information. It will, however, be implemented through a series of products and services from the carriers and vendors of hardware solutions for the end user. These products and services, as we have seen, will handle a variety of applications on the basis of the users' needs.

In many cases the concept of ISDN has been introduced as a feature of Centrex services. Users are being lulled into the false impression that if they want ISDN, they must use an LEC Centrex offering. Nothing could be further from the truth.

Centrex services make ISDN implementation a lot easier and less risky today. The reason is that the carriers, and in particular the LECs, have the digital switching capability in their exchanges to provide the transport across the network. Because Centrex is also a CO-based service, that is, the Centrex is a software partition of the CO, it would make sense that the ISDN capabilities are merely a logical extension of what the LEC can readily provide to its Centrex users.

However, the capability is also a transport mechanism from an end user's PBX, located at the customer's premises. We are dealing with the ability to deliver digital connectivity to the PBX and interface at the BRI or PRI level. No magic, just connectivity. So, for now let's dispel the rumor that it is a Centrex service.

In the United States, telephony service falls into different categories (Table 26.2).

Centrex

Centrex is a service typically found in North America, but it is now making its way into the international marketplace, where there is no equivalent service at this time. The two main suppliers of Centrex, through switches and offered by the LECs, are Northern Telecom (NORTEL) and Lucent Technologies. Each has a slightly different approach to offering ISDN through its products; however, they are coming closer and closer together as the standards near completion.

NORTEL markets its product, known as Meridian Digital Centrex, offering a full range of features in the Centrex world using a proprietary digital technology. In 1986, NORTEL first offered BRI ISDN on a limited basis as part of the Meridian Digital Centrex line.

Lucent Technologies markets the 5 ESS and 5E-ESS that offer digital Centrex services with a full array of features and capabilities. However, Lucent has gone to a full ISDN environment. Therefore Lucent only offers ISDN Centrex in 5 ESS; no other type of digital Centrex is available.

Both vendors have caused a lot of confusion in the marketplace as they compete for market share in the ISDN arena. Each has published various reports on its implementation of ISDN. Yet, looking more closely at the real world, many of the customers were actually installing Centrex service. The manufacturer, depending on philosophy, stated that it was ISDN. This is especially true in Lucent's case, because ISDN Centrex is all you can get.

This is where the user gets confused. Is ISDN really proliferating, or are the vendors just forcing the numbers to show greater acceptance?

TABLE 26.2

Summary of Telephony Services in the United States

Basic dial tone. Standard telephone service from the LEC to the customer locations (business or residential), terminated on a single line set.

Key systems. The ability to bring multiple lines to the customer premises and terminate in a key service unit (KSU) for access to multiple lines by multiple lines. Usually, in a small office environment, a one-for-one line to user fits.

Centrex. Comes in a variety of names but still does the same thing. Centrex is a CO-based technology that employs a line to every user in the location over a twisted pair of wires. Additional features are available over the basic and key system. All service and maintenance of the Centrex is provided by the LEC.

PBX. Also called by many names, but extends the dial tone from the CO to a DEMARC point at the customer premises. The LEC provides trunks that multiple users compete for. The mix here is far more users competing for fewer trunks, an economic issue. Fully featured, similar to Centrex, but the end user provides all of the administration and maintenance for the PBX.

Hybrids. Some mix of features and capabilities that match PBX, Centrex, and/or key systems.

To use ISDN Centrex, the process is straightforward. Simply order digital Centrex from an LEC running a 5 ESS, or order Centrex in a digital multiplexing switch (DMS) world (NORTEL product) and get the BRI capability. This would provide 2B+D to every extension.

ISDN is a reality today, but it is not limited to Centrex services. The 2B+D capability to the extension and the 23B+D at the access through the CO as a gateway to the interexchange network, using the B channel to access packet services, provide automatic identification of outward dialed calls (AIOD) and user-to-user information through SS7 connections.

ISDN PBX Capabilities

The PBX hasn't been neglected during the evolution of the central office product lines. Many manufacturers have been introducing support of the PRI (23B+D) interface at the PBX level. Every PBX manufacturer has some plan to introduce products and services used in an ISDN, depending on the maker.

The CCITT standards never really addressed ISDN at the PBX level, so many of the manufacturers have created their own "steak-and-sizzle" marketing campaigns to compete for the market share of ISDN PBX capabilities. Primarily, these manufacturers are providing interfaces for the PRI, but they can also access the BRI at the PBX level. Many have introduced BRI services on a station-to-station basis, and support for signaling and control within the PBX for use across public or private (switch-to-switch) networks. Most of these products are proprietary inside the PBX, so the buyer must beware of potential incompatibilities.

Telephone terminal equipment made by one manufacturer will not work in a PBX made by another manufacturer. Typically, when a PBX is selected and installed, the buyer is making a sole source purchase for new or add-on equipment. This is also true in the Centrex environment, where a digital Centrex telephone set from Lucent will not work on a NORTEL ISDN Centrex. So much for standards!

Each of the PBX manufacturers is now trumpeting the capabilities of its own ISDN-compatible PBX to deal with private networking arrangements. Once again, the manufacturers will have some proprietary software in dealing with PBX-to-PBX connectivity through a private network. The users must be aware of how the system will integrate into their networks before making a purchase/lease decision.

Messages and Frame Formats

D Channel Protocols

The D Channel adds a new dimension to digital networking. It provides for an out-of-band common-channel signaling facility for ISDN. Common-channel signaling uses a single channel to convey the necessary signaling information through special messages. This information is used to identify called and calling stations, set up connections, maintain connections, release connections, identify line and network status, provide billing and other operational data, and provide input for network management.

The D Channel for the basic rate interface (BRI) is a 16-Kbps derived channel that is shared for signaling, low-level packet switching, and telemetry. The D Channel for the primary rate interface (PRI) is a 64-Kbps derived channel that is used exclusively for signaling.

The D Channel uses a specific bit-level time slot for the BRI and a specific byte-level slot for the PRI. The network termination (NT) function builds the message-oriented signaling procedure from these bits. The NT uses layer 2 and 3 functions that will use the services of the layer 1 to create a standard signaling channel.

CCITT recommendation I.400 describes terms as link access procedure on the D Channel (LAPD). LAPD is a protocol that operates at the data link layer of the OSI model. LAPD is independent of transmission rate and requires a duplex, bit-transparent D Channel. The D Channel protocols are shown in Table 26.3.

The LAPD layer 2 functions include:

- The provision of one or more data link connections on a D Channel
- Discrimination between the data link connection via a data link connection identifier in each frame

TABLE 26.3

Summary of the Protocols Used on the D Channel

	D Channel Protocols
Layer 3	Q931/I.451
Layer 2	LAPD/I.441
Layer 1	BRI/I.430 PRI/I.431

■ Frame delimiting, alignment, and transparency, allowing recognition of a sequence of bits transmitted over a D Channel as a frame

■ Sequence control to maintain the sequential order of the frames

■ Detection of transmission, format, and operational errors on the data link

■ Recovery from detected transmission, format, and operational errors with notification to the management facility of unrecoverable errors

■ Flow control

The LAPD layer 3 functions are as follows:

■ Providing a means to establish, maintain, and terminate network connections across an ISDN between communicating application entities

■ Routing and relaying

■ Network connections

■ Conveying user-to-user information

■ Network connection multiplexing, segmenting, and blocking

■ Error detection and error recovery

■ Sequencing, flow control, and reset

D Channel Message-oriented Signaling

The D Channel handles the transfer of both user data and signaling information. The signaling information facilitates the establishment, maintenance, and clearing of ISDN channels. ISDN uses out-of-band signaling over the D Channel for a more efficient transfer of data and the overall efficiency of the bearer channels (B channels). The D Channel is presently defined as a three-layer protocol set. These are the bottom three layers: the physical, data link, and network layers.

The network layer protocol, CCITT Q931/I.451, uses messages to convey signaling information between ISDN layer 3 entities. The components of the messages are referred to as *information elements*. Many types of messages exist at layer 3, depending on the network connection. ISDN will support circuit mode, packet mode, and frame relay mode connections. At layer 3, the information field from the LAPD frame is processed. The LAPD frame was constructed at layer 2 from the D Channel layer 1 time slots. I.451 defines the message-oriented signaling application for the information field. The messages transported on the D Channel are as follows.

Call establishment messages:

- Alerting
- Call proceeding
- Connect
- Connect acknowledge
- Setup
- Setup acknowledge

Call information phase messages:

- Resume
- Resume acknowledge
- Resume reject
- Suspend
- Suspend acknowledge
- Suspend reject
- User information

Call disestablishment messages:

- Detach
- Detach acknowledge
- Disconnect
- Release
- Release complete

Miscellaneous messages:

- Cancel
- Cancel acknowledge
- Cancel reject
- Congestion control
- Facility
- Facility acknowledge
- Facility reject
- Information
- Register
- Register acknowledge
- Register reject
- Status

27

Asynchronous Transfer Mode

ATM Capabilities

Asynchronous transfer mode (ATM) is one of a class of packet switching technologies that relays traffic via an address contained within the packet. Packet switching techniques are not new; some (such as X.25 and the original ARPANET) [now called the Internet] have been around since the late 1960s. However, when packet switching was first developed, the packets used variable lengths of information. This variable nature of each packet caused some latency within a network because the processing equipment used special timers and delimiters to ensure that all of the data was enclosed in the packet. The X.25 packet switching techniques were covered in Chap. 18. As a next step toward creating a faster packet switching service, the industry introduced the concept of *frame relay* (covered in Chap. 25). Both of the packet switching concepts (one [X.25] a layer 3 and the other [frame relay] a layer 2 protocol) used variable-length packets. To overcome this overhead and latency, a fixed cell size was introduced. In early 1992, the industry adopted a fast packet or "cell relay" concept that uses a short (53-byte) fixed-length cell to transmit information across both private and public networks. This cell relay technique was introduced as ATM. ATM represents a specific type of cell relay that is defined in the general category of the overall broadband ISDN (B-ISDN) standard. In fact, when we talk about B-ISDN we are actually talking about ATM.

ATM is defined as a transport and switching method in which information does not occur periodically with some reference, such as a frame pattern. All other techniques used a fixed timing reference; ATM does not. Hence the name *asynchronous*. With ATM, data arrives and is processed across the network randomly. There is no specific timing associated with ATM traffic, so the cells are generated as data needs to be transmitted. When no traffic exists, idle cells may be present on the network, or cells carrying other payloads will be present.

What Is ATM?

ATM is a telecommunications concept defined by ANSI and CCITT standards committees for the transport of a broad range of user information including voice, data, and video communication on any user-to-network interface (UNI). Because the ATM concept covers these services, it might well be positioned as the high-speed networking tool of the 1990s and beyond. ATM can be used to aggregate user traffic from multiple existing applications onto a single UNI. The current version of the UNI is 4.0 which specifies the rates of speed and the agreed to throughput at the user interface. As shown in Fig. 27.1, the ATM concept aggregates a myriad of services onto a single access arrangement. Some of these applications include:

- PBX-to-PBX TIE lines/trunks
- Host-to-host computer data links
- Video conferencing circuits
- LAN-to-LAN bridged or routed traffic
- Multimedia networking services between high-speed devices
- Workstations
- Supercomputers
- Routers
- Bridges
- Gateways

All of these services can be combined at aggregate rates of up to 155 Mbps today for user traffic. However, the end-user future rates of speed will be 622 Mbps, accelerating into the 1.2 to 2.4 Gbps class. Currently the carriers are using speeds of 622 Mbps across their backbone networks, but will be deploying the 2.4 Gbps soon. In the future the carriers will step up to the 10 Gbps and higher rates.

Figure 27.1
Multiple inputs are
aggregated onto a
single UNI.

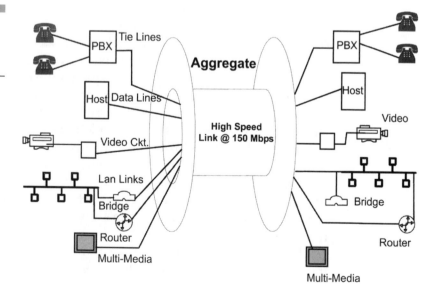

Broadband Communications

The use of fixed statistical (and synchronous) time-division multiplexing services on the telecommunications networks of old had certain limitations. One of these was the restrictive bandwidth available to perform functional transmission. Fixed time slots were used, requiring that amount of bandwidth to be allocated to a specific service at a specified transmission rate. When considering the evolution and migration of these network services, remember that these came from the telephony world. In telephony, a 4-kHz bandwidth was used, and a digital time-division multiplexing scheme was applied to this bandwidth. The result was fixed time slots of 64-Kbps baseband services delivered to a user network. Where larger amounts of throughput were required, these fixed slots were concatenated together to achieve higher rates of speed (i.e., 128, 256, or 384 Kbps, etc.).

However, the fixed slot was just that: fixed. If a user only needed 128 Kbps for 1 hour, the time slots would go unused for the rest of the day. Newer multiplexers called *inverse multiplexers* or *I muxes* are available to allocate time slots together by groupings of channels (i.e., 128, 256, 384, 512, and 768 Kbps). However, some of these are not intelligent enough to pull noncontiguous channels together. For example, if a user wanted to pull together six channels and conduct a videoconference at 384 Kbps, the I mux would look at channels 1 through 6. If one of these six channels was already in use (for a voice or data transfer), then the I mux would have two choices: either knock the existing connection down or wait until the channel became free. This means that the I mux also has to reserve the other five channels so that another connection does not set up. The more expensive I muxes can allocate noncontiguous channels for this video call, but the price for this type of service became prohibitive. Choices had to be made. The choices are less cost and less convenience or more cost for more flexibility. Either way, this is not consistent with the way we run our everyday business. Broadband communications service was introduced as a breathable demand network concept. This allows an organization to use as much as required when required. On release of 384 Kbps of capacity, the networking equipment would reset the values and create a resource that was available. Users, therefore, could select and utilize the required throughput rate as needed. This "breathing" concept makes the service far more usable.

ATM contrasts with synchronous transfer mode (STM). Synchronous transfer mode, which sounds as though it has something to do with synchronous optical network (SONET), actually describes the way the digital telephone networks have worked since the inception of time-division multiplexing. Most synchronous transfer mode signals run on the standard asynchronous DS1 and DS3 used in today's telephony networks.

ATM, on the other hand, is an outgrowth of B-ISDN standards that are intended to run on SONET. Therefore, asynchronous transfer mode is designed to run on synchronous facilities, whereas synchronous transfer mode deals with asynchronous traffic. Talk about confusion! However, STM is based on the fixed time slot that can carry asynchronous traffic, such as voice and video signals. The transmission is based on fixed time slots to prevent delays and latency on these time-sensitive communications. Whoever thought these terms up had to have a perverse sense of humor.

Thus, the need for broader communications channel capacities emerged. *Broadband* is defined as a higher throughput rate of bursty traffic than is traditionally available on the telephone companies' channel capacities. Instead of a 64-Kbps (DSO) channel, the broadband communications handle multi-megabit transport starting at the very low end at 1.544 Mbps, progressing upward in "chunks" or increments up to 155 Mbps, 622 Mbps, and 1+ Gbps.

The characteristics of broadband communications will be the transport of higher-speed communications of a continuous or bursty nature from a variety of inputs to a mix of outputs. All of these techniques and characteristics are based on strict TDM/PCM and SONET. Before going any further, here's a quick review of both TDM and SONET so that we are all on the same wavelength.

Time-Division Multiplexing

Time-division multiplexing and pulse code modulation were designed as the first steps of a digital carrier and transport system. In 1958, the Bell System had created this concept to replace the older analog system (Chap. 14).

A Review of the Digital Multiplexing Function

First, to create a TDM/PCM scheme, the techniques were evolved from the 4 kHz of bandwidth typically allocated to a telephone circuit. Remember that this 4 kHz has already been referred to earlier in this book. See Fig. 27.2 for a graphic representation of the 4-kHz channel.

Figure 27.2
The 4-kHz channel is
used to create a
digital data stream.

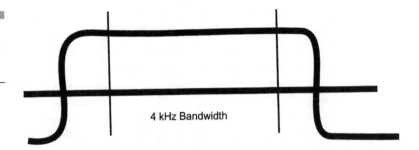

Figure 27.2
The 4-kHz channel is
used to create a
digital data stream.

4 kHz Bandwidth

Using the 4-kHz channel, sampling of the introduced signal will occur at twice the highest range of frequencies on the line. This means that a sampling rate of 8000 times per second is achieved as follows:

$$\text{Sampling rate} = \text{Bandwidth} \times 2$$

$$8000 = 4000 \times 2$$

8000 samples per second yields a standard 125-μs sample time

$$1 \text{ second}/8000 \text{ samples} = 0.000125 \text{ second}$$

Time-Division Multiplexing/ Pulse Code Modulation

Once the sampling rate (8000 samples per second) is established, the value of the sample is established as well. Under normal conventions, a signal is converted from the analog wave into a digital pulse stream. The pulses, therefore, must relate to some value along the curve of the wave. This is done by considering the coarseness of the analog wave. We use a set of values based on 256 combinations, which is derived by placing sensitivities and using binary coding based on an 8-bit sample. This is derived by:

$$2 \text{ states in binary} = 0/1$$

$$2^8\text{-bit sample} = 256 \text{ combinations}$$

Using an analog wave with amplitudes peaking on both the positive and negative sides of the zero line results in 128 possible values on the positive side of the wave and 128 values on the negative side of the wave. This is pulse amplitude modulation. Note also that when representing the values, a 0 in the most significant bit (MSB) slot (the farthest to the left) represents a negative value. Alternately, a 1 placed in position 8 (the MSB) represents a positive value. Zero has two values then: 0000 0000 and 1000 0000. We see that a 0 can exist on the positive side of the wave and on the negative side of the wave.

The final step is to digitally encode and transmit the signal in pulse code modulation. Using the typical sine wave, if we select the value of one sample at 5, the value of the sample of this wave is digitally encoded into a data stream of:

0000 0101

This is then transmitted in digital form. Refer to Fig. 27.3 for a graphic representation of this consolidated onto a single drawing for ease of viewing. TDM/PCM using the 8-bit value and 8000 samples per second therefore operates at a channel speed of 64 Kbps. This is the basis for the digital carrier system.

Using the four-wire circuit for transmitting the digital signal, the time-division multiplexer can take 24 inputs and multiplex them onto a single-carrier system called the T1 carrier. Therefore a circuit carries 24 simultaneous channels multiplexed together with an extra overhead of 8 Kbps (1 bit per sample of 24 channels × 8000 samples per second). The resultant yield is 1.544 Mbps. This was covered in much greater detail in Chap. 14; refer back to that chapter if you need more detail on how this works.

$$[24 \times (8000 \times 8)] + [1 \times (8000)] = 1.544 \text{ Mbps}$$

$$(24 \times 64{,}000) + (8000) = 1.536 \text{ Mbps} + 8 \text{ Kbps} = 1.544 \text{ Mbps}$$

Figure 27.3
The PCM technique transmits a binary data stream.

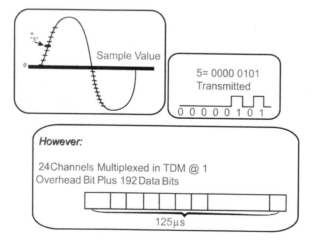

Sample Value

5= 0000 0101
Transmitted

0 0 0 0 0 1 0 1

However:

24 Channels Multiplexed in TDM @ 1 Overhead Bit Plus 192 Data Bits

125 μs

The Digital Hierarchy

Using these TDM/PCM values, a North American digital hierarchy was developed. Everything is based on a standard of 8000 samples per second and a digital multiplexing scheme. The hierarchy used below set the rules for future levels of multiplexing. This North American digital hierarchy is shown in Table 27.1.

Each of the previous rates was established on the same rules. A frame of information that carries an 8-bit sample from each of the channels within the frame is transmitted in 125 microseconds. Thus a DS1 frame consists of 193 bits (192 bits of data and 1 framing overhead bit), whereas a DS3 frame (called an M13 frame) consists of 4760 data and overhead bits. In a DS3 multiplexing scheme, the timing is adjusted by adding bits to make up the 125-microsecond time slots (called justification). The M13 frame is shown in Fig. 27.4, depicting the 4760 time slots. A problem exists with the M13 asynchronous protocol used to transmit this information. The frame is too small to multiplex 28 T1 frames into a single M13 frame. The thought behind this is that 28 T1 frames will require 5404 bit slots. The M13 only has 4760 (only 4704 of these are available because of the overhead added, as shown in Fig. 27.4). Thus the M13 frame must be transmitted 9366 or 9367 times per second. This makes it asynchronous, because the arrival rate of the frames will be different. Further, the 125-microsecond clock will be lost. The bit sequencing within this frame is also very ambiguous. You cannot trace a channel through the M13 asynchronous protocol, because the bits are moving from frame to frame into different bit slots. In order to perform any diagnostics or troubleshooting, you must demultiplex the entire DS3, a process that is inconvenient and disruptive.

TABLE 27.1

Summary of the North American Digital Hierarchy Values

Name	Capacities	Equivalent	Yield
DS 1	24 channels @64 Kbits/s	One T1	1.544 Mbits/s
DS 1C	48 channels @64 Kbits/s	Two T1s	3.152 Mbits/s
DS 2	96 channels @64 Kbits/s	Four T1s	6.312 Mbits/s
DS 3	672 channels @64 Kbits/s	Twenty-eight T1s	44.736 Mbits/s

Figure 27.4
The DS3-M13 frame uses a 4760-bit framing sequence.

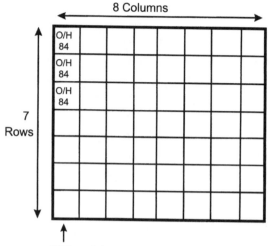

8 Columns

7 Rows

85 time slots composed of 1 overhead bit plus 84 data bits yield 4760 bits per frame.

SONET

To rectify the problem with the M13 frame format, SONET was created. All of the digital transmission speeds were initially set to be carried on the existing wired systems in place at the telephone company and inter-exchange carrier levels. These systems included microwave, coaxial, and twisted-pair copper wires. As newer fiber-optic systems were introduced, a new synchronous digital hierarchy (SDH) evolved. Throughout the world, SDH was accepted, yet in the United States the term *synchronous optical network* (SONET) was adopted. Although the international and United States versions of SDH/SONET are very close, they are not identical.

SONET specifies a synchronous transport signal (level 1) of 51.84 Mbps, which is a DS3 with extra overhead. The overhead allows for diagnostic and maintenance capabilities on each synchronous transport signal (STS). The optical equivalent of the STS-1 is an optical carrier level 1 (OC-1). This is the basic building block for chunks of bandwidth as they are multiplexed together to form much higher capacities. Increments of SONET include the capacities shown in Table 27.2. These transport systems and carrier levels work from the base of a T3 plus overhead creating an STS-1.

SONET goes further than just defining the multiplexed values of speed. It breaks the architecture of the link into three separate steps for

Chapter Twenty-Seven

TABLE 27.2

Typical SONET
Specified Speeds

SONET Name	Speed
OC-1*	51.84 Mbits/s
OC-3*	155.52 Mbits/s
OC-9	466.56 Mbits/s
OC-12*	622.08 Mbits/s
OC-24†	1.244 Gbits/s
OC-36	1.866 Mbits/s
OC-48†	2.488 Gbits/s
OC-96	4.976 Gbits/s
OC-192	9.953 Gbits/s
OC-255	13.92 Gbits/s

*Common speeds for ATM.
†Future speeds for the broadband ISDN networks.

purposes of defining the interfaces and defining responsibility. These layers include (a fourth is added here to describe the layer-by-layer responsibilities):

■ *The Photonic layer.* Deals with the transport of bits and the conversion from an STS electrical pulse into an OC signal in light pulses.

■ *The Section layer.* Deals with the transport of the STS-N frame across a physical link. The functions of the section layer include scrambling, framing, and error monitoring.

■ *The Line layer.* Deals with the reliable transport path of overhead and payload information across a physical system.

■ *The Path layer.* Deals with the transport of network services such as DS1 and DS3 between path terminating equipment.

SONET incorporates each of these into an architecture such that overhead and responsibilities are clearly defined. This layered architecture is shown in Fig. 27.5. Note that the four layers are shown in this graphic. Actually, the photonic and section layers are one, but have been subdivided for clarity in showing the structure. The photonic and section layers deal with the physical medium, or the OSI layer 1 protocols.

The SONET architecture actually consists of three parts:

Figure 27.5
The four layers of the SONET protocols.

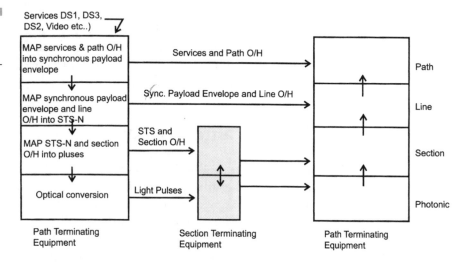

| Path Terminating Equipment | Section Terminating Equipment | Path Terminating Equipment |

- The *section* is defined as the transport between two repeating functions or between line-terminating equipment and a repeater.

- The *line* is defined as the transport of the payload between two pieces of line-terminating equipment.

- The *path* is defined as the transport of the payload between two pieces of path-terminating equipment (multiplexers, etc.) or the end-to-end circuit.

This is shown in Fig. 27.6, where the pieces are laid out on a plane. The graphic shows how the pieces all work together in a fiber link. The fiber specification is what was originally used in SONET (and SDH interna-

Figure 27.6
The SONET architecture is laid out to define responsibilities for the signal.

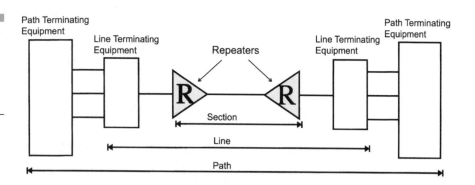

Section, line and path definitions

tionally); however, newer microwave and satellite termination equipment is being introduced to provide the same capacities and formats. SONET defines the responsibility of the carriers in multiplexing the signal onto the physical medium, as well as the points of demarcation for the various carriers involved with the circuit.

SONET Frame Format

In each of the layers listed, overhead is added to the data being transported to assure that everything is kept in order and that errors can be detected. This overhead allows for testing and diagnostics along the entire route between the functions specified in the SONET layers. A new frame size is also used in the SONET architecture. The older M13 asynchronous format just couldn't cut the mustard. In order to get the network back to a synchronous timing element, a frame is created that consists of 90 octets (8-bit bytes) across and 9 rows (810 bytes) down. This is the OC-1 frame, or the equivalent of the DS3 frame plus the extra overhead. However, with the change in size, the frequency of frames generated across the network is now brought back to 8000 per second. The 125-second clocking and timing for the digital network can be reestablished. The frame shown in Fig. 27.7 is 90 bytes (columns) wide and 9 rows high, yielding the 810-byte frame format.

Of the 90 columns, the first 3 ($3 \times 9 = 27$ bytes) are allocated for transport overhead. The transport overhead is divided into two pieces: 9 bytes (3 columns, 3 rows) for section overhead and the remaining 18 bytes (3 columns by 6 rows) for line overhead. This is for maintenance and diagnostics on the circuit.

The 783 remaining octets (87 columns, 9 rows) are called the *synchronous payload enveloper* (SPE). From the SPE, an additional 9 bytes (1 column, 9 rows) are set aside for path overhead. The path overhead accommodates the maintenance and diagnostics at each end of the circuit—typically, the customer equipment. Thus the leftover 774 bytes are reserved for the actual data transport.

(774×8 bits = 6192 bits $\times$ 8000 frames per second 49.536 Mbps)

ATM uses the STS payload as the transport of information by inserting 53 cells into the STS payload. This allows for the fluid and dynamic allocation of the bandwidth available. As it is advertised, ATM starts at 50 and 155 Mbps. To achieve the 155-Mbps rate, three OC-1s are concatenated (kept together as a single data stream) to produce an OC-3C. This works out rather nicely.

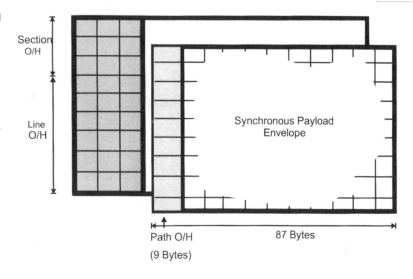

Figure 27.7
The SONET frame is
810 bytes long; it will
be represented 8000
times a second.
Overhead is allowed
for maintenance.

While using the SONET specification for carrying the data, cells are mapped horizontally into the SONET frame shown in Fig. 27.7. However, newer things have occurred since the initial SONET standards were developed. Now SONET operates at speeds of up to the OC-192, which is the equivalent of 192 T3s. OC-192 is 9.958 Gbps of throughput. However, using a dense wave-division multiplexing concept (sending different colors of light onto the fiber) we can achieve approximately 32 OC-192s on a single piece of fiber, or 320 Gbps. Moreover, the manufacturers (like Nortel Networks and Lucent Technologies) are rapidly pursuing data rates at 1.6 Tbps* on a single fiber. With these speeds being generated on the fiber, we shall see nearly unlimited bandwidth in the future. This, of course, creates the scenario to offer voice, data, and video on a single access link using ATM and SONET at a fraction of current costs.

One can only imagine where this will all lead. If we can get to 1.6 Tbps, what about 10 Tbps? If we get to 10 Tbps, then what about 100 Tbps? This is the way we have been pushing the carriers and manufacturers alike. They, of course, have been responding in kind.

The use of the higher-speed communications methods also leads to a newer transmission rate. Nortel Networks recently announced that using SONET OC-192, they will deliver 10-Gbps Ethernet speeds across the wide area networks. This becomes a battleground for the various suppliers; the telcos want to deliver gigabit speeds over their ATM backbone, whereas the newer carriers (like Qwest Communications and Level 3 Communi-

---

*Terabits (trillions of bits) per second.

cations) want to deliver across SONET backbones and they do not care what form is provided here.

The Cell Concept

In the past, using the traditional time-division multiplexing arrangement, bandwidth was provided on a fixed time slot basis. If a user wanted a high data rate transfer, this fixed time slot limitation got in the way. Using the standard 64-Kbps transmission speeds, you would be able to derive rates based on standards of 64-Kbps chunks. However, this only worked for a channel capacity of 64 Kbps or 1.536 Mbps.

If a user wanted 384 Kbps (six channels multiplexed together), the service was not ubiquitous. In many cases, the user had to either derive 64 Kbps or use a full T1/E1 service. There just wasn't much in between. In addition, even if these services were available, the price might have been too expensive. The network did not lend itself to user needs, so the user had to fit the application to the capacities offered by the network. Obviously, this impeded the ability to use bandwidth efficiently and cost the user either money or throughput performance. Something had to be done!

Network suppliers began to look at the inefficient use of capacity. They found that they were consistently overbuilding their infrastructure of links or trunks only to lease it off at lower rates than they should have. These low rates were supposed to make the service more attractive to the end user. However, this overbuilding is expensive because the suppliers were providing service that was only minimally used and they could not reclaim the bandwidth that sat idle for 8, 12, or 16 hours a day.

Looking at user demands and the suppliers' own financial picture, a new concept emerged: rather than forcing the user to adapt to the constraints of the network, why not let the network adapt to the needs of the user? Thus ATM was born. ATM gets around the inefficiency of the fixed time slot, rates, and formats of the TDM world by allocating whatever is necessary to the user whenever the user wants it. To do this, the network suppliers looked to the packet switching world and settled on a service that mimics packet switching. Differences exist, however. Instead of using the processor-intensive slow-speed services of X.25 packet switching (which tops out at 64 Kbps), or the speedier frame relay (2 Mbps or a deviation up to 50 Mbps), a mix of packet/frame technology evolved using a fixed-size cell that offers higher throughput due to efficient use of bandwidth without the overhead of X.25 and frame relay.

The Importance of Cells

These cells can get around the waste of frame relay or other frame concepts. In frame relay, the frames are larger, but variable in length. Admittedly, the variable amount of data in a frame accommodates the burstiness of the data. However, the use of the timers and delimiters places extra constraints on the use of packets and frames. Therefore, if a frame size is set to accommodate LAN traffic, such as an Ethernet frame of 1500 bytes, the network deals with a frame of the same size. Consequently, the 1500-byte frame is used, even if only 150 bytes are to be transmitted. In many implementations, a pad function (filler) is used to fill the frame for transmission. The network has been used for a much longer period to send an insignificant amount of data. The use of this frame was only 10 percent efficient. This was done through the fills and the buffering of data until the network switch was sure that all of the data was received before sending it on to the next location in the network. While the network was handling this partially empty frame, other users who wanted to transmit were delayed, all because the network had to recover timing, framing, and formatting on the fixed time slots. Moreover, when a variable frame size is used, the buffers across the network have to work accordingly. In many cases the buffers in the frame relay switches are set to expect a full frame of 1610 bytes (plus the associated overhead). If a switch has a nearly full buffer, and a frame arrives, the switch must allocate the full-size buffer for the receipt of the frame. If a full frame buffer cannot be allocated, the switch throws the frame away. Although the frame may have been partially filled, the maximum buffer cannot be allocated. Therefore, the network is inefficiently used because the data is discarded when it did not have to be. Ultimately the frame size was extended to a maximum of 4096 bytes because of efforts and improvements by Cascade Communications* to accommodate IBM token and other frames. This just adds to the complexity of the needed buffer space. The switches on the network have to receive the beginning (opening) flag, buffer the frame until it has everything, then look for the closing flag before the network can process the data.

On the other hand, a much smaller cell allows the transmitter to break down large blocks of information into more manageable pieces. If the frame used (in this example) is only partially filled, the network need not be concerned. It will only be required to send the 150 bytes of information plus any associated overhead in a couple of cells. Thus, the network

---

*Cascade Communications was later acquired by Ascend. Ascend was then acquired in 1999 by Lucent Technologies.

performs more efficiently. The processor speed can be used to maximize the throughput and minimize delays across the network. Cells can now be processed very quickly in silicon (chipsets). Many of the ATM switch manufacturers are now using application-specific integrated circuits (ASICs) in the systems to process the cells efficiently and quickly.

Deriving Bandwidth

When these fixed cells are used, another benefit is achieved. Users who needed to transmit 8 Mbps were stuck. Either they would have to settle for a T1 (1.544 Mbps) transmission, which would ultimately cause congestion, or they would have to lease a T3 (44.736 Mbps), which would be too expensive and mostly unused. Again, the fixed time slot arrangement of the TDM world got in the way.

Using a cell concept, the user would transparently send interleaved cells across the network regardless of the amount of bandwidth needed, and consequently get an effective throughput of 8 Mbps. This is based on the availability of the fiber distribution to the user's door, because it is the fiber (using SONET) that will deliver the capacity necessary to derive the bandwidth being discussed.

As cells are transmitted, they will be stacked up in the SONET frame (STS-N frame) to derive the necessary bandwidth that the user needs. Once the transmission of 8 Mbps is completed, a new user might only need 384 Kbps, so the cells will be used to derive that amount, making the rest available to others.

Cell Sizes and Formats

The standard fixed cell size for ATM is 53 bytes (octets). This comprises 5 bytes of overhead for activities such as addressing, and 48 bytes of payload. The cell is shown in Fig. 27.8. The 48-byte payload will be a variable, depending on the information and control necessary. There are several types of cells used that could take 4 bytes of the 48 away for control, leav-

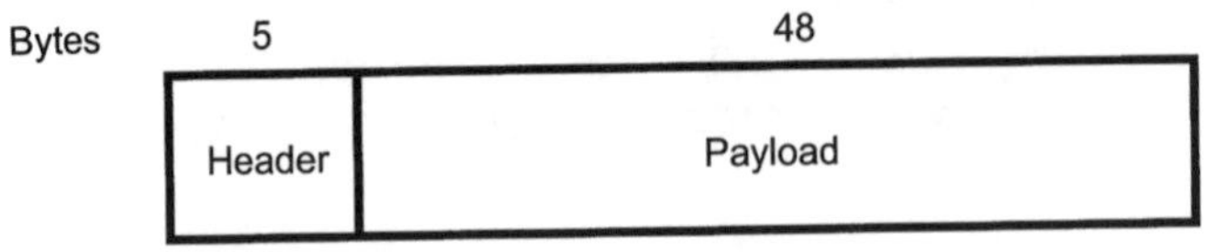

Figure 27.8
The ATM cell consists of 5 bytes of header and 48 bytes of payload.

ing the user 44 bytes of effective data; another option gives the user all 48 bytes of payload for data. It depends on the implementation and the service run on ATM. Either way, the cell is still fixed at 53 bytes long.

The cell is broken down as follows. Five bytes of overhead are shown in Fig. 27.9.

General Flow Control Identifier

The first 4 bits of byte number 1 contain what is known as a *general flow control identifier* (GFC). This is used to control the flow of traffic across the user-to-network interface (UNI) out to the network. The standards bodies are still deciding how to best use the GFC, and are not explicit in their definition of its use. Remember that this is only used at the user interface, because once the cell goes onto the network-to-network interface (NNI) that is between network nodes, these extra 4 bits will be reassigned for network addressing. The GFC is only used at the entry to the ATM backbone network between the user device and the rest of the network.

Virtual Path Identifier

The next 8 bits of the header are called the *virtual path identifier* (VPI). A VPI is part of the network address. A virtual path is a grouping of channels between network nodes.

Virtual Channel Identifier

The virtual channel identifier (VCI) is a pointer on which channel (virtual) the system is using on the path. The combination of the virtual path and virtual channel make up the data link running between two network nodes. The VCI is 16 bits long, using the second 4 bits of byte 2, all 8 bits of byte 3, and first 4 bits of byte 4.

Figure 27.9
The overhead uses 5 bytes to keep addressing in order.

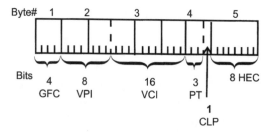

Payload Type

Three bits in byte 4 are allocated to define the payload type (PT) which indicates the type of information contained in the cell. Because these cells will be used for transporting different types of information, the network equipment might have to handle it differently. Several types of payload type indicators have been defined:

■ Types 0 to 3 are for the data types inside the cell.

■ Types 4 and 5 are for network maintenance and control.

■ Types 6 and 7 are being defined.

Cell-Loss Priority

Bit number 8 of byte number 4 is a cell-loss priority (CLP) bit. The user can define whether or not to discard the cell if congestion occurs on the network. If congestion occurs and the bit is set to 1 by the user, the network can discard the cell. If the bit is set to 0, then the cell may not be discarded.

Header Error Control

The 5th byte of the overhead is used as a header error control (HEC). This is an error-correcting byte that is conducted on the first 4 bytes of the header. It is used to correct single-bit errors and detect multiple errors in the header information. If a single-bit error occurs, the HEC will correct it. However, if multiple errors occur in the header, it will discard the cell so that cells will not be routed to the wrong address because of errors occurring on the network. The HEC only looks at the header information; it does not concern itself with the user data contained in the next 48 bytes.

The Cell Format for User Data

Once the header is completed, the user information is then inserted (Fig. 27.10). As already mentioned, the user field is either 44 or 48 bytes of information, depending on the process used. Here's how this works.

Figure 27.10
Once the header is
computed, the
payload is attached.

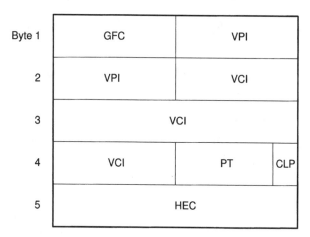

ATM Cell Overhead

Figure 27.10
Once the header is
computed, the
payload is attached.

The Adaptation Layer

Called the ATM adaptation layer (AAL), this is probably the most significant part of ATM. The adaptation layer provides the flexibility of a single
communications process to carry multiple types of traffic such as data,
voice, video, and multimedia.

Each type of traffic has varying needs. Voice needs constant data traffic;
LAN is bursty in nature; video is time sensitive; data transfers from host to
host can be delayed without a problem. It is in this adaptation process that
the network can deal freely with varying types of information and only
route cells on the basis of the routing information in the header.

Just about any type of transmission will require more than a single cell
of information (48 bytes), so the ATM adaptation layer divides the information into smaller segments that are capable of being inserted into cells for
transport between two end nodes. Depending on the type of traffic, the
adaptation layer functions in one of five ways. These are shown in Table
27.3, with various types of information and adaptation layers involved.

The Adaptation Layer Process

The AAL is broken down into two sublayers; first is the convergence sublayer, the second is the segmentation and reassembly sublayer (SAR). This

TABLE 27.3

Summary of the
Five Different
Adaptation Types

Type	Name	Description
1.	Constant-bit-rate (CBR) services	Allows ATM to handle voice services at DS0, DS1, and DS3 levels. Recovers timing and clocking for voice services.
2.	Variable-bit-rate (VBR) time-sensitive services	Not finalized but reserved for data transmissions that are synchronized. Also will address packet mode video in a compressed mode using bursty data transmission.
3.	Connection-oriented VBR data transfer	Bursty data generated across the network between two users on a prearranged connection. Large file transfers fall into this category.
4.	Connectionless VBR transfer	Transmission of data without a prearranged connection. Suitable for LAN traffic that is bursty and short. Same reasoning as X.25, where dial-up connection and setup take longer than data transfer.
5.	Simple and efficient adaptation layer (SEAL)	Improved type 3 for data transfer where higher-level protocols can handle data and error recovery. Uses all 48 bytes as data transmission and handles message transfer as sequenced packets.

SAR is similar to X.25 PADs. The purpose of the process is to break the data down into the 48-byte payloads, yet maintain data integrity and pointers for ID purposes. This process of two sublayers produces a protocol data unit (PDU). The convergence sublayer PDU is of a variable length that is determined by the AAL type and the length of the higher-layer data passed to it. The SAR-PDU is always kept at 48 bytes to fill an ATM cell data stream. This is shown in Fig. 27.11 as it goes through the process.

As the user data, which can be multimegabyte files, is passed down to the convergence sublayer (CS) process, the data is broken down into variable block lengths. A maximum of 64 Kbytes is used in this process. The large user file is broken down into the 64-Kbyte segments. A header and trailer describing the type and size of the CS-PDU are added. This is then passed on to the next sublayer process.

The SAR then receives the CS-PDU and breaks it down into 44-byte cells (if less than 44 bytes, the rest is padded). Additional overhead (2 bytes of header and 2 bytes of trailer) is added to the SAR-PDU. The simple and efficient adaptation layer (SEAL) instead uses all 48 bytes for user information, and therefore uses the bandwidth more efficiently.

Figure 27.11
The adaptation process goes through two steps to produce the ATM payload.

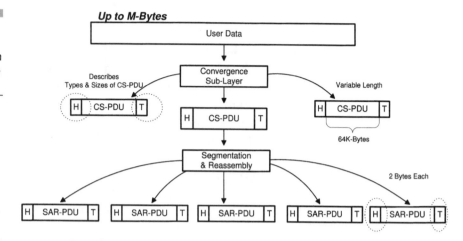

Finally, the PDU is then inserted into an ATM cell with the 5-byte header. Following this, the cell is handed down to the physical cable system. These cells are then delivered across the physical media into SONET or other transmission. Thus the speeds used involve continuous or variable bit rates, depending on the application. Being carried in a SONET frame, the cells can be strung out to get the highest bandwidth possible. Unlike DS1 or DS3 services, ATM cells are slotted into the SONET frame horizontally rather than vertically.

ATM Standards Protocols

The development of communications standards takes on a global perspective and coordination. There are many standards bodies involved with the often lengthy task of generating a consensus on implementing a broadband communications technology. These bodies are continually refining and extending the standards, adding more advanced features and capabilities. At each revision of a standard, the cumulative experience with the older implementation is used to update and improve the standard. The older standard implementations remain useful; they are simply built on. The standards process has always been a fine balance between technology and commercial interests. A recent phenomenon among standards groups is the emergence of the forum from various commercial groups. The purpose of the forum is to impact positively the standards process for a particular communications technology. The forum consists of a group of interested carriers and vendors supporting education and implementation activities.

CCITT

The International Telegraph and Telephone Consultative Committee Communications Standards Organization (CCITT) provides communications standards for the International Telecommunications Union (ITU), which is supported by the United Nations. The 15 study groups of CCITT have specific technical responsibilities for developing international standards. For example, ATM and SONET/SDH standards are being developed in Study Group XVIII. Every four years, the study groups submit recommendations to the CCITT plenary for approval. The CCITT standards are law in some European countries, but in most countries, they are treated as recommendations. The clout of CCITT in setting standards comes from coordination with national standards groups, such as ANSI, the thoroughness of its standards process, and its worldwide stature.

CCITT has recommended that ATM be used in worldwide broadband networks. The first standards were produced in the mid-1980s and provided a basic outline of the service. ATM was chosen as the transfer mode for B-ISDN in 1988, and an initial set of recommendations was agreed on in 1990. These recommendations specify ATM among the protocol suites at the lower layers of the OSI reference mode. Most ATM standards are already specified by the international CCITT committee. However, work continues in the area of signaling, call setup definitions, and network management functions.

The B-ISDN protocol reference model shown in Fig. 27.12 is defined in CCITT recommendation I.121 into multiple planes:

- *The U plane.* The *user plane* provides for the transfer of user application information. It contains the physical layer, the ATM layer, and multiple ATM adaptation layers required for different service users (e.g., continues bit rate and variable bit rate service).

- *The C plane.* The *control plane* protocols deal with call establishment and release and other connection control functions necessary to provide switched service. The C plane shares the physical and ATM layers with the U plane. It also includes the ATM adaptation layer procedures and higher-layer signaling protocols.

- *The M plane.* The *management plane* provides management functions and the capability of exchanging information between the U and C planes. The M plane contains two sections: layer management, which performs layer specific management functions, and plane management, which performs management and coordination functions related to the complete system.

Figure 27.12
The B-ISDN protocol
reference model
(CCITT).

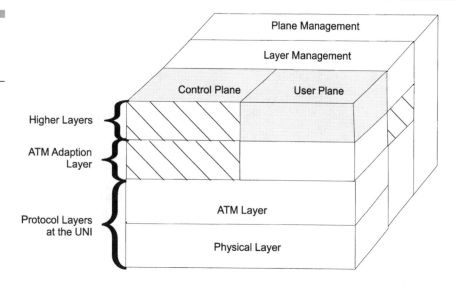

Protocol layers involved for initial deployment (PVC).
Extra layers necessary for switched service (SVC).

Each of these protocols are contained in the user-network interface (UNI) protocol architecture. The UNI specification involves the protocols that are either terminated or manipulated at the user-network interface. ATM bearer services, defined by both ANSI and CCITT, provide a sequence for preserving connection-oriented cell transfer service between the source and the destination, with an agreed-on quality of service (QOS) and throughput. ATM bearer services involve two lower protocols, the ATM and the physical, of the B-ISDN protocol stack. These two layers are service independent and contain functions applicable to all upper-layer protocols. The ATM bearer service at the public UNI is defined as the point-to-point, bidirectional virtual connections at either a virtual path (VP) level and/or a virtual channel (VC) level. The UNIs (private and public) are limited to the physical and ATM layers and higher-level protocols required for UNI management.

Physical layers can include DS3 and SONET for both public and private UNIs. Additional physical layers can be specified for the private UNI.

Rates

Supported U.S. access rates to a public wide-area network supporting cell relay service (ATM) include those shown in Table 27.4.

TABLE 27.4

Basic Rates
Supported on
ATM Under U.S.
Standards

Supported Rates
DS3
STS-1
STS-3c
STS-12c

The physical access channel can be a T3/DS3 or SONET-based (STS-3c or STS-12c) facility. ATM cells must conform to ITU-T recommendation (CCITT) I.361. Cells are transported to their destination with a level of assurance consistent with a prespecified quality of service (QOS). The actual services provided vary, and service is delivered in a transfer rate of cells per second. Supported transfer rates in the United States range from 150 to 1,470,000 cells per second. The cell/channel control protocols are based on CCITT Q93B, which is further defined by the ATM Forum.

ATM is a standard based on the overall B-ISDN reference model, as already stated. It is based on a layered architecture concept similar to that used by the International Standards Organization (ISO) seven-layer open systems interconnection (OSI) model. Basically, these models divide any communications process into subprocesses, called *layers*, arranged in a stack. Each layer provides services to the layer above to aid in communications between the top-layer process or applications at the end of the connection.

The ATM-relevant portions of the B-ISDN model comprise the bottom three layers of the B-ISDN protocol stack; the ATM adoption layer (AAL), divided into two sublayers; the ATM layer and the physical layer, also divided into two sublayers. There is not a one-to-one relationship between the B-ISDN and OSI stacks. However, the three layers (the ATM portion of the B-ISDN protocol stack) are roughly equivalent to the ISO layers one (physical) and two (data link), as shown in Fig. 27.13. Be careful to keep these separate; they are not identical.

ANSI

The American National Standards Institute (ANSI) was created in 1918 to coordinate private sector standards development in the United States. ANSI is the U.S. representative to international standards groups and includes 300 standards committees, as well as associated groups, such as the Exchange Carriers Standards Association (ECSA). The ANSI subcommit-

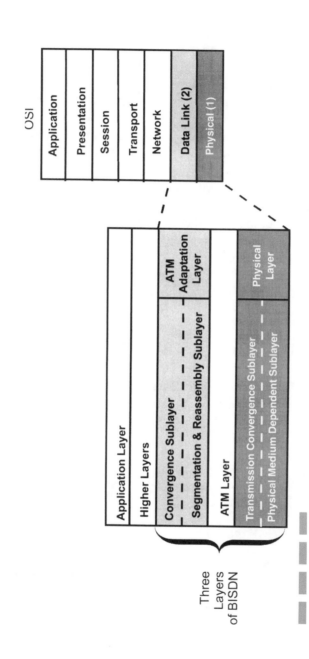

Figure 27.13

The three layers of B-ISDN closely resemble the bottom two layers of the OSI.

tees active in broadband network standards are T1X1 and T1M1. T1X1 plays a role in SONET rates and format specification, and T1M1 guides the effort to define the standard for operations, administration and maintenance, and provisioning (OAM&P).

The physical-media-dependent sublayer deals with aspects that are dependent on the transmission medium selected. The PMD sublayer specifies physical medium and transmission (e.g., timing, line coding) characteristics and does not include framing and overhead information. Physical characteristics of the UNI at the user, subscriber, and terminal interface for broadband services (U_B, T_B, and S_B, respectively) reference points are defined in ANSI T1E1 2/92-020 and shown in Fig. 27.14. Other SONET physical medium specifications are defined later. This is a moving and emerging target as the standards bodies move toward the broadband specifications.

The transmission convergence (TC) sublayer specification deals with physical layer aspects that are independent of the transmission medium specifications. B-ISDN-independent TC sublayer functions and procedures involved at the UNI are defined in ANSI T1.105-1991 and T1E1-2/92-020.

ATM Forum

The ATM Forum is an international consortium of members that is chartered to accelerate the acceptance of ATM products and services in local,

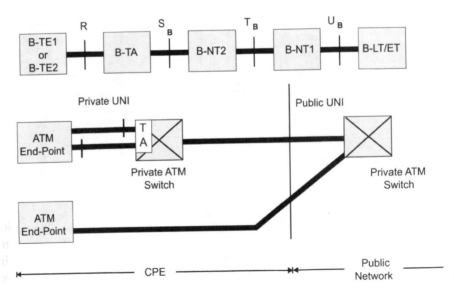

Figure 27.14
The physical access reference points of the UNI developed by ANSI and CCITT (ANSI).

metropolitan, and wide-area networks. Although it is not an official standards body, the Forum works with the official ATM standards groups (ANSI and CCITT) to ensure interoperability among ATM systems. The consortium accelerates ATM adoption through the development of common implementation specifications. The forum's first specification on the ATM user-network interface (UNI) provides an important platform on which vendors can design and build equipment. It defines the interface between a router or a workstation and a private ATM switch, or between the private ATM switch and a public ATM switch.

The ATM Forum has created interest in private uses of ATM in LANs. LAN management standards, such as SNMP, are being promoted for ATM network management and miniature switched ATM networks used as LANs.

Equipment

At this point in your path to understanding ATM, it is logical to consider how we'll access these services. Several products have been introduced, with others announced for future delivery. The real gist of the connection process is how we can migrate from one platform to another without scrapping everything and starting over. Many of the early products were quite expensive and only attractive to very large corporations.

Consequently, the 500+ members of the ATM Forum—which is comprised of carriers, equipment suppliers, consultants, and end users—have all been active in trying to specify the best approaches to allow ATM evolve in the workplace. They are discussing the ability to interconnect equipment, such as:

- Bridges
- CSU/DSU
- Gateways
- Multiplexers
- Routers
- Servers

Industry experts foresee a time when ATM technology will be used from the desktop, across a LAN, and through a WAN to serve an entire enterprise. Today, ATM vendors expect the ATM switching technology to overtake conventional switching techniques over the next 10 years.

Bridges

As we know them today, bridges link similar LANs together across a short (departmental) or a long (WAN) boundary. Overall, bridges will likely continue to be used as an interconnection device for LANs. However, newer bridges will emerge that perform bridging and routing into a high-speed hub or concentrator. The bridge will act as the filter or buffer device onto the higher-speed hubs.

As each of the high-speed networking technologies continues its evolution, the ability to link services together will become more crucial. The bridge will be required, as more networks are broken into smaller pieces to improve throughput performance. Thus, as network segmentation continues, the bridge will become one of the major interfaces to the hub or controller (Fig. 27.15). The hub is an ATM format service, whereas the controller could be an intelligent switching concentrator at raw or native speeds. The bridge's significance will be in providing a high-speed access capability into the hub. This will use a high-speed serial interface (HSSI) or digital exchange interface (DXI).

Figure 27.15
The bridge uses a high-speed interface into an ATM hub. The hub created the ATM interface.

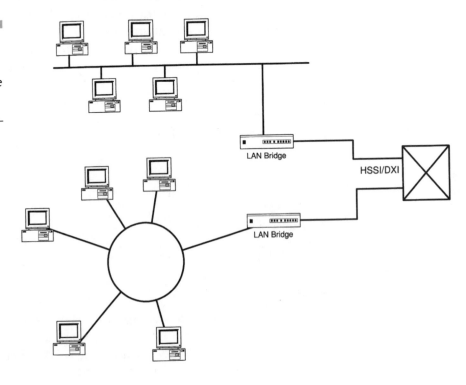

Routers

Routers are becoming more sophisticated every day. They can be used to interconnect LAN segments for high-speed access and provide enhancements to network security. Firewall routers allow for the screening, access, and denial of data transfer between segments.

Upgraded port speeds, increased numbers of LAN and WAN ports, and software will all add to the router's capabilities. The routers will also be tied to the high-speed hubs and bridges to access high-speed ATM services. However, the router user will most likely be looking for interim steps to access ATM, as shown in Fig. 27.16. Not every user can access ATM via T3 because of the cost for this service. Therefore, routers will be available to connect to ATM via a simple T1 interface. Although this can only be viewed as a near-term solution, it is the first step toward accessing ATM at a reasonable price. Router hardware and software changes will allow this device to create the cells at the local level to attach and forward the cells to a switching system. As a connectionless device, the router will deliver data packets to the ATM device, which will forward them to a cell processor.

Figure 27.16
Initially, the router will use a high-speed interface to high-speed ATM hubs. Later, the router will become fully ATM compatible.

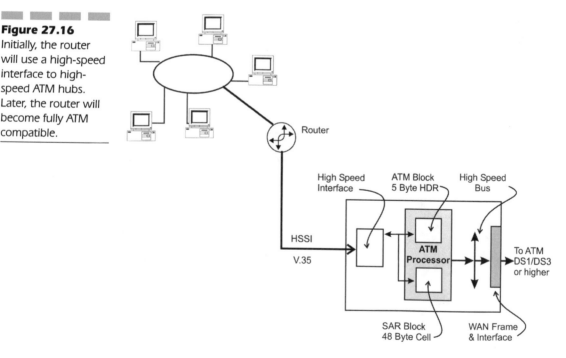

Gateways

These devices are designed to interconnect a potpourri of dissimilar connections or networks. Working up the OSI model, the gateway performs the full seven-layer service. As a connection to the network, the gateway will use ATM as the interface at layers 1 and 2 of the OSI model.

Because ATM will be protocol independent, the gateway will provide high-speed connections to an ATM hub or switch (Fig. 27.17). This will require upgrades of hardware and software, but the impact will be somewhat minimal. Such traffic services as TCP/IP, IPX/SPX, LU6.2 APPN/APPC, and SNA tunneling will still be services provided via a gateway. High-speed interconnectivity across these services at 45, 100, and 155 Mbps will be achieved at the hub or switch. But the gateway will function at native speeds of 4, 10, or 16 Mbps today, or a subset at 64-Kbps to 1.544-Mbps transport. Several LAN and WAN ports will be available through the gateway offering aggregates of bandwidth at much higher speeds.

Figure 27.17
Even gateways will interface to a high-speed ATM hub, providing protocol independent service.

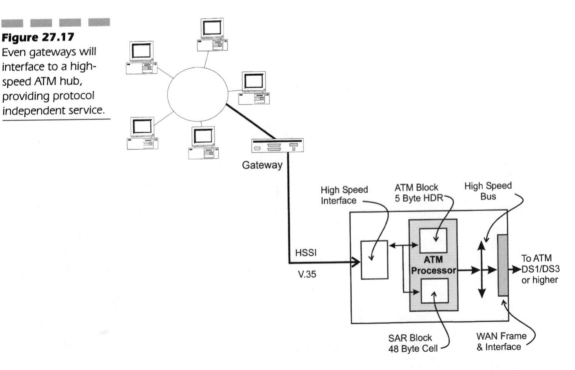

Servers

Servers in this regard will provide access to the services offered on a network via ATM transport. The public switched and private network arenas will have a series of server functions (Fig. 27.18) to provide the end user functionality and features. These high-speed multitasking servers will offer:

- Remote video and cable services
- Plain old telephone services (POTS)
- Pretty amazing new stuff (PANS)
- Multimedia
- Video services
- Remote database access
- Dense graphics and CAD transfer and access

The fabric of the inner working will require dual bus parallel/serial interfaces at the high end. The high end would include the 100-/155-/622-Mbps throughput range and would lend itself to a series of channel attachments to dual fiber-optic links for server access. These dual fiber links will provide a level of fault tolerance and an aggregate throughput in the gigabit spectrum. The end user need only request the high-speed service and the server does the rest.

Figure 27.18
Servers will be accessible for a variety of applications and multitasking.

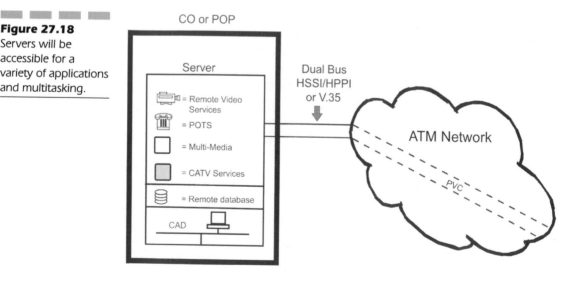

Switches

ATM switches will come in a variety of flavors, speeds, and access ports. At the heart of the switch will be a dual bus architecture with throughput starting at 622 Mbps on up to 2.4 Gbps and more (Fig. 27.19). Several inputs (starting at 4 and currently up to 16 in the user environment) will match across a matrix of outputs (ranging from 4, 10, or 16 Mbps and more). The ATM switch will process the AAL 3-, 4-, and 5-level services in the enterprise network.

The local switch will handle the switching and routing of cells in a building or department. This local switch will connect to a network backbone switch via an OC-3 (155 Mbps) or less service to start. Faster, higher throughput will become available through upgrades at the backplane level and the number and speed of ports attached. This requires that the switch be scalable and migrateable to the gigabit ranges. Traditional switches support 4 to 90 ports at speeds of 45 to 155 Mbps. Parallel processing of an ATM switch will allow concurrent switching among many parallel paths, providing each port full access to the allocated bandwidth. Cell switching breaks up data streams into very small units that are independently routed through the switch. The routing occurs mostly in hardware through the switching fabric. The combination of cell switching and scalable switching fabrics is a key ingredient of ATM.

Figure 27.19
ATM switches at the user side will have 4 to 16 ports and support various data streams.

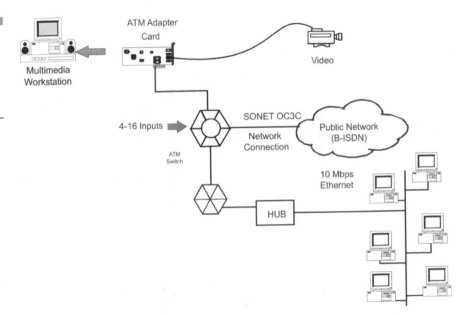

Public Switches

The public network ATM switch is a larger, more intelligent version of the customer premises switch. A public switch is capable of handling hundreds of thousands of cells per second and has thousands of switch ports, each operating at rates of up to 622 Mbps. All cell processing functions are performed by the input controllers, the switch fabric, and the output controllers. In an ATM switch, cell arrivals are not scheduled. The control processor resolves contention when it occurs, as well as call setup and teardown, bandwidth reservation, maintenance, and management. The input controllers are synchronized so that cells arrive at the switch fabric with their headers aligned.

The resulting traffic is said to be *slotted,* and the time to transmit a cell across the switch fabric is called a *time slot.* All VCIs are translated in the input controllers. Each incoming VCI is funneled into the proper output port as defined in a routing table. At the output controllers, the cells are formatted in the proper transmission format. For example, a broadband ISDN output controller provides an interface that consists of a line terminator to handle the physical transmission and an exchange terminator for cell processing.

Multiplexers

The slowest defined rate for ATM today—45 Mbps—is faster than most routers, bridges, and video codecs operate. This leaves the network manager with the choice of dedicating a 45-Mbps circuit to each lower-speed device, possibly wasting bandwidth, or multiplexing several lower-speed signals onto an ATM circuit. An ATM multiplexer might be considered similar to a switch, but it does more than just provide connectivity to multiple users. It handles isochronous data such as voice and video. The multiplexer accepts data from routers or bridges through a standard interface (V.35 or HSSI), segments the data into cells, addresses each cell, and maps the cells into a WAN framing structure (T3/E3 or SONET/SDH).

CSU/DSU

The purpose of the CSU/DSU is to encapsulate information into the proper framing before it enters the WAN. The CSU/DSU converts between one communications technology and another by providing the interface between the on-premises hub, switch or router formats, and the broadband network. The CSU/DSU regenerates the signals received from the network as well as the sent data. Thus it can also serve as a way to troubleshoot the transmission line. The CSU/DSU automatically monitors the signal to detect violation and signal loss. When problems are detected, the CSU/DSU allows remote network testing from the central office, including loopback testing of the transmission line.

ATM in the LAN Environment

LAN bandwidth continues to increase. Although fiber distributed data interface (FDDI) pushed the bandwidth of the LAN to 100 Mbps, it only delayed the bandwidth congestion problems that are inherent in a shared LAN. The capacity of shared media networks is only as large as the speed of the common line (or bus).

ATM-based LANs allow users to connect to the network at their required speed, adding bandwidth as needed. ATM removes the limitations of shared LAN backbones by collapsing them into the ATM switching fabric. The typical extended ATM LAN will employ an ATM switch to link slower LANs (Fig. 27.20). To interconnect a series of LANs, multiple switches can be concatenated together.

The switch will be more powerful than today's intelligent LAN hubs in terms of processing power, routing capability, and network management. In addition, the ATM switch is distinguished from today's hubs in that its total bandwidth is the sum of the bandwidth from its input ports rather than a fixed bandwidth. Thus, ATM switches allow information to travel from any port to any other port without the risk of blockage. In contrast, traffic on a bus-based LAN or hub can be blocked if there are too many sources on the bus.

Evaluating the Need for ATM

As for any system, the need for ATM will be based on traffic loads and future demand for bandwidth. In general, if congestion is occurring on a LAN today, the obvious solution is either to segment the LAN into smaller pieces or add the necessary capacity to support the traffic.

As new applications emerge, the drain on available transport will be greater. Particular attention must be exercised in areas of heavy traffic loads. The applications will include:

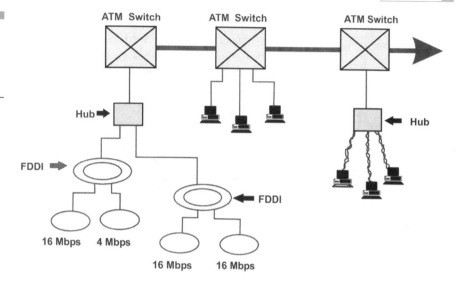

Figure 27.20
ATM switches will provide connections to multiple LANs in the office.

- CAD
- Magnetic resonance imaging (MRI)/electronic media imaging (EMI)
- Multimedia
- Voice on LANs (future)
- Video on LANs (future)

As these applications begin to emerge, the need for ATM might well become evident more quickly. The use of the ATM switch will relieve the congestion problem, but at a cost. What is your motivation to grow the network? If service and function are important, then ATM can serve the demand.

LAN to LAN via ATM

If all goes according to the plans of vendors, carriers, standards bodies, and users alike, then ATM should have plug-and-play capability. The use of shared wire, single-bus technology with all stations tapped in might become a thing of the past. As the cable originally snaked from office to office, distances became the bottleneck. When the cable had to be made longer, two pieces were joined together with repeaters. However, a single break in the cable could bring down the entire LAN. The idea of a hub started with the application of a star wiring topology for Ethernet LANs. Now hubs are becoming the central switching point for the entire enterprise. In the near future, hubs will embrace internetworking switching and multimedia.

The next-generation ATM-based hub will provide an integrated framework (a star-shaped backbone with a high-capacity switch) that provides support for multimedia applications (Fig. 27.21). This design eases the task of routing video streams and simplifies network management. As the capacity of a single switch is exceeded, switches will be added to the backbone, and the high-speed intelligent hub will provide a structure to interlink networks.

Local Exchange Carriers

Just about all of the local exchange carriers have begun using ATM in their backbone networks. Their plan is to convert the existing technologies they have now into the broader-bandwidth transport services. However, this will likely be a slow migration path unless these carriers see the use of ATM as a quick ramp-up at a reasonable cost.

Because ATM is really based on the network being shaped to fit the customer's needs rather than the age-old philosophy of the customer fitting the networks needs, we will likely see the LECs offering their early services as connection-oriented services for virtual private data networks (VPDN) with the intent of bringing fiber to the door and displacing copper T1 circuits. Literally, the LECs will have to start moving the capacity closer to the customer's premises. A combination of services will have to emerge to draw the customer's data out to these virtual private data networks. Much as the LECs offered VPN/SDN voice networks, the application and bursty data will have to be fully and transparently supported.

Figure 27.21
Hubs and switches in the backbone will support future bandwidth needs.

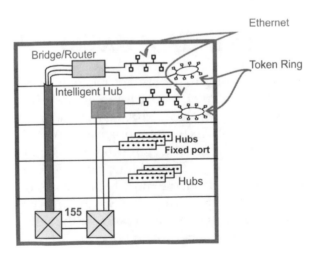

The LECs might have to initially offer a meshed private line network to get the customer to change over from the older leased line services. Such a network might be expensive at first, but the longer-range strategy is what's at stake here. As the LECs gain a larger customer base, they can begin to move ATM switches closer to the customer and offer high-speed switching at a more reasonable price.

Interexchange Carriers

Just as the LECs are cautious in the provisioning of ATM, the IECs are being equally careful. Major vendors (three to five) will likely begin offering ATM in their backbones in major metropolitan areas only.

Each of the major players has been quickly trying to deploy the range of services that it feels the customer is looking for. Internetworking has been an issue with many of the past services, because different technologies manufactured by different vendors have been used. In many of these situations, it was the proprietary handling of the information that restricted ubiquitous access across carrier platforms. We should hope that this won't occur with ATM.

Each of the IECs has a software or virtual defined network (SDN/VPN) for voice. In addition, the same deployment of ISDN has been used for the IEC and the LEC. Consequently, separate networks might still exist for the short term.

CATV Companies

The battle over bandwidth is never done until the last round is fired. CATV companies have also been quiet about ATM. Yet, most of the major providers are in the process of upgrading their coaxial-based networks with fiber to the door. Could it be that these suppliers have a larger vision than just delivering TV signals across the fiber? You might think that this is not a good application for delivering bandwidth. However, looking at the cost to upgrade the cable and the limitations on customer services, this could prove to be a logical extension of the metropolitan area network (MAN) and access to the WAN. Many of the CATV suppliers have been acquiring ATM switching equipment and voice switches for their networks. Do they know something we don't?

After all, the CATV companies go to every door. ATM to the door is what this is all about. Further, ATM will support voice, data, video, and multimedia, so the CATV suppliers might well be in a position to transport door-to-door applications. Then, if needed, they can access the long-distance WAN arena through a simple connection to the IECs.

Application Needs Driving

ATM is designed to meet the requirements of next-generation networks on a more global scale for LAN and WAN: for data, voice, video, and image for both public and private networks, and for various countries worldwide.

As enterprise networks continue to grow in complexity and diversity, newer demands are being put on the existing network infrastructure. Crucial issues that must be addressed include such areas as network management and ease of use. Users continually seek out higher-speed communications, cost efficiency, the ability to truly scale bandwidth as needed, the capacity to aggregate services in a super or subrated format, and independent protocols on the network. Further, as needs change, users desire the ability to reconfigure the network on the fly.

Consequently, the TDM schemes of the past did not meet the needs of the bandwidth-hungry applications that were emerging. Because voice and video networks grew apart from data applications, something had to be done. The industry sought out and found the transport system that should support such applications as:

■ *Enterprise networking on a wide area or global area.* As new techniques are developed, the applications can be migrated onto the WAN or global area network (GAN). Local area and campus area networks can be quickly added and tied to the wide-area network. Dynamics will foster, rather than inhibit, the interconnectivity of corporate computing and human resources. It is possible that by the turn of the century such mundane terms as *local, metropolitan,* and *wide-area networks* will be history. Instead, the network manager will manage and control all resources on a single domain, that being the enterprise. Regardless of where a resource exists, the addressing scheme will likely be available to route requests to and subsequently receive responses from the desktop of any authorized user. Virtual/dynamic workgroups, regardless of their location geographically (whether next door or around the globe), will have instantaneous access to the corporate data and to each other. Simple forms of electronic mail, irrespective of complex addressing schemes, will be defined as entities encased within the network. As more organizations continue to downsize, skill sets in various locations can easily be assembled and linked together for the culmination of organizational projects and goals. Long response times associated with the traditional networking techniques will become passé. The speeds to be offered can allow for response times to improve with latency across the network valued at less than 80 milliseconds. Through this dynamically assembled workgroup, the physical boundaries of distance become obsolete. The users are virtually connected as though they were in the same room.

- *Joint development on various R&D or engineering project work through the ability to conference engineers from various parts of the country or the world.* As complex documents, such as those required for design and engineering, are used, the workforce can physically be anywhere. Simple text manuscripts or complex documents, such as CAD files, video clips, voice notation, and database or spreadsheets, can all be amassed together as a single document. These can easily be ported from site to site across the ATM network for revisions, comments, or new input as an original, then ported right back into their original host system. Thus duplication of effort can be reduced, if not totally eliminated. Through this ability, these enormous (200+ Mbytes) files can easily be accommodated by the network. This is also the beginning of multimedia application deployment. Today's transport systems make these files unmanageable because they take too long to transport and cause severe congestion on the network. Distributed computing using the future client/server architecture will surely cause network congestion and stress the current levels of bandwidth availability. As networks (local, metropolitan, and wide-area) continue to grow because of the proliferation of servers and superservers, the need for added bandwidth grows almost exponentially. Databases that are truly distributed will require large throughput capacities because the actual program might reside in one site while a shell interface resides at another. To update these distributed databases, huge amounts of data will be competing for bandwidth. Furthermore, as massive file transfers are required for journaling or vaulting electronically to a remote site, the network will be called on to deliver more capacity. Using ATM in any of these scenarios, the network could be very flexible. When 155 Mbps is not sufficient, users can upgrade to 622 Mbps by increasing ports and speeds without any undue hardship or impact to the rest of the network.

- *Multiwindow desktop conferencing via video or computer conferencing.* Cost-efficient services with reliable transport will be the key operatives here. Teams of business administrators or senior staff can review plans and strategies for new product introduction or development while a team of researchers can review current results of testing and diagnostics through electron microscopes or a group of physicians in various departments or buildings around town can confer on the real-time output of a medical resonance imaging system across town. Files can be created that include video conferencing on one screen (or window), an MRI process on the other, the patient's historical files on a third, and other diagnostics (i.e., EKG, EEG) on the next. These pulled together in a window environment can give a total picture of the patient's current and past medical condition. Better-quality health care can result without disrupting a physician's schedule via electronic viewing.

■ *Telelearning at colleges, universities, and even K-12 locations.* This of course will be handled through multimedia applications. Many schools are currently downsizing their staffs, and resultant decreases in course offerings have occurred. Rather than transporting staff from school to school, which is wasteful and time consuming, or transporting students to remote schools, with the ensuing insurance and safety liabilities, the schools can be linked together for real-time compressed video on an interactive basis. Many of the past video training experiments have been reviewed as too limiting because students really are in delay and one-way modes from remote campuses. Many of the limitations were financially restricted because dedicated lines of sufficient capacity were too expensive and could not be justified for limited (1 to 2 hours per day) scheduling and use. With the ability to connect as much bandwidth as necessary on demand, these limitations should soon be past history.

■ *Full-service networking.* Voice, data, video, text, and imaging can all be truly integrated onto a single network platform. This totally integrated solution requiring hundreds of megabits to gigabits per second will have to be accessible and price sensitive. Thus, the network will finally be used to match the customer needs, rather than the other way around.

Transparency across the globe will exist from every desktop to any other desktop or service at any location and at any time. The network will have delivered real communication, that being transparent, native (not compressed or throttled) throughput across the network at native speeds (4, 10, 16, 100, and now 1000 Mbps) and linking diverse speeds to the single-network platform that is capable of handling multiple inputs to single threads at on-the-fly speeds.

ATM and Frame Relay Compared

One way to understand the interworking functionality of the two sets of protocols is to compare and contrast their capabilities. Table 27.5 gives an overall summary of the two techniques in use today. One can see a summary of the characteristics of the two service offerings in this comparison. This is not the whole story, but it does give a visual means of seeing the reasons for the popularity of the two techniques.

Network Interworking

The network interworking function provides the transport of frame relay user traffic transparently across the ATM link. It also handles the PVC signaling traffic over ATM. As already discussed, this is sometimes called *tunneling through the network.* Other times it is called *encapsulating the traffic.* Regardless of the name it is given, the function provides for the transparent movement of end-user frame relay information across the ATM network. The benefit is that with the tunneling or encapsulation formatting, the service is as good as though the end-user had a leased line service between the two end points. The benefit of this tunneling approach is connecting two frame relay networks across an ATM backbone. This is shown through the use of the network interworking unit in Fig. 27.22.. The interworking function is shown as a separate piece of equipment between the frame relay and ATM networks, and in some cases it is just that. However, newer implementations of this architecture place the network interworking function and interfaces inside an ATM switch. Regardless of where it resides, it is the functionality that is really important, not the location of the box. The interworking function will allow for each Frame Relay PVC connection to be mapped on a one-to-one basis over an ATM PVC. In other cases, may frame relay PVCs can be bundled together across a single higher-speed ATM PVC. The one-to-one or one-to-many services allow more flexibility.

Service Interworking Functions

The use of a service interworking function takes away some of the flexibility and transparency across the network. It actually acts more like a gateway (protocol converter) to facilitate the connection and communications between different disparate pieces of equipment. Figure 27.23 is a rep-

TABLE 27.5

Comparing
Frame Relay and
ATM Characteristics

Characteristic	Frame Relay	ATM
Connection type	Connection oriented	Connection oriented
Connection mechanism	PVC	PVC
Switched access (SVC arrangement)	Yes, but not widely implemented	Yes
Multiplexing arrangement	Statistical TDM	Statistical TDM
Current speeds available	Typically 56 Kbps–2 Mbps; implementations up to 50 Mbps	1.544 Mbps (T1) 12.3 Mbps 25.6 Mbps 34 Mbps (E3) 45 Mbps (T3) 51 Mbps (OC-1) 155 Mbps (OC-3) 622 Mbps (OC-12)
Area served	WAN	LAN CAN MAN WAN
Sequencing of data	No	No
Protocol data units	Variable	Fixed
Protocol data unit size	<4096 bytes	53 bytes
Flow mechanisms	Circuit by circuit	Circuit by circuit
Traffic congestion management	DE bit	CLP bit
Congestion notification	FECN/BECN	Payload type field
Bursty	Yes; defined by B_c and B_e rates	Yes; defined by PCR (maximum burst sizes)
Addressing method	DLCI	VPI and VCI
Address size	10 bits (normal)	24 bits
Standards	Joint development with Frame Relay Forum and ATM Forum, ANSI specifications, ITU standards, and others	

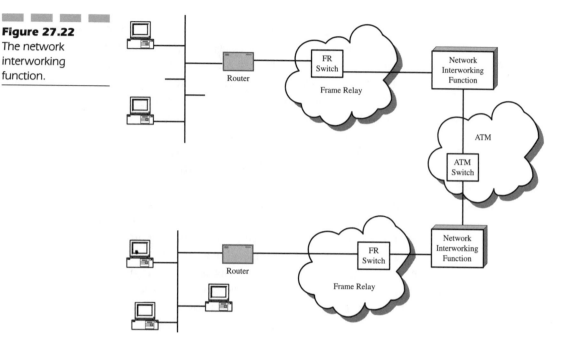

Figure 27.22
The network interworking function.

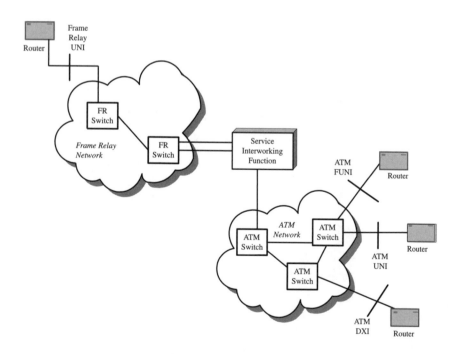

Figure 27.23
Frame/ATM service interworking function.

resentation of the interconnection of the service interworking devices across the network. The end user actually sends traffic out across a frame relay network on its own PVC; then it gets passed through the frame relay network to the service interworking function, where the data is then mapped to an ATM PVC. The IWF functionality provides the mapping of the DLCI to the VPI/VCI, as well as other optional features. The IWF is shown as a separate box, whereas newer implementations will have the dual mode functionality inside an ATM switch.

Frame relay and ATM can interwork and interoperate through several different techniques. This allows the carrier a sense of comfort, knowing that the legacy systems of the past can still be accommodated. The frame relay investments made in the early 1990s are still viable, and the ATM investments will be around for some time to come. Through the ATM protocols such as FUNI, end users have the comfort of knowing their networks are not obsolete. One can now see why the networking and service interworking functions are so important. Millions of dollars of investment can still be used, and newer protocols can be deployed without making the entire network obsolete. This is what internetworking is all about.

28

Digital Subscriber Line (DSL)

While the industry has been scurrying around looking for new ways of transmitting data from the consumer's door, the ILECs found new life in their copper cable plant. Clearly, the telcos had a personal stake in finding something that would allow them to continue to use the unshielded twisted pairs of wire at everyone's door. The movement to a data communications driver gave them the impetus they were looking for.

What Is *x*DSL?

One of the major problems facing the ILECs was the ability to maintain and preserve their installed base. Ever since the Telecommunications Act of 1996 there has been mounting pressure on the ILECs to provide faster and more correct Internet access. Therefore, a new form of communications was needed to work over the existing copper cable plant. One of the technologies selected was the use of *x*DSL. The DSL family includes several variations of what is known as *digital subscriber line*. The lowercase *x* in front of the DSL stands for the many variations. These will include:

- Asymmetric digital subscriber line (ADSL)
- ISDN-like digital subscriber line (IDSL)
- High-bit-rate digital subscriber line (HDSL)
- Consumer digital subscriber line (CDSL)
- Rate-adaptive digital subscriber line (RADSL)
- Very high speed digital subscriber line (VDSL)
- Symmetric or single digital subscriber line (SDSL)

One can see that the variations are many. Each digital subscriber line (DSL) capability carries with it differences in speed, throughput, and facilities used. The most popular of this family under today's technology is asymmetrical digital subscriber line (ADSL).

ADSL is a technology primarily provided by the ILECs because the existing cable plant can support the speeds, which can vary depending on the quality of the copper. The CLECs are fast becoming involved in providing *x*DSL now. However, the most important and critical factor in dealing with ADSL technology is the ability to support speeds from 1.5 up to 8.192 Mbps. The ILECs can also support POTS for voice or fax communications on the same line. What this means is that the ILECs do not have to install all-new cabling to support high-speed communications access to the Internet, which is burning up the wires today.

Asymmetric Digital Subscriber Line (ADSL)

Asymmetric digital subscriber line is the new modem technology to converge the existing twisted-pair telephone lines into the high-speed communications access capability for various services. Most people consider ADSL as a transmission system instead of a modification to the existing transmission facilities. In reality, ADSL is a modem technology used to transmit speeds of 1.5 to 6 Mbps under current technology. In the future ADSL is expected to support speeds up to 8.192 Mbps. This definition of the higher range of ADSL speeds is one that is yet to be proven; however, with changes in today's technology one can only imagine that the speeds will be achievable.

ADSL Capabilities

Many of the capabilities being considered with the DSL family are the services for converging voice, data, multimedia, video, and Internet streaming protocols services. The carriers see their future in the rollout of products and services to the general consuming public so they can access the Internet. Table 28.1 shows the theoretical speeds and distances of various ADSL technologies.

TABLE 28.1

Data Rates for ADSL

Current Data Rate, Mbps	Wire Gauge	Distance, ft	Distance, km
1.5–2.048	24	18,000	5.5
1.5–2.048	26	15,000	4.6
6.3	24	12,000	3.7
6.3	24	9,000	2.7

Remember that the speeds and distances shown here are theoretical. If the copper has been damaged or impaired in any way, the speed and distances will change accordingly (downward), and we all know what condition the wiring is in from earlier discussions in this book. Reality and the actual distances and speeds very likely will be less than shown here. What is most important is the assumption that these speeds can be established and maintained on the installed base of unshielded twisted pair (UTP) wire. As long as the ILECs can approximate these speeds today, the consumer will most likely not complain.

Modem Technologies

Before proceeding too far in this discussion, a quick review of modem technology is probably in line. Modems, or *modulator/demodulators*, were designed to move data across the voice communications network. Users still struggle to transmit data across the voice networks at speeds up to 33,600 bps. Even with the newer modems called the *V.90*, which are supposed to operate at 56 Kbps, we still see significant reductions in speed. Although this may seem like high-speed communication, our demands and needs for faster communications have quickly outstripped the capabilities of our current modem services, making the demand for newer services more evident. Higher-speed modems could be produced, but the economics and variations on the wiring system prove this somewhat impractical. Instead, the providers looked for a better way to provide data communications that mimic the digital transmission speeds we readily accept.

Using the telephone company's voice services, the end user installs a modem on the local loop. This modem is the data circuit-terminating equipment (DCE) for the link. A modem is used to communicate across the wide area networks as shown in Fig. 28.1. The ILEC installs a voice-grade line on the copper cable plant and allows the end user to connect the modem. The modem then converts the data into an analog signal. There is no real magic in modem communications today, but in the early days of data communications, this was considered voodoo science. The miracle of data compression and other multibit modulation techniques quickly expanded the data rates from 300 bps to today's 33.6 Kbps. Newer modems are touted to handle data at speeds of up to 56 Kbps, but few come close to these rates. So, the reality of all the pieces combined still has the consumer operating at approximately 33.6 Kbps.

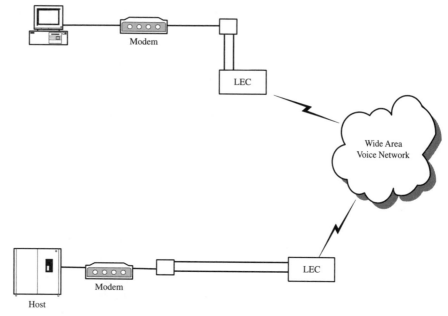

The Analog Modem History

In the early days of modem communications, the Bell Telephone companies provided all services across North America. A customer desiring data needed to call the local supplier, who would install the telephone line, modem, and associated services. Leased lines were used when specific speeds or volumes were anticipated. Regardless of the modem and lines used, the main provider was the important ingredient. Local providers supported only very low speeds. If the customer had a leased line and needed faster data, special equipment was installed on the line. Moreover, the technological advancement of modem technology was not a priority for the local providers, because they owned the installed base.

In 1968, things began to change. Court decisions allowed competitive devices on the network. Demands escalated! The customer-provided modems were interconnected through a data coupler (called a *data access arrangement* [DAA]) provided by the local regulated carriers for a fee.

Later, the U.S. Federal Communications Commission (FCC) and the Canadian Communications Radio and Television Commission (CRTC) allowed changes in the way the interconnection was handled. Modem manufacturers were allowed to produce their products according to a set of specifications and registrations, eliminating the need for the telephone company protection equipment and the fee associated with the monthly rentals.

Soon the market began to swell with modem products that could take advantage of the voice network to transmit data. However, limitations still existed on the speeds and services allowed by these newer devices. The telephone network was designed to carry a voice call with reasonable quality. The market spurred the development of modem technologies from the original 300 bps speeds to the current speeds (28.8–33.6 Kbps). In 1997, the 56-Kbps modem was introduced. However, even at 56 Kbps, users were looking for more. The modems just did not satisfy the demands for higher-speed Internet access and video demands.

ISDN-Like Digital Subscriber Line (IDSL)

DSL refers to a pair of modems installed on the last mile of line, facilitating higher data speeds. Network providers use the existing wires and add the DSL modems to increase the throughput. DSL modems offer duplex operations. The speed of a DSL modem may be 128 Kbps on copper at distances up to 18,000 feet using the twisted-pair wires. The bandwidth used is from 0 to 80 kHz. IDSL uses the 128-Kbps full-duplex basic rate interface (BRI). As shown in Fig. 28.2, the IDSL technique is all digital, operating at two channels of 64 Kbps for voice or nonvoice operation and a 16-Kbps data channel for signaling, control, and data packets. ISDN was very slow to catch on, but the movement to the Internet created a whole new set of demands. Now more ILECs and the CLECs offer ISDN services. As the deployment of IDSL was speeding up on the local loop, the providers developed a new twist, called *always on ISDN,* mimicking a leased set of channels that are always connected. Bonding the channels together, users can surf the Net at speeds of 128 Kbps. Note that this is a *symmetrical* digital subscriber line.

High-Bit-Rate Digital Subscriber Line (HDSL)

In 1958, Bell developed a voice multiplexing system that uses the 64-Kbps pulse code modulation (PCM). Using the PCM techniques, voice calls were sampled 8000 times per second and coded using an 8-bit encoding. These samples were then organized into a framed format, using 24 time slots to

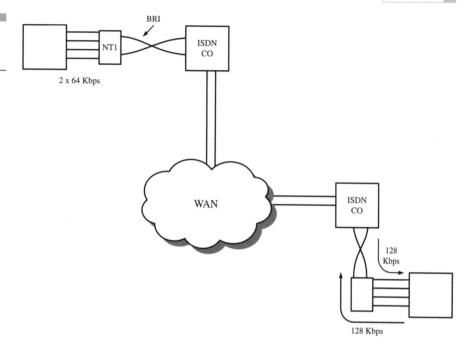

Figure 28.2
IDSL line connection
allows 128 Kbps.

bundle and multiplex 24 simultaneous conversations onto a single four-wire circuit. Each frame carries 24 samples of 8 bits and 1 framing bit, 8000 times a second. This produces a data rate of 1.544 Mbps. We now refer to this as a *Digital Signal Level 1* (DS-1) at the framed data rate. This rate of data transfer is used in the United States, Canada and Japan.

Throughout the rest of the world, standards were set to operate using an E1 with a signaling rate of 2.048 Mbps. The differences between the two services (T1 and E1) are significant enough to prevent their seamless integration.

However, in the digital arena, T1 required that the provider install the circuits to the customer's premises. The local provider installs a four-wire circuit. Repeaters are spaced at every 5000 to 6000 feet. When installing the T1 on the local loop, limitations of the delivery mechanism get in the way. Alternate mark inversion (AMI) consumes all the bandwidth and corrupts the surrounding cable spectrum quickly. Consequently, the providers can only use a single T1 in a 50-pair cable. Figure 28.3 is a representation of this cable layout. This inefficient use of the wiring makes it impractical to install T1s to small office and residential locations. Further limitations require the providers to remove bridge taps, clean up splices, and remove load coils from the wires to get the T1 to work.

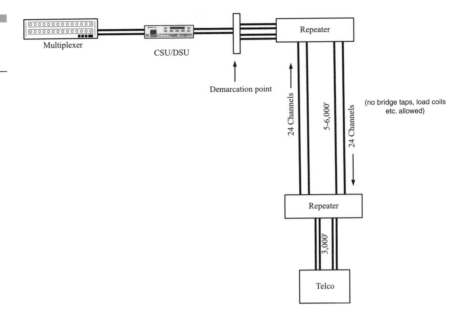

Figure 28.3
The typical layout of
the T1.

To circumvent these cabling problems, HDSL was developed. HDSL does not require the repeaters on a local loop of up to 12,000 feet. Bridge taps will not bother the service, and the splices are left in place. This means that the provider can offer HDSL more efficiently for 1.544 Mbps. The modulation rate on the HDSL service is more advanced. Sending 768 Kbps on one pair and another 768 Kbps on the second pair of wires splits the T1. This is shown in Fig. 28.4.

Originally, HDSL used two pairs at distances of up to 15,000 feet. HDSL at 2.048 Mbps uses three pairs of wire for the same distances. The most recent version, HDSL-2, uses only one pair of wire and is more acceptable to the providers. Nearly all providers today deliver T1 capabilities on some form of HDSL.

Symmetric or Single Digital Subscriber Line (SDSL)

The goal of the DSL family was to continue to support and use the local cable plant. Therefore, providing high-speed communications on a single cable pair became paramount. Most local loops already employ single cable pair today; thus, it is only natural to assume that providers would want this

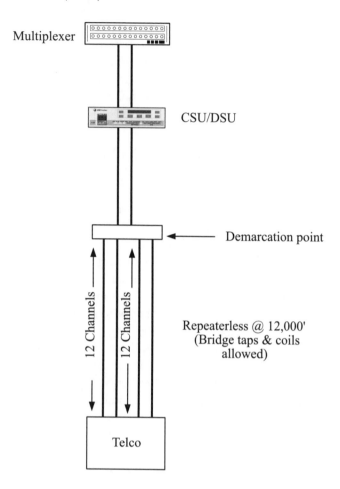

Figure 28.4
HDSL is impervious
to the bridge and
splices. The T1 is split
onto two pairs.

capability. SDSL was developed to provide high-speed communications on that single cable pair but at distances no greater than 10,000 feet. Despite this distance limitation, SDSL was designed to deliver 1.544 Mbps on the single cable pair. Typically, however, the providers provision SDSL at 768 Kbps. This creates a dilemma for the carriers because HDSL can do the same things as SDSL.

Asymmetric Digital Subscriber Line (ADSL)

SDSL uses only one pair of wires to provide duplex high-speed communications but is limited in distance. Not all users require symmetrical speeds at the same time. ADSL was therefore designed to support different speeds in each direction at distances of up to 18,000 feet. Because the speeds requested are typically to access to the Internet, most users look for higher download speeds and lower upload speeds. Therefore, the asymmetrical nature of this service meets those needs.

Rate-Adaptive Digital Subscriber Line (RADSL)

Typically, when equipment is installed, assumptions are made based on minimum performance characteristics and speeds. In some cases, special equipment is used to condition the circuit to achieve those speeds. However, if the line conditions vary, the speed will be dependent on the sensitivity of the equipment. In order to achieve variations in throughput and be sensitive to the line conditions, rate-adaptive DSL was developed. This allows the flexibility to adapt to changing conditions and adjust the speeds in each direction to potentially maximize the throughput on each line. In addition, as line conditions change, one could see the speeds changing in each direction during the transmission. Many of the ILECs have installed RADSL as their choice given the local loop conditions. Speeds of up to 768 Kbps are the preferred rates offered by the incumbent providers.

Consumer Digital Subscriber Line (CDSL)

Not all consumers need symmetrical high-speed communication to access the Internet. Furthermore, ADSL speeds are more than the average consumer may be looking for. Lower-speed communications capability was developed using CDSL. With other forms of DSL, splitters are used on the line. CDSL was designed to eliminate the splitter on the line. Moreover, speeds of up to 1 Mbps in the download direction and 160 to 384 Kbps in the upward direction are provided. It is expected that the speeds and CDSL will meet needs of the average consumer for some time to come. A universal ADSL working group developed what is called *ADSL-lite*, also called *G.lite*. This specification was ratified in late 1998, using the working group's specifications for service delivery to the average consumer. An example of this DSL-lite service is provided by the Nortel Networks 1-Mb modem.

Very High Speed Digital Subscriber Line (VDSL)

It was only matter of time until some users demanded higher-speed communications than was offered by the current DSL technologies. VDSL was introduced to achieve the higher speeds. In fact, speeds ranging from 13 to 52 Mbps are available, but the distance limitations of the local cable will be a big factor. In order to achieve the speeds, one can expect that a fiber feed will be used to deliver VDSL. This technique will most likely carry ATM (cells) as its primary payload. We can expect some hybrid arrangements to deliver this speed to the door for high-speed data at up to 52 Mbps downward and 1.5 to 6 Mbps upward.

Table 28.2 summarizes the speeds and characteristics of the DSL technologies discussed here. These are the typical installation and operational characteristics; others will certainly exist in variations of installation and implementation.

TABLE 28.2

Summary of DSL Speeds and Operations Using Current Methods

Service	Explanation	Download	Upload	Mode of Operation
ADSL	Asymmetric DSL	1.5–8.192 Mbps	16–640 Kbps	Different up and down speeds, one pair of wire.
RADSL	Rate-adaptive DSL	64 Kbps–8.192 Mbps	16–768 Kbps	Different up and down speeds. Many common operations on 768 Kbps. One pair of wire.
CDSL	Consumer DSL	1 Mbps	16–128 Kbps	Now ratified as DSL-lite. No splitters. One pair of wire.
HDSL	High-bit-rate DSL	1.544 Mbps in North America; 2.048 Mbps in rest of world	1.544 Mbps 2.048 Mbps	Symmetrical services Two pairs of wire.
IDSL	ISDN DSL	144 Kbps (64 + 64 + 16) as BRI	144 Kbps (64 + 64 + 16) as BRI	Symmetrical operation. One pair of wire. ISDN BRI.
SDSL	Single DSL	1.544 Mbps 2.048 Mbps	1.544 Mbps 2.048 Mbps	Uses only one pair, but typically provisioned at 768 Kbps. One pair of wire.
VDSL	Very High Speed DSL	13–52 ± Mbps	1.5–6.0 Mbps	Fiber needed and ATM probably used.

The Hype of DSL Technologies

The local providers are extremely excited at the possibility of installing higher-speed communications and preserving their local cable plants. No one wants to abandon the local copper loop, but getting more data reliably across the local loop is imperative. Therefore, the ability to breathe new life into the cable plant is an extension of the facilities in place. This also means that they can create a new form of revenue streams from the old copper. Consumers are looking for higher-speed access (primarily to access the Internet) for whatever the application. Yet, at the same time, consumers are looking for a bargain. They do not want to spend a lot of money on their communications services.

The providers are trying to bump up their revenues, without major new investments. They would like to launch as many new service offerings on their existing cable plant and increase the costs to the end user. This is a business decision, not a means of trying to rake the consumer over the coals. Yet, there has to be a happy medium of providing services and generating revenues with limits on expenses. To do this, the *x*DSL family offers the opportunity to meet the demands while holding down investment costs. The key ingredient for success is to minimize costs and satisfy the consumer. Make no mistake, if the local provider does not offer high-speed services, someone else will.

*x*DSL Coding Techniques

Many approaches were developed as a means of encoding data onto *x*DSL circuits. The more common are carrierless amplitude phase (CAP) modulation and discreet multitone (DMT) modulation. Quadrature with phase modulation (QAM) has also been used, but the important part is the standardization. The industry as a rule selected DMT, but several developers and providers have used CAP. It is therefore appropriate to summarize both of these techniques.

Discreet Multitone (DMT) Modulation

DMT uses multiple narrowband carriers, all transmitting simultaneously in a parallel transmission mode. Each of these carriers carries a portion of the information being transmitted. These multiple discrete bands—or, in the world of frequency-division multiplexing, subchannels—are modulated independently of each other, using a carrier frequency located in the center of the frequency being used. These carriers are then processed in parallel form.

In order to process the multicarrier frequencies at the same time, a lot of digital processing is required. In the past this was not economically feasible, but integrated circuitry has made this more feasible.

The American National Standards Institute (ANSI) selected DMT with the use of 256 subcarriers, each with the standard 4.3125-kHz bandwidth. These subcarriers can be independently modulated with a maximum of 15 bps/Hz. This allows up to 60 Kbps per tone used. Figure 28.5 shows the use of the frequency spectrum for the combination of voice and two-way data transmission. In this representation, voice is used in the normal 0 to 4 kHz band on the lower end of the spectrum (although the lower 20 kHz are provided). Separation is allowed between the voice channel and the upstream data communications, which operates between 20 and 130 kHz. Then a separation is allowed between the upstream and the downstream channels. The downstream flow uses between 140 kHz and 1 MHz. As shown in Fig. 28.5, the separation allows for the simultaneous up and down streams and the concurrent voice channel. It is on this spectrum that the data rates are sustained. Each of the subchannels operates at approximately 4.3125 kHz, and a separation of 4.3125 kHz between channels is allocated.

Using DMT for the Universal ADSL Service (G.Lite)

Provisions for the high-speed data rates of full ADSL are good, but not every consumer is looking for the high data rates afforded on ADSL. Therefore, the Universal ADSL Working Group decided to reevaluate the need for the end user. What they determined is that many consumers need downloads of 1 to 1.5 Mbps and uploads of 160 to 640 Kbps. Consequently, ADSL-lite specification was designed with these speeds for the future. Initially introduced in early 1998, the specification was ratified in late 1998 to facilitate the lower throughput needs of the average consumer. DMT is the preferred method of delivering G.lite service. There is no way to know if the network providers can support hundreds of multimegabit ADSL up- and download speeds on their existing infrastructure. However, using the G.lite specification can support lower-demand users more efficiently. Similar to the DMT used in the ANSI specification, the carriers are divided as shown in Figure 28.6. Note that in this case the high end of the frequency spectrum tops out at approximately 550 kHz instead of the 1 MHz range with ADSL.

Carrierless Amplitude Phase (CAP) Modulation

CAP is closely aligned to quadrature amplitude modulation (QAM). QAM as a technique is widely understood in the industry and well deployed in older modems. Both CAP and QAM are single carrier signal techniques. The data rate is divided in two and modulated onto two different orthogonal carriers before being combined and transmitted. The main difference between CAP and QAM is in the way they are implemented. QAM generates two signals with a sine/cosine mixer and combines them onto the analog domain.

CAP was one of the original proposals for use with ADSL technology. Unfortunately, this was a proprietary solution offered by a single vendor, which turned heads away from acceptance. CAP's use of the frequency

Figure 28.5
The ANSI DMT specification.

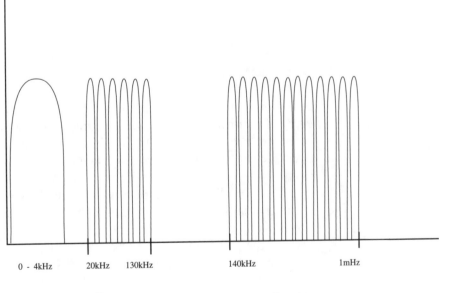

0 - 4kHz	20kHz 130kHz	140kHz 1mHz
Voice	Upstream	Downstream

Figure 28.6
The ANSI and UAWG
G.lite spectrum.

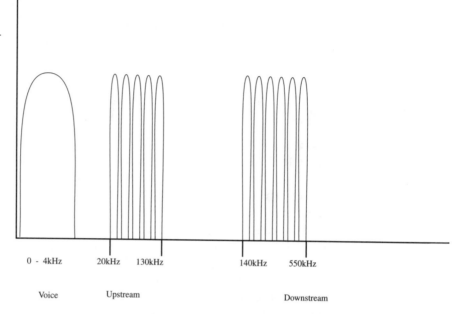

Figure 28.6
The ANSI and UAWG
G.lite spectrum.

0 - 4kHz   20kHz   130kHz   140kHz   550kHz

Voice   Upstream   Downstream

spectrum of the line is shown in Figure 28.7. Most industry vendors agree that CAP has some benefits over DMT, but also that DMT has more benefits over CAP. The point here is that two differing technologies were initially rolled out for ADSL (and the other family members) which contradict each other in their implementation.

Figure 28.7
The spectral use of
CAP.

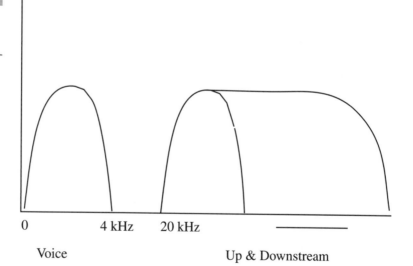

0   4 kHz   20 kHz

Voice   Up & Downstream

CAP uses the entire loop bandwidth (excluding the 4-kHz baseband analog voice channel) to send the bits all at once. There are no subchannels as found in the DMT technique. The lack of subchannels removes the concern about and problems with individual channel transmission.

Comments on Deployment

ADSL service is catching on. However, the ILECs and CLECs are dragging their feet. As of late 1998, there were only about 150,000 ADSL modem pairs installed in the United States. In contrast, there are over 800,000 cable modems installed in residences and businesses across the country. The local owners of the copper loop have to take a more aggressive approach to delivering high-speed services, or consumers will go somewhere else. As the market continues to mature and standards continue to develop, the local providers must preserve their infrastructure.

Whereas consumers are reluctant to proceed with ADSL, the HDSL and SDSL services are still very attractive alternatives, offering 1.544- to 2.048-Mbps symmetrical speeds or some variation, as already discussed.

In the future, when high-speed media are installed to the door or to the curb, the logical stepping stone will become the VDSL service, perhaps sometime in 2002 to 2003. Although trials are already under way, too much time passes until the results are complied and analyzed. Therefore, the reality of VDSL for the masses is still a long way off.

29

Cable Modem Systems and Technology

In the late 1970s a major battle arose in the communications and the computer industries. Convergence of the two industries was happening as a result of the implementation of the local area networks. In the local networking arena, users began to implement solutions to their data connectivity needs within a localized environment. Two major choices were available for their installation of wiring: *baseband coaxial cable* and *broadband coaxial cable*.

The Ethernet Cable

The baseband cable was based on Ethernet development, using a 20-MHz, 50-Ω coax. Designed as a half-duplex operation, Ethernet allowed the end user to transmit digital data on the cable at speeds of up to 10 Mbps. Clearly the 10 Mbps was maximum throughput, but was attractive in comparison to the technology of twisted pair at the time (telephone wires were capable of less than 1 Mbps bursty data). Moreover, the use of the baseband technology allowed the data to be digitally applied directly onto the cable system. No analog modulation was necessary to apply the data. It was dc input placed directly onto the cable. The signal propagates to both ends of the cable before another device can transmit. This is shown as a quick review in Fig. 29.1. To control the cable access, the attached devices used carrier sense multiple access with collision detection (CSMA/CD) as the access control. CSMA/CD allowed for the possibility that two devices may attempt to transmit on the cable at the same time, causing a collision and corruption of the actual data. Consequently, the cable had to be very controlled.

A second alternative at the time was to use the broadband coaxial cable, operating with a bandwidth of approximately 350 MHz on a 75-Ω cable. Broadband systems were well known because this is the same as CATV, which had surfaced in the early 1960s. Therefore, the technology was well deployed and commodity priced. Moreover, the 350-MHz capacity was attractive to the computer industry and the communications industry partisans. The issues began to surface quickly regarding the benefits and losses of using each technique. This is shown in Fig. 29.2.

What the issue really boiled down to was one of analog versus digital and the baseband versus broadband implementations to achieve this goal. This was a hot issue throughout both industries. The issue included using a broadband cable under the turf of the voice communications departments, whereas the baseband cables were under the primary con-

Figure 29.1

CSMA/CD cable networks are collision domains.

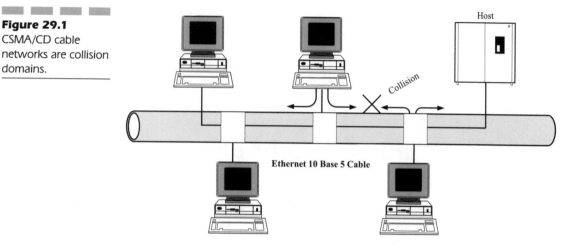

Ethernet 10 Base 5 Cable

trol of the data processing departments. If one technology was chosen over another, the lines in the sand would be washed away and the convergence of voice and data would force the convergence of the two groups.

The issue was therefore not whether to use a cable, but what *type* of cable so that the LAN would fall under the correct jurisdictional authority within the organization. Unfortunately, control is not the goal of

Figure 29.2

Broadband coaxial cable.

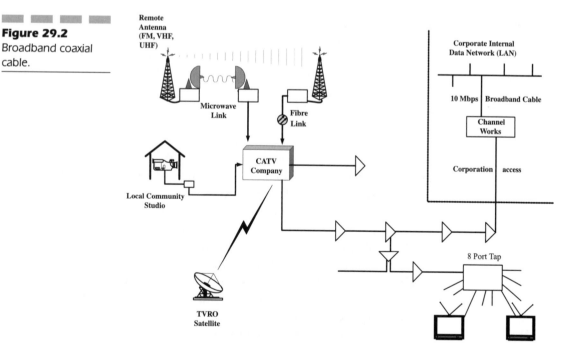

organizations, but access and profitability are. As an industry, too much time was wasted over semantics. However, what ultimately rolled out of the bandwidth argument was that the baseband cable systems were better for the LAN. This was the decision of the 1980s, when all traffic on the LAN was geared to data only at speeds of 10 Mbps and less.

Cable TV Technology

CATV has been around since the early 1960s. It is proven technology. In the early days of Ethernet, Digital Equipment Corporation (DEC) rolled out many systems using baseband (Ethernet) cable. However, some organizations needed more than just data on a large localized network. They worked with two major providers at the time to develop the interfaces for the broadband cable systems to attach an Ethernet to the CATV cable.

DEC developed several working arrangements with various suppliers to provide a frequency-agile modem (FAM) to work on the cable TV systems. The CATV companies did not necessarily own the broadband cable. Instead, this cable was locally owned in a high-rise office or a campus complex by the end user. The cable system provided a high bandwidth, but was very complex for the data and LAN departments to understand. The reason is obvious: the broadband coax operated using frequency-division multiplexing (analog techniques), which was beyond the scope of the LAN administrators and the data processing departments. The voice people knew of analog transmission but had a hard time with digital transmission in those days. There was a silent department in the crux of all the arguments—the video departments within many organizations stayed out of the fight.

As DEC began to roll out various choices, the average user had to justify the connection of the analog technologies (used as a carrier) with the digital data demands of the LAN. What many organizations did on a campus was consolidate voice, data, LAN, and video on a single cable infrastructure. What the industry came up with was a specification for 10Broad36 to satisfy the LAN needs over a coax cable. *10Broad36* stands for 10 Mbps, on a broadband cable 3600 meters long. A classic representation of the combined services on 10Broad36 is shown in Fig. 29.3.

The data industry was distraught because this encouraged the use of an analog carrier system to move digital data. Over the years, however, this has been revisited several times. Wang Computer Company developed a proprietary cable system for connecting Wang systems using two broad-

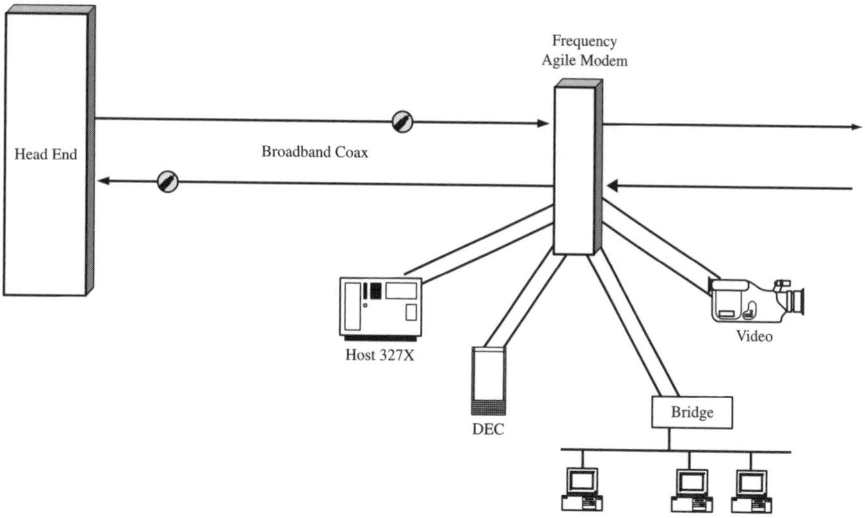

Figure 29.3
A 10Broad36 cable.

band coax cables. Technologically, the system was sound. However, the price and the proprietary nature of the Wang system forced its demise.

Later in the evolution of this service, the term *broadband LAN* became popularized. Ethernet grew to 100 Mbps, and then on to the gigabit range. Justifying this high-speed communication met with resistance until the use of the various fiber and coaxial systems emerged. By taking a quantum leap in the industry, the data and voice departments saw the benefit and need of converging the two services to the desktop and offering voice and video over the LAN. The 10-Mbps Ethernet and coaxial cables could not handle this offering. Moreover, access to the Internet continued with demands to add speed and capacity (voice and video on the Internet). The industry began to seek a new method of bypassing the telephone companies' local loops. A technology already at the door, of course, was CATV. So a new idea emerged: use CATV to support high-speed Internet access and bypass the local loop from telephone companies. Hence, cable modem technology changed the way we will do business in the future.

The New Market

The cable television companies are in the midst of a transition from their traditional core business of entertainment video programming to a position as a full-service provider of video, voice, and data telecommunications services. Among the elements that have made this transition possible are technologies such as the cable data modem. These companies have historically carried a number of data services. These have ranged from news and weather feeds, presented in alphanumeric form on single channels or as scrolling captions, to one-way transmission of data over classic cable systems.

Information providers are targeting upgraded cable network architecture as the delivery mechanism of choice for advanced high-speed data services. These changes stem from the commercial and residential data communications markets. The PC and LAN explosions in the early 1980s were rapidly followed by leaps in computer networking technology. More people now work from home and depend on connectivity from commercial online services (i.e. AOL, CompuServe, and Prodigy) to the global Internet.

Increased awareness has led to increasing demand for data service, and for higher speeds and enhanced levels of service. Cable is in a unique position to meet these demands. There appear to be no serious barriers to cable deployment of high-speed data transmission.

System Upgrades

The cable platform is steadily evolving into a hybrid digital and analog transmission system. Cable television systems were originally designed to optimize the one-way, analog transmission of television programming to the home. The underlying coaxial cable, however, has enough bandwidth to support two-way transport of signals. The hybrid network is shown in Fig. 29.4.

Growth in demand for Internet access and other two-way services has dovetailed with the trend within the industry to enhance existing cable systems with fiber-optic technology. Many cable companies are in the midst of upgrading the HFC plant to improve the existing cable services and support data and other new services. Companies are taking different approaches to online service access. For some applications, customers may be accessing information stored locally at or near the cable headend or regional hub such as the @Home services being offered in many cities. This may be temporary until wide area cable interconnections and expanded Internet backbone networks are in place to allow information access from any remote site.

Figure 29.4
The hybrid data network.

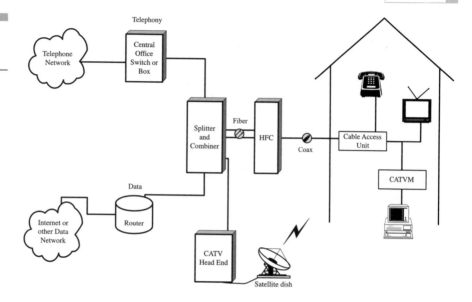

Cable Modems

Digital data signals are carried over radio frequency (RF) carrier signals on a cable system. Digital data utilizes cable modems, devices that convert digital information into a modulated RF signal and convert RF signals back to digital information. The conversion is performed by a modem at the subscriber's premises, and again by headend equipment handling multiple subscribers. See Fig. 29.5 for a block diagram of the cable modem.

A single CATV channel can support multiple data streams or multiple users using shared LAN protocols such as Ethernet, commonly in use in business office LANs today. This is where Ethernet networks can be applied to the broadband coaxial networks. Different modulation techniques are being tried to maximize the data speed that can be transmitted through a 6-MHz channel. Comparing the data traffic rates for different types of modems shows why the cable modem is so popular under today's environment. Table 29.1 shows a comparison of a file download of 500 Kbytes using different techniques.

Careful traffic engineering is being performed on cable systems so that data speeds are maximized as customers are added. Just as office LANs are routinely subdivided to provide faster service for each individual user, so too can cable data networks be custom tailored within each fiber node to

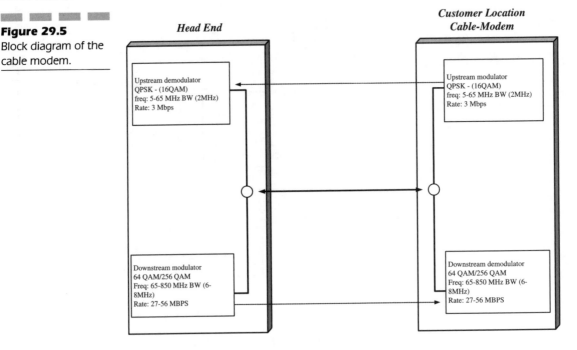

Figure 29.5
Block diagram of the cable modem.

meet customer demand. Multiple 6-MHz channels can be allocated to expand capacity as well.

Some manufacturers have designed modems providing asymmetrical capabilities, using less bandwidth for outgoing signals from the subscriber. CATV companies in some locations may not have completed system upgrades. Therefore, manufacturers have built migration strategies into such modems to allow for eventual transmission of broadband return signals when the systems are ready to provide such service and customers demand it. A representative sample of the way data speeds are provided on cable modems is shown in Table 29.2.

TABLE 29.1

Comparison of Transmission Speeds—Time to Transmit a Single 500-Kbyte Image

Method	Rate	Time
Telephone modem	28.8 Kbps	6–8 min
ISDN	64 Kbps	1–1.5 min
Cable modem	10 Mbps	Approximately 1 s

SOURCE: CableLabs

TABLE 29.2

Representative
Asymmetrical Data
Cable Modem
Speeds

Manufacturer	Upstream	Downstream
General Instrument	1.5 Mbps	30 Mbps
Hybrid/Intel	96 Kbps	30 Mbps
LANcity	10 Mbps	10 Mbps
Motorola	768 Kbps	30 Mbps

Standards

Modems are available today from a variety of vendors, each with their own unique technical approach. These modems are making it possible for cable companies to enter the data communications market now. In the longer term, modem costs must drop and greater interoperability is desirable. Customers that buy modems that work in their current cable system need assurance that the modem will work if they move to a different geographic location served by a different cable company. Further, agreement on a standard set of specifications allows the market to enjoy economies of scale and drives down the price of each individual modem. Ultimately, these modems will be available as standard peripheral devices offered as an option to customers buying new personal computers at retail stores. The cable companies and manufacturers came together formally in December 1995 to begin working toward an open standard.

Leading U.S. and Canadian cable companies were involved in this development toward an open cable modem standard. Specifications were to be developed in three phases, and then be presented to standards-setting bodies for approval as standards. Individual vendors were free to offer their own implementations with a variety of additional, competitive features and future improvements. A data interoperability specification will comprise a number of interfaces. The resultant specification is called the Data Over Cable Service Interface Specification (DOCSIS).

Some interfaces reside within the cable network. Several of these system-level interfaces also will be specified to ensure interoperability.

Return Path

The portion of bandwidth reserved for return signals (from the customer to the cable network) is usually in the 5- to 40-MHz portion of the spec-

Figure 29.6
Frequency spectrum
allocated to the cable
modems.

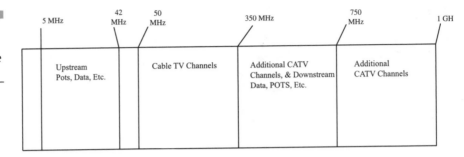

trum. This portion of the spectrum can be subject to ingress and other types of interference, so cable systems offering two-way data services have been designed to operate in this environment.

Industry engineers have assembled a set of alternative strategies for return-path operation. Dynamic frequency agility (shifting data from one channel to another when needed) may be designed into modems so that data signals may avoid unwanted interference as it arises. Other approaches utilize a "gate" that keeps the return path from an individual subscriber closed except for those times when the subscriber actually sends a return signal. Demarcation filters, different return laser types, and reduced node size are among the other approaches, each involving tradeoffs between capital cost and maintenance effort and cost.

Return-path transmission issues have already been the subjects of two years of lab and field-testing and product development. The full two-way capability of the coaxial cable already passing most U.S. homes is now being utilized in many areas, and will be available in most cable systems soon. Full activation of the return path in any given location will depend on individual cable company circumstances ranging from market analysis to capital availability.

The spectrum used for the forward and reverse paths is shown in Fig. 29.6 as an indication of the frequencies available and the overall management of the system. This also shows that additional 6-MHz channels can be set aside to handle the data traffic on the cable modems and the cables themselves.

Applications

Cable modems open the door for customers to enjoy a range of high-speed data services, all at speeds hundreds of times faster than telephone modem calls. Subscribers can be fully connected, 24 hours a day, to services without interfering with cable television service or phone service. Among these services are:

- Information services—access to shopping, weather maps, household bill paying, and so forth.

- Internet access—electronic mail, discussion groups, and the World Wide Web.

- Business applications—interconnecting LANs or supporting collaborative work.

- Cable commuting—enabling the already popular notion of working from home.

- Education—allowing students to continue to access educational resources from home.

The promises of advanced telecommunications networks, once more hype than fact, are now within reach. Cable modems and other technology are being deployed to make it happen. Regardless of the technology selected, the main goal is to get the high-speed data communications on the cable adjacent to the TV and entertainment. This gives the CATV companies the leverage to act in an arbitrage situation, competing with the local telephone companies who have dragged their feet in moving high-speed services to the consumer's door.

The Combined Corporate and End-User Networking Strategies

The use of a single PC on a cable system is okay for the telecommuter (or cable commuter now), but what of the small office or home office where more than a single PC is connected? Figure 29.7 shows an example of various ways the CATV connection can be accomplished. This figure uses an example of local home networking with two PCs connected to a single cable modem. Most of the providers have instructions on how to accomplish this and require a home user (or small user) to download additional software to accommodate the dual connection on a single modem. The second alternative is to have a router connected to the cable modem, such as a *branch office router*. The network is attached to the router, and the

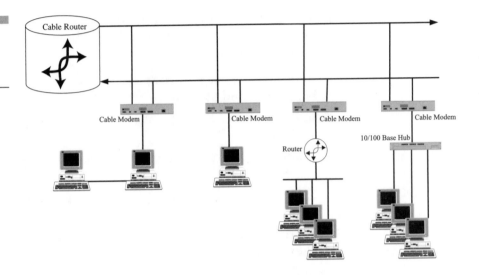

Figure 29.7
Multiple ways of
connecting to the
cable modem.

router is responsible for handling the dispersing of the traffic onto the cable system. This can be a very effective use of the link. Next in the figure is a connection to a hub, such as a 10- or 100BASE-T connection into a LAN hub. CATV providers state this is not supported and will likely not work. However, it has been done and works well for a small office or home office connection. Using the connection directly into the hub from the cable modem makes the modem available to more users instead of just a single PC. The hub will act as a bridging function onto the modem and concentrate the traffic through the individual devices. These configurations all work, but the providers do not support problems if they arise. You are on your own if it does not work.

Security on CATV

When CATV systems are used, they are shared high-speed Ethernet backbone access to the Internet or other connections. One must be aware that on a shared cable, the PC is a peer to all others on the same cable, although they are in physically different locations. With 500 connections, there will be many people who acquire the service from the CATV suppliers. The CATV company installs according to the appropriate technology, not according to security parameters. This is okay because they are merely providing the bandwidth to gain access. It is the end user's responsibility to turn off all the leaks in the local system (the PC). By default, when you run the Microsoft Windows environment and the appropriate networking software, the *shares* on the PC are turned on. The end user must therefore go in and turn them off. This means that if the shared services are not turned off, a user down the street, or across the town, can double-click on the "network neighborhood" icon and see all the other PCs connected to the cable. Not only can they see the devices, they can double-click on the PC and see the resources available on that PC. From there, when a remote device has double-clicked on your PC, they can open your drives and see your files. Unless some provisions have been taken to block this access, the intruder (used as a method of entry only) can read, write, edit, or delete your files. Worse yet, while the intruder is on your system, you do not even know they are there.

Many users who have cable modem service from the cable companies are not aware of the risks. Worse, the installation personnel on these systems do not totally understand or forget to point out the risks. Therefore, the user leaves the PC on 100 percent of the time, day and night, leaving access to the computer totally available. The cable modem is available 100 percent of the time, making the computer a target for hackers and the mischievous, without the permission or the knowledge of the penetrated computer owner.

Be aware that the risk is there and find out how to shut these open doors before leaving your computer on the network. Do not assume you are secure just because you shut your system off when not using it. When you do log on, you are exposed, and the perpetrator can get on your system while you are on it too.

30

Signaling
System 7 (SS7)

The preceding chapters throughout this book have discussed various ways of getting a connection across the network and hosting a voice or a data transfer call. The ability of a caller to go off-hook anywhere in the world today, dial some digits, and then miraculously talk to someone continues to be a mystery. The network's ability to set the call up almost instantly, then tear it down just as fast is what really carries the mystique. How can the network figure out where to send the call so quickly, get the connection, and ring the telephone on the other end? All of this happens in under a second and the user is oblivious as to the intricacies of what occurs. What happens behind the scenes constitutes the backbone of the signaling systems. The networks are now dependent on the ability to handle subsecond call setup and teardown.

Evolution of Signaling Systems

Several signaling systems have been introduced into the telecommunications networks. The current one in use is called *Signaling System Number 7* (or SS7) in North America. In the rest of the world this is referred to *Common Channel Interoffice Signaling System* (CCITT) *Number 7* (CCS7 for short). Although the names are different, the functions and the purposes of the two systems are the same. As always, the North Americans do things one way, and the rest of the world does things a different way. This is an age-old problem, but one that we have learned to deal with and adjust to. The essence of the signaling system boils down to many different factors, but one of the most significant reasons the carriers employ these systems is to save time and money on the network. Following that fact, the carriers are also interested in introducing new features and functions of an intelligent network, as previously discussed. The best signaling systems are designed to facilitate this intelligence in the network nodes designated as signaling devices, separate and distinct from the switching systems that carry the conversations.

Pre-SS7

Prior to the implementation of common channel SS7, per-trunk signaling (PTS) was used exclusively in the networks. The PTS method was used for setting up calls between the telephone companies' exchanges. This method continues to be used in some parts of the world where SS7 has not yet been implemented. Admittedly, the number of exchanges using the PTS method is declining. SS7 is gaining in its deployment worldwide. However, the network is always in a state of change and this is no exception. PTS sends tones or multiple frequencies (MFs) to identify the digits of the called party. The trunk also provides all of the intelligence for monitoring and supervision (call seizure, hang up, answer back) of the call. Telephone systems at the customer's location (PBXs) that are not Integrated Services Digital Network Primary Rate Interface (ISDN PRI) compatible use the PTS method.

On a long distance call, when a call setup is necessary, each leg of the call repeats the MF call setup procedure until the last exchange in the loop is reached. In essence, the call is being built by the signaling as the progress is occurring on a link by link basis. As each link is added to the connection, the network is building the entire circuit across town or across the country. Each leg of the call setup takes approximately 2 to 4 seconds, using the configuration shown in Fig. 30.1, with a total call setup taking approximately 6 to 12 seconds (at a minimum) from end to end.

This method is an inefficient use of the circuitry. Although the call gets to its end destination, several complications could arise. Regardless of the complications, the outcome is the same; the carrier ties up the network and never completes the call. Hence, there is no revenue generated.

Figure 30.1
Per-trunk signaling preceded SS7 but was slow.

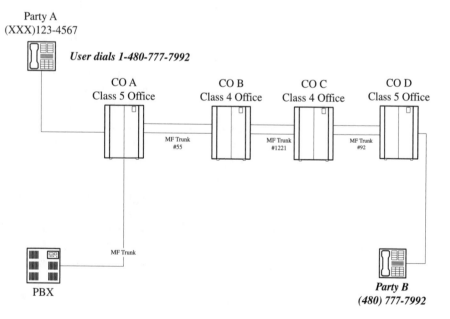

This inefficiency costs the carriers a significant amount of money. Therefore, something had to be done to improve this method of call establishment. The call-establishment part of the connection could take as much as 24 seconds, then time out and never get to its destination. However, the carrier tied up parts of the network without getting a completion. This is no big deal when discussing one call. However, when a network carries hundreds of millions of calls per day, this accumulated lost time is extensive and expensive.

Introduction to SS7

The ITU-TS developed a digital signaling standard in the mid-1960s called *Signaling System Number 6* (SS6) that would revolutionize the industry. Based on a proprietary, high-speed data communications network, SS6 later evolved to SS7. SS7 has now become the signaling standard for the world.

The success of the signaling standards lies in the message structure of the protocol and the network topologies. The protocol uses messages, much like the X.25 and other message-based protocols, to request services from other entities on the network. The messages travel from one network element to another, independent of the actual voice and data that they pertain to, in an envelope called a *packet*.

The first development of the SS6 in North America was used in the United States on a 2400-bps data link. Later these links were upgraded to 4800 bps. Messages in the form of data packets were used to request connections on voice trunks between central offices. Placing 12 signal units (of 28 bits each) assembled packets into a data block. This is similar to the methods used today in SS7 architectures.

SS6 used a fixed-length signal unit (28-bit signal units) but SS7 uses variable-length signal units. The most recent version of SS7 uses a 56-Kbps data link throughout North America, whereas in the rest of the world SS7 runs at 64 Kbps. The differences in the speeds between 56 and 64 Kbps results in the fact that the local exchange carriers have not yet fully deployed the use of B8ZS* on the digital circuits. Consequently, the 56 Kbps is an anomaly in the SS7 networks. Further, SS6 was still being installed by the North American carriers up through the mid-1980s (even though it was invented in the 1960s) while SS7 deployment began in 1983, leaving two separate signaling systems in use throughout North America.

---

* For a discussion of B8ZS, see Chap. 14.

SS6 networks are very slow, whereas SS7 is much faster. The use of a full DS-1 (1.544-Mbps) data link is still being considered in the North American marketplace.

Purpose of the SS7 Network

The primary purpose of SS7 was to access remote databases to look up and translate information from 800* and 900 number calls. There were several benefits to using this lookup process so that the carriers do not have to maintain a full database at each switching node, but know how to get to the remote database and find the information quickly. The second purpose of the SS7 network and protocols was to marry the various stored program controlled systems throughout the network. This allows the quick and efficient call setup and teardown across the network in 1 second. Moreover, this integration provides for better supervision, monitoring, and billing systems integration. Additional benefits of the SS7 network were geared to replacing the SS6 network, which as of today is well over 30 years old. Like anything else, the networks have served us well, but need upgrading on a regular basis due to technology changes and demands for faster, more reliable services.

SS7 networks allow the introduction of additional features and capabilities into the network. This makes SS7 attractive to the carriers so they can generate new revenues from the added features. SS7 also allows the full use of the channel for the talk path because the signaling is done out of band on its own separate channel. This is more efficient in the call setup and teardown process.

What Is Out-of-Band Signaling?

Out-of-band signaling is signaling that does not take place in the same path as the conversation. We are used to thinking of signaling as being in band. We hear dial tone, dial digits, and hear ringing over the same channel on the same pair of wires. When the call connects, we talk over the same path that was used for the signaling. Traditional telephony used to work this way as well. The signals to set up a call between one switch and

---

* Now the addition of 888 and 877 area codes are included.

another always took place over the same trunk that would eventually carry the call.

In early days, out-of-band signaling was used in the 4-kHz voice-grade channel. In Fig. 30.2 we see the 4-kHz channel. The telephone companies used band-pass filters on their wiring to contain the voice conversation within the 4-kHz channel. The band-pass filters were placed at 300 Hz (the low pass) and at 3300 Hz (the high pass). The range of frequencies above the actual filter is 700 Hz (4000 – 3300 = 700). In this additional spectrum, in-band signaling was sent down the wires outside the frequencies used for conversation. Actually, the signals were sent across the 3500- and 3700-Hz frequencies. Although these worked, and were not in the talk path (out of the band) they were limited to the number of tones that could be sent. The result was also a limit to the states that could be represented by the tones.

Out-of-band signaling evolved to a separate digital channel for the exchange of signaling information. This channel is called a *signaling link*. Signaling links are used to carry all the necessary signaling messages between nodes. Thus, when a call is placed, the dialed digits, trunk selected, and other pertinent information are sent between switches using signaling links, rather than the trunks which will ultimately carry the conversation.

It is interesting to note that while SS7 is only used for signaling between network elements, the ISDN D channel extends the concept of out-of-band signaling to the interface between the subscriber and the switch. With ISDN service, signaling that must be conveyed between the user station and the local switch is carried on a separate digital channel, called the *D channel*. The voice or data that comprise the call are carried

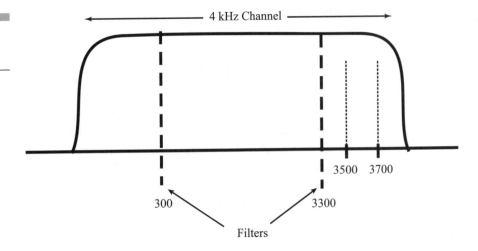

Figure 30.2
Out-of-band
signaling.

on the B channel. In reality, the out-of-band signaling is *virtual* because the signaling information is actually running on the same path as the B channels. The signaling and the conversational data flows are separated by time slots on the same physical paths. Therefore, the signaling is virtually out of band, while it is physically in the same bandwidth.

Why Out-of-Band Signaling?

Out-of-band signaling has several advantages that make it more desirable than traditional in-band signaling:

- It allows for the transport of more data at higher speeds (56 Kbps can carry data much faster than MF outpulsing).
- It allows for signaling at any time in the entire duration of the call, not only at the beginning.
- It enables signaling to network elements to which there is no direct trunk connection.

The SS7 Network Architecture

If signaling is to be carried on a different path than the voice and data traffic it supports, then what should that path look like? The simplest design would be to allocate one of the paths between each interconnected pair of switches as the signaling link. Subject to capacity constraints, all signaling traffic between the two switches could traverse this link. This type of signaling is known as *associated signaling.* Instead of using the talk path for signaling information, the new architecture includes the connection from the *signal switching point* (SSP) to a device called the *signal transfer point* (STP). It is then the responsibility of the STP to provide the necessary signaling information through the network to effect the call setup

When necessary, the STP will send information to the *signal control point* (SCP) for translation or database information on the routing of the call. The pieces combined to form the architecture of the SS7 network are described in Table 30.1 and shown in Fig. 30.3 with the connection of the overall components.

Figure 30.3 shows a typical interconnection of an SS7 network. Several points should be noted:

TABLE 30.1

Components of SS7
Networks

Component	Function
Signal switching point (SSP)	SSPs are the telephone switches (end offices and tandems) equipped with SS7-capable software and terminating signaling links. They generally originate, terminate or switch calls.
Signal transfer point (STP)	STPs are the packet switches of the SS7 network. They receive and route incoming signaling messages toward the proper destination. They also perform specialized routing functions.
Signal control point(SCP)	SCPs are the databases that provide information necessary for advanced call processing capabilities.

- Paired STPs perform identical functions. They are redundant. Together, they are referred to as a *mated pair* of STPs.

- Each SSP has two links (or sets of links), one to each STP of a mated pair. All SS7 signaling to the rest of the world is sent out over these links. Because the STPs are redundant, messages sent over either link (to either STP) will be treated equivalently.

- A link (or set of links) joins the STPs of a mated pair.

- Four links (or sets of links) interconnect two mated pairs of STPs. These links are referred to as a *quad*.

- SCPs are usually (though not always) deployed in pairs. As with STPs, the SCPs of a pair are intended to function identically. Pairs of SCPs

Figure 30.3
SS7 Architectural
beginnings.

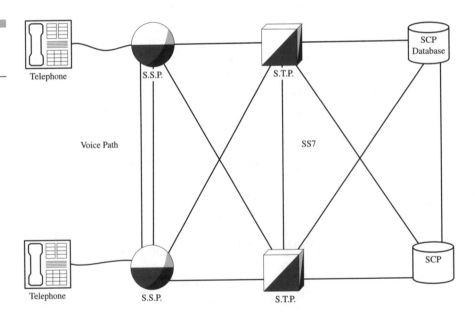

are also referred to as mated pairs of SCPs. Note that a pair of links does not directly join them.

■ Signaling architectures such as this, which provide indirect signaling paths between network elements, are referred to as providing *quasi-associated signaling*.

SS7 Interconnection

The actual linkage allows the local exchange offices to send the necessary information out of band across the signaling links. SS7 therefore uses messages in the form of packets to signal across the network through the STPs. This allows the full use of the talk path for information exchange, and the messaging paths for informational dialogue between the switching systems and the transfer points.

The links are used to pass control and billing information, network management information, and other control functions as necessary without interfering with the conversational path.

Basic Functions of the SS7 Network

The basic functions of the SS7 network include some of the following pieces of information:

■ The exchange of circuit-related information between the switching points along the network.

■ The exchange of non-circuit-related information between the databases and the control points within the network.

■ The facilitation of features and functions by marrying the stored program control systems together throughout the network into a homogeneous network environment.

Further, the SS7 network allows these features to be put into place without unduly burdening the actual network call path arrangements.

■ It handles the rerouting of network traffic in the event of circuit failures by using automatic protection switching services, such as found in SONET or alternate routing information.

■ Because it is a packet-switching concept, the SS7 network prevents misrouted calls, duplication of call requests, and lost packets (requests for service).

- It allows the full use of out-of-band signaling using the ITU Q931 signaling arrangements for call setup and teardown.

- It allows growth so that new features and functions can be introduced to the network without major disruptions.

Signaling Links

SS7 signaling links are characterized according to their use in the signaling network. Virtually all links are identical in that they are 56-Kbps (or 64-Kbps) bidirectional data links that support the same lower layers of the protocol; what is different is their use within a signaling network. The bidirectional nature of these links allows traffic to pass in both directions between signaling points. Three basic forms of signaling links exist, although they are physically the same. They all use the 56-Kbps DS0A in North America and 64-Kbps DS0C data facilities in nearly every other portion of the world (except Japan, where they still use a 4.8-Kbps link). The three forms of signaling links are:

- *Associated.* The simplest form of signaling link is referred to as the *associated signaling link,* as shown in Fig. 30.4. In associated signaling, the link is directly parallel from the end office with the voice path for which it is providing the signaling information. This is not an ideal situation, because it would require a signaling link from the end office to every other end office in the network. There are some associated modes of signaling in use, but they are rare.

  Where one will most often find associated signaling deployed is at an end-user location using a single T1 and common channel signaling. Channel 24 on a T1 is the associated out-of-band signaling channel for the preceding 23 talk channels.

  In some cases, it may be better to directly connect two SSPs together via a single link. All related SS7 messages to circuits connecting the two exchanges are sent through this link. A connection is still provided to the home STP using other links to support all other SS7 traffic.

- *Nonassociated signaling links.* In the nonassociated signaling link arrangement there is a separate logical path from the actual voice

Figure 30.4
Associated signaling.

Signaling and Talk Path

S.S.P.   S.S.P.

path, as shown in Fig. 30.5. There are usually multiple nodes involved to reach the end destination, while the voice may have a direct path to reach the final destination. Nonassociated signaling is a common occurrence in many SS7 networks.

The primary problem with this form of signaling is the number of signaling nodes that the call must use to progress through the network. The more nodes used, the more processing and delay that can occur. Nonassociated signaling involves the use of STPs to reach the remote exchange. To establish a trunk connection between the two exchanges, a signaling message will be sent via SS7 and STPs to the adjacent exchange.

■ *Quasi-associated signaling links.* In quasi-associated signaling, a minimum number of nodes are used to process the call to the final destination, as shown in Fig. 30.6. This is the preferred method of setting up and using an SS7 backbone because each node introduces additional delay in signaling delivery. By eliminating some of the processors on the setup, the delay can be minimized.

SS7 networks favor the use of quasi-associated signaling. In quasi-associated signaling, both nodes are connected to the same STP. The signaling path is still through the STP to the adjacent SSP.

Figure 30.5
Nonassociated
signaling.

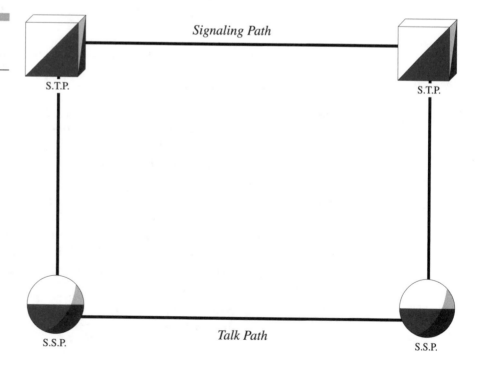

Figure 30.6
Quasi-associated
signaling.

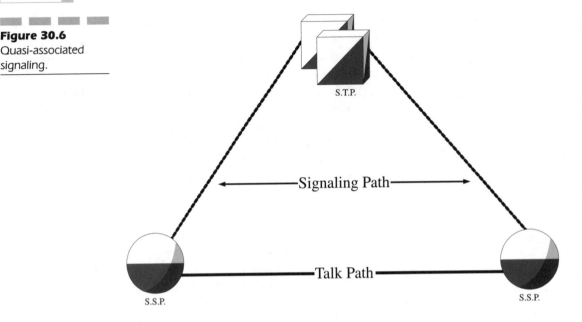

The Link Architecture

Signaling links are logically organized by link type (A through F) according to their use in the SS7 signaling network. These are shown in Fig. 30.7 with the full linkage in place.

A Link. An A (access) link connects a signaling end point (SCP or SSP) to a STP. Only messages originating from or destined to the signaling end point are transmitted on an A link.

B Link. A B (bridge) link connects one STP to another STP. Typically, a quad of B links interconnect peer (or primary) STPs (the STPs from one network to the STPs of another network). The distinction between a B link and a D link is rather arbitrary. For this reason, such links may be referred to as *B/D links.*

C Link. A C (cross) link connects STPs performing identical functions into a mated pair. A C link is used only when an STP has no other route available to a destination signaling point due to link failure. Note that SCPs may also be deployed in pairs to improve reliability; unlike STPs, however, signaling links does not interconnect mated SCPs.

D Link. A D (diagonal) link connects a secondary (local or regional) STP pair to a primary (internetwork gateway) STP pair in a quad-link con-

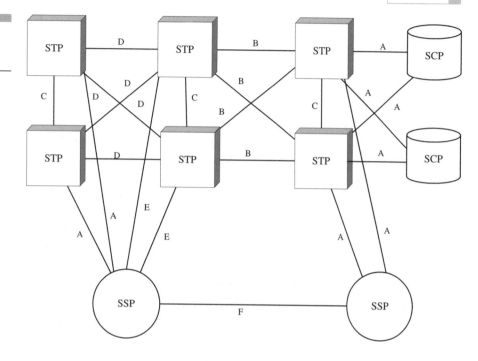

Figure 30.7
The signaling link
architecture.

figuration. Secondary STPs within the same network are connected
via a quad of D links. The distinction between a B link and a D link
is rather arbitrary. For this reason, such links may be referred to as
B/D links.

E Link. An E (extended) link connects an SSP to an alternate STP. E
links provide an alternate signaling path if a SSP's home STP cannot
be reached via an A link. E links are not usually provisioned unless
the benefit of a marginally higher degree of reliability justifies the
added expense.

F Link. An F (fully associated) link connects two signaling end points
(SSPs and SCPs). F links are not usually used in networks with STPs.
In networks without STPs, F links directly connect signaling points.

Links and Linksets

A linkset is a grouping of links joining the same two nodes. A minimum
of 1 link to a maximum of 16 links can make up the linkset. Normally
SSPs have 1 or 2 links connecting to their STPs, based on normal capacity

and traffic requirements. This constitutes a 1- or 2-link linkset. SCPs have many more links in their linksets to handle the large amount of messaging for 800/888/877, and 900 numbers, calling cards, and AIN services.

Combined Linksets

Combined linkset is a term used to define routing from a SSP or SCP toward the related STP, where two linksets are used to share the traffic outward to the STP and beyond. The requirement is not that all linksets be the same size, but the normal practice is to have equally sized groupings of Linksets connecting the same end node. Using a linkset arrangement, the normal number of links associated with a linkset is shown in Table 30.2.

Linksets are defined as a grouping of links between two points on the SS7 network. All links in a linkset must have the same adjacent node in order to be classified as part of a linkset. The switches in the network will alternate traffic across the various links, to be sure that the links are always available. This load spreading (or balancing) serves many functions. Some of them are:

- To be aware when a link fails
- To recognize when congestion is occurring in the network
- To use the links when traffic is not critical and be aware when a link is down before it becomes critical

Routes and Routesets

The term *routeset* refers to the routing capability of addressing a node within the SS7 network. Every node within the network has a unique address. This address is referred to as a *point code*. The addressing scheme or point code is the major routing characteristic of the CCS7 (SS7) network. The terms *routeset* and *point code* are synonymous.

TABLE 30.2	Link Type	Number of Links
Configuration of Linksets	A links	Maximum of 16 links
	B/D links	Installed in quads up to a maximum of 8 links
	C links	Installed individually up to a maximum of 8 links

The point code is made up of 9 digits broken down into three 3-digit sequences. An example of this is 245-100-000. Reading the point code from left to right, we find that:

- The first three digits refer to the *network identifier* (245)
- The next three digits refer to the *cluster number* (100)
- The final three digits refers to the *member number* (000)

In any given network, there can be 256 clusters and each can have 256 members. The network number in this case is for Stentor Communications in Canada.

The routing of SS7 messages to a destination point code can take different paths, or routes. From the SSP perspective, there are only two ways out from the node, one toward each of its mated STPs. From that point on the STPs decide what routes are appropriate, based on time, resources, and status of the network. From the SSP, various originating and terminating (destination) addressing scenarios are defined as follows:

- If the route chosen is a direct path using a directly connected link (SSP1-STPA), then the route is classified as an *associated route*.
- If the route is not directly connected via links (SSP1-SSP2), the route is classified as a *quasi-route*.
- All routing is controlled by nodal translations, providing flexible and network-specific routing arrangements. This is shown in Fig. 30.8.

SS7 Protocol Stack

The SS7 uses a four-layer protocol stack that equates to the seven-layer OSI model. These protocols provide different services depending on the use of the signaling network. The layers constitute a two-part functionality: the bottom three layers are considered the communications transmission of the messages, whereas the upper portion of the stack performs the data processing function. Refer to Fig. 30.9 for the protocols.

The stack shows that the bottom three layers make up the *message transfer part* (MTP), similar to the X.25 network function. At one time SS7 messages were all carried on X.25. Now newer implementations use SS7 protocols, whereas in older networks or third-world countries X.25 may still be the transmission system in use.

The SCCP is used as part of the MTP when necessary to support access into a database, and occasionally for the ISDN user part. This extra

Figure 30.8
Routes and routesets.

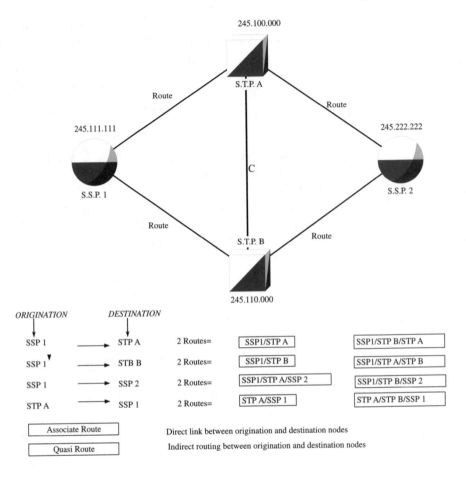

ORIGINATION		DESTINATION				
SSP 1	→	STP A	2 Routes=	SSP1/STP A		SSP1/STP B/STP A
SSP 1	→	STB B	2 Routes=	SSP1/STP B		SSP1/STP A/STP B
SSP 1	→	SSP 2	2 Routes=	SSP1/STP A/SSP 2		SSP1/STP B/SSP 2
STP A	→	SSP 1	2 Routes=	STP A/SSP 1		STP A/STP B/SSP 1

Associate Route	Direct link between origination and destination nodes
Quasi Route	Indirect routing between origination and destination nodes

link is the equivalent of the transport layer of the OSI model supporting the TCAP.

The SS7 network is an interconnected set of network elements that is used to exchange messages in support of telecommunications functions. The SS7 protocol is designed both to facilitate these functions and to maintain the network over which they are provided. Like most modern protocols, the SS7 protocol is layered. Functionally, the SS7 protocol stack can be compared to the Open Systems Interconnection (OSI) reference model. Although OSI is a seven-layer stack designed to perform several communications and transparent functions, the SS7 protocol stack is similar but different.

Like any other stack, the SS7 protocol stack is specifically designed for reliable data transfer between different signaling elements on the network. Guaranteed delivery and the prevention of duplication or lost

Figure 30.9
SS7 protocols.

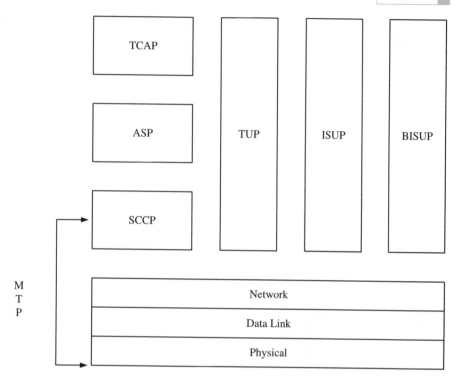

packets are crucial to network operations. To satisfy differing functions, the stack uses various upper-layer protocols, but consistently uses the same lower layers.

Basic Call Setup with ISUP

The important part of the protocols is the call setup and teardown. This next example is shown in Fig. 30.10.

When a call is placed to an out-of-switch number, the originating SSP transmits an ISUP *initial address message* (IAM) to reserve an idle trunk circuit from the originating switch to the destination switch (1a). The IAM includes the originating point code, destination point code, circuit identification code dialed digits, and, optionally, the calling party numbers and name. In this example, the IAM is routed via the home STP of the originating switch to the destination switch (1b). Note that the same signaling links are used for the duration of the call unless a link failure condition forces a switch to use an alternate signaling link.

Figure 30.10
Call setup with ISUP.

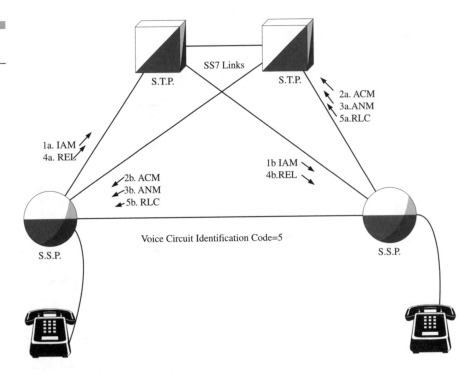

Figure 30.10
Call setup with ISUP.

The destination switch examines the dialed number and determines that it serves the called party, and that the line is available for ringing. The destination switch transmits an ISUP *address complete message* (ACM) to the originating switch (2a) via its home STP to indicate that the remote end of the trunk circuit has been reserved. The destination switch rings the called party's line and sends a ringing tone over the trunk to the originating switch. The STP routes the ACM to the originating switch (2b), which connects the calling party's line to the trunk to complete the voice circuit from the calling party to the called party. The calling party hears the ringing tone on the voice trunk.

In this example, the originating and destination switches are directly connected with trunks. If the originating and destination switches are not directly connected with trunks, the originating switch transmits an IAM to reserve a trunk circuit to an intermediate switch. The intermediate switch sends an ACM to acknowledge the circuit reservation request and then transmits an IAM to reserve a trunk circuit to another switch. This process continues until all trunks required to complete the connection from the originating switch to the destination switch are reserved.

When the called party picks up the phone, the destination switch terminates the ringing tone and transmits an ISUP *answer message* (ANM) to the originating switch via its home STP (3a). The STP routes the ANM to the originating switch (3b), which verifies that the calling party's line is connected to the reserved trunk and, if so, initiates billing.

If the calling party hangs up first, the originating switch sends an ISUP *release message* (REL) to release the trunk circuit between the switches (4a). The STP routes the REL to the destination switch (4b). If the called party hangs up first, or if the line is busy, the destination switch sends an REL to the originating switch indicating the release cause (e.g., normal release or busy).

Upon receiving the REL, the destination switch disconnects the trunk from the called party's line. It next sets the trunk's state to idle, and transmits an ISUP *release complete message* (RLC) to the originating switch (5a) to acknowledge the release of the remote end of the trunk circuit. When the originating switch receives (or generates) the RLC (5b), it terminates the billing cycle and sets the trunk state to idle in preparation for the next call.

ISUP messages may also be transmitted during the connection phase of the call (i.e., between the ISUP ANM and REL messages.

SS7 Applications

At this point we switch gears and look at some of the applications that are possible because of SS7 implementations. The use of advanced intelligent network (AIN) features, ISDN features, and wireless capabilities all became a reality because of the functions of SS7 integration. Some of the features are listed here; remember, they are formulated and possible because of SS7, although they may be part of other systems or concepts. These include:

- 800/888/900 services
- Enhancements in 800/888 services within call centers
- 911 enhancements
- Class features
- Calling card toll fraud prevention
- Credit card approval and authentication
- Software/virtual defined private networks
- Call tracing
- Call blocking

It is with the SS7 protocols and signaling systems that all these features and functions are possible. Many of the features are possible through the stored-program CO switches. However, when we activate features that work across the network (or world), SS7 facilitates the delivery mechanism. Other systems will be introduced in the future to implement more intelligence in the networks, such as the advanced intelligent networks (AINs) but they will require the infrastructure of the signaling systems to enact and carry their messages.

31

Local Number Portability

In 1984 when the breakup of the industry began, the FCC and others had a vision that personal communications services would become a reality. The Telecommunications Act of 1996 took this one step closer to the expectations of the regulators. The ultimate goal was to provide seamless and transparent communications for the consumer, with a choice of providers. No one ever really expected that the local loop would remain in the control of one supplier. Thus, the local competitive market took hold. However, the number plan assignments were rigidly set.

An essential stipulation of the Telecommunications Act of 1996 requires that customers must be able to change local carriers while maintaining the same telephone number. The ability to make numbers *portable* across service providers is driving significant investment in the network infrastructure for both the wireline and wireless carriers. Although the carriers were not required to reroute calls from their networks to ported numbers until June 1999, the advent of wireline number porting also had an immediate and significant financial impact. This also presents an impact on the ability to complete outgoing calls between wireline and wireless numbers.

All calls to entire NPA-NXX blocks will require local number portability (LNP) processing to determine if the specific number has actually been *ported,* increasing the demand on the network resources. Because there will be a technical necessity to use the number portability (NP) databases to complete wireless-to-wireline calls, both providers need to address both the cost and the service issues.

Several other issues are in play here at the same time. LNP is only one of the major impacting changes. The wireless and wireline carriers are dealing with the regulatory issues for other scenarios, such as right-of-way issues, physical interconnection problems, the needs of integrating operational support systems, and settlements between the carriers for access and cocarrier charges.

The Three Phases of LNP

The ultimate goal of the Telecommunications Act of 1996 includes the ability of the end users (customers) to take their numbers anywhere. This constitutes full location transportability. Of course, we cannot just make such a quantum leap of technology without considering interim steps to minimize confusion.

Consumers do not have a clue about LNP issues. They hear about the ability to use LNP, but do not have any concept of what it means to the carriers. Polls of end users reveal that 80 percent of residential customers would not be willing to change service providers if they have to change telephone numbers. Moreover, 90 percent of business subscribers would be unwilling to change service providers if it involves changing telephone numbers. To foster competition the idea of LNP was embraced and pushed to the public.

Consequently, we look at the three steps of LNP to get to an end. The three steps include:

- *Service provider portability.* The first step in LNP permits the subscriber to keep their current telephone number when they change service providers. Changes in service providers can mean between and among the ILEC, CLEC, wireless cellular, wireless PCS, wireless SMR, and the cable TV providers, to name the most common. Applying initially to the wireline providers, service provider portability will eventually extend to the wireless networks.

- *Service Portability.* When a subscriber desires services unavailable through the local end office, service portability allows the customer to secure these services from another switching system without changing telephone numbers. This allows the customer to change the mix of services available for use.

- *Location portability.* The final step in LNP allows customers to keep their telephone numbers when they move away from a service area covered by the current central office. The farther away the customer moves, the more the technological challenge posed by LNP. Let's use an example of Bud moving from Phoenix to Boston, but wants to keep the number (480) 777-7992. The ramifications of this solution change the entire numbering plan and the geographic boundary of the service providers.

LNP for Cellular and PCS Suppliers

Few carriers reveal their final plans in advance. They will say that they will meet the mandates of the law, but not exactly how. However, let's look at the impacts on their systems.

Cellular Providers

The cellular industry has been a success story to behold. Over the past 14 years, it has grown from nothing to over 75 million customers. With very few exceptions, the cellular carriers see themselves as a complementary carrier to the ILEC business. Some of this comes from the fact that two parts of the cellular network started when the licenses were issued: wireline carriers meant the ILEC-operated, regulated portion of the cellular network, and nonwireline meant the competition. In most cases, the ILECs that operated the wireline side of the business spread out across the country as part of the nonwireline provider (competing for wireless services with their cousins). Therefore, the cellular industry is influenced heavily by the ILEC side of the business. The cellular providers see themselves as complementing the wireline business instead of competing with it.

PCS and SMR Providers

These carriers seem to be the ones seeking a niche in the market. As they offer services to their customers, they view themselves as an alternative to the ILEC wireline services and at the same time as complementary. Depending on how the customer reacts, the PCS providers especially will move toward one position. The providers offer the same features and functions as the wireline service providers, including:

- Call forwarding (busy or don't answer)
- Voice messaging
- Three-way calling
- Caller ID
- Call transferring

The PCS providers using LNP in their potpourri of offerings are now saying that the customer can use the same number for the home number, the business number and the traveling number. Why pay for two different lines and service offerings when you can do it all on one telephone (and number)?

Their speech is somewhat convincing, but they fail to mention the cost of the airtime for receiving calls and the added cost for making local calls. However, they are now packaging these services in such a way that they are invisible.

The LNP implementation is designed to occur in the top 100 MSAs by March to June 2000. The implication is that the wireless carrier implementations will be fully IS-41 compliant, using an LNP trigger mechanism. Newer triggerless methods will be implemented to meet the first phase of LNP by March 31, 2000.

LNP Differences

Wireless carriers handle their customers differently than do the wireline carriers. In general, major differences arise in the way they handle customer care. Some are very good at managing the customer to provide interface; others never see the customer and rarely get involved with solving problems.

The way the infrastructure is established is also a potential problem. The wireless carriers have completely different overlays in the network for rating calls. The wireline carriers have geographically bounded rate centers, but the wireless carriers have different boundaries. It is normal for a wireless carrier to backhaul a call several miles (or a couple of hundred miles) to process out of its MSC. The calling area for a wireless carrier is set in a different way.

The wireless carriers' billing systems are different too. In the wireline networks, customers do not pay to receive calls (with the exception of a collect or an 800/888/877 call), whereas in the wireless network airtime is billed for every minute of usage. Moreover, in the wireline network the caller gets unlimited outbound local calls (for most network providers, except where message units are billed), whereas in the wireless networks airtime is billed for all local calls. The wireless carriers have been attempting to minimize the visibility by offering packages of X minutes for Y dollars per month. However, in revenue generation the wireless carriers have the advantage because they prebill for the service whether the customer uses it or not. Now, with LNP, they must know where the call is going and how it is actually terminated in order to capture the usage. Most important, they need the originating and terminating minutes to bill when the customer exceeds the allotted time in the package. If the wireline carriers offered these packages, we'd be burning down the COs.

Somehow the interface between the wireless and wireline carriers must come together to make this all transparent. Further, the wireless carriers need an interface to the number portability administration center (NPAC), which requires more updates and more SS7 links to facilitate the LNP operation.

What's It All About, Anyway?

LNP is a far-reaching mandate. Because the Telecommunications Act of 1996 ensured competition could be allowed to grow in the local and long distance markets, we have seen competition in the long-distance market grow and prosper. Therefore, the FCC mandated that in order for new entrants to move into the business of local access, and to foster the competitive spirit, LNP was a requirement.

Few consumers or businesses will take a chance with a new provider, if they have to change numbers. LNP was made available to wireline customers in the 100 top MSAs by the end of 1998 and to the wireless subscribers by March 2000. Although the deadline was pushed back to the year 2000, this is not a trivial matter. The CTIA petitioned the FCC for a five-year moratorium. They want a reprieve for the broadband PCS suppliers to build out their basic infrastructure. They feel that if the LNP issue is to be addressed while these carriers are still building their infrastructure, it will prove to be negative in their market segment. Most wireless carriers say they will have something in place by 2000, but the choices are varied in the way they will implement this.

Wireless customer care and billing systems are different from those of the wireline carriers. Even the way they define their rate centers differs widely from that of the wireline providers. The wireless carriers must now make the wireline LNP systems work within the wireless systems, which is no simple matter. There are still many unanswered questions about how the systems will work for the wireless networks and providers. These include:

- How will the wireless carriers transfer the line information to the number portability administration centers (NPACs)?
- What happens to the roaming discussed in the earlier chapters?
- How will over-the-air provisioning be handled when LNP is added?
- What will become of the mutual compensation billing systems they use?
- When calling party pays becomes a reality, who gets billed when the wireless carriers don't have LNP information telling them where the call was actually terminated?

Where's the Money?

One of the big issues in complying with the LNP mandates is the fact that there has to be a return on investment for the carriers. The cost implications of building out the network to accommodate the LNP requirements are significant. The FCC recognized this issue when mandating that LNP be implemented between the local providers (ILECs), the CLECs, and the wireless providers.

Several mechanisms will help the carriers recoup their costs. These include the ability of the carriers to bill back to the consumer starting in February 1999. The consumer will pay for the overall implementation, but carriers will never really recoup all their costs because of other regulatory and tariff issues. The local utilities commissions are always dragging their feet when approving any costs passed through to the consumer. Carriers subject to a rate-of-return or a price-cap constraint will be allowed to recover their carrier-specific costs directly related to the installation and operation of LNP. A federally approved monthly charge was enacted, to last no more than five years. Costs related to the N-1 querying protocol (used to determine the last carrier before the incumbent carrier) may be recovered from noncompliant carriers or default carriers, through a federal tariff for query services. The wireless carriers have considered the option of just paying the ILECs for the LNP services. However, this can be too expensive for them, leaving the other possibilities still available.

There is more, though, not just about recouping costs, but also in gaining revenue. The Gartner Group conducted a study on LNP and found that a substantial amount of money can be made. The domestic local telephone services market, after opening the market to competition with LNP, will result in basic service opportunities of $4.9 billion in 1999 and up to $6.2 billion in 2000. The network access market potential will total $4 billion in 1999 and $5.4 billion in 2000. These estimates are based on an assumption that 75 percent of the U.S. population lives in the 100 top MSAs and that 75 percent of all the service revenues will come from the 100 top MSAs. Market-share rates are based on the ability of the customer to churn.

The Wireless Carriers' Decision

Carriers have choices on how they will handle the LNP situation. Not every carrier is affected by the services (such as the IECs and the payphone providers) but those who are affected can choose their own destiny. They can choose to do one of the following:

- *Do nothing.* Carriers can remain noncompliant within the guidelines of the Telecommunications Act of 1996 and allow the terminating carrier to perform the N-1 database query for them. A carrier without LNP has no way of knowing whether outgoing calls are dialed to numbers that are ported; therefore the calls are automatically sent to the carrier who owned the number originally. If it is a number in a ported region, the destination carrier must dip into the LNP database to correctly route the call to the final end point. ILECs are expected to charge $0.001 to $0.008 for this. For a wireless carrier this can be substantial, with estimates set at $250,000 to $1 million per year, based on very conservative averages. That being the case, the wireless carriers are likely to get on the LNP bandwagon quickly if this is their choice. They will still need 6 to 9 months to implement a decision.

- *Choose a Service Bureau.* This is one that many of the smaller wireless carriers have expressed the most interest in today. Most of the cellular carriers are already using clearinghouse functionality from service bureaus. The most common are GTE and Cincinnati Bell Information Services (CBIS), who provide reciprocal billing and accounting functions. For a fee, the service bureau will offer the LNP service. If the wireless carrier outsources to a service bureau, minimal upgrades to the wireless carrier networks are required. Some errors are likely to occur because this manual system requires entries into multiple OSSs. Carriers that choose this option have LNP-enabled call routing, but lack the other competitive advantages that go with LNP in full operation.

- *Choose their own LNP solution.* This is the most expensive of the choices in the short term, but for the long term it will be of real benefit. For both wireline and wireless carriers, this is an expensive and complicated solution. The benefits of LNP are that it will take full advantage of the intelligent network (IN), the advanced intelligent network (AIN) features, and/or the wireless intelligent network (WIN) infrastructure. This will give the wireless carrier a competitive edge over the competition by bringing a new level of intelligence into the wireless telecommunications industry and networks.

Basic LNP Networks

The components of LNP are not that much different from the original SS7 networks used for years. The pieces serve different functions; looking at the components in Fig. 31.1 we see:

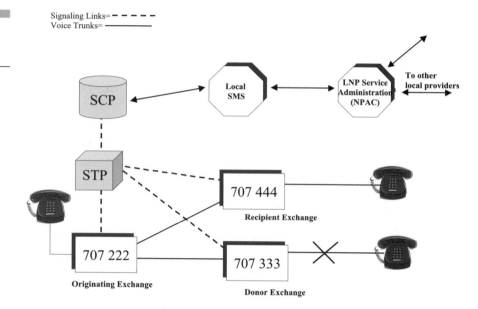

Figure 31.1
Basic LNP network
components.

- The switching service point (SSP) is the local CO or tandem.

- The signal transfer point (STP) is a packet-mode handler that routes data queries through the signaling network.

- The signal control point (SCP) is the database for features, routing, and global title translation (GTT).

- The local service management system (LSMS) and service order administration (SOA) can be provisioned separately, but when combined they are referred to as the number portability administration center (NPAC) connectivity.

For wireless providers, the LNP capabilities depend on the MSA served. Even if a wireless network does not support LNP, the wireless provider's methods and procedures should be updated to support troubleshooting for calls placed to ported wireline numbers.

Once wireline networks in the top 100 MSAs have implemented LNP, wireless providers serving the same MSAs need to address wireless end users making calls to portable wireline numbers. Wireless SPs are not required to support LNP routing at this time. Calls for wireless end users to a ported wireline number will complete successfully through the donor network.

LNP Services Administration

At issue are a few minor points:

- The LECs administer the numbering plan in their regions. The North American Numbering Council (NANC) issued a requirement for a neutral third-party administrator.

- LNP number administration was limited to regional implementations. Therefore, the NANC directed the INC to develop the guidelines for LRN assignment.

These issues were surmountable but did cause minor delays in getting the NPAC services up and running. Many of the wireless providers bought the 10,000 numbers in an NPA-NXX from their LECs. They therefore feel they have ownership and proprietary rights to the numbers assigned to them. Furthermore, the inconsistencies in which the plans were applied called for a third party to manage the assignments of numbering plans as well as the LRN assignments.

Location Routing Number (LRN)

LRN depends on intelligent network (IN) or advanced intelligent network (AIN) capabilities deployed by the wireline carriers' networks. LRN is a 10-digit number to uniquely identify a switch that has ported numbers. The LRN for a particular switch must be a native NPA-NXX assigned to the service provider for that switch.

LRN assigns a unique 10-digit telephone number to each switch in a defined geographic area. The LRN now serves as the network address. Carriers routing telephone calls to end users that have changed from one carrier to another (and kept their same number) perform a database dip to obtain the LRN corresponding to the dialed telephone number. The database dip is performed for all calls where the NPA-NXX of the called number has been flagged in the switch as a portable number. The carrier then routes the call to the new provider based on the LRN.

Figure 31.2 shows the flow of LRN information, which is the same for both a wireline and a wireless provider.

When the caller dials the number (333-3333 in this case), the originating switch (612-222) sends its signaling information (info dialed 333-3333)

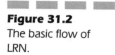

Figure 31.2
The basic flow of
LRN.

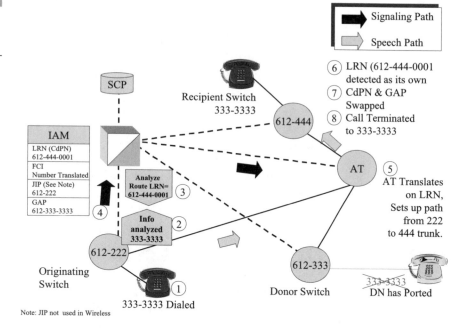

Note: JIP not used in Wireless

through the STP to the SCP, which analyzes the route and returns the
LRN (612-444-0001). Next, an IAM message is forwarded from 612-222
through the STP to the access tandem. The AT translates the LRN and sets
up a speech path (trunk) from 222 to 444. Switch 612-444 detects the LRN
(612-444-0001) as its address; therefore, the called number and the generic
address parameter are swapped. From there, the call is connected (terminated) at 333-3333. The donor switch in this scenario has been uninvolved
throughout the process.

LNP Impact on Routing and Rating

As already stated, there are changes for the wireline and the wireless
providers. Wireless providers have had different billing and rating measures to satisfy their network demands. The ILECs were tied to the regulatory boundaries of the local PUCs and the FCC. The impacts for both
providers will mean that changes in the way calls are routed, rated for

access, and rated for end-user billing will include the use of the LRN in the future. This means that the wireless providers will have to use some inclusions of the LRN differently than they did in the past. The simplest way of doing this (as represented by three major wireless providers) is to use the standard V&H coordinates and then apply from-and-to billing and access ratings to the call.

Using the N-1 architecture, the carriers use the next-to-last service provider to perform the LNP database dip. This will pertain to both wireline and wireless providers. When two IECs are involved, the N-1 query is launched by IEC number two. When only two networks are involved (ILEC and IEC), then the originating service provider is responsible for the LNP query.

Demand for LNP

Local LNP issues will first be addressed in service provider portability. However, as we wait for this to happen, we know that 90 percent of wireless users say they would choose a different carrier if they could keep the same number. This percentage is stated differently than others in this book. All the other studies indicate that more than 80 percent of customers would not be interested in a change if it involved a number change. The wireless carriers are fraught with the fear that the statistic for their networks is that 90 percent *would* change.

This is a dramatic number in an industry whose biggest problems are churn and fraud. Churn is killing these suppliers in light of the packages that are being offered. The possibility of getting free minutes and/or free local calling from a new supplier is enough to make the wireless carriers' customers jump providers. Long-term contracts are outdated, but were a way to hold the customer for at least some extended period.

Location NP is becoming a bigger issue and demand is becoming a hotter topic for the carriers and customers alike. However, location portability raises many other issues that have not been fully fleshed out, such as:

- Added database dips because of the number of ILECs and IECs that can be involved in a call process.

- 911 PSAP administration will become more complicated.

- Rating and billing issues. How do you process a call? Who pays?

- Loss of the NANP geographical significance. In the old days we knew where a caller was by the telephone number.

- CPE impacts in the ability to process local versus long-distance calls from a database that must be kept updated. This puts more burdens on the CPE (PBX, key systems, etc.).

New local calling plans will be needed, but separate from the rate center plans in place and regulated today. Possibly, rate center consolidation will be an alternative for number portability from a centralized location and a better way of handling the local versus long-distance calling. Toll alerts may be needed in the signaling network. When a number has been flagged as ported to a new LRN, the network will intercept and advise the calling party that it will be a toll call. This involves more database dips and more equipment to interface in the network. Who pays for this?

Another issue will be that callers will be relegated to making a 10-digit call for all calls, local and long distance, because of the lookup that will be required. Users are already screaming about that; what do providers do to appease this issue?

LNP Operations, Administration, and Maintenance Impacts

Clearly, we can see that all parts of the network are being affected by the implementation of LNP. Many of these issues and impacts are quite costly for the wireline carriers. Looking at all the pieces when we add the MSC in place of the SSP, the impact on the integration becomes far more complex because of the infrastructure differences between the providers. There are many places where the wireless providers do things differently.

One area is over-the-air provisioning by the wireless carriers. Getting that information to customers who port their numbers to a wireline provider will have to be done through the NPAC. STP and SCP interfaces are using IS-41 and SS7 combinations that work differently from the wireline carriers.

The point here is that it can be done at a greater cost to the wireless carrier because of the mobility involved, and the database dips that will be required.

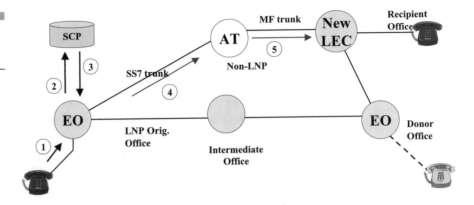

Figure 31.3
Scenario with a non-LNP office.

Scenarios

These scenarios look at the different choices that may appear in the implementation of LNP by the wireline or wireless carriers. The first, as shown in Fig. 31.3, is a non-LNP intermediate office using MF trunks between suppliers.

1. Call is placed to ported subscriber.

2. Originating switch performs LNP query.

3. SCP replies.

4. LNP IAM packet is sent to non-LNP tandem.

5. Non-LNP tandem dials over MF trunk, only LRN is sent (in the called-party ID parameter), dialed number is lost, and call fails.

The next scenario, shown in Fig. 31.4, deals with a dialed number pulsed to a non-LNP tandem via MF signaling.

Figure 31.4
Scenario with non-LNP tandems.

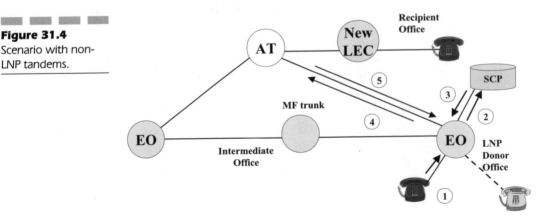

1. Call is placed to a ported subscriber from a donor switch.

2. Donor switch performs the LNP query.

3. SCP replies with the LRN.

4. Donor switch sends the dialed digits over an SS7 trunk to a non-LNP tandem (if the donor is provisioned to send only the dialed digits to non-LNP switches, the signal ported number option).

5. The non-LNP tandem sends the call based on the dialed number back to the donor switch. The call will keep going through steps 2 to 5 until the call is ended by the network management. (This is *message looping,* a possible big problem in a wireless network.)

In the third scenario, shown in Fig. 31.5, we see the operator services involved in a dial 0 call.

1. The subscriber (708-232-1111) dials 0 and is connected to the TOPS operator.

2. The operator receives a request to connect to 708-828-2222 and bill to the originator's station.

3. TOPS determines that the requested number has been ported and sends the LNP query to the SCP.

4. The LNP SCP sends back the response with the LRN (312-225-0000) of the recipient switch.

5. TOPS routes the call to the recipient switch.

6. The recipient switch terminates the call to the ported number, 708-828-2222.

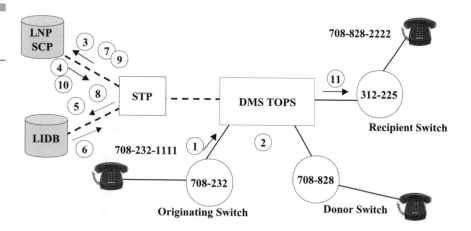

Figure 31.5
Scenario with the operator model.

Occasionally, a database file must be accessed. Using a line information database (LIDB), this scenario follows the flow into and out of the database to determine the location of the called party.

1. Subscriber A (708-232-111) dials 0-708-828-2222 and call is routed to TOPS.
2. TOPS sends a "bong" to the subscriber, requesting the calling card number.
3. TOPS routes calling card number to LNP SCP to obtain correct LIDB point code address.
4. LNP SCP returns LIDB point code to STP.
5. STP queries LIDB for calling card validation.
6. LIDB returns calling card validation to TOPS.
7. TOPS routes billing number to LNP SCP to obtain correct calling party service provider (LRN).
8. LNP SCP returns LRN of billing number and TOPS populated appropriate AMA module.
9. TOPS routes called party number (708-828-2222) to LNP SCP for LRN.
10. LNP SCP returns LRN of recipient switch.
11. TOPS routes call.

Wireless and Wireline E-911

When we think about LNP in a wireline environment, the process is straightforward. The caller places an emergency call from a fixed location, which is easily identifiable. Routing of the E-911 call is to the proper public safety answering point (PSAP) based on a known prearranged location from each telephone. ANI is mapped one to one for all wireline calls using the appropriate callback number and a location from an automatic location information (ALI) database. This is shown in Fig. 31.6.

911 calling and routing is built on the limited capability of the CAMA trunks (trunks that were originally used for cost accounting and messaging call information), and the CAMA will remain in use for some time to come. This does cause some difficulty for the wireless companies, because the need to send more information for wireless 911 calls cannot be met the same way as in wireline calls. Figure 31.7 shows the wireless 911 process using LNP.

Is 911 Important? Nearly 25 percent of all the calls (80,000) per day are from wireless subscribers. This places a great deal of problems on the network.

Figure 31.6
Wireline E-911.

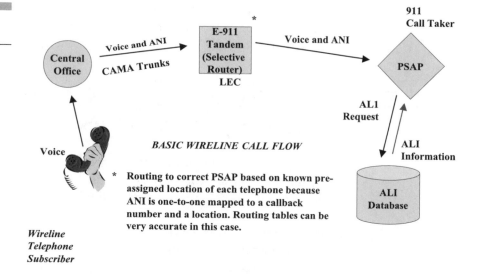

BASIC WIRELINE CALL FLOW

* Routing to correct PSAP based on known pre-assigned location of each telephone because ANI is one-to-one mapped to a callback number and a location. Routing tables can be very accurate in this case.

Wireline Telephone Subscriber

The Ultimate Goal—2001

The goal by the year 2001 is to have a fully transparent network that can handle the wireless 911 calls and get to the caller very quickly. Upgrades across the network require that a satellite global positioning system (GPS) installed at the wireless network interface will help in locating the caller and will route to the appropriate PSAP every time. Using the GPS, the feeling is that the caller can be located 67 percent of the time within 125 meters (410 feet), because the call will be routed to the PSAP with the

Figure 31.7
Wireless E-911.

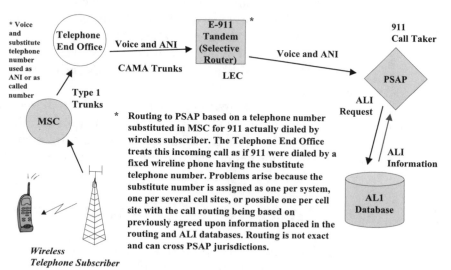

* Routing to PSAP based on a telephone number substituted in MSC for 911 actually dialed by wireless subscriber. The Telephone End Office treats this incoming call as if 911 were dialed by a fixed wireline phone having the substitute telephone number. Problems arise because the substitute number is assigned as one per system, one per several cell sites, or possible one per cell site with the call routing being based on previously agreed upon information placed in the routing and ALI databases. Routing is not exact and can cross PSAP jurisdictions.

Wireless Telephone Subscriber

appropriate *x, y* locator coordinates. The ALI will be updated to support both ANI and *x, y* lookup. The call to the PSAP will arrive with a mapping system showing the *x, y* coordinates of the calling party.

Benefits of Using LNP and Roaming Capabilities

The benefits of making and receiving calls in a wireless network or a wireline network are the ultimate revenue generation. When a user on the wireless network is receiving a call today from the wired world (or vice versa), the network uses its own appropriate databases. In this case the wireless networks rely heavily on their own SCPs, which house the database information of their callers.

Home location registers (HLRs) keep track of all network suppliers' users, based on their own network ID. The HLR database has all the appropriate information to recognize the caller by user ESN and mobile telephone number. The database controls the features and functions the user has subscribed to. When a call comes into the network today, the called number is sent to the HLR responsible for the number dialed. These are SS7 messages. The HLR then looks into its database and determines where the caller is located, or if the user is on the air.

If the user is located somewhere else, a visiting location register (VLR) entry exists at a remote MSC. This entry has been updated by the remote end when the user activated the telephone (powered it on and registered) or when an already powered device rolled into range of a cell site from the new location. Then an SS7 message is sent out to the HLR that the called party is now in someone else's ALI. See Fig. 31.8 for the registration process.

If a call is coming in from the network, the SS7 inquiry comes into the HLR, which then sends a redirect message to the remote MSC (the VLR) serving the end user.

Using this roaming capability, database dips are occurring on a frequent basis. Many of the wireless networks are already taxing their databases. They are trying to move more information out to the STP or the local switches rather than constantly hammering into the SCP.

Incoming Calls to an Idle Mobile

Using the IS-41 and SS7 linkage, when a call comes in to a mobile number, the immediate link is to the HLR. The HLR then sends out a query to the

Figure 31.8
Mobile registration.

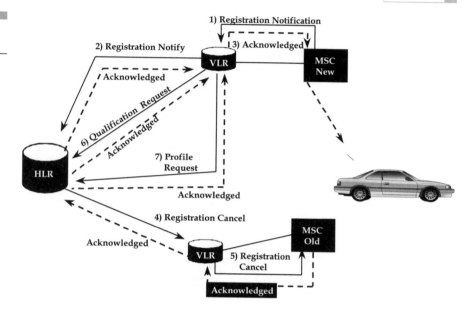

VLR indicating that there is a call coming in for the mobile. See Fig. 31.9 for the process of incoming to idle sets.

The VLR verifies the information and sends back confirmation to the HLR indicating that the call should be steered to the temporary local directory number (TLDN), which is effectively a call-forwarding arrangement. The call is then forwarded to the end-user device, using the dynamically assigned DN.

Once the call is received at the remote MSC, the call is logged into the appropriate billing and accounting system for reciprocal payment and processing. At the end of the month, the billing systems are polled and processed by a service bureau for peering arrangements.

Calls to a Busy Set or a Set That Doesn't Answer

Assume that after the TLDN has been assigned by the VLR and the call has been redirected to the set, the set is busy or does not answer. Now the VLR sends an SS7 message back to the HLR and requests additional information, such as "What do I do now?"

The VLR then sends out the page to the mobile, but if no answer comes back, then the SS7 message to release the call is sent back to the HLR.

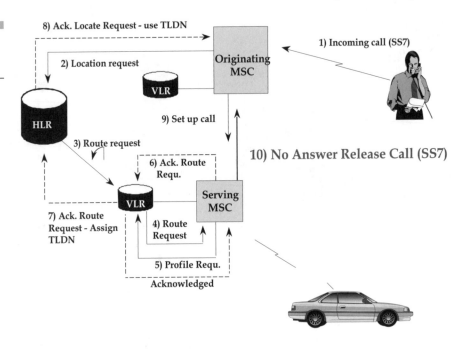

Figure 31.9
Incoming call to an
idle mobile set.

No Answer with Call Forwarding

Assume that the VLR assigned a TLDN to process the call to the called telephone user, then sends out the page, but the caller does not answer. This time the VLR goes back to the HLR for a clarification on the next step. The HLR does a database dip and finds that the called party has call forwarding to another number, such as a business wireline number or a voice-mail system. This information is then fed back to the VLR for processing. The VLR then sends the message back to process the call to the forwarding number, as shown in Fig. 31.10. The VLR is completed, and the HLR has provided all the information necessary. The SS7 network now sends the incoming call request to the wireline end office, alerting it that a call is coming in for the forwarded number, and the process continues.

Calls to a Busy Set with Message Waiting

Assume that the call is coming in for a wireless user but the wireless user is on the phone. Now the VLR goes into react mode and assigns a second

Figure 31.10

Call forwarding to a mobile set.

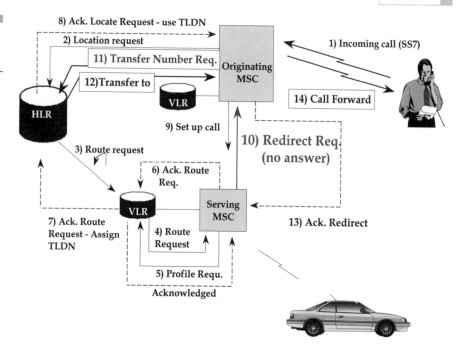

- 8) Ack. Locate Request - use TLDN
- 2) Location request
- 11) Transfer Number Req.
- 12) Transfer to
- Originating MSC
- VLR
- HLR
- 1) Incoming call (SS7)
- 14) Call Forward
- 9) Set up call
- 10) Redirect Req. (no answer)
- 3) Route request
- 6) Ack. Route Req.
- Serving MSC
- 7) Ack. Route Request - Assign TLDN
- VLR
- 4) Route Request
- 13) Ack. Redirect
- 5) Profile Requ. Acknowledged

TLDN back to the HLR, indicating to forward the call to the new number. The call is then sent to the serving MSC on a separate telephone number reserved for this call. Then a message is sent through the airwaves to the air interface indicating that a new call is coming (in the form of a beep). The caller can then choose to answer the call or ignore it.

32

Cellular and Personal Communications Systems

Radio Transmission

Cellular transmission is the latest and greatest front in the industry. In 1946, Illinois Bell Telephone Company introduced a mobile telephone service that would allow users driving in and around the Chicago area to communicate directly to and from vehicles. Using a radio-to-telephone interface, the telephone companies had a lock on this service. Users lined up by the hundreds if not the thousands to request mobile telephone service.

Unfortunately, this was not as simple as driving up to the telephone company and getting the service requested. There was a 2-year wait in many areas of the country as the service rolled out. The reason was simple: radio waves are not intelligent! The telephone companies mounted high-gain antennae on top of high-rise office buildings in the major downtown areas. From there, they would boom out a signal at approximately 250+ watts. This is a very strong signal in any transmission. The antenna was on top of the highest building, and the power output was so great because the telephone companies wanted to get the best coverage possible. The FCC had only issued a limited number of frequencies for use in the Advanced Mobile Phone Services (AMPS) and Improved Mobile Telephone Services (IMTS) networks. Twelve channels were typically allocated in each of the metropolitan areas. Therefore the channels were limited, and the telephone companies needed to provide the greatest area of coverage possible. All calls within the city had to be routed through the centralized tower because of the limited amount of channels. In this particular network, 12 channels were available. Each channel was used on a high-powered radio transmitter to provide the required coverage. Further, the system operated on a one-way (simplex) basis. Only one side of the call could speak at a time. This was analogous to the older two-way "push-to-talk" radio systems. The telephone companies were trying to provide coverage in approximately a 25-mile radius from the downtown area. Back in the late 1940s that was sufficient, because the work force was less mobile.

Problems with the AMPS/IMTS

This arrangement worked for short periods, but, as newer demands were placed on the system, channel capacity was very limited. Only 12 users could be on the network at one time. The limited number of channels (frequencies assigned by the FCC for this service) led to the frequency reuse planning in the telephone companies.

However, because the telephone companies were also trying to get the greatest coverage, the output signal did not stop at the 25-mile marker. Radio-based communications will travel greater distances as a function of the frequency used and the power output. If you want greater coverage, you need to use a higher power output system and antennae or you have to use more equipment. Even though the signal pattern was designed to provide coverage only in a 25-mile radius, it kept on going beyond the distance planned. This meant that if another system was set up in the nearby area, using the same frequencies, the telephone company system would cause interference. As clearly shown in Fig. 32.1, the overlap areas

Figure 32.1
A map of IMTS coverage areas and frequency reuse.

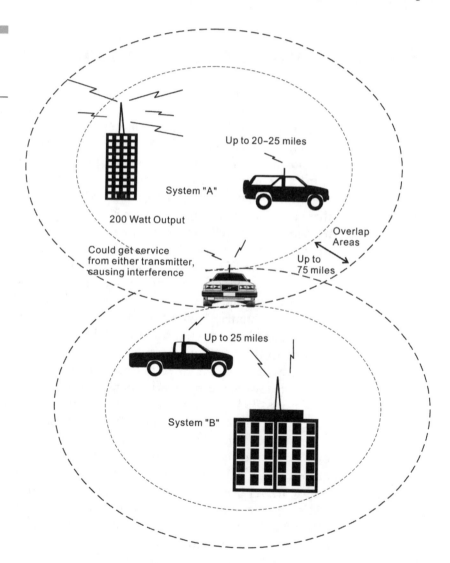

Up to 20–25 miles

System "A"

200 Watt Output

Could get service from either transmitter, causing interference

Overlap Areas

Up to 75 miles

Up to 25 miles

System "B"

were bounded by another 50 to 75 miles. This means that the radio-based system could only use the assigned channels every 100 miles apart. The areas around the circle in this figure—the *buffer zones*—exist to prevent interference. You might recall from the bandwidth discussion earlier in this book that sidebands and frequency bandpass filters are used in the wired telephone world. Essentially the same holds true in the wireless world, where the telephone company allocated the buffer zones so that two conversations would not occur on the same channel. The figure shows that the signal was used in the center (thicker) core, but the overlap areas are shown in the lighter core.

Cellular Communications

Prior to the introduction of an improved telephony service through radio-based systems, the industry standards bodies faced a dilemma. They wanted to meet the demand for more services, but they just did not have the frequencies available. Therefore the engineers went back to the drawing board and created a new technique called *cellular communications*. Using a frequency pattern from radio transmitters, the concept allows a honeycomb pattern of overlapping "cells" of communication. Because these cells can be minimal in size, it is possible to reuse the frequencies repeatedly. Thus cellular communications was born. The goal of cellular was to make more service available to vehicular users.

In 1981, the FCC finally set aside 666 radio channels for cellular use. These frequencies were assigned or set aside for two separate carriers. The lower frequencies were reserved for *wireline companies*. Wireline companies are the regulated providers (the local telcos). The higher frequencies were assigned or reserved for nonwireline carriers, which are the competitors to the telephone carriers. Both operating carriers (the wireline and nonwireline) are licensed to operate in a specific geographic area. The areas are classified as the metropolitan service areas (MSA) and rural service areas (RSA). Each carrier uses approximately 312 frequencies for voice/data communication and 21 frequencies for control channels.

Another major difference in cellular communications is the use of control supervision and switching of calls to serve the cellular user adequately. This is particularly true in a mobile environment when the vehicle moves from one cell to another. The dynamic switching and control necessary to facilitate a smooth and seamless handoff from one cell to another is paramount. If this does not work properly, all communication may be terminated.

The signal on conventional mobile radio telephones degraded as the vehicle moved farther away from the base station. This degradation frustrated both the caller and the called party. As the user went beyond an area of coverage, the call would be cut off, further frustrating both calling and called parties.

With cellular communications, when a call is in progress and the caller moves away from the cell site toward a new cell, the call gets "handed off" from one cell to another. In Fig. 32.2, this process begins to take place. Here's how it works. Once a call is in progress and the handoff from cell to cell becomes necessary to keep it that way, the system initiates a change. As the cellular telephone approaches the imaginary line, the signal strength transmitted back to the cell site will start to fall. The cell site equipment will send a form of distress message to the mobile telephone switching office (MTSO), indicating that the signal is getting weaker. The MTSO then orchestrates the passing of the call from one cell site to another. In Fig. 32.3, the initial sequence begins as a cell site notifies the MTSO that something is going on.

Immediately after receiving the message from the cell site, the MTSO sends out a broadcast to the other cell sites in the area. It requests a determination of which site is receiving the cellular user's signal the most strongly; this is called a *quality of service measurement*. Each site responds accordingly. In Fig. 32.4, the MTSO's initial broadcast goes out across the network.

Figure 32.2
A call in progress will be handled by the cell in which the user is located.

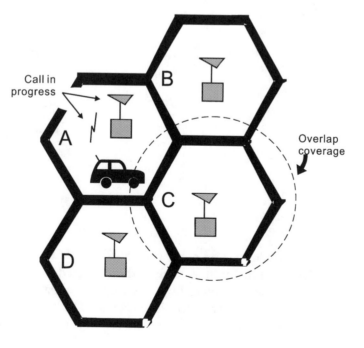

Figure 32.3
The cell site senses
the drop in signal
strength and alters
the MTSO.

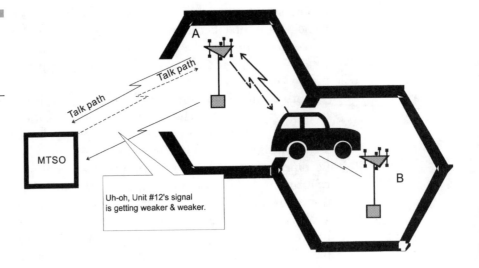

Figure 32.4
The MTSO begins its
initial sequence to
prepare for the
handoff.

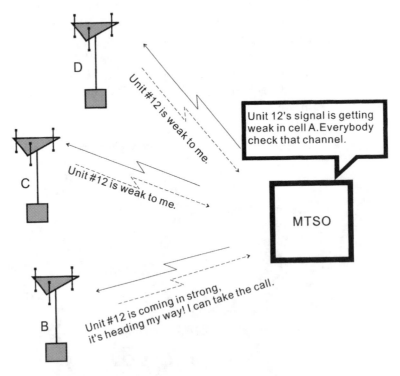

As the responses come back to the MTSO, the particular cell receiving
the signal the most strongly is then selected to accept the call. The MTSO
directs the targeted cell site to set up a voicepath in parallel to the site losing the signal. When this is ready, the MTSO sends a command to the cellular set to resynthesize, or go to the new frequency. The cellular set then

retunes itself to the new frequency assigned to the receiving cell site, and the handoff takes place. This takes approximately 100–200 milliseconds to enact. Fig. 32.5 shows the handoff.

This might sound a little more complicated than it really is. The designers recognized that the original radio systems put out way too much power. Therefore, frequency reuse was not possible in a 100-mile radius. Using a lower-power output device, the radius of the radio transmission system is smaller. In fact, the cellular system was designed to operate in a range of approximately 3 to 5 miles. If the power output is reduced to 10 watts, the radio transmission will travel less distance. Therefore, the user must be closer to the equipment to receive the call. To accommodate this, the cellular "cells" include equipment placed (typically) in the center of each of these overlapping cells. The distance from the radio equipment to the user will be approximately 1 to 2 miles. Therefore, the system should work more efficiently. With only a 10-watt output, the frequencies used in each of the cells can be reused repeatedly. A separation of at least two cells must exist, but that was taken into account. Figure 32.6 is a representation of a cell network showing the honeycomb pattern.

Figure 32.5
The alert goes out to the cellular site, advising it to go to a new channel.

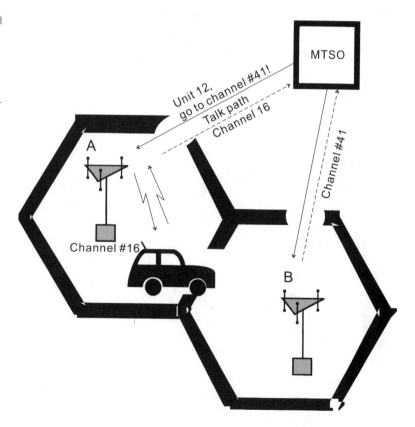

Figure 32.6
The cellular honeycomb pattern is a series of overlapping cells.

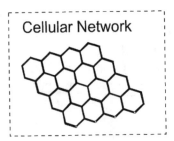

Cellular Network

The carriers have been getting better at providing this coverage as the years have gone by. Many of the cellular suppliers now have provided for a seamless transition from cell to cell anywhere in the country. To do this, they have developed an interface into another telephone- or landline-based technique called signaling system 7 (SS7). Using SS7, the carriers can now hand a call off across the country without user involvement. It just happens transparently. The SS7 linkage is shown in Fig. 32.7, where the MTSO equipment is tied into a computer system called a signal transfer point (STP). In this figure, the MTSOs are tied to duplicate STPs as a means of providing redundancy in the network. With this interconnection, additional features were also provided in the cellular networks similar to the telephone networks.

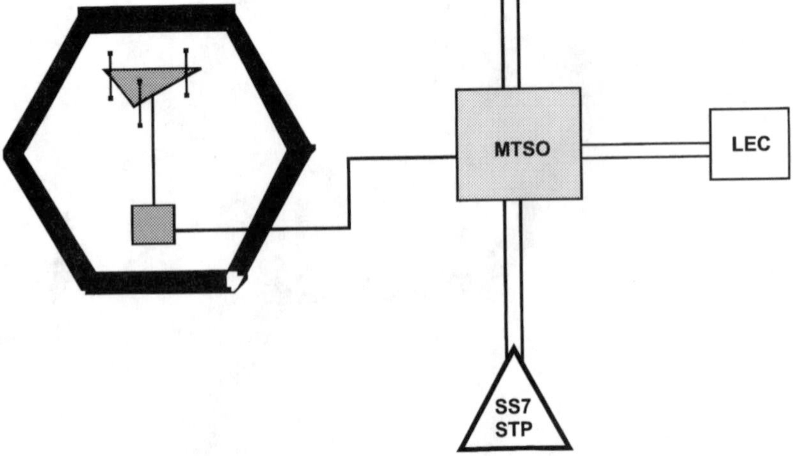

Figure 32.7
Cellular equipment now links to the backbone network through SS7 linkage.

SS7
STP

MTSO

LEC

SS7
STP

Meeting the Demand

The capacity to provide communications from a car telephone was easily met but expensive to start. A vehicle-mounted cellular telephone in 1984 retailed for $3000. However, since then the cost has declined dramatically.

The Telephone (Mobile) Set

The set houses a transceiver capable of tuning to all channels within an area. These are frequency-agile units capable of receiving/transmitting on all 666 frequencies, as opposed to the fixed-frequency units of old. The mobile set is shown in Fig. 32.8. The major components of the unit are:

- Handset
- Number assignment module (NAM), an electronic fingerprint (a 32-bit binary sequence)
- Logic unit
- Transmitter

Figure 32.8
The cellular unit sets up the calls to the cell site using a frequency-agile capacity to synthesize the necessary frequency.

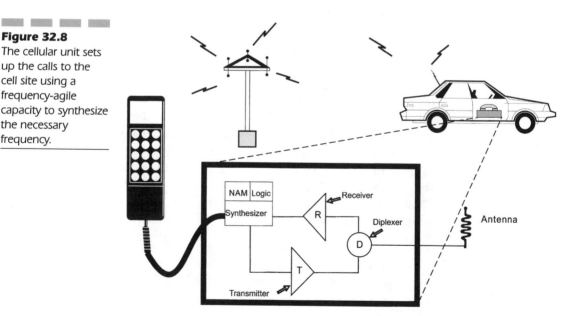

NAM= Numeric Assigned Mode

- Receiver
- Frequency synthesizer (generation of frequencies under control of the logic unit)
- Diplexer—separates transmitter/receiver functions
- Antenna

Cellular's Success and Loss

In 1984, cellular communications became the hot button in the industry. This was the original attempt to add capacity to several systems around the country. In the preceding years, the Advanced Mobile Phone Services (AMPS) and the Improved Mobile Telephone Services (IMTS) were very limited. Part of the problem stemmed from the fact that all systems were analog. Frequency-division multiple access (FDMA) is an analog technique designed to support multiple users in an area with a limited number of frequencies. An analog radio system will use analog input, such as voice communications. Because these systems were designed around voice applications, no one had any thought of the future transmission of data, fax, or packetized (X.25) data from a vehicle. When cellular was first introduced, the industry experts were predicting that by the year 2000, approximately 900,000 users would be on the network conducting voice telephone calls. After all, it was a voice network designed to carry dial-up telephone calls.

No one was sure what the acceptance rate would be of cellular radio-telephone services. Currently, there are approximately 47 million cellular users in the United States. This was a surprise to the industry experts, who had no idea of the pent-up demand. Approximately 100,000 to 150,000 new users sign up monthly. The problem is not one of acceptance and signing up new users but retaining users and encouraging them to use the service more. The churn ratio has been as high as 15 to 30 percent. This is a costly process and one that the providers are trying to overcome. Frustration exists on three counts:

- *The cost of the service can be quite high.* Depending on the location, the basic monthly service will range from $19 to $39, and the per-minute charge for incoming or outgoing calls is $0.35 to $0.45.

- *The possibility of getting poor transmission or reception on the network.* This is a design problem that must be overcome before total acceptance is achieved.

■ *The lack of available service.* The network can get congested very quickly, and if you are paying for the service, you obviously want it available all the time. Users typically don't want to hear about the providers' problems. To overcome this problem, many carriers have split their cells into half and more, just to be able to use the frequencies repeatedly. Figure 32.9 shows where cell splitting has been performed. The graphic shows multiple smaller cells inside a normal cell. This might be overaccentuated, but it is primarily a means of showing how this can be accomplished.

Many users sign up for the service but rarely use it. Users might buy a cellular telephone and keep it for emergency purposes only. This is fine for the user because the costs are minimal, but for the supplier, it means having to prepare to serve more users than will actually use service. This causes concern for the carriers because their revenue is based on usage. Without usage, there will be little to no revenue. The capital costs of building the network are fixed and must be recuperated through usage. The carriers are trying to figure out how they can:

■ Accommodate more users

■ Encourage subscribers to use more price-sensitive services

■ Generate new revenue

We face a quandary in the industry:

■ The carriers need more users to generate higher revenues to pay off the investment.

■ The carriers must begin an evolution from analog to digital systems that will allow more efficient use of bandwidth (frequency spectrum).

Figure 32.9
Cell splitting is required when constant congestion occurs. This helps to use more frequencies in smaller areas.

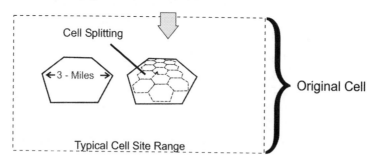

Average 3 miles; however, in major metropolitan areas, cell splitting will occur. Average can be a few blocks.

■ The need for security and protection against theft is putting pressure on carriers and users alike.

Analog systems do nothing for these three needs per se. Using either amplitude modulation or frequency modulation techniques, which were discussed in the data communications section of this book, to transmit voice on the radio signal uses all of the available bandwidth (or most of it). Distance, noise, and other interference demand power, amplification, and bandwidth to deliver a quality service. This means that the analog cellular carriers can support a single call today on a single frequency (specific frequency). As such, the limitations of the systems are in channel availability on a given set of technologies. If you consider the noisy nature of analog signals, you can also determine that quality is a matter of multiple factors, not the least of which will be congestion and atmospheric conditions, as covered in other sections of this book.

Other concerns with analog systems rise from both security and fraud issues. The systems using a single frequency are subject to monitoring, but this is no different from the situation with any other airborne technique. Furthermore, when the analog network can be penetrated fraudulently, we must be concerned. This is not strictly an analog systems problem.

As mentioned earlier in this chapter, the analog system was designed to provide the benefit of quick communications while on the road. Because this was considered a service that would meet the telephone needs of users on the go, the thoughts of heavy penetration were only minimally addressed. However, as the major metropolitan service areas (MSA) began expanding, the carriers realized that the analog systems were limited. With only a single user on a frequency, congestion in the MSA began to be a tremendous problem.

When cell splitting takes place, other issues become problematic for the carriers. The more cells in use because of splitting, the more critical the arrangement of frequencies available to use. Furthermore, as smaller cells are created through splitting, other issues (Table 32.1) become problematic.

This put an added financial burden on the carriers as they attempted to match need with return on investment. These problems have plagued the industry for quite some time now. As the industry begins to search for new solutions, other uses of cellular communications began to crop up, putting an added load on the service.

As mentioned, the cellular systems were designed for the road warrior driving around with a vehicular telephone. Initial installations were handled in cars and trucks. However, as the technology improved, trans-

TABLE 32.1

Some of the
Problems Carriers
Will Experience
with Cell Splitting

Issue	Problem/symptom
More equipment	Cost issues associated with buying the equipment. A cell site costs in the range of $650,000–800,000.
Real estate	Getting the space to mount added equipment plus local ordinances in the area. Consumer pressures leading to the "not-in-my-backyard" syndrome.
Power	Remote generation equipment required that is a cost and a security problem.
Logistical problems	Managing the new sites, new frequencies, and other ancillary services for a much larger equipment base.

portable telephones became nouveau for cellular users. Why should they be required to remain in the vehicle to use the service? Units could be mounted in the vehicle, and when the user got out, the set could be removed and used for several hours of "standby operation" and 1 hour of on-line operation. This opened a whole new avenue for service opportunities. The transportables were initially expensive, but prices have dropped to a far more reasonable rate (in the range of $300 to $600). Mobility became more flexible for the cellular user, who was no longer relegated to the vehicle. The transportable unit is shown in Fig. 32.10.

Portable handheld units began the next wave for cellular users. If a set could be removed from the vehicle, why then should it even be tied to the vehicle? If a set could be designed with a rechargeable battery pack, a briefcase model could be made available. Better yet, why not come up with a flip-type telephone or a shirt-pocket telephone? As prices were dropping for the vehicle telephone, so, too, were those for the handheld telephones. Recently, ads in the trade magazines and local newspapers

Figure 32.10
The transportable set took the set out of the vehicle. Up to several hours of standby operation are available, depending on battery and technology.

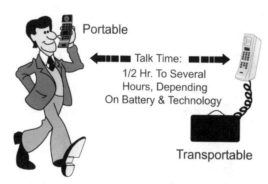

Portable

Talk Time: 1/2 Hr. To Several Hours, Depending On Battery & Technology

Transportable

around the United States have been offering the handheld devices for a mere $0 to $15 with a 1-year subscription to the carrier's network. The retail value of these telephones is as low as $360 to $1500. This has come a long way from the initial days of the cellular networks in 1984.

Features exploded also, with the introduction of:

- Voice messaging
- Redial
- Display
- Call forward
- Memory
- Hands-free operation

Subcompact or shirt-pocket flip telephones caught on. Users became enamored with the technology, and the price was almost too good to pass up. These latest and greatest models have between 0.3 and 0.6 watts of power output; they are totally portable and accessible. Professionals meeting with doctors and lawyers outside their normal office environment like the idea of using a cellular flip telephone. It gives them the ability to be reached at a moment's notice but frees them from the vehicle and respects their briefcase space, which is valuable. These telephones became the talk of many social events.

As mentioned earlier, these evolutions in the analog arena were all driving forces in the use and acceptance of cellular communications. However, as the congestion problem continued to go unsolved, the carriers entered a new era. The whole world has been evolving to digital transmission systems on the local and long-distance scene. This evolution to digital has many benefits for the standard wire-based carriers. These include:

- Higher usage
- Better quality
- Multiple techniques
- Digital speeds

Digital Transmission

The cellular carriers are looking to step into the digital world. This stems from the perspectives of compatibility and frequency utilization. If users can share a frequency or range of frequencies, then more users can be accommodated on less bandwidth. This has a definite financial impact on the carrier investment strategies. They could support more users on fewer frequencies and require less equipment.

Digital

Digital transmission will introduce better multiplexing schemes so that the carriers can get more users on an already strained radio frequency spectrum. Additional possibilities for enhancing security and reducing fraud can also be addressed with digital cellular. Again, this appears to be a win-win situation for the carriers. A summary of the key benefits for the cellular carriers' migration to digital communications is shown in Table 32.2.

Once the decision was made to consider digital transmission, the major problem was how to achieve the switch, what flavor to use, and how to transfer the existing customer base seamlessly to digital. The digital techniques available to the carriers are:

- Time-division multiple access (TDMA)
- Extended time-division multiple access (ETDMA)
- Code-division multiple access (CDMA)
- Narrowband advanced mobile phone service (N-AMPS)

Carriers and manufacturers are testing and evaluating each of these systems. There are several pilots and installations in progress. Many have been successful, while others have been extremely disappointing. This creates a split in the industry, depending on the technique chosen and the carrier support. We must wait and see what happens here. But users who have recently purchased portable telephones should also be kept in mind. When the network goes digital, those new telephones just purchased or subscribed to will have to be changed. Can you imagine the hundreds of thousands of users who are going to rebel against the change of equipment?

TABLE 32.2	**Key Benefits of Digital**
Summary of the Key Benefits for the Cellular Carriers' Migration to Digital Communications	Less costs overall. Initially conversion will be expensive, but longer range will be less.
	More users on the same or fewer frequencies and spectrum.
	Better security than offered by the analog transport systems.
	Less risks of theft and fraud.

Voice Technology and Applications

Because cellular was originally designed for voice communications, all of the development was in the ability to service voice transmissions and applications. Many of the technological aspects have already been addressed. However, the applications for voice are just as many. Some of these include the following:

- *Dial-up calls from and to the user on the road.*   This is what the cellular network was designed to do. The primary goal of cellular is facilitation of the voice dial-up call without wires.

- *Dispatch services for maintenance and repair crews.*   Many municipalities are now using cellular communications to notify the crews on the road. Very specific two-way communications can be handled. This certainly beats the disruptive two-way radio systems that were used in the past.

- *Trucker availability.*   Because truck drivers need to be reached anywhere and anytime, the use of a portable telephone keeps them in touch with the dispatchers. This saves valuable time and money for the trucker and the company, a savings that hopefully also gets passed along to the customer.

- *Safety and security (accident reporting and 911 services).*   The books are full of cases where lives have been saved and emergency response services have been dispatched more quickly thanks to the cellular telephone. The use of a telephone in a vehicle to call 911 (999 in some parts of the world) can get ambulances, fire trucks, and police agencies on the roll much quicker than ever before. No single benefit has proved the value of cellular more than this one.

- *Secondary services for the Coast Guard.*   Whereas two-way radio-based services were limited and subject to extensive monitoring, the Coast Guard now has a more reliable means of communication without having to go ashore and find a landline.

- *Voice service for police vehicles.*   Using cellular telephones, police agencies can eliminate the radio patch system that was predominant in day-to-day operation. Motor vehicle registrations, discussions with other officials, and private communications on closed channels all have their unique benefits.

- *Disaster recovery for office or manufacturing locations.*   Quick reestablishment of voice communications can be handled on a limited basis with cellular telephones. Call forwarding arrangements are easily made to a portable telephone after a building is destroyed. Many of the key players in an organization are highly mobile and therefore benefit from the ability to use the telephone on the run.

■ *Replacement for two-way radio applications (such as taxi, ambulance, and other on-demand services).*   Two-way radio is difficult to use for the novice. It is noisy, sporadic, and congested in most metropolitan areas around the country.

■ *Air phone services.*   As the nomadic user moves from the office to the vehicle to the airplane, being reachable is still the name of the game. Therefore, a special form of cellular communications will make the traveling executive or salesperson more available.

■ *Vehicle scheduling/diversion for dispatchers in a storm, major traffic jam, etc.* The vehicle can be notified to use alternate routes.

■ *Fire departments as a backup technology.*   At a fire scene, things can get quite hectic. Fire fighting personnel must always compete for resources and get second opinions from experts. Safety ties in closely here. The ability to reach a chemical expert regarding the mixing of water with certain materials to extinguish flames makes critical decisions more reliable, as the mixture of water with specific chemicals could produce an explosive reaction. This is not easily discussed on a radio, but it makes sense to use a portable telephone.

■ *Supplementing the wires for the nonwired user in an organization.*   Within the walls of an organization, catching anyone on the telephone these days is a haphazard proposition. Meetings, social gatherings, and group informal discussions all have a propensity to draw us out of our office areas and away from our PBX or key system. However, taking the cellular telephone with us means that there should be no missed calls in the future.

Facsimile Technology

It was only a matter of time before the users of cellular communications started looking for and finding new applications. A voice application is fine, but supplementing that capability with fax is a natural progression.

Facsimile transmission has been used in the trucking industry for some time. A bill of lading or customs form can be shipped right into the cab of the truck. The advantages of this technique are varied. Other applications, such as sales literature, quotations, medical records, and so on, could have strategic and savings benefits. An example of a fax transmission across the network to a trucker is shown in Fig. 32.11. This particular case involves the arrival of a trucker at a border crossing without all of the appropriate paperwork. Assuming that perishable goods are on the truck, the dispatcher needs to get the truck rolling as soon as possible. However, the government needs to see that the appropriate permits, fees, and inspections have been taken care of. Rather than delaying the movement of the vehicle, the trucker can immediately call the dispatcher and request the appropriate forms. When these forms are ready, the dispatcher can simply fax them into the cab of the truck. What used to be an overnight and very expensive proposition has been reduced to a few minutes of far less expensive time.

Cellular systems transmit information from sender to receiver. As a mobile unit is transmitting or receiving a fax, handoffs from cell to cell might take place. This handoff might cause a loss of information during the cycle. However, because pixels (dots) are being transmitted, the loss of information will be minimal.

As a facsimile is being scanned, a Huffman and Read (pronounced *reed*) code set is being used. Actually, the facsimile machine is scanning a page of information on a line-by-line basis. During handoff, a 100- to 200-ms delay could occur while the cellular user is handed from one cell to another. For data communications, this is a long time, but fax technology is far more forgiving. The facsimile machine waits for carrier detect, then continues to send. However, any data that was sent during the handoff could be lost. Because the line-by-line scan and transmit is used, all that will be lost is a line of dots. This shouldn't impact the information very much, yielding a high-quality facsimile transmission. This is shown in Fig. 32.12, where a line of dots, or one scanned line, has been lost. This is not a major problem because, although the line is lost, the rest of the document is still legible. So, using a fax modem allows for a wait time of up to 400 ms—a data communications modem will drop

Figure 32.11
The post-fax scenario allows information to be shipped directly into the vehicle.

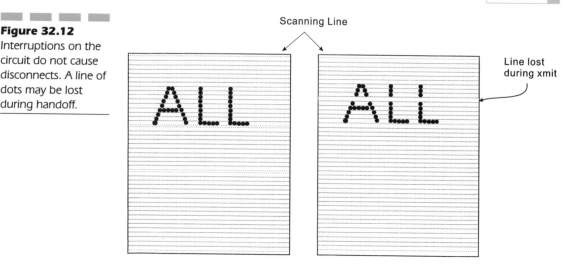

Figure 32.12
Interruptions on the circuit do not cause disconnects. A line of dots may be lost during handoff.

the line if the carrier is lost for greater than 80 ms. A handoff can take from 100 to 200 ms in an average cell. The information should, therefore, be usable and understandable. Further, as the fax modems communicate to each other, they have the ability to monitor line quality and fall back to slower speeds, if necessary, to get the transmission through.

Data Transmission

Many users say that data transmission does not work on cellular communications. Although this was limited in the early stages, things have gotten much better. The first problems occurring were with handoff. As a vehicle using data services rolled from cell to cell, the handoff would disrupt the call for 100 to 200 ms. This was just enough to disrupt the carrier detect (CD) cycle, so the modem assumed that one of the callers had hung up. Thus the modem would hang up, too! Obviously, this problem could be overcome by having data transmission take place only while the vehicle is stationary. However, that becomes a limitation of communications. As Fig. 32.13 represents, when data is disrupted, the modem will fall off and disconnect. This graphic is designed to show the early problems with the wireless transmission services.

Newer data modems for cellular communication use a technique similar to the fax modem. The modem will delay for 400 ms before hanging up. Because the handoff only takes 100 to 200 ms, the modem has time to

Figure 32.13
Early implementations
of wireless data
caused disconnects
during cell-to-cell
handoff.

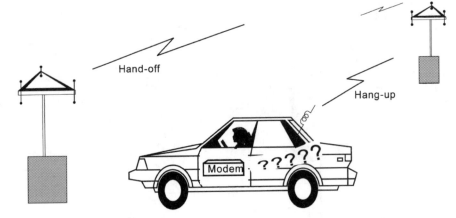

spare without losing the call. Some data might be affected, but error detection, ACK/NAK, and CRCs all will come into play in detecting and correcting the errors that occur on the cellular network. See Fig. 32.14 for a representation of the data correction problem. Conventional data was always regarded with suspicion on cellular airwaves. Because the modems had to hang onto the line for 400 ms while waiting for the carrier to return, everyone thought that data would only be effective at 2400 bps. Motorola has introduced a 9600-bit-per-second cellular modem, and another company has a 19,200-bit-per-second cellular modem. As these techniques become more common, we can expect to see more users clamoring for the capabilities. Unfortunately, across a data link via cellular (analog) technology, the bandwidth constraints will limit this service to casual, bursty data needs. If large amounts of data are to be transmitted, the costs and the available frequency spectrum might get in the way.

Figure 32.14
Improvements in the
data communications
prevent the
disconnects as
handoff occurs.

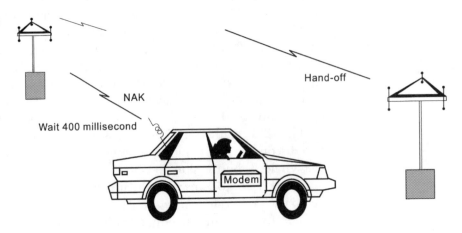

Digital Cellular Evolution

As the spectrum for radio cellular users becomes more congested, two primary approaches are in the offing:

- Time-division multiple access (TDMA)
- Code-division multiple access (CDMA)

However, standards bodies and manufacturers are all coming up with variations on the use of the frequency spectrum. TDMA had been discussed and was well settled on. Nevertheless, manufacturers have been developing an extended TDMA (ETDMA) that will bring more use to the system. TDMA will derive a 3-fold increase in spectrum use, whereas ETDMA will produce 10- to 15-fold increases. The concept is a digital speech interpolation (DSI) technique that uses the quiet times in normal speech, thereby assigning more conversations to fewer channels, gaining up to 15 times more capacity than an analog channel. Each of the moves to digital requires newer equipment, which means more capital investments for the cellular carriers.

TDMA uses a time-division multiplexing scheme where time slices are allocated to multiple conversations. Although TDMA usually deals with an analog-to-digital conversion using a typical pulse code modulation technique, it performs differently in a radio transmission. PCM is translated into a Quadrature phase-shift keying technique, thereby producing a four-phased shift, doubling the data rate for data transmission (actually driving a voice call at 8 Kbps).

Spread spectrum, which was first used back in the 1920s, has evolved from its initial military security applications. Spread spectrum uses a technique of organizing the radio frequency energy over a range of frequencies instead of modulating the frequency. The system uses frequency hopping (similar to ETDMA) with time-division multiplexing. One minute the transmitter is operating on one frequency, the next on another, according to some code. The receiver is synchronized to switch frequencies in the same pattern. This is effective in preventing detection (interception) and jamming. Thus, additional security is derived. These techniques should produce 10 to 20 times increased capacity over existing analog systems.

Another possible solution for bandwidth or user expansion is narrow-band advanced mobile service (N-AMPS), a technique developed by Motorola. N-AMPS is a channel splitting technique where channel bandwidths are allocated in chunks of 10 kHz instead of the traditional 30 kHz. However, industry experts look at this technique as being more complex and requiring stringent filtering, a problem not associated with the TDMA, ETDMA, and CDMA solutions.

Personal Communications Services

Why require a person to be tied to a pair of wires or to a fixed location with an associated number when the provider can offer a tie to the individual through an intelligent network that is capable of finding a called party no matter where he or she is? This concept will appear on the mass market at a time when the population is far more mobile than it is now. The use of cordless telephone technology has already made its niche in the industry, particularly in the home, where teens use a cordless telephone as the norm. Security or convenience is not a concern. Service is an issue because the set must be kept directionalized with the base station. The limitation of distance to a 300-foot radius around the base station works fine in the residence. However, this does not work as well in the business environment because of all the concrete and steel around the set.

Therefore, it is only natural that the next logical extension of this service will be to provide unrestricted service and access regardless of where the recipient of the call might be at any time. Consider the ability to use cordless telephony anywhere in the world and the reality of personal communications becomes less myth, more fact. Throughout the evolution of wireless communications, users have always wanted the ability to be reached at any time. Unfortunately, the use of earlier systems was always limited to very specific regions, such as a major metropolitan centers and the surrounding areas (typically, a 40- to 50-mile radius from the user's home base). This distance limitation led to many complex machinations in the use of paging and radio handheld systems. The industry finally responded with a more robust application that used satellite and radio repeaters. Now an individual can move out of the home area and be reached virtually anywhere around the country.

Pockets of dead space always exist, but these are limited. Even when two-way improved mobile telephone service (IMTS) was deployed, several limitations were inherent. Limited frequencies, long waiting lists, and expensive service all kept the use of this technology to a minimum. Cellular technology improved the situation, but the cost of the airtime was considered very high. At a time when this cost should have been dropping, carriers and providers of cellular service were raising their rates to compensate for losses because of the high churn (turnover) rates of customers and because of a new phenomenon: theft and fraud. The legitimate user was penalized for the losses being sustained by the providers.

Now the industry wants to deliver wireless communications in the form of telephony, data, paging, and E-mail applications to the mass market. To do this, a newer concept is needed that will allow for rapid deployment of equipment and more diverse coverage of all areas around the country, then the world. This will make major investments far less lucrative, so the proposed carriers and providers of personal communications services are looking for ways to introduce the technology to the masses with limited investments and large-scale frequency reuse. The resultant discussions being held by these carriers and providers focus on the purchase and installation of microcells and picocells (Fig. 32.15). Primarily, the newer carriers are looking at providing much smaller systems and equipment to reuse the frequency spectrum increasingly. The obvious goal is to produce the equivalent of a wireless dial tone service to the masses (both businesses and residences) for a reasonable, affordable price. When the technology and service are first introduced, the industry pricing might be as high as or higher than the monthly costs for cellular communications and wired dial tone from the local telephone company. However, as the proliferation of the service takes place, the service costs will begin to fall dramatically. One can now expect to see this form of service renting for $0 to $10 per month and $0.10 per minute of usage. It will take some time to arrive at this point, but it has already begun to happen. PCS is still beginning to roll out, and will take time to provide the necessary coverage. Spotty services and coverage still prevail during the deployment of these services, causing some dissatisfaction by users

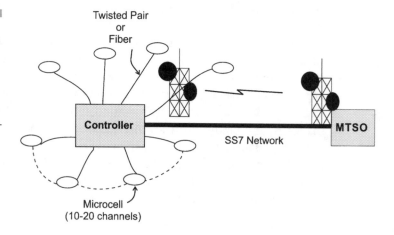

Figure 32.15

A single controller will be able to manage up to 96 microcells, with each cell handling 20–25 calls.

who sign up for the service early in its implementation stages. This, too, will pass as the providers gain footholds in their market penetration and service areas.

This equates to a catch-22 situation. The carriers and providers need users to sign up and use the service so that they can deliver more benefits and coverage. Yet, many new users will be frustrated as the deployment takes place and will return to wired dial tone because this is an older, proven technology.

Technology

The technology decisions in this arena are equally perplexing. Where the use of typical frequency division services (an analog transmission system) in normal multiuser wireless communications is proven, it will not allow for the frequency reuse, as described. Furthermore, any attempt to use a narrower band for transmission would only provide a limitation to the services anticipated for use. Consequently the industry has been wrestling with the question of what technology to deploy for the future. Again, the costs will be astronomical when a nationwide or worldwide network service is installed. Therefore time-division multiple access (TDMA), enhanced time-division multiple access (E-TDMA), and code-division multiple access (CDMA) are all being considered for the various future offerings. Each of the vendors or purveyors of service could theoretically take a different approach. These various technologies are summarized in Table 32.3. In this table, the various options are shown with their respective advances in frequency reuse increases, the comparison of digital versus analog capabilities, the services that might well be provided, and the potential number of suppliers that will exist.

One further added comparison includes the use of narrowband advanced mobile telephone service (N-AMPS), which is being considered and deployed on a trial basis by some of the cellular carriers. This adds a degree of complexity, because the variations will not be fully compatible. Therefore, a change in the transmitted signal or the format of the information will be required if an internetworking application is used.

Cost penalties will exist with the various degrees of technology that are selected. The most expensive solution will be the use of CDMA, a technology that is still new and that will require the total change of the equipment and techniques used.

TABLE 32.3	Technology capability	TDMA	E-TDMA	CDMA	N-AMPS
Comparison of the Technologies Being Considered for PCS/PCN	Frequency reuse gains	3×	10–15×	20×	3–5×
	Digital	Yes	Yes	Yes	No
	Services to be offered	Voice Data Paging	Voice Data Paging E-mail	Voice Data Paging E-mail Radio determination	Voice Data Paging
	Costs	Low–Medium	High	High	Low
	Quality	Fair	Good–Excellent	Good–Excellent	Good

The Concept of PCS/PCN

Visualize how this system might work! The local wired telephone services from the local exchange carriers will still exist. Therefore the use of the wireless local loop in the dial tone arena will still use a twisted pair of wires, as has always been the case. However, the personal communicator will now act as a cordless telephone in the business or residence.

As shown in Fig. 32.16, a cordless telephone arrangement will be set up with a smart terminal. This smart box will allow the user to place a personal communicator in the charging unit at the location. While in the charging unit, the set will act as any cordless telephone. A call coming into the telephone will travel across the local telephone company wires and ring into the network-attached box. When the user picks up the telephone and answers the call, the set will now be cordless. In Fig. 32.17, the same scenario will be used, but the most significant difference will be the lack of a pair of telephone wires into the business or residence. In this scenario, the set is 100 percent wireless—with a smart interface in the location, a call coming into the business or residence will come in across the airwaves rather than over the wires. The primary difference here is the absence of a telephone wire.

Taking this one step further, when the owner of the telephone set chooses to leave the primary location, the set will be taken from the charging unit and possibly clipped onto a belt loop or placed into a

Figure 32.16
The personal communicator placed into a charger or smart box acts like a standard cordless phone when used with telco wires.

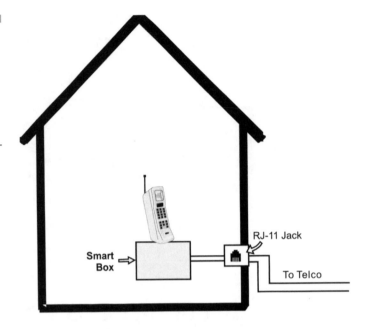

pocket (purse, jacket, or other). Wrist-mounted telephone sets might also appear. AT&T Wireless has a wrist mount that is spring-loaded. When the user opens the set, the hand is cupped around the ear to act as the receiver. Actually, the spring-loaded receiver slips into the palm of the hand and allows for a simulation of the normal telephone set. The transmitter is

Figure 32.17
In the wireless arena, the communicator is connected to the telco via microcell technology.

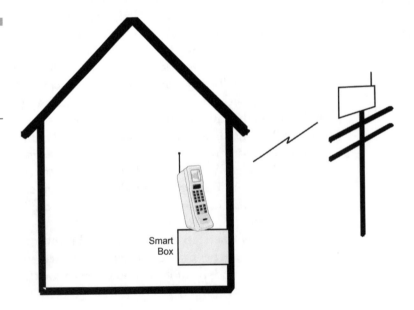

located on the wristband. This is just one variation of the telephone set that might appear in the personal communications world.

A conceptual rendering of this is shown in Fig. 32.18. As the user leaves the building and steps out into the open space of the neighborhood, the set will immediately begin communicating with a local cell. This local cell might be provided by the local telephone company or other supplier. However, the cell will be a microcell, possibly mounted atop a light stanchion or telephone pole. In Fig. 32.19, the cells are located on top of a telephone pole line. These cells will be wireless to and from the personal communicator set, but the individual cells will be hard-wired along the pole line via fiber, coax cable, or twisted pairs of wire. These wired facilities will be carried back to the telephone company central office. In this particular area, the telephone companies have an edge over their competitors because they already have the poles installed and own the "rights-of-way" to these poles.

Continuing with the scenario, the user's device will continue to communicate with the local microcell as the user proceeds down a path or walkway. As the user gets close to the boundary of the cell, a handoff will be required from one cell to another. The microcells will only cover a limited distance to begin with, possibly in the range of 80 to 200 feet. This implies that handoffs will happen on a far more frequent basis. As the

Figure 32.18
The spring-loaded wrist phone places the receiver in the cupped palm to accommodate full handset operation. The microphone stays at the wrist.

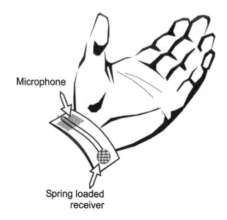

Microphone

Spring loaded receiver

Figure 32.19
Once the set is
removed from the
smart box and the
user leaves the locale,
an instant
communication is
established via the
microcell to alert the
network that the user
is mobile.

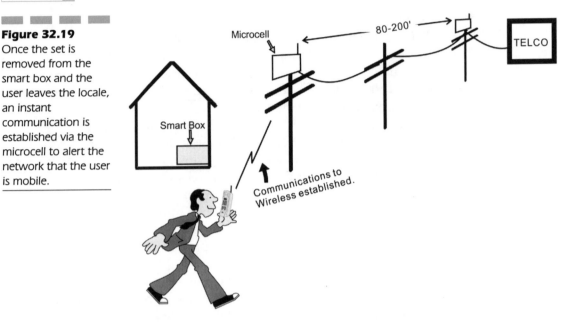

user continues down the path, the handoffs continue every 80 to 200 feet along the way. As the user enters an area where there are no poles, the microcells are now mounted as telepoints along the way. A telepoint is a wireless access mounted atop or on the side of a building. All that is necessary is a close proximity to the telepoint or microcell. In Fig. 32.20 the telepoint is mounted on the side of a building, and in Fig. 32.21 the microcells are mounted on top of the buildings, each cell serving a radius of 200 feet. Each of these cells will be controlled by a central cell controller, with a cell serving 20 to 25 simultaneous calls and a controller managing between 32 and 64 cells. At some future point in time, these controllers will likely manage up to 96 cells, each controlling 20 to 25 simultaneous calls. This amounts to a single control device managing between 1900 and 2400 concurrent calls. Imagine the efficiency of this concept!

Why Personal Communications?

In the original concept of business communications as well as its connection to personal communications, the use of cordless telephones emerged. Serious efforts have always been made to eliminate the telephone tag problem within the business community. Instead of requiring users to leave

Figure 32.20
When no pole lines are available, telepoints can be mounted on the side of buildings.

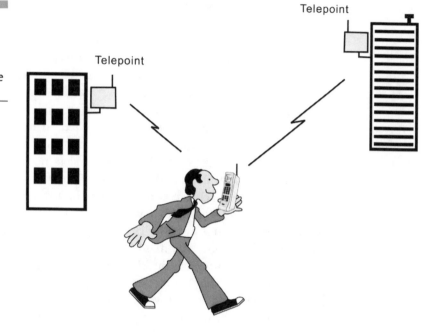

Figure 32.21
Microcells can also be mounted on the tops of buildings, with each cell serving a radius of 200 feet.

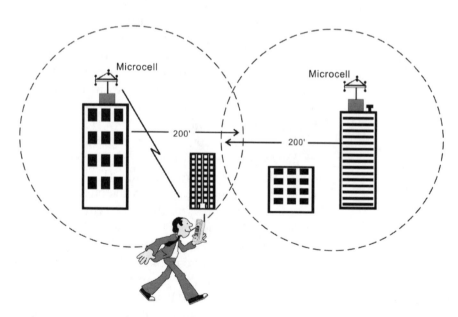

messages and wait for a return call, the industry has been attempting to provide the connection directly. However, whether in an office environment or with a mobile user, the limitations of the telephony world still prevailed. Too often, the user was relegated to a pair of physical wires. To overcome some of these problems, cordless telephones were introduced in the business and the residence to allow for added flexibility. Unfortunately, the same problems of distance limitations reigned. With a wireless arrangement, the amount of wiring needed in an office environment decreased, while the use of paging systems (meaning loudspeaker paging within an office) and personal pagers (wireless notification services on a radio-based system) increased. Businesses felt that added flexibility and mobility could be achieved. Yet, human nature crept into this picture: individuals could plead that they were out of range and did not receive a page or that the batteries to the portable pager or telephone set were dead.

Hence, the problem was magnified. The frustrations of management, customers, and clients were building up. The real intent of a call is to connect two people who want to talk to each other, not to result in a page or voice mail message. Using a voice mail arrangement only allowed us to leave a message, not to actually speak to the person desired. Nothing has really been gained through this technology. Furthermore, many organizations have learned that, by introducing voice mail as a tool to assist in the passing of information, the technology quickly erodes to the point where it is used solely as an excuse not to answer the telephone (at least not in real time). An individual selectively decides whether to return a call. Pagers with alphanumeric displays allow callers to leave more information while trying to reach a party, but again, this only lends itself to the same kinds of abuses.

Clearly, the attempts to put two people together are only creating logical steps toward the ultimate goal. Therefore the next logical step is to replace the pager and voice mail with a device used for real-time connectivity. The use of the cordless telephone, as mentioned, was the first attempt at accomplishing this replacement. Despite what many feel, the cordless telephone was very successful, although limited. In particular, the cordless telephone was extremely successful in the residential and small business market. Overcoming industry concerns of interference, lack of security (or eavesdropping), and lack of total privacy was the unabated desire to be untethered. Retail organizations, distribution and warehousing facilities, and small branch offices found the use of a cordless telephone an improvement over the hard-wired environment. This movement occurred around the world, but legal issues of eavesdropping, smuggling, and interference had to be dealt with. The evolution from single-line cordless telephones that operate in the shared, unlicensed frequency ranges led to the need for a licensed (or controlled device) technology that would eliminate the fear of eavesdropping and interference. Hence the evolution to the personal communications arena.

Personal communicators have undergone experiments under license agreements from the regulatory bodies around the world. The results have been so mixed and the offerings so varied that there had to be another round of discussion about the packaging and delivery of the services. Personal communications is the attempt to rectify the ills of all the past experiments and trials to deliver the call to the originally intended recipient.

Clearly, a portable communicator will not solve all these ills. *Personal* means that the individual, not a company or department, will be reachable. Yet the decision still remains with the individual as to whether to accept a call, forward it to another location, allow it to route to an answering machine or voice mail system, or just ignore it. The fact that *communications* implies that the recipient can accept or reject the call is only another part of an equation that has to account for a much broader range of services and a much wider area of coverage. The device should allow for connection anywhere in a building, city, state, or country. Furthermore, the services should include other connections, such as paging, E-mail, data transmissions, facsimile traffic, and other types of communications that would normally be provided in the wired office environment. Thus the personal communications concept must include any form of telecommunications.

The Evolution of Personal Communications

Since its inception, the wireless personal communications concept that started early in the 1980s has mystified even the industry experts. What services could be provided at a reasonable cost, be ubiquitous enough to serve the masses, and still produce a profit? Further, although voice telephony is the primary goal of providing wireless services, newer applications are constantly surfacing. New systems should include a provision for the following capabilities, based on user-selectable options:

- Voice dial-up services

- Data communications at various speeds, but minimally at 9.6 Kbps

- Paging and alphanumeric messaging on a pager

- E-mail access and file transfer capabilities

- Radio determination and vehicle location services

- Personal digital assistant services for calendaring, mail, contact management, and other service options

- Facsimile at Group III speeds or higher

- Future access to higher bandwidth applications such as multimedia

- Message center operations to leave or retrieve messages

- Telemetry services for process control and alerting

- Dial-up video in a slow-scan mode (7 to 10 frames per second)

Comparing PCS to Cellular Networks

As this chapter continues, you might begin to ask, "What are the differences between PCS/PCN networks and the already existing cellular networks discussed?" There probably aren't going to be a lot of technological differences between the two networks, but the functionality and the population served will likely be the differentiating factors. PCS has been described (hopefully) as the replacement for the local loop or the hard-wired service.

Understand that this section is not predicting that all local wired facilities will disappear or be abandoned. Nothing like that will likely take place in our lifetimes. However, as a secondary means of providing the single-number telephone concept with the ability to be totally mobile yet still reachable, PCS will fit the niche. However, on a higher quality, 64-Kbps standard premium sustained service for the fast-moving, vehicular-based user, cellular technology will offer the appropriate connectivity. Therefore, we must take a position here that the two technologies and services should be viewed as complementary rather than competitive. The cellular network will continue to support the office-in-the-car concept for voice, data, video, fax, and image services. The costs will continue to be higher for this service because it will be more robust and dynamic.

For the casual user or the mobile person who needs to be near the residence or the office, PCS will suffice in most cases. Therefore, two separate target markets will still be evident. This is not to say that the providers of the service will necessarily be different. The cellular provider of today might well be the cellular and PCS provider of the future. At an inter-

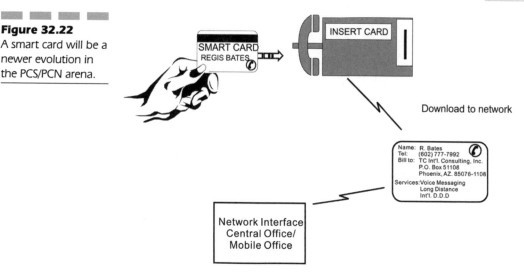

Figure 32.22
A smart card will be a newer evolution in the PCS/PCN arena.

national level, this has always been true, as the local Post Telephone and Telegraph (PTT) organizations were the only authorized providers. However, on a global scale, the shift to competition in this local loop bypass is already under way. What was once the sacrosanct monopoly of the government around the world is now breaking down to a competitive service offered by many. At this point, the market will likely be two major cellular players and up to five PCS suppliers. However, as time goes by, newer services are offered, and newer technologies used, the number of players in this market will be wide open.

Consider that, in October 1993, the FCC announced its intentions to provide PCS services through a series of auctioned licenses to approximately 2500 licensees. There are likely to be 11 national PCS providers followed by 49 regional ones. The rest will be more localized in a specific metropolitan or rural area. The market penetration now will be wide open. Furthermore, the smart telephone, smart card telephone (Fig. 32.22), or wrist telephone shown earlier will emerge as the wave of the future to allow total connectivity anywhere and anytime. This is the vision of how the world will work in 1998 and beyond. To set this in motion, several experimental steps were conducted in 1995.

33

Radio Systems

Chapter Thirty-Three

Radio-Based Systems

Satellite and microwave communications are radio-based systems that have been around for a number of years. At one point, the bulk of the long-distance services in the network were served primarily via microwave, and the international long-distance services via satellite. A good number of these systems are still in effect, serving long-distance networks, rural areas, and international long-distance connectivity. Many organizations have built their own private networks by launching their own capacity into a satellite orbit, buying into their own rights-of-way for microwave towers and radios across the country, or just renting the capacity from other suppliers. This is by far not a dead service or technology.

Much has changed with the introduction of fiber optics. Yet, both of these techniques still have a use for the high- and low-bandwidth services for:

- Voice
- High-speed data
- Low-speed data
- Video
- Facsimile
- LAN-WAN connectivity

Satellite

The satellite method uses a microwave transmission system, with radio connectivity being the primary means of broadcast. This technique has long been considered useful in very long-distance communications. Satellite transmissions were initially begun in the early 1960s, but several enhancements have been made over the years.

Typically, an earth station (uplink and downlink) is used to broadcast information between receivers. Current technology uses a geosynchronous orbiting satellite. *Geosynchronous orbit* means that a satellite is launched into an orbit above the equator at 22,300 miles. This orbit distance means that the satellite is orbiting the earth as fast as the earth is rotating. Therefore, it appears to earth stations that the satellite is stationary, thus making communications more reliable and predictable—and earth stations less expensive, because they can use fixed antennas.

This communications technique uses a frequency (uplink) broadcast up to the satellite, where a transponder receives the signal, orchestrates a conversion to a different frequency (downlink), and transmits back to the earth. As the communication is transmitted from the satellite toward the earth, a 17° beam is used. This produces a pattern of reception known as a *footprint*. The footprint is shown in Fig. 33.1, with the entire United States covered by a single satellite. There are obviously many more than just three satellites in orbit around the earth, but only three are needed to provide global coverage. The satellites used for commercial applications are in a geosynchronous orbit.

Frequencies

Currently used satellite frequencies are categorized in bands. These bands are divided into the RF spectrum according to capacity and reception variation. The frequency bands are divided into two separate capacities:

Figure 33.1
The satellite footprint covers ⅓ of the earth.

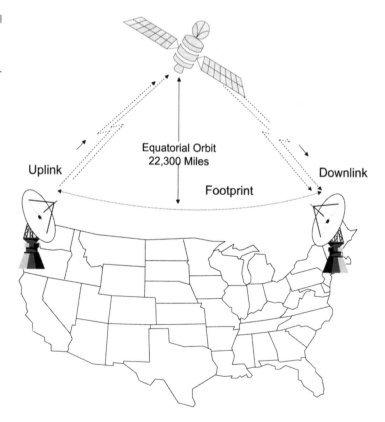

Equatorial Orbit
22,300 Miles

Uplink

Footprint

Downlink

uplinks and downlinks. The service provides for full duplex operation, so a pair of frequencies is used. The uplink is used for transmission, and the downlink is used for receiving (Table 33.1).

Satellites are good for broadcast communications—one transmission source to many receiver locations. This method is *distance insensitive,* because once the signal travels up (22,300 miles) and back down (22,300 miles), the position of the receiver doesn't matter on a landline basis. We do not rent or lease the channel based on distance but on the capacity of the channel. If two sites are using a satellite to communicate, the price is the same, whether they are 10 miles or 3000 miles apart. The round trip from earth to the "bird" and back is essentially the same distance.

Advantages of Satellites

Obviously, there are some distinct advantages to the use of satellite, or no one would want the capacity. These advantages include:

- *Distance insensitivity.* As stated, the distance between the two end points is not a consideration in the pricing.

- *Single-hop transmission.* With just a single shot up and back from the satellite, most communications coverage should be easily addressed. There is only one repeater function taking place, which helps to eliminate some of the errors in data transmission whenever a signal must be repeated.

- *Solutions for remote areas and maritime applications.* Ships at sea have a problem in their communications needs. They cannot see a microwave dish because they are too far from land; thus no line of sight is available. Furthermore, it is not practical to use any landline systems, because there are no wires available to run from the ship back to the land-based telephone networks. For remote areas where it is impractical to get wiring because no major development has occurred, a satellite link might be the only true solution.

TABLE 33.1

Summary of the Frequency Bands Used for Satellite Transmission

Band	Uplink Frequency	Downlink Frequency
C	5.925–6.425 GHz	3.700–4.200 GHz
Ku	14.0–14.5 GHz	11.7–12.2 GHz
Ka	27.5–31.0 GHz	17.7–21.2 GHz

■ *Good error performance for data.* Once a signal has penetrated the earth's atmosphere and gotten into space, little can corrupt a data stream. Therefore, after traveling approximately 5 to 20 miles in the atmosphere, the signal is relatively immune to any other problems. The bit error rate of a satellite transmission is approximately 10^{-11}, which is very good compared to those of other transmission systems.

■ *Broadcast technology.* Because the footprint is so large, the transmission can be sent to many locations simultaneously. There is no need to mesh a private line network or transmit serially and sequentially when using satellites. The signal broadcast, once up to the satellite and once back down, can be sent to as many locations as are tuned to the specific frequencies. This can be a time savings factor and a cost savings opportunity.

■ *Large amounts of bandwidth.* An average satellite has 24 transponders. A transponder is a transceiver (transmitter + receiver). Each of these transponders can handle approximately 36 MHz of bandwidth, which, as you will recall from earlier chapters, is significant. Using some various modulation techniques, such as QAM from Chap. 13, we could achieve approximately 3.4 Gbps of overall data throughput. Of course, this varies depending on the modulation technique used and the channel spacing (separation). The idea of massive amounts of bandwidth is what counts, however. Many of the TV broadcasters use satellite communications to serve their distribution of broadcast-quality video communications at 45 or 90 Mbps.

Disadvantages

No system offers only great advantages, so the opposite side of the transmission system must be viewed for fairness' sake. Satellites have some distinct and peculiar disadvantages. Many can be overcome, but they do surface whenever the technology is discussed.

■ *One-way propagation delay.* This delay is about $\frac{1}{4}$ to $\frac{1}{2}$ second and can be disruptive to voice and asynchronous data protocols or any protocol that requires an ACK or a NAK before the next block of data can be transmitted. The turnaround on a data stream would be double the $\frac{1}{4}$ second, or an average turnaround of $\frac{1}{2}$ second. In using a bisynchronous protocol, a 4800-Bps transmission speed could be reduced to 400 Bps because of the transit delays.

■ *Multihops increase delay, detrimentally impacting voice.* If one cannot transmit on a single satellite link (for example, Boston to the Middle Eastern countries), then a double hop is required. This means that the signal will be transmitted up to one satellite and back down to the earth station, then retransmitted up to a second satellite, then finally back down to the earth station around the horizon. The delay will be at least one second from the time a message is originally sent until it is finally received. The disruptive nature of this for voice makes it virtually unusable. For data, however, protocols can be "spoofed" or faked into thinking that the data got there sooner, and the problem can be overcome.

■ *High path loss in transmission to satellite.* The transmission path will introduce more loss to the signal while it is making its way to the satellite. Therefore more power is required, and signal-to-noise ratios must be carefully administered, or else the information could become unusable.

■ *Rain absorption affects path loss.* Any radio-based medium (especially in the microwave frequencies) is subject to a certain amount of absorption from water. Rain in a channel path can absorb much of the signal strength, leaving a far weaker signal to find its way to the receiver. Once received, the signal (now weaker and noisier) will have to be amplified. Remember the analog discussion in Chap. 6 about the impact of analog amplification.

■ *Congestion buildup.* The satellites were originally designed to be spaced 4° apart, based on a 360° circular orbit. However, as more countries in emerging worlds have sought to use radio rather than wire-based network access, the satellites have been placed in closer orbital slots. They are now being positioned at 2° increments. This can lead to congestion on the transmission paths, cause occasional interference, and (of course) limit future parking places.

Very Small Aperture Terminals (VSATs)

Very small aperture terminals came into being in the mid-1980s, when the size of the satellite terminal shrank. Since then, hundreds of networks have cropped up. The pricing schemes have not yet made even greater numbers totally feasible, but as technology continues to shrink, the use might expand. Figure 33.2 is a graphic representation of the VSATs located around the country.

Figure 33.2
V-SATs are located
around the country
to provide coverage.

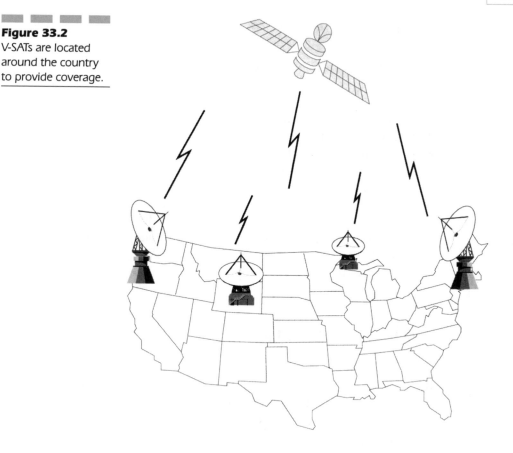

Costs are continually falling, and benefits derived from VSAT networks are rising, breathing new life into this technology. No longer do users have to own and operate their own hubs now that shared hubs are more realistic. Prices from suppliers have dropped; VSAT units can cost as little as $250 per month in quantities of 1000 units. Many specific data applications are now being switched over to satellite via VSAT. In 1991, more than 35,000 VSATs were on record. Newer regional operators are now approaching smaller network users (10–65 nodes). The dishes (VSATs) can be as small as 1.0 to 2 meters, thus making it more realistic to place units in remote sites.

However, these smaller terminals are limited in their applications; voice and video (motion) applications aren't generally feasible. VSATs do, however, provide sufficient bandwidth capacities for data transmission (host to host, file transfer LAN to LAN, and some receive-only broadcast video).

Microwave

The local telephone companies, long-distance carriers, and users have all been successful in the use of microwave radio systems. Microwave was the predominant means of handling long-haul transmission here in the United States. The advantage is that rights-of-way and natural barriers are more easily overcome. A single radio channel can carry as many as 6000 voice channels in 30 MHz of bandwidth.

Microwave, like any radio technology, is designed around space and the frequency spectrum, which are finite resources. Thus, coordination between paths is extremely important to prevent interference. Microwave radio can use either an analog or a digital modulation method. Each operates differently. Analog radio uses either amplitude or frequency modulation, usually frequency modulation. In 30 MHz of bandwidth, 2400 voice channels can be carried, equating to a voice channel of 12.5 kHz.

Digital Microwave

Digital microwave has been available since the mid-1970s, with the direct modulation of an RF carrier. Because a modulation technique uses 1 bit per hertz, using a 64-Kbits/s channel would consume far too much of the radio frequency spectrum. Therefore a better modulation technique is needed. A microwave installation is shown in Fig. 33.3. Phase shift keying or Quadrature amplitude modulation techniques are more common. Using 16 QAM, a total of 1344 voice channels can be carried on 30 MHz of radio, yielding 90 Mbps (3 bits/Hz). Newer 64 QAM supports 2014 channels at 135 Mbps (4.5 bits/Hz). Digital microwave allows for the direct interface of T1/T3 carrier circuits.

Service in the 18- to 23-Hz radio spectrum is called *short-haul microwave*. Typically, this is a point-to-point service operating between 2 and 7 miles and supporting up to four T1s. Newer systems support up to eight T1s.

Digital termination services that operate in the 10-GHz range began in the early 1980s. This short-haul service is designed for metropolitan area digital services, similar in concept to cellular radio. This TDMA technology has access to the system for 24 users at up to T1 speeds. As when dealing with any radio frequency spectrum, the need for licensing exists. Satellite services are usually coordinated by the network suppliers. Microwave can be coordinated by either the network suppliers or user. You will need:

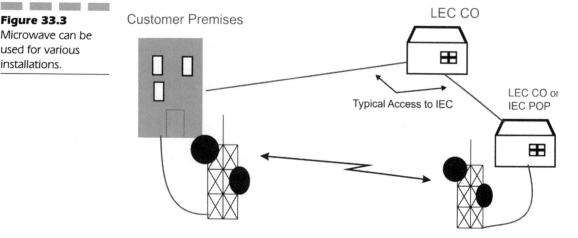

Figure 33.3
Microwave can be
used for various
installations.

- Line of sight
- Path coordination
- Frequency/licensing
- FCC-trained personnel
- Tower/antenna location
- Power/backup power
- Time

Future Use of Microwave and Satellite Systems

Because the radio systems can carry such broadband capacities over radio frequency, we will see several opportunities to use these technologies.

- *CATV.* Coaxial broadband cable delivers compressed video to users. These systems use digital modulation at 90 Mbps. However, as high-definition TV becomes less expensive and more widespread, 135 Mbps of bandwidth throughput can be achieved on microwave/satellite systems to a coax or fiber interface.

- *DBS.* Direct broadcast satellite systems send a receive-only (one-way) transmission for individual reception. Using a very small aperture antenna (1 meter or less), a viewer will receive TV signals through a personal earth station (or some small component). DBS will be available to CATV, MATV, and broadcast users alike. This is perceived as a supplementary service, not a replacement. New uses of DBS include 400-Kbps reception of data from an ISP.

■ *Wireless local loop (WLL)* technology using multipoint multichannel distribution systems (MMDS) and local multipoint distribution system (LMDS) capabilities.

■ *Interconnection already exists between cellular systems and microwave.* However, as iridium comes about in the 1998, satellites in low polar orbit will be used for worldwide cellular connections. This will go further into the personal communications networks (PCNs), using a personal communications server (PCS) system to deliver calls to and from the cellular arena.

■ *VSATs will continue to get smaller, and newer modulation/time-sharing techniques will provide greater access to bandwidth.* Prices will continue to drop as these services continue to compete with fiber and microwave alternatives.

CHAPTER 34

Light Systems

Another form of wireless communications is provided by transmitting infrared light through the air. This can be an exciting opportunity to provide basic connectivity in a localized area at a very reasonable cost. Infrared light systems have been around for three decades but have never really caught on until recently. Clearly, the need to provide connectivity is the issue in any telecommunications environment. Infrared or laser beams can be considered in the overall corporate strategy if the bandwidth needs are limited to a few channels or if the communications system is being complemented or supplemented with this technique. Operating high in the frequency spectrum of laser beams in the tera-hertz (trillions of hertz) range, an invisible light beam is focused from a transmitter to a receiver over a very short distance, usually under 1.5 miles. Similar to radio waves, infrared light is subject to specific curtailments and impairments in transmission paths, such as:

- Very limited distances, typically under 1.5 miles.
- The necessity of a clear line of sight in order to communicate with the other end.
- Limited bandwidth of approximately 155 Mbps. Unguided light systems were introduced in 1999 by Lucent Technologies that will operate at 2.488 Gbps, and a promise that we will see 10 Gbps in the near future.
- Susceptibility to environmental disturbances and limitations, such as movement and electrical disruptions.
- Susceptibility to disruption from fog, dust, heavy rain, and other objects in the path, because the fine beam of light might not be able to penetrate these impairments.

With this in mind, the use of laser or infrared light still has its benefits. One would certainly have an opportunity to gain benefit under the right conditions. No single technique can be 100 percent effective; each has its own limitations and advantages. This is why so many options are available. Organizations can choose the transport systems that best meet their needs.

Despite the current technical bandwidth limitations, there are certain cost advantages and ease-of-use considerations. These factors keep this technology high on the acceptability scale of many users for many applications. The availability of low-powered solid-state laser diodes has opened up many possibilities. Producing cost-effective through-the-air transmission capabilities for voice, data, video, and now LAN traffic at native speeds will enhance a network immensely. Transmission through the air without a cable system infrastructure reduces the need for the rights-of-way that would be necessary with a cable system. Furthermore, the infrared system falls into the category of nonlicensed technology. This removes many of the licensing requirements and associated delays and costs incurred with other systems. For distances of under 2 miles, this technique holds a wide range of exciting possibilities. To use such a system, however, three criteria must be considered:

- The basic geometry of the system
- Atmospheric conditions that will affect the transmission
- Site selection for the installation of the link

System Geometry

As already mentioned, the infrared system is generally used in a point-to-point, line-of-sight application. Some multipoint and repeated systems have been developed, but these are rare. There are three elements that make up the basic system, including:

- The transmitter, where the signal is modulated from its original electrical form into a light beam. Modulation was covered in detail in Chap. 5.
- The medium used—in this case, air.
- The receiver, which demodulates the light back into its original electrical form.

These basic components are shown integrated into a complete laser system in Fig. 34.1. In this figure, the components look like bars: this is for representation only.

The data is input into the transmitter, where it is modulated into a light beam. The raw laser beam is a very narrow beam of light (the actual light is referred to as monochromatic) that has very little divergence or spreading. Theoretically, the beam is usable by the receiver in its raw form, but in reality, it is extremely difficult to aim and maintain this raw light beam. Therefore, the raw light beam must pass through a lens that will produce a divergent beam. This helps to solve the initial limitation of the beam itself. However, it also imposes some distance limitations on the transmitted signal. Because the lens creates the divergence, it reduces the distance from the transmitter to the receiver by diffusing the strength of the beam. Safety requirements also limit the output power of the beam. The trans-

Figure 34.1
The basic components of an infrared system.

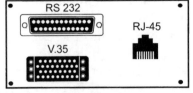

Rear Panel

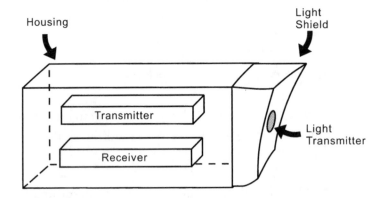

mitter generally uses a constant diameter of light, called a *footprint,* regardless of the distance being used.

The incoming light is focused by a collector, or collecting lens, that sends the received beam to an optical detector. The receiver uses an angle of acceptance (RAA) of approximately 3 to 5 mR to provide a degree of selectivity. The transmit beam divergence angle (TBA) is always greater than the receiver angle of acceptance. The receiver needs a minimum level of incident light in order to demodulate the light back to the original electrical input. There are trade-offs in the size, complexity, and cost of the

Figure 34.2
The transmitted beam divergence angle compared to the receiver acceptance angle.

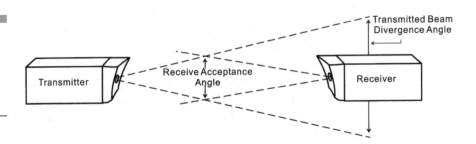

receiver lens that limit the range. The receiver angle of acceptance and the transmitter beam divergence angle are shown in Fig. 34.2.

The setup in the previous paragraph is a one-way, point-to-point system (called a simplex system). However, most applications need two-way communications; for this, a duplex system can be used. To create a duplex system, two simplex systems are coupled together. The creation of a duplex system is shown in Fig. 34.3. Both systems will operate with lasers of the same wavelength (typically 830 nm). This is an advantage of a light-based system, because the lasers do not switch from transmitting to receiving like a radio system would. (Remember that in radio systems, two separate sets of frequencies are used to keep the transmit frequency apart from the receive frequency.) A further feature of focused light transmission systems is that multiple systems can be mounted close to each other without causing interference as long as the angle the systems make with each other is greater than the original angle of the receiver. The use of multiple systems is shown in Fig. 34.4. Two light beams are transmitted

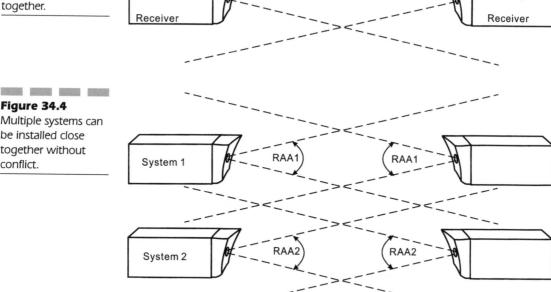

Figure 34.3
A duplex system is created by linking two simplex systems together.

Figure 34.4
Multiple systems can be installed close together without conflict.

across the same path, yet they are kept separate because the angles of acceptance are different.

Atmospheric Conditions

The effects of atmosphere on light transmission will influence the total overall performance of the system. The severity and the duration of these conditions will affect the distance and performance. All electromagnetic radiation used in any communications system is affected by the atmosphere. However, each of the transmission systems is affected differently, be it low/high-frequency radio, microwave radio, or laser transmission. The wavelength of each transmission determines the effects on the actual transmission. The three most significant conditions that affect laser transmission are:

- Absorption
- Scattering
- Shimmer

All three conditions can reduce the energy at the receiver, which will then affect the reliability and the bit error levels. These three conditions are covered in more detail.

Absorption

Absorption is caused primarily by water vapor (H_2O) and carbon dioxide (CO_2) in the air. The density of water and carbon monoxide are conditioned by humidity and altitude. Gases that form in the atmosphere have many resonant bands, called *transmission windows*, that allow specific frequencies of light to pass through. These windows occur at various wavelengths. The window we are most familiar with is that of visible light. The near-infrared wavelength of light (830 nm) used in laser transmission occurs at one of these windows; therefore absorption is not generally a big concern in an infrared laser transmission system.

Scattering

Scattering has a greater effect than absorption. The atmospheric scattering of light is a function of the wavelength of the light and the number and size of scattering elements in the air. The optical visibility along the path is directly related to the number and size of the scattering particles. The three most common scattering elements in the air that will affect laser beam transmission are as follows.

FOG AND SMOG. Fog appears when the relative humidity of air is brought to an appropriate saturation level. Some of the nuclei then grow through condensation into water droplets. Attenuation by fog is directly attributable to water droplets less than a few microns in radius. The result of the scattering is that a reduced percentage of the transmitted light reaches the receiver.

Visualize the situation of driving a vehicle in dense fog. The headlights on the car are somewhat useless. Try turning on the high beams and see what difference this action makes. The resultant scattering of the visible light from the headlights makes it almost impossible to see. The light is being reflected back at the vehicle as well as being scattered in all directions. This is exactly how fog affects the infrared light beam.

Smog's effects are similar to those of fog, but not as severe. This is because of the difference in radius of the particulate matter. Consequently, fog is the limiting factor in laser transmission performance.

RAIN. Although the liquid content of a heavy shower is 10 times the density of a dense fog, the radius of a raindrop is approximately 1000 times that of a fog droplet. This is the primary reason that attenuation via rain is 100 times less than that via fog. Think about this for a moment; when fog is created, the saturation level in the air is very high, but the droplets are much finer. In rain, although more liquid is falling, the coarse drops don't affect the transmission as much. Visualize that there is more space between the drops of rain than between the droplets in fog; therefore the light can get through rain much more easily.

When you are driving in a heavy rainfall, the headlights still penetrate the raindrops and give greater visibility over a greater distance. Even when you turn on the high beams, the distance visibility is better than with the low beams. A certain amount of the light is still being reflected back at the vehicle and scattered in all directions, but more is penetrating the rain. The same results occur with the transmission of the laser beam through the rain. It would take an extremely heavy rain (greater than 5 inches per hour) to significantly reduce the transmission of the laser beam.

SNOW. The effects of snow on laser transmission fall somewhere between those of fog and rain, depending on the level of water particles in the snow. A very wet snow is close to rain, and an extremely dry snow is similar to fog. Visualize driving through a dense snowstorm. Driving through a heavy, wet snow is like driving through a heavy rain. It is not all that pleasurable, but at least the light penetrates the snow and visibility is achieved. However, when a whiteout occurs, the ability to get the light through the snow is significantly reduced. The light appears to be reflecting right back at the vehicle, rendering all attempts to increase visibility useless. The same effect occurs with a laser transmission.

Shimmer

Picture driving down a long road on a hot day. As you look toward the horizon, it appears that water is on the road. As you look closer, you can see what appear to be heat waves shimmering off the ground. This is the direct result of a combination of factors including:

- Atmospheric turbulence
- Air density
- Light refraction
- Cloud cover
- Wind

These factors combined cause disturbances similar to a laser beam passing through the atmosphere. As local conditions begin to combine, shimmer begins to affect the transmission. Shimmer imposes a low-frequency variation on the amount of light reaching the receiver. This variance could result in excessive data error rates or video distortion on a laser communication system. By using simple, proven techniques such as frequency modulation (FM) and automatic gain control (AGC) in the system, the effects of shimmer can be minimized. Moving the equipment several meters above the heat source will also reduce the effects of shimmer.

Site Selection

Proper site selection is an equally important factor. To assure successful installation and use, several factors must be considered, such as:

- Clear line of sight
- Mounting conditions

- Avoiding adverse conditions
- Mounting structures

Each of these factors is important. Diligence must be applied.

Clear Line of Sight

When setting up a transmission system, a clear line of sight is needed. This means that nothing can obstruct the light. The two ends must physically "see each other" or the system will not work: it is as simple as that. The laser beams should not be directed through or near wires (i.e., electric wires, telephone, cable TV). Trying to force the light through tree limbs is not a good idea. As Fig. 34.5 shows, getting over cables and wires is important. Allowances must be made for the growth of tree branches over time (Fig. 34.6).

The system can be mounted either indoors or out. Figure 34.7 shows the system mounted indoors. The light beam passes through glass. The glass must not have any reflective coating or tinting; these might impair the ability of the beam to pass through.

Adverse Mounting Conditions

Few precautions must be dealt with when mounting the unit. It is important to ensure that the maximum signal strength is received and that the beam is not inadvertently misdirected. These considerations include avoiding heat sources, water obstructions, east-west orientations, and vibration.

Figure 34.5
Towers may be necessary to keep the path above electric or telephone wires.

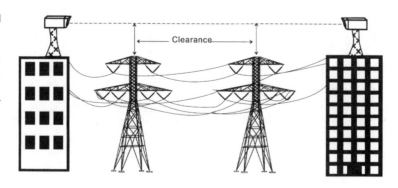

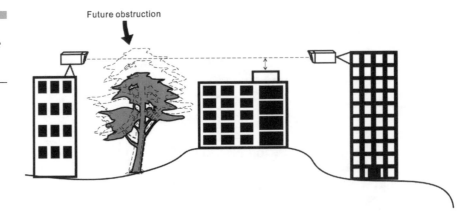

Figure 34.6
A clear path must be ensured for the transmission.

HEAT SOURCES. Heat scintillation (or shimmer) occurs as a result of differences in the index of refraction of air caused by temperature. The air might act as a lens that incorrectly steers the laser beam. The beam should not be directed over exhaust vents, cooling towers, chimneys, smokestacks, or large air-conditioning units.

WATER OBSTRUCTION. Water droplets on glass can redirect the light. To prevent this when using the system indoors, use an awning

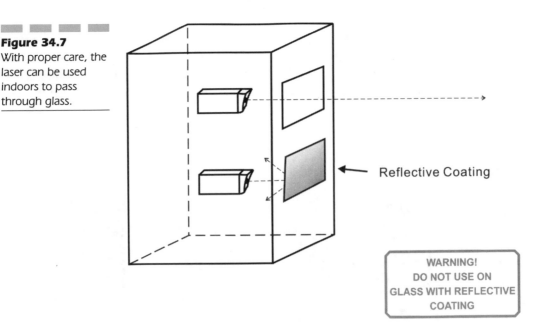

Figure 34.7
With proper care, the laser can be used indoors to pass through glass.

Reflective Coating

WARNING!
DO NOT USE ON
GLASS WITH REFLECTIVE
COATING

above the window. If building codes preclude using an awning, mount the unit as high as possible on the glass. This will prevent buildup of water on this section of the glass, because the water will run down quickly.

EAST-WEST ORIENTATIONS.   The angle of acceptance is designed to be quite narrow. Any extraneous infrared light passing directly into this angle of acceptance will be accepted, thus causing errors. East-west shots that have direct sunlight on the same axis will cause interference. The sun emits infrared light, and this cannot be filtered out. Therefore, a system will be "blinded" for periods when the sun's axis is the same as that of the beam from the system.

VIBRATION.   Both transmitter and receiver deal with very narrow beams of light. The smallest movement of the transmitter or receiver can severely limit transmission or reception. The system could even lose its angle totally, resulting in an outage.

Licensing Requirements

Because infrared and laser-beam technology employs wavelengths in the invisible light spectrum, no licensing is required. The use of light fortunately falls into the frequency spectrum where there are no regulatory guidelines. The U.S. Government has not figured out how to charge for the use of light, nor can they determine how much we use. Therefore, they can't tax it on a usage-sensitive basis. Because the government cannot figure this out, neither can the telephone companies. This, of course, makes it much easier to deal with the technology itself, allowing an organization to place a system in operation quickly.

Unlike radio technology, where the bulk of the frequency spectrum requires licenses and path clearances (with limited exceptions) from the regulatory bodies around the world, the use of light theoretically causes no interference between systems. Thus, use of this spectrum is allowed with little to no coordination effort. So long as the acceptance angle of one receiver does not coexist on the exact same axis with that of another, there should be no problems. This is an attractive feature of light-based systems.

Bandwidth Capacities

The system's bandwidth can cover a myriad of capacities, depending on the configuration purchased. The specific application must be considered in terms of the bandwidth necessary as well. At the high end of the bandwidth, the systems today can deliver as much as 16-Mbps throughput for LAN-to-LAN connections on a token-passing ring network. Point-to-point systems at 155 Mbps are used for LANs and MANs. Thus, one could say that the current bandwidth is 16–155 Mbps. However, other systems are designed to meet the needs of the user interface. An example of the capacities is summarized in Table 34.1. In this table, the needs are met by different systems at the capacities required to meet the application. In several of these systems, the pricing is dependent on the actual needs.

Applications

As noted in Table 34.1, infrared and laser technologies have several applications. These are addressed below as they pertain to a need for connectivity from site to site. In the following scenarios, a typical customer with two sites (buildings) located approximately 1000 meters apart, whether on a single campus or in two buildings in the same town, needs to provide various forms of connectivity between these sites. After the customer looks at the alternatives, the choice is between:

■ A direct burial cable (copper, coax, or fiber)

TABLE 34.1

Summary of Various Systems Available on the Market and the Application Used

Bandwidth	Application
10 Mbps	Ethernet data link
4 and 16 Mbps	Token-passing Ring
1.544 /6.312 Mbps	T1/T2 capacity data/video
4×1.544 Mbps	Quad T1 voice/data
RS-232 to 19.2 Kbps	Data transceivers
RS-422A to 2.048 Mbps	Data transceivers
Audio 20 to 20 kHz	Audio services

SOURCE: Laser Communications Inc., Lancaster, PA.

- A microwave radio system
- An infrared laser system

Although more considerations and information are needed to arrive at any conclusion, the following might play as an active comparison.

The Cable Decision

On looking at the cable scenario, the customer is faced with the extraordinary costs of digging a trench, laying a conduit, and running a cable connection to the two buildings. Additionally, the necessary right-of-way might be extremely difficult to obtain. The customer decides that a cable-based solution is not appropriate in this situation. A note here: another alternative exists: the use of local leased line facilities from the local exchange carrier (LEC) or a competitive access provider (CAP). However, the long-term leased line rental costs are not attractive to the customer.

The Radio Decision

Looking at the area and the potential ordinances and variances that would be required in his scenario, the microwave decision is also determined to be feasible only if another alternative cannot be found. Given that certain forms of licensing are required and that a clear path is necessary, in a very heavily installed radio area, the costs and time constraints are greater than the customer wishes to endure. The customer is not even sure that investing all this time and effort in the process will result in the granting of a license. Hence the microwave choice is ruled out.

The Infrared Decision

Because the cable and the microwave choices are not that attractive, the logical choice is to use an infrared system. No licensing requirements are necessary, so the system can be put into action as soon as the acquisition is made. Because no right-of-way constraints impede the process, it can be handled appropriately and quickly. An added benefit is that there will be no monthly recurring charges for the use of the facilities as there would with leased lines. Finally, implementation requires no major tower, power equipment, cable entrances, or other construction projects.

The cost of buying the system is also very attractive, because a one-time outlay of less than $20,000 is all that is required. The customer also takes into consideration the fact that if a move from one of the buildings is ever necessary, the system can be packed up and moved along with the rest of the office equipment. Assuming that a clear line of sight can be obtained and that the distance does not exceed the specifications of the system, it can be back into action in a matter of hours. This is not the case with the other options. The clear winner in this particular scenario is the infrared laser system. However, what of the application? Regardless of the application, the same rules apply. However, in this particular scenario the applications might include the following:

The Use of Infrared Technology to Link Two LANs Together

The technology can link a token ring-to-token ring network together at native speeds. Either a 4- or 16-Mbps capacity can be applied here. In Fig. 34.8, two LANs are linked together at native speeds. Here this means that the LANs are connected at the same speed at which the LAN operates. This allows for transparent connectivity at the full rated speed of the network itself for a price of approximately $35,000 for a full 16 Mbps, or $26,000 for 4 Mbps, rather than slowing the transport down to some lower-level speed. If a bridging arrangement is used here, the total throughput is limited to a speed of 1.544 Mbps or less.

Figure 34.8
Linking two token ring LANs together at a native speed of 4 or 16 Mbps provides transparent connectivity.

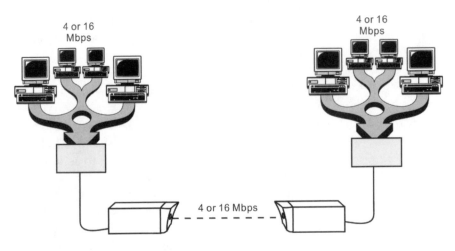

Infrared Laser Technology Used to Link Two Ethernet LANs

The use of infrared laser technology to link two Ethernet LANs together at a native speed of 10 Mbps is shown in Fig. 34.9. A different system is used here, with the native throughput operating at the full rated 10 Mbps, rather than using a slower-speed bridging arrangement that runs at 1.544 Mbps or less. The system is priced at around $15,000.

In both of these cases, the alternatives to these connections could be leased T1/E1-speed lines at connections slower than the LAN operating speed. This could potentially cause network congestion because the throughput is buffered to accommodate the line speed. Further, the leased line of up to 1 mile from a local operating company (telco) could cost as much as $700 to $1000 per month. The payback on this arrangement would obviously be dramatic—gaining the additional speed and still paying for the system in less than 2 to 3 years, depending on the system selected. Furthermore, if the native speeds were leased from the local telco, the costs for the services could be as much as $1800 to $2400 depending on the speed selected, thereby producing a return on the investment in less than 12 to 16 months. This is for the close-in services. Complete installation of a microwave system could cost as much as $80,000 to $100,000, three to four times the cost of infrared. As can be seen for this environment, the use of the infrared system has some definite advantages.

Connecting PBX to PBX Using a Digital Transmission System

Connection of PBX to PBX using a digital transmission system for tie line capabilities between systems using a quad T1 capability for voice connectivity is another application. Figure 34.10 shows this arrangement with a digital trunk interface connection at each end of the system. In this particular situation, 56/64 Kbps of voice connections are provided with 24 channels on each of the four T1s, resulting in 96 digital voice connections. This would be a prime case for connecting two PBXs in very close proximity to a single network, or a means of providing other voice networked services such as voice messaging, off-premises extension, and so on. For a quad T1 service, the infrared

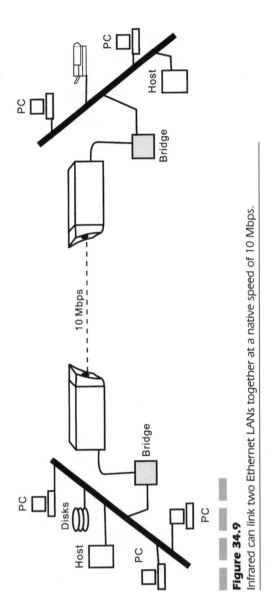

Figure 34.9

Infrared can link two Ethernet LANs together at a native speed of 10 Mbps.

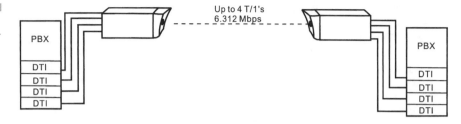

Figure 34.10
Linking PBXs together
with up to four T1s
for tie lines. DTI cards
are used in the PBX
to yield up to a
96-PCM circuit at
64 Kbps.

system would price out at approximately $23,000, yielding a net payback time of less than 1 year, given four T1s at $1000 per month rental from the local telco.

T1/T2 Service Between Two Sites

The provision of a T1/T2 service between two sites to provide a 6-Mbps channel or four 1.544-Mbps channels for video conferencing capabilities at compressed video standards is shown in Fig. 34.11. These compression rates deliver superior-quality video compared to the more compressed speeds currently being offered in the industry (i.e., 256-, 384-, or 512-Kbps video services). This service would cost approximately $15,000, and would certainly deliver a return on investment in less than 1 year.

Infrared High-Speed Digital Dataphone Services

The use of an infrared service to provide for high-speed digital dataphone services using a host computer is shown in Fig. 34.12. In this sce-

Figure 34.11
The infrared system
can deliver speeds of
up to 6 Mbps (T2) for
video conferencing or
four links at 1.544
Mbps (T1).

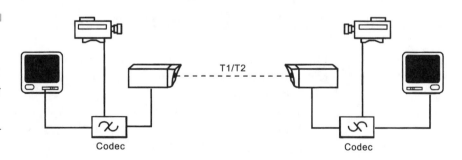

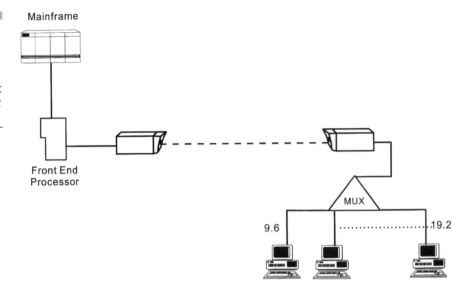

Figure 34.12
The infrared system can replace multiple DDS service for data connections to a host computer at 9.6–19.2 Kbps each.

nario, a DDS service can be replaced with the system between these two buildings, whereby a statistical time-division multiplexer (stat-mux) or a subrate multiplexer can be used to provide several 9.6- to 19.2-Kbps data channels at the remote site. This system would cost approximately $10,000, but a DDS service between these two sites at 56 Kbps could cost as much as $1800 per month. This would yield a payback in less than 6 months.

These examples are merely representative of the types of connections that can be provided at a 1:1 ratio. Clearly, mixing and matching the services on these infrared systems would offer even greater returns on investment.

Can the System Carry LAN Traffic Transparently?

Obviously, the question always comes up as to whether an infrared system can really carry true LAN native speeds. For this to occur, several conditions must be met.

First, the system must have the bandwidth capacity to carry the native LAN traffic. This has been evidenced in these scenarios, where the system throughput is at least 4, 10, or 16 Mbps.

Second, the system might not add significant latency to the network because the entire LAN is based on very specific timing for the delivery of the frames on the network. Should the latency become a problem, the frames will not be delivered in time, causing a discarded frame or a network time-out, which could require resynchronization of the network to recover timing. In either case, transmission speed of the infrared system across the air would approximate 3.3 nanoseconds per meter. A 1000-meter distance between these systems would add only a 3300-nanosecond latency to the transport of the frame. Further, the system must modulate the signal onto the medium with some added delay based on the equipment. If a standard delay of approximately 30 nanoseconds is added for the modulation and the demodulation process, then an additional 60 nanoseconds of delay is added. These delays should fall well within the defined tolerance levels of the networks.

Third, the connection arrangement must either fall within the scope of an access to a bridge or repeater for the cable system in an Ethernet world, or else be connected as a station or at a multistation access unit (MAU) for a token ring network. Once again, these connections are easily made with the use of the infrared interfaces and the network components.

Advantages of Infrared

The use of infrared technology brings advantages along with it. The benefits are quite numerous and should be continually considered whenever a connection is required, rather than immediately ruling this technology out. Many organizations automatically discount this as an old technology not worthy of consideration. The more astute telecommunications person keeps it in mind as an option, at least until some other technology proves to be either better suited for the job or more cost effective. Some of the major advantages of infrared are:

- Immune to radio frequency interference (RFI) and electromagnetic interference (EMI)
- Reasonably high bandwidth available
- Secure technology, more so than a radio-based technique
- Cost effective
- Easy to install
- Efficient when dealing with a potential move of the system

- No license required
- Systems can traverse similar paths without interference
- Can be up and running in a matter of hours
- Supports multiple standard interfaces
- Handles voice, data, video, and LAN traffic either separately or on the same system, depending on application

Disadvantages of Infrared

As with any technological invention and use, there are always pros and cons. The primary disadvantages of infrared are not as dramatic, but should be factored into the equation. Because the use of a light-based system such as this is primarily based on actual need, these disadvantages exist:

- Very limited distances, 1.5 miles at most, depending on the manufacturer and the rate of speed being used
- Concern over the use of a laser in an office environment
- Negative effects of vibration
- Impairment by atmospheric disturbances of reception of the beam from the transmitter
- Risk of exposing cleaning personnel to the light (laser) beam
- Needs for a clear path in a line of sight, regardless of any other condition
- Need for stable mounting because of the sway of buildings, mounts, poles, or trees—very little tolerance can be allowed as a result of the small footprint of the received signal

Fiber-Optic-Compatible Systems

A newer product has emerged on the market that utilizes a fiber-optic system of wave-division multiplexing services with up to four clear channels of pulse frequency modulation video or digital data signals per wavelength window via one multimode fiber. Although this is not a wireless technique, the concept is similar to a development that has been touted as *free-space optics.* In the conceptual model of this newer service, vendors are creating the bandwidth of a fiber-optic channel capacity across free airwaves. Hence the name *free-space* as opposed to capacity inhibited by a piece of glass cable. A standard fiber-based system uses different wavelengths to carry the channels of communication across a glass fiber. The free-space optics system uses a different color of light in a different wavelength to carry various streams of voice, data, or video transmission over short distances. Obviously, because this concept uses laser beams in free space, the need exists to use low-power output devices. Distances of approximately 3 to 5 miles are being discussed at this time. Some of the applications for this free-space optic system will include, but not be limited to:

- Security and surveillance systems
- Multimedia learning over a short distance
- Video multiplexing using NTSC, PAL, and SECAM standards
- Digital data at native speeds
- Digital voice transmission

Low power will be a requirement to prevent any damage from the effects of the laser output in close confines, such as major metropolitan downtown areas. The system will also have to meet the requirements for infrared systems, such as not being too close to pedestrian traffic. However, because the free-space optics system operates in the visible light spectrum, the signal will be less impeded by the various atmospheric conditions that affect the operation of an infrared system.

The system will also be fully modular and easily transportable, as with an infrared system. This will allow the flexibility to set up and operate a transmission quickly. Because this system will use light beams traveling across the airwaves, the licensing issue will not come into play. The bandwidth of these systems will range from approximately 2 to 40 MHz, yielding data rates of up to 16 Mbps for native LAN traffic at the LAN speed. Wavelengths will be 750, 780, 810, and 840 nanometers.

The use of this form of communications medium will obviously offer some very attractive options to end users for access to the long-distance carrier's point of presence (POP) and ultimate access to the long-distance networks. Local area, campus area, metropolitan area, and wide area networks will all have some use of the free-space optic systems. Consider the use of such a system as an alternative to access the long-distance network or as a means of bypassing the local loop in the telephone company arena.

CHAPTER **35**

Videoconferencing

Videoconferencing Systems

Up to now, this book has covered the use of the older analog dial-up voice network and the newer techniques used to produce digital transport systems to carry voice and data transfers. However, another use of this network is coming of age: conducting face-to-face communications between and among sites in the business organization. The concept of jumping on a plane and traveling to a remote location is fast becoming passé. The reason for this is the wear and tear on traveling executives, professionals, sales forces, and others.

Timing is also a critical concern. To be competitive in today's marketplace, decisions must be made quickly and firmly. Many times, obtaining the information necessary to make these decisions requires a face-to-face meeting or the ability to see a product. In the engineering world, the ability to see a component or piece of equipment that needs modification is also crucial. In the past, the only way to handle this was to go to the location and conduct the meeting or review. The options were limited and many times they were too expensive. Now this is all changing quickly. The use of cameras that can project an image of the remote site across the network is far more prevalent. Equipment has improved, the network has increased in reliability and performance, and the technology needed to transmit such information has become far more reasonably priced. Therefore more and more organizations are deploying videoconferencing systems throughout their hierarchies. This technology is becoming one of the tools of the late 1990s and will carry over into the turn of the century.

The use of videoconferencing systems is not new; they appeared in the industry back in the early 1960s. The concept is a simple one: conduct a face-to-face meeting between members of an organization anywhere in the world without the travel associated with such a meeting. This takes advantage of the leased line or dial-up capabilities of the telecommunications industry techniques that use a device to convert the analog signal of a videoconference into digital pulses. This device is called a coder/decoder (codec).

Videoconferencing has been technically possible for several years, but at a steep price that made it a practical alternative only for very large organizations with specially designed facilities. In the late 1970s and early 1980s, an organization might have spent $1 million just to prepare a site for the video capability. This was particularly true in boardroom conferencing, which was therefore considered a senior-level management tool because of the cost. Now, the arrival of universal standards and improvements in equipment and signal transmission has led to much smaller systems with significantly lower costs. The costs of the equipment were $200,000 to $300,000 in the early days. New systems are available in the $20,000 to $60,000 range. These developments have made videoconferencing an affordable alternative for businesses of all sizes.

Rollabout and tabletop videoconferencing systems have rapidly replaced their larger and more expensive predecessors, and a new market for videoconferencing between groups of two or three users has emerged. The use of videoconferencing systems is no longer reserved for the very expensive boardroom environment; it can now be easily accommodated at the desktop or in a regular conference room. The special facilities and lighting necessary for the larger systems have been dispensed with by the newer, more sophisticated systems that overcome the environmental situation. Small groups of engineers, designers, financial personnel, or managers can conduct ad hoc meetings. These can be impromptu meetings, because the equipment can operate virtually anywhere. Desktop units that cost as little as $10,000 to $15,000 make this a more viable and palatable alternative to the cost and time associated with travel.

The logistics of transmitting a video signal across the leased line or dial-up telephone network are no longer as complex as in the past. For the average business, the bandwidth (or pipe) is already in place to support the videoconferencing systems. Therefore video equipment transmission need only be added and piggybacked on the existing voice or data communications links. Digital leased services, such as T1 or switched 56 Kbps, are readily available. Additionally, the emergence of integrated services on a digital network (ISDN) has opened the door for the impromptu dial-up connection at speeds of 56 Kbps up to 1.544 Mbps. The most commonly used speed is in the range of 128 to 384 Kbps. This is easily accomplished through an inverse multiplexer (I mux). (You might remember the discussion of the I mux in Chap. 27.) The lower economic barriers to entry in videoconferencing and the managerial efficiency sought in corporate America make videoconferencing the opportunity of the 1990s.

What Is Videoconferencing?

Back in the early 1960s, AT&T introduced the concept of videoconferencing at the New York World's Fair. The idea at the time was a product known as the Picture Phone. In this early introduction, callers could see a motionless snapshot and hear the voice of the party on the other end of the connection at the same time. Even though this exhibit generated a lot of excitement and opened the eyes of many organizations, it was really just a demonstration of how far the technology could go. The service and the technology would have to evolve and mature before videoconferencing could be taken seriously as a business tool. Early attempts at videoconferencing were stymied by several factors.

The systems had to be connected to similar devices on both ends, and only a handful of suppliers existed at the time. Standards for interconnectivity were nonexistent. The special circuits required at the time were extremely expensive and not universally available. The operating costs for such a system were astronomical.

Eventually, AT&T's Picture Phone died a slow death, but the spirit and concept survived in the minds of the engineers. Research continued and improvements were made. During the 1970s, full-motion two-way closed-circuit video emerged in the boardroom environment. Also, one-way non-interactive private TV networks that used satellite transmissions or very high bandwidth began to appear. Although firms made good use of their networks, others were not willing to invest in the buildout of special rooms and circuits at each of their sites. Despite the improvements that were evident, the financial burden slowed the development and installation of video systems.

During the 1980s, things began to change. Several major breakthroughs opened the door to reasonably priced videoconferencing. The more significant accomplishments included the following:

- Improved video compression techniques made the bandwidth constraints less difficult to deal with because the systems could operate at less than full T1 speeds.

- Divestiture of the AT&T and Bell Companies opened the door for competition in the telecommunications industry, reducing transmission costs by 40 percent and more.

- The evolution and rollout of digital transmission systems occurred throughout the industry rather than just in the major metropolitan areas.

- Computer chip technology progressions led to the decline in the cost of the electronic componentry.

Prior to 1986, a typical videoconferencing system cost in excess of $200,000. Because these systems required special acoustics, lighting, and other environmental conditions, they were installed in the special video rooms. The treatment of these rooms could cost as much as $700,000. This was obviously too expensive for the average organization.

A mere 1 year later, the cost of the equipment had dropped to $120,000. By the end of the decade, a standard videoconferencing system bore a $60,000 price tag. Price competition increased significantly in the early 1990s, as shown in the graph in Fig. 35.1. Although some companies are still spending over $100,000 to build and equip lavish video rooms, the market trend is toward a functional, accessible, modular system that can be used by all levels within the organization.

Videoconferencing Overview

When we think of videoconferencing, the image of high-tech multi-media conference rooms comes to mind. Although this is true in some environments, the reality of the situation is that, more and more, lower-end services are what drives this application. Videoconferencing has a significant potential as an efficient, productive means of conducting point-to-point or point-to-multipoint meetings. Identifying and achieving this goal depends on the organization's acceptance and knowledge of videoconferencing systems. Teleconferencing in itself offers many benefits. Some of these can be summarized as follows.

Increasing Managerial Productivity

The time for travel to and from sites is normally nonproductive. Regardless of whether the individuals involved with travel are dedicated, the problem is putting that time to productive use. This is becoming increasingly more difficult when air travel is involved. Most of us have been crammed into an airplane at least once—perish the thought of winding up in the center seat. There just isn't any space to move or spread documents out. Coupled with this is the fact that many of the airlines are now stuffing every seat they can get onto a plane to produce more revenue. The lack of space is compounded when the plane is chock-full. The qual-

Figure 35.1
The costs of videoconferencing equipment have dropped dramatically.

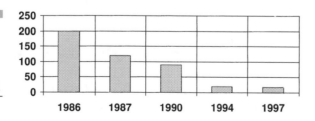

ity of air travel has decreased over the years, despite what the airlines want us to believe. Constant interruptions, constant passenger fidgeting on the plane, and passengers climbing over each other disrupt work. Even the road warriors who have computers (laptops and notebooks) are prevented by the lack of space from conducting much useful business on the plane. So, if travel time is nonproductive, the counter to this is to replace the travel with a technique that allows the executive to attend the meeting for as long as necessary, then walk down the hall back into his or her office and get right back into the thick of things.

Quality of Meetings

A funny thing happens when videoconferencing systems are used: the quality of the meetings held usually increases. This is in part because of better planning and execution of the meetings. When someone needs to travel to a meeting, the inevitable happens: they get to the meeting but find they have left something behind. Or they still have to get back to the office to check with others. Clearly, an entire organization or department cannot travel to a meeting. When video is used and we need to check with another department, we just go down the hall. Decisions and discussions can be facilitated much more expeditiously.

Planning a videoconference usually makes the use of time more efficient. If everyone has an agenda and premeeting materials, the time is used better. With video we tend to do more of this type of planning. Because the cost per minute used to be so much higher, everyone tried to be as quick and to the point as possible. Otherwise, the expense of the conference would have been exorbitant.

Time and money savings mount up. The planning mentioned above causes savings by making better use of time. Further, the cost of travel is escalating at a very rapid pace, despite what the travel industry tells us. Consequently, we can involve more people, bringing in conferees only when needed rather than having them sit around for hours waiting for a turn to speak and otherwise participate. This all adds up to benefits not just in terms of money but also of valuable time.

Improvements in the Quality of Life

We travel a lot (Bud runs up 350,000 to 400,000 miles a year, and Don is no slouch either). The amount of time dedicated to road warrior events constantly demands five to six days a week of our time. This means that travelers like us get only one day a week at home. This puts a lot of pressure on the home life as well as situations in the office. We have to be more accessible, and we have to work much longer hours. There are no substitutes for person-to-person communications. Even though techniques such as E-mail, voice mail, network access from remote all exist, they still do not completely eliminate the need to speak with a human. As travel gets reduced or replaced by videoconferencing, the amount of free time that the individual retrieves is phenomenal. This is not to imply that some limited amount of travel will not be required, but when one day per week is compared to six days per week, there is no contest. Unfortunately, senior-level people in organizations wind up in situations where they are on the road five and six days per week. Such people will be prime candidates for videoconferencing applications.

The Parts of a Video System

The components of a videoconferencing system vary depending on budgets, applications, and personal preferences. However, certain basic parts are required. These include the following:

- A camera
- One display screen or monitor—minimum. Many systems come with two.
- A codec (the heart of the video system is the coder/decoder)
- Audio input/output capabilities (unless you are very good at reading lips, you should have an audio service integrated)
- Control, either manual or automatic
- The communications link

The Camera

Crucial to a video system is the ability to see what's going on at the other end. The camera is the basic element of this service, and one that should not be overlooked: the better the camera, the better the presentation. Hopefully, the camera system will be unobtrusive in a conference environment. In the early days, the cameras were a constant source of confusion. As needs dictated, the camera had to be shifted from one speaker or object to the next, then back again. This required that a controller in the room be assigned to move the camera from position to position. Unfortunately, one of two scenarios constantly got in the way. First, the controller got so engrossed in the meeting that the camera never got moved. Therefore people would see blank walls and not the body language of the speaker. Second, the controller got so involved with the positioning and movement of the camera that he or she lost all interest and input in the meeting. This should be taken into consideration right up front. Most systems now come with simple-to-use camera arrangements that allow fixed and variable movements on the camera so that positioning can be simplified. In many systems, the camera might be mounted inside a cabinet so it is not visible. The need for a quality camera cannot be overstated. Transport systems and codecs cannot make up for poor-quality input from a cheap camera.

The Monitors

Every videoconferencing system includes at least one monitor. Larger systems typically offer two or more, using standard TVs. One monitor is for the graphics camera, the other displays the participants. Some systems use the two monitors to display the participants at each end. Still others might include picture-in-a-picture (PIP) so that both ends of the transmission can be seen at once. The variables are many, depending on the manufacturer, the system, and the depth of the budgetary pockets. Monitors should also be a concern so that sharp images can be projected, but much of the sharpness on the monitor will be determined by the speed of the link. Poor quality of reception might not be caused by a problem with the monitor. As we move videoconferencing systems to the desktop, variations in monitor size increase exponentially. Typical monitors for conference room units have 35-inch screens, whereas the rollabout systems today use a 25- to 35-inch monitor. Desktop units are now using a 19-inch monitor. The size is not a function of the quality but of the overall display that must be viewed by the number of participants.

Audio Capabilities

Most people take the audio portion of a videoconference for granted. Unfortunately, this part of the connection is as important as, if not more important than, any other. Most dissatisfaction with videoconferencing usually involves poor audio quality. There is a reason the audio information is so critical: research has proven that 60 to 70 percent of all information in a videoconference is still transferred by audio. Although we supplement the audio portion of the meeting with visual input, the bulk of what happens is verbal.

Other ideas state that the combination of the visual and the verbal communications enhances the retention level to approximately 85 percent. Therefore the audio portion of the conference is as important as the video. Another way of looking at the situation: if a videoconference is established and the video portion goes out, a good part of the meeting can still take place. Attendees might be able to refer to handouts or other materials and follow along with the audio presentation. However, if the audio is lost, the connection will be rendered useless. This points further to the total need for all pieces to work, but also underlines the need for the audio connection. Typical audio systems allow for up to 15 kHz of voice quality in a video system to give the best-quality voice communications possible.

The Control System

The control system and the software portion of the system constitute the "brain" of the overall videoconferencing system. These components help to orchestrate the overall performance of the videoconferencing capabilities. Every part of the system interacts with the control and software functions. The audio mixers use the control to allow for full duplex communications; the angle, pan and tilt, and zoom functions of the cameras are controlled as well. The control portion developed in the software makes the system transparent to the participants in a conference. This part of the system executes the preset functions for scanning the audience or locking into a specific camera. Also, adjustments to focus and volume controls can be accomplished here.

The Coder/Decoder

Using compression techniques to send a lot of information over the communications link is part of the function of the coder/decoder. This is the chip set that will take the analog picture of the participants and digitally encode it for the transmission link. At the receiving end, the decoding function will take the digitally encoded picture and convert it back to an analog form for delivery to the monitor. The codec acts similarly to a modem; it is the change agent in the videoconferencing system. The codec handles the digital compression of the information so that we can get away with using a much lower data rate on the actual physical link. Compression algorithms are becoming so sophisticated that the codecs can do with 128 Kbps or better what used to take a 90-Mbps transmission. This is a big savings in terms of the availability and acceptability of the videoconferencing deployment.

The Communications Link

A less visible but very important component of the videoconferencing systems is the communications link. Improvements in videoconferencing, both from a quality and cost standpoint, closely parallel advances in the transmission technology. The demands for speed have decreased over the years from a full T1 speed to today's technology that requires as little as 64 Kbps (Fig. 35.2).

Transmission or data rates vary, but most common systems use either 56,000 bits per second (56 Kbps), 64, 112, or 384 Kbps, or 1.544 Mbps at the high end of the spectrum. In comparison, these data rates are a dramatic reduction from the 90 Mbps of bandwidth needed to support the original Picture Phone at the World's Fair, which actually produced a very hazy picture at best. That 90-Mbps transmission rate in 1960 was considered quite reasonable. This is also the speed and throughput used by the major networks to bring us one-way broadcast-quality transmission services on our TVs. Compare a still frame of video at that rate to what you see on TV, and you have the differences in the world of video.

The higher the data rate, the better the picture quality and the clarity of the motion. A data signal transmitted at 56 Kbps has difficulty handling a lot of motion and picture detail. Even though strides have been made to achieve tolerable videoconferencing at 56 Kbps, the average participant would not be happy with the results. Signals sent at 128 Kbps (or multiples of 64 Kbps higher) produce detail that increases proportionately with the increments of bandwidth added. From a personal perspective,

Figure 35.2
The overall improvements in technology have reduced the bandwidth for video over the past decade.

Decreases in Transmission Speeds over Time

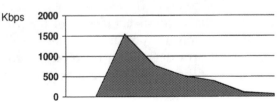

the only starting point that really works in a dynamic meeting is transmission at 384 Kbps. However, this is not a general service, and it is very pricey. The average organization will start with very high expectations, but as the pricing starts to emerge, expectations in picture quality start to drop. It is amazing what the financial picture will do for a person's tolerance level. There are two options for transmission. One is dedicated or private line services that include high-capacity T1 circuits between and among locations. This can be very expensive. Switched carrier services are also available from the long-distance companies. Switched services are often referred to as *bandwidth-on-demand* services because the user only pays for the usage when it is needed. These switched carrier services can also fall into two different categories: switched digital services (such as switched 56 and 64 Kbps) or ISDN (which is a digital switched service that provides two 64-Kbps channels). Although this service is not available everywhere, it might be needed, and differences exist in the implementation of ISDN around the country and the world. ISDN is offered at two interface rates. *Primary rate interface* is a 23-bearer service and a data link channel, all operating at 64 Kbps each. This is essentially the dial-up version of a T1, with a signaling system 7 channel (the D channel) for the call setup and teardown. *Basic rate interface* is a two-bearer service operating at 64 Kbps each, plus a D channel operating at 16 Kbps. One BRI can support a dial-up videoconference at up to 128 Kbps. If more is needed, an I mux can be used to pull together the increments of 64 Kbps, such as 256 or 384 Kbps, that are needed. This is a very effective tool in enabling use of the videoconferencing systems of today.

Many of the systems on the market today use the BRI as the interface for their video dial-up services. A low-end I mux costs approximately $10,000 to $20,000 and allows the user to pull together the capacity of the BRI or multiple BRIs.

Standards in Videoconferencing

Although videoconferencing has been around for so many years, only recently has it taken the industry by force. Why now? What is different? Actually, 30 years of pent-up demand might be one of the reasons that video is such a "hot button" in the industry. However, the major contributor to the most recent interest and hype is the result of newer and better standards for interoperability in video systems.

The ITU has made significant strides in the videoconferencing standards arena. The standards fall under the H.320 specification, which details the way video systems communicate with each other. Recommendation H.320, developed by ITU Study Group 15, still has some hurdles to overcome, especially in the LAN arena because many organizations are now considering the use of video on the LAN. However, the outlook is a positive one. The interoperability between differing systems marks a removal of many of the obstacles that kept a number of organizations away from the use of video. The P × 64 standard permits equipment from different manufacturers to speak the same language based on the bandwidth used. In these standards, the *P* stands for the multiples of 64 Kbps being used. The number can range from 1 to 30 (or 64 to 1920 Kbps) and conforms to international standards.

Prior to the adoption of the standards, choosing a videoconferencing product was a risky proposition. To guarantee compatibility, users were restricted to purchasing from a sole provider of every point in the video network. As a result, the market was filled with diverse product lines that generally were incompatible unless a mirror image was installed on the other end of the line.

The most notable of the standards is the H.261, which specifies how the codecs (coder/decoders) decipher digitally transmitted signals from dissimilar systems. This standard stipulates a minimum design to ensure that all codecs will interoperate at the lowest level. H.261 covers transmission from the basic ISDN rate of 64 or 56 Kbps up through the full T1 rate. In addition, H.261 defines two resolution standards:

■ Quarter common intermediate format (QCIF), a format used mainly in desktop and videophone applications.

■ Common intermediate format (CIF), a format that is optional under the H.261 standard used for room systems.

CIF requires a coder/decoder to transmit video at a rate of 30 frames per second, with 288 lines of lumina pixels per frame, 352 lumina pixels per line, 144 chroma lines per frame, and 176 chroma pixels per line. This produces a picture with lower resolution than broadcast TV. QCIF has even lower resolution because the pixel and line numbers are both halved, yielding 144 lines of lumina pixels per frame, 176 lumina pixels per line, 72 chroma lines per frame, and 88 chroma pixels per line.

Flexibility is built into this standard to allow each manufacturer to fine-tune the performance of its codec. Although most manufacturers have developed their own proprietary algorithms that translate the analog information into digital data, the standards ensure that divergent systems will be able to communicate on a generic level. In early 1992, the three largest manufacturers successfully achieved interoperability using the P × 64 standard, proving that once-noninteroperable systems could internetwork. This has led to a significant increase in the demand for the video-conferencing systems, because one of the major barriers was removed.

Other standards are also included in the use of videoconferencing systems. Some of the more common are shown in Table 35.1 with a representative summary of what each is designed to accomplish. These fall under the umbrella of the H.320 standards.

Most manufacturers now offer videoconferencing equipment that is H.320 compliant. Therefore, a room system and a desktop system that are

TABLE 35.1

Summary of the Standards in Videoconferencing

Standard	Description	Status
H.261	Uniform coding of a signal. Outlines how to compress digital video information, the syntax and semantics of a video bit stream (P × 64).	Adopted
H.221	Communications framing. Specifies what information is in a bit stream so each codec can keep track of video frames. Outlines whether a bit is audio, video, or control as well as how to weave bits together.	Adopted
H.230	Specifies how commands between codecs are exchanged during a session for control and diagnostics.	Adopted
H.242	Transmission protocols for call setup, transfer, and teardown.	Adopted
H.320	Technical requirements for low-bandwidth audiovisual systems. The overall specification for videoconferencing.	Adopted
H.331	Half duplexing multiplexing. Tells the codec how to string together data for transmission over half duplex channels such as satellite.	Adopted
G.711*	64-Kbps pulse code modulation for audio.	Adopted
G.722	48/56/64-Kbps variable adaptive differential pulse code modulation (VADPCM) audio.	Adopted
G.728	16-Kbps audio.	Adopted
H.233	Transmission confidentiality. Defines the method for identifying and negotiating encrypted data.	Adopted
H.243	Multipoint handshake. Defines communications between a multipoint control unit (MCU) and the codec. Multipoint videoconferencing.	Adopted
Not named	Still-frame graphics coding	
	Data transmission	

*The G.711, G.722, and G.728 standards specify the audio compression parameters that range in bandwidth speeds of 3.5–7 kHz.

both H.320 compliant should interoperate. However, while virtually all group systems are H.320 compliant, less than half of the desktop video-conferencing systems on the market support the standard. The buyer must beware. As the H.320 continues to roll out, it must be expanded to address video on the LAN and the interactive sharing of documents. Study Group 15 is currently working on standards for screen and document sharing, known as collaborative computing.

The major emphasis today in the industry is on the desktop and roll-about video systems. Sales of these systems are projected to skyrocket over the next few years. These projections are shown in the graph in Fig. 35.3. Clearly this is going to be a big movement in the industry if the sales skyrocket according to the projections. The use of video will become another tool in the conduct of our everyday business, regardless of where we reside in the chain of command.

Multipoint Control Units

The use of multipoint control units (MCUs) is beginning to increase as more organizations are bringing in videoconferencing systems and the declining cost is making it easier to use these systems in more sites. Once a videoconferencing program begins, the next logical step is to begin the multipoint conferences. What was once just a point-to-point service now needs to adapt to the changing business needs. Ad hoc meetings among sites are becoming more prevalent. The collaborative and dynamic workgroup formulation requires three or more sites to communicate with each other simultaneously. To perform this function, the user is faced with a decision to either bring the MCU functionality in house; use a bridging function from an outside resource (there are limited-service bureaus such as Sprint and AT&T that perform this function); or use a combination of the two.

Figure 35.3
The sales projections
for video systems.

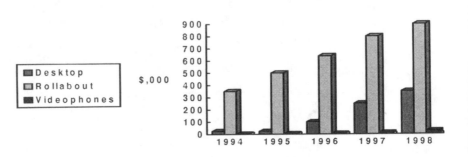

MCUs are devices that will allow the bridging and concatenation of the necessary sites into a single conversation. Costs for these devices will range from $10,000 to $100,000, depending on the system configuration. Typical sites use a device that will allow up to 8 to 20 ports. These range from $10,000 to $20,000 on average.

Many of the organizations that use a mix of these services will start off slowly with a four- to eight-port MCU. When larger meetings are required, or when a boardroom-type meeting is scheduled, these organizations use the services of the bridging companies. Unfortunately, the use of these systems results in mixed reaction and varying satisfaction levels. Normally the quality of the connection is not the issue. Rather, it is attitude and scheduling conflicts that result in user dissatisfaction. Forgotten appointments, uncooperative bridging operators, and even cost issues have created a demand for the MCU to be moved in house. More organizations are looking at using these systems internally. This decision actually is happening at a time when many organizations are considering outsourcing all communications to a service provider. This issue is yet to be resolved, but a significant amount of activity will spur a new business opportunity. When there was little need for such a service, there were few players. When the need increases as the number of units expands, then the number of service providers also increases. Many of the carriers have introduced various flavors of the multipoint conferencing systems. To do this, however, they have purchased an MCU from each of the players in the industry to maintain compatibility. Very few will discuss the fact that they try to accommodate a user's needs through a system that matches the end user system rather than adapting to the standards. We know that this will change over time, but the costs associated with using a public bridging system are kept artificially high because so many pieces of equipment from various manufacturers are required.

The MCU is typically provided as a dial-up service by the major players (Fig. 35.4). Here an end user can be provided with the connectivity needed for a multilocation teleconference.

Connecting the Parts

Once the decision to use a videoconferencing system is made, the next choice to be made is what system to use: the boardroom version, the rollout unit, or the desktop model. Depending on the needs of the organization, this could involve multiple units. One of the authors has had several clients who have purchased multiple units in the three cate-

Figure 35.4
Multiple locations can
be connected
through the carrier
MCU.

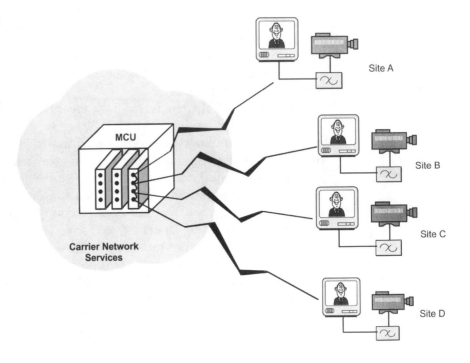

Figure 35.4
Multiple locations can be connected through the carrier MCU.

gories at each of their sites. This is a financial and a business decision. Whatever can be accomplished with a single system is great. However, if users have to wait in line for a system to become available, they might try to avoid using this technique and revert back to the old travel arrangements. A view of each of these systems is presented in the next few pages.

The Boardroom Version

This is the larger of the systems. It is not a mobile unit by any stretch of the imagination. To accommodate very high-end conferences and presentations, organizations still use the fixed room location. This is a unit that can start at $60,000 and run up into the $100,000+ range. Furthermore, this unit will require a specially treated facility. Although the special treatment is not a system requirement, people at the senior levels of an organization will want everything to be perfect. This will take some doing. First, in the design of the equipment, the following pieces will likely be used (Table 35.2). These are the basic pieces that will accommodate this type of a conferencing system. Added pieces can be brought into the equation, depending on the amount of functional systems needed to satisfy the business need.

	Description	Number/type
TABLE 35.2 Summary of Equipment in a Boardroom Environment	Monitors (presentation)	1 at 27" or 35", autoscanning, multiple inputs at composite video, RGB, S-video, high resolution at 640 × 480.
	Cameras	Two of high quality with the following features: 8–10× zoom; autofocus; motorized pan, tilt, and zoom; lens focal points at 6–48 mm; zoom ratio at 8 × 1; >400,000 total pixels; >380,000 effective usable pixels (494H × 768V); horizontal resolution of 450 lines; vertical resolution of 350 lines, fade capable.
	Codec	One each, using H.261 plus video coding specifications, H.320 compatible, providing picture-in-picture service, RS-449 and G.704 standards, echo cancellation, G.728 audio output, encryption capability, public or private line service.
	Controller	One with the following features: pan, tilt, and zoom on cameras; preset camera positions; separate control buttons for camera, video, and graphics input; variable volume controls.
	Document camera	One CCD single-chip design.
	Speeds supported	Variable at P × 64 from 128 Kbps to 1.544 Mbps.
	Monitor	Two at 35" for large rooms with dual RCA jack for audio and video delivery to the monitor.
	Microphones	Multiple omnidirectional with desk or table mounts.
	FAX machine	For reception of documents from remote locations, this should be CCITT group III standard.
	VCR	Used for recording or playback of videoconferencing sessions.
	I mux	One with variable inputs to draw up various channels at 64 Kbps, from 2 to 30 using international standards.

A graphic representation of this type of room is shown in Fig. 35.5. Large amounts of money will be spent to support this environmental conditioning and conferencing equipment. The systems are shown as standing alone, although there are several ways to connect the video service. This can be through the I mux and ISDN lines from the outside world at the BRI or PRI rates. If BRIs are used, the I mux can support pulling the three BRIs together into a single data stream at 384 Kbps. If the I mux is connected to the PRI service, it can allocate as many channels as necessary to support the full P × 64 channel capacity (Fig. 35.6). Another way to connect this service is to use an I mux behind a PBX (Fig. 35.7). This is a generic connection that allows various opportunities to integrate the

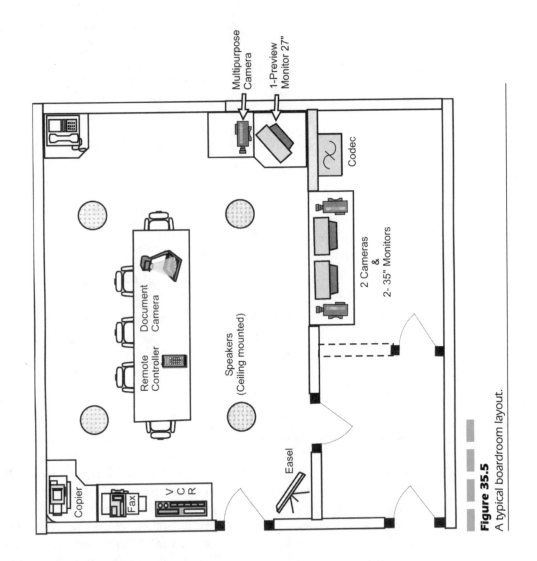

Figure 35.5

A typical boardroom layout.

Figure 35.6
The I mux is used to
pull channels
together for video.

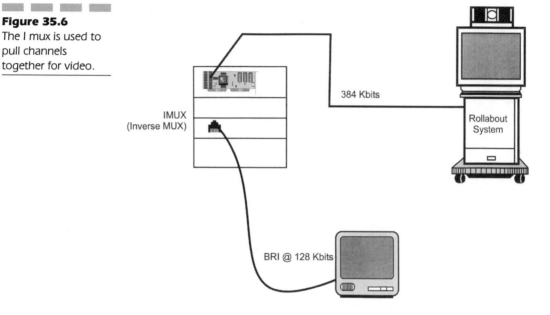

Figure 35.7
The I mux can sit
behind a PBX, or a
BRI card in the PBX
can deliver desktop
video.

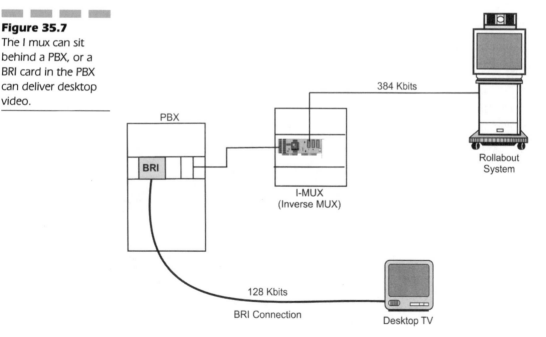

communications technologies into an enterprise strategy. No one design will work in all cases, but the general layout will at least provide a basic understanding of how it all works.

The Rollabout System

The rollabout is the system that can be used in between multiple conference rooms. This is particularly attractive when the organization does not have the resources or the ability to dedicate a specific room for videoconferencing systems. The users merely book the system and bring it to the most convenient conference room or individual office (Fig. 35.8). This system costs less than the boardroom configuration, so there might be many of these purchased in an organization and rolled about from room to room. One must be careful that trained personnel know how to move this equipment from room to room to avoid tipping and breakage. Although these units are designed to roll around, and they do, care must still be exercised in moving them. A sample checklist of equipment concerns for any videoconferencing system is provided in Table 35.3. This checklist should include pieces that would be considered for any system

Figure 35.8
A rollabout system can be moved from office to office without dedicated rooms set aside.

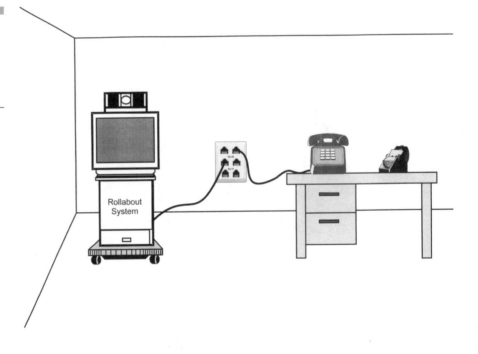

TABLE 35.3

Generic Checklist of
Options and
Pricing for Video
Systems

Description	Standard vs. Optional	Pricing
1. Codec		
2. Monitors • Small (12–18″) • Medium (20–25″) • Large (26–35″)		
3. Cameras • Autofocus • Document		
4. Multiplexer or inverse multiplexer • Speeds • Number ports • Channels that can be banded		
5. Microphones • Voice actuated • Lapel • Desktop • Built in • Upgraded available		
6. Controller • Handheld for remote pan, zoom, and tilt • Central desktop mounted		
7. Input for other devices • VCR record and playback • Slide presentation • White board • Electronic blackboard • PC input for playback		
8. Interfaces available • RJ45s (a number available) • TNC/BNC • RS449 • V.35 (dual) or RS366 • ISDN (PRI or BRI) (number available) • Dual DSU • I mux (ports, capacities, speeds)		

TABLE 35.3
(Continued)

Description	Standard vs. Optional	Pricing

9. Carts
  • Self-contained
  • Custom or standard		

10. Outputs
  • NTSC
  • PAL
  • SECAM		

11. Other associated pieces
  (vendor list of options)
  •
  •
  •		

12. Installation		
13. Wiring		
14. Testing		
15. Training (number of personnel)		
16. Total system price		

17. Maintenance
  • Warranty period
  • Parts and labor 5 days
    × 8 hours on site
  • Parts and labor 7 days
    × 8 hours on site
  • Extended periods on site
  • Factory return		

and their status: standard or optional. Along with this information, the checklist also should include the vendor's pricing information so a comparison can be made of the various systems.

The PC-Based System

The initial systems were mounted in the boardroom; then they evolved into rollabout systems. The most recent innovation is to use a system that is desktop mounted or PC-based. Cards that have the chip sets built into them can now be installed in various forms of PCs. The standards include boards for ISA, Enhanced Industry Standard Architecture (EISA), MAC, and S-Bus architectures. The Macintosh lines also have cards available for the NU-Bus. The convenience of placing the videoconferencing right at the desktop facilitates impromptu meetings and the ability to share documents on a computer. Some of these cards might take up a single slot in the PC, whereas others require two slots. The expansion slots on a bus are prime real estate; therefore the movement is to find a system that will occupy only a single slot. Regardless of the number of slots, these systems are designed to work at the 56/64- to 112/128-Kbps rate. They will not go any higher than that today. The small camera is a fixed-focus "lipstick" camera mounted atop the PC's monitor. Two dial-up lines are available, specifically with the BRI/ISDN service. The picture on the set can occupy the full screen or a small corner of the screen. The software in these systems will permit document and video sharing simultaneously. This allows for collaborative development and work group formation on the fly. If you can get used to seeing a very small and grainy image, then this system will fit the bill for most impromptu meetings. However, experience shows that the number of participants that can easily work together on these systems is one or two. Any more than two and the people involved in the meeting will be fighting to see what is going on. This being the case, the rollabout system will probably be a better solution. The cost for the PC-based systems has been moderate, with prices ranging from $2000 to $6000. For a

Figure 35.9
A PC-based system serves the impromptu meeting well. A single or dual card is used.

desktop system, this might well be the price that we'll see for the next couple of years. However, as the system becomes more mature and more applications develop to run the impromptu meeting, the prices should fall. The only significant problem with these systems is the control software needed to orchestrate the entire hardware platform simply. In numerous demos, the authors have seen these software systems lock up or fail to perform as advertised. This is because the systems are new and the bugs have not all been worked out yet. You can expect these problems to become nonissues in the near future. The next issues are just getting the bandwidth to the desk, the available connectivity through the building, and the applications necessary to share with other workers.

The cost of this system will be a factor in the acceptance rate. With a base entry into videoconferencing at this price range, more and more small branch offices, small businesses, and home-based businesses will be looking to this system. The PC-based system is shown in Fig. 35.9. The number for the system is above and beyond the cost of the regular PC hardware and monitors that are required.

CHAPTER **36**

Finances for Telecommunications

Introduction

Up until about 1980, most communications, both data and voice, were regarded by companies as "costs of doing business." There were many reasons for this:

- For the first half of the century, only voice communications existed in electronic form.
- Voice communications capability was supplied by the telephone company.
- The telephone company really did not offer a great variety of options to its customers.
- What it did offer worked well, but management could make few choices.
- The telecommunications manager's main responsibility was to place and coordinate orders with the telephone company and to track the bills to ensure that they were correct.
- Because there was little choice in what to order, "managing telecommunications" was almost an oxymoron.

Also, people who managed communications tended not to have a management point of view. They were not oriented toward providing management enough information to make business decisions. After all, presenting a telecommunications budget as a fait accompli was much easier than justifying each project individually. This orientation is still seen in many telephone bills rendered by Postal Telephone and Telegraph authorities (PTTs); the level of detail seen in a normal bill in the United States is often, outside the United States, either a separate (priced) service or simply unavailable.

In the late 1960s, choices began to proliferate. Data communications began to grow as a percentage of the total communications budget. Still, the mentality that regarded communications as an overhead function with little potential to be managed—certainly not a possible profit center—remained in most companies.

However, as more and more companies in the United States felt the cold winds of international competition, and as each market niche started to fill up with more and more competitors, every area of expenditure in organizations came under scrutiny. Telecommunications (used here to include both voice and data) constitutes a significant expenditure of funds in many companies—an expenditure that tends to increase steadily. In many organizations, this is the second or third largest cost on the balance sheet. As an overhead function, it is a prime candidate for cost control and optimization, especially with the increasing number of ways to satisfy any given communications requirement.

Cost is always an issue no matter what financial position your organization is in. Management expects that the best pricing and value for its dollars are being achieved. Yet, management still sees telecommunications as an overhead function, a necessary evil. This image must be dispelled. The use of telecommunications as a strategic resource will:

- Increase productivity
- Reduce costs
- Improve the bottom line

All you have to do is convince your management of these benefits. Moreover, companies are increasingly becoming aware that telecommunications can be used as a competitive weapon in the marketplace—a method of improving attractiveness to customers, and in some cases a way of directly generating additional revenue. Of course, turning a new set of technologies into a "competitive weapon" can be much more difficult than it sounds. Nonetheless, it has been done often enough that the very idea no longer sounds as implausible as it once did.

In fact, notwithstanding the brouhaha about the "information highway," modem communications technologies should indeed be regarded as means to improve revenues and productivity. But expenditures on these technologies could be wasted if they do not rest on a sound financial basis. As in any other major capital or expense decision, the financial effects of such choices should be considered before, during, and after the choices are made. In some cases, the dollars involved might drive a decision. It is important to understand when and how to cost justify a decision, or perform a comparative financial analysis among several alternatives, in order to prove to upper management that the investment and/or monthly expenditures will appropriately benefit the organization.

Sometimes the benefits are so obvious or overwhelming, or the costs of not making the investment are so horrendous, that no analysis need be done. This chapter, in contrast, focuses on the financial aspects that can influence the decision of whether or not to make the investment:

- How do you approach this tender subject without alienating management?
- How do you persuade management that the communications expenditures will generate greater returns than many other forms of expense?

The answer is different in every organization. First and foremost, you have to begin to speak in management's terms. All too often, telecommunications and telephony people want to describe technologies, applications, and state-of-the-art concepts to their management. Unfortunately, jargon and the use of technical terms turns management off. Consequently, when vying for the same corporate dollars as all the other departments, telecommunications tends to lose ground. If you learn the language of business and apply business principles to your purchasing decisions, you will stand a better chance of getting the funds needed to support the organization.

In order to analyze or justify an expenditure, two key aspects of the transaction and technology must be identified:

- Benefits
- Costs

The normal hurdles such decisions face in an organization must also be understood in order to perform the appropriate cost/benefit analysis.

Benefits

When considering any expenditure, the first step is to identify the expected benefits. Such benefits are normally categorized into *hard* and *soft* benefits.

Hard Benefits

Hard benefits are those that can easily have objective and predictable dollar figures assigned to them. Examples of hard benefits might include some of the following ideas:

- Reduce dial-out voice communications costs by $3000 per month
- Decrease PBX maintenance costs by $500 per month
- Avoid a one-time PBX acquisition cost of $150,000 in the next fiscal year
- Allow direct charge-back of appropriate calls to clients (versus absorbing them as overhead) in the amount of $3500 per month
- Avoid the requirement of acquiring a new T1 (costing $4000 per month)
- Reduce travel costs (airline, hotel, meals) by $8000 per month

■ Control overnight package shipment costs by avoiding $2500 in expected monthly bills

■ Reduce the cost of dial-in data communications by $5000 per month

Every one of these examples shares a few key elements. It states a particular type of expenditure, a specific dollar amount to be saved, and a (usually recurring) time period (typically per month). Most communications and maintenance bills are stated in terms of months, so your savings will usually also be stated that way initially. Ultimately, as you work with the figures, they will be translated into yearly figures, as management normally thinks in terms of yearly budgets. This translation also benefits you, because yearly savings are normally much more impressive, at least in the case of savings that recur monthly.

Soft Benefits

Soft benefits are results that you expect to be good for the company, but that are more difficult to quantify, and sometimes even to identify. The object of this exercise is to make the soft benefits as hard as possible. This will make more sense later on.

Identifying soft benefits is typically a two-pass activity. On the first pass, one simply makes a general identification of areas that are expected to benefit from the planned expenditure. Examples might include:

■ Improve productivity of call handlers in the customer service area

■ Reduce the time required to add items into inventory

■ Increase the productivity of our consultants by giving them quicker access to reference information

■ Reduce the cycle time required to develop a new product (or troubleshoot a problem in an existing one)

■ Diminish the number of errors experienced in information retrieved by our customers when they dial in to our network

■ Optimize the routing of our delivery vehicles

■ Minimize the time to respond to trouble calls

Any organization would jump at the chance to achieve any of these goals, right? Well, perhaps, but a few questions will probably arise before management signs on the dotted line. Those questions are, in summary:

- How much money will this save or generate?
- How much do I have to pay now to experience this saving later?
- When is later?

In the following, we will address how much must be paid. But it is your job to calculate probable savings to be derived from the expenditure you propose. How do you do it?

After all, it is quite possible that the real reason you are proposing an expenditure is to achieve a soft rather than hard saving. That is, you really do not know exactly how much will be saved—but you believe (or your internal client believes) that the change will be a beneficial one. But you must quantify this belief—"wishing doesn't make it so"—or convince management to loosen the purse strings.

An additional pass is required to refine the benefit statements. Soft benefits often refer, in one way or another, to improvements in the efficiency of human beings. In the list of benefits, the first, second, third, and sixth items express the hope that such a benefit will occur. Quantifying such benefits is fairly straightforward, although a number of assumptions must be made. Always state your assumptions. You will most likely use a spreadsheet for your analysis. If someone questions—or corrects—an assumption, you can easily change it and redo the analysis. If your assumptions are not clearly stated, management cannot evaluate how realistic your predictions are.

To quantify the benefits expected from an improvement in productivity, you need to determine or assume the following:

- Exactly what activity is to be shortened? Examples include length of time on a phone call, the average time to travel from one customer location to the next, or the length of time to check one item into inventory.
- How many times does that activity occur per year?
- How much time is to be cut off each occurrence of the activity? For example, if an average new order call takes 5 minutes, the average length of such a call might be reduced by 30 seconds.
- What is the cost (average salary, hourly rate, etc.) of a person that typically performs this activity? This should be a "fully loaded" figure; that is, it should include medical benefits, retirement fund, vacation, and so on, if applicable. The place to start to get this information is the supervisor; however, a trip to human resources might be required.

Once you have established this, it is time to hit the spreadsheet. Figure 36.1 illustrates a sample calculation based on a proposal to add a computer-

Figure 36.1

A sample calculation on a proposed CTI solution.

1	Expected time savings per call (in seconds)	30
2	Annual telephone orders	250,000
3	Expected seconds saved (row 1 × row 2)	7,500,000
4	Expected hours saved (row 3/3600)	2083
5	Customer rep salary (fully loaded)	$40,000
6	Customer rep cost per hour (row 5/2000)	$20
7	Potential savings (row 4 × row 6)	$41,667

integrated telephony function to a PBX. It is assumed that if many of the customers' records can be brought up on screen automatically for customer representatives, then an average of 30 seconds will be saved during each order by speeding the recording of address, telephone, and other customer information that is usually static.

It would be nice to be able to point to row 7 and say, "See, this will save us over $40,000! Let's do it!" But this is a raw number. Much remains to be considered before the analysis is ready for management's review. Does management expect to hire additional customer reps? If so, then the number of calls is probably expected to increase. Should you do a 5-year analysis, incorporating a volume growth factor? Would the use of this improvement forestall the additional hires? Would the organization be able to make do with the same number of agents if a 30-second reduction per call is achieved? (This is possible: if you are handling tens of thousands of calls per day and you save 30 seconds on each, it can equate to significant hours of savings.) Refer to Chap. 8 on traffic engineering, which had a similar example of the savings of time. This includes the ability to save more than just lines, trunks, and other ancillary equipment. It can also account for substantial people savings.

How sensitive is this analysis to changes in the assumptions? You should be able to get accurate figures on an average salary. But how precise is your 30-second estimate? Be prepared with either studies done in your industry that bear out your calculations or perhaps the results of a study done with a stopwatch showing how much time is used by taking down the information to be avoided. Do not forget that not all calls fit your paradigm; new customers will not be in your database, and many will call from locations other than their own telephone (thus not allowing a matchup of their telephone number with the customer records in your database).

Costs

The benefit figures, whether soft or hard, are "gross"; that is, they identify changes in or avoidance of expenditures, but they do not identify what must be done or spent to realize those savings.

Cost is a relative term. We think of cost as an outlay of dollars. Yet, true costs include both the long-term investments and support, coupled with the returns on investment. This is important to selling concepts to management. Systems in raw dollars cost approximately $700 to $1000 per station when dealing with PBXs. The long-term dollars over a 5-year period can be three times that. However, if the telephone set is used to produce revenue of $1 million, what is the cost?

Cost/Benefit Calculations

As should be obvious from the name, a cost/benefit analysis balances the two key aspects of an opportunity (also known as an investment or, more commonly, an expenditure). As discussed, there are hard costs and soft benefits. If your goal is to justify an expenditure, you should do your best to turn soft benefits into hard benefits. But once you have reduced everything to dollars, your work has just begun.

Life Cycles

Contrary to popular opinion, a life cycle is not (only) a piece of equipment found at your local health club. Rather, it is a key concept in the analysis of acquisitions of both communications and data processing equipment. It is not applicable to items paid for on a monthly basis, but only to equipment with an expected useful lifetime of several years.

What is an item's life cycle? The answer is not so obvious. It is not the length of time the equipment could successfully operate (if properly maintained). It is not even necessarily the length of time the equipment could provide service more economically than other alternatives. Rather, it is the length of time that an organization can realistically expect to use the item in its planned role before discarding or replacing it. But, you might say, coming up with such a time frame is only slightly more accurate than fortune-telling with a crystal ball! Indeed. The process of determining the life cycle of a piece or category of equipment is one that involves forecasting technology trends, the expected health of the business, possibly interest rates, and other factors. In short, it is largely a matter of opinion.

Nonetheless, you must try to nail down a figure—for financial, operational, and career reasons.

FINANCIAL. Assume that someone is developing an annual budget for desktop personal computer purchases. For simplicity, we'll assume that a group of 10 users needs to be equipped. A reasonable approximation of the cost of one desktop unit, with software, is about $5000. If you further assume that each machine will serve productively for an average of 4 years, what should the annual budget be? Ten people times $5000 is $50,000 in the first year. After that, we'll spend $50,000 every 4 years, or $12,500 per year on average.

But wait! It has just been determined that all of the computers will have to be replaced an average of every 3 years rather than every 4. The startup cost remains the same, but now $50,000 will be spent every three years—$16,667 per year.

PBXs typically serve for a long time, often as long as a decade or more. But technology is changing more rapidly than in the past. Assuming that a PBX bought now will still be in service 10 years from now could be a big mistake.

PAYBACK PERIODS. One of the simpler methods of evaluating investments is payback analysis. This isn't tripping someone that hit you in the schoolyard 30 years ago. Rather, it is an approach to taking into account the time value of money without actually doing the requisite calculations. However, it does give an initial rough feel for an investment's viability as compared to alternatives.

Notice that we are using the term *investment*, rather than *expenditure*, in this section. To have a payback period at all, an investment must bring in, over time, revenue greater than that spent to get the revenue. Otherwise, there is a negative payback, suggesting that this would be a poor investment indeed!

Payback analysis, also sometimes described as *break-even analysis*, is straightforward. A revenue stream or periodic savings figure associated with an investment is estimated (for example, $10,000 per year in reduced calling costs achieved by installing a virtual voice network). The costs associated with that stream of revenue are totaled (for example, $20,000 in one-time costs, plus a marginal increase in the cost of access lines to the network of $5000 per year). A little work with a spreadsheet, and voilà! The break-even point is 3 years from the date of investment. Table 36.1

TABLE 36.1

Summary of
Calculations in a
Payback Analysis

Year	Savings	Cumulative Savings	Cost	Cumulative Costs
0	$10,000	$10,000	$25,000	$25,000
1	$10,000	$20,000	$5,000	$30,000
2	$10,000	$30,000	$5,000	$35,000
3	$10,000	$40,000	$5,000	$40,000
4	$10,000	$50,000	$5,000	$45,000
5	$10,000	$60,000	$5,000	$50,000

shows the numbers used to perform the calculations on the spreadsheet.
From this table we then graphed the information shown in Fig. 36.2. Note
that the graph shows the cumulative figures.

In today's business climate, a 3-year break-even point will most likely
not generate intense management excitement.

Notice that with break-even or payback analysis, the time value of
money is typically disregarded. When comparing alternative projects
with similar time horizons, this is not unreasonable. But when management begins to focus more carefully on your numbers, you most likely
will need to create a more rigorous analysis. One way to do that is with
return on investment calculations.

Figure 36.2
The savings versus
costs graphed to
show B/E.

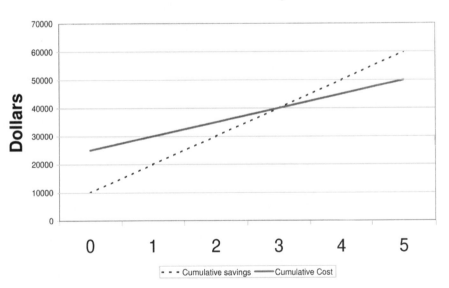

Break-Even Analysis

RETURN ON INVESTMENT CALCULATIONS. Return on investment (ROI) is where the rubber meets the road in investment analysis. One way to approach ROI is to use an internal rate of return (IRR) analysis. To quote the help available with Microsoft's Excel spreadsheet, the IRR

> Returns the internal rate of return for a series of cash flows represented by the numbers in values. These cash flows do not have to be even, as they would be for an annuity. The internal rate of return is the interest rate received for an investment consisting of payments (negative values) and income (positive values) that occur at regular periods.

In English, the IRR is comparable to the interest rate your company might get on the money it is considering investing if it put the funds in a bank instead of into your tender clutches. The IRR does take into account the time value of money, so if you include all of the factors it does provide a good decision tool when selecting among alternative investments.

Any modern spreadsheet should be able to do this for you. Using the previous example, the net initial cost in the first year is $15,000. Each year thereafter there is a net savings of $5000. The IRR for this stream of outgo and income, during the first 5 years, would be about 13 percent. Better than passbook savings—but, of course, considerably more risky.

When the analysis of the life cycle concept is mixed in, the results become much more interesting. That is because ROI is intimately tied to the total length of the expected revenue stream. It is a statement of the average interest rate compared to the expected cash flow over the duration of the analysis. Table 36.2 shows how much better the investment looks if a 6-year period is assumed: a 24 percent internal rate of return!

TABLE 36.2

Internal Rate of
Return Analysis

Year	In/Out	IRR
0	−$15,000	
1	$ 5,000	−66.7%
2	$ 5,000	−23.2%
3	$ 5,000	0.0%
4	$ 5,000	12.6%
5	$ 5,000	19.9%
6	$ 5,000	24.3%

On the other hand, if the technology is supplanted in 2 years, we lose money at an annual rate of 23 percent.

Pricing Considerations

If you had to plan for a new system today, the world would be at your beck and call. Vendors and carriers alike are in a slump. The sales of equipment, peripherals, and services are all slowing down. Increased pressure is being applied on the sales forces to increase revenue. This implies that a buyer's market exists—which is true if you know what your needs are and how to negotiate with the vendors.

Unfortunately, many telecommunications and telephony personnel might be good technically but lack the skills to deal with other issues. Thus you have to understand the pricing schemes, and be prepared to wait them out, if necessary, in order to get your best deal. For equipment purchases, many of the vendors use a one-third rule. One-third of the quoted price is for equipment, one-third is for installation, and one-third is for the profit margin. The final third is where you have room to negotiate.

For example, a digital PBX system would cost approximately $750 to $1000 per line. If you use a 1000-line system, you can assume that you will pay between $750,000 to $1 million. Request a budgetary price from your vendors and see what they quote. You might be pleasantly surprised, or you might get sticker shock when you read the budgetary price. First, the price might be quoted on a rudimentary scheme of 1000 lines, 50 percent single-line sets and 50 percent digital multiline sets. So, let's use $750,000.

However, the vendor might also assume that you'll want features and peripheral equipment above and beyond a rudimentary system. Therefore you might see duplicate CPUs, battery backup, enhanced features, station message detail recording (SMDR), enhanced networking, least-cost routing, among other features. The cost might come in at $1.5 million. How did it get so expensive? What happened? If the vendors start in a position of strength (at $1500 per line), they have plenty of room to negotiate. They will offer you a 50 percent reduction on initial station equipment and features right up front. The price dropped only 15 percent to approximately $1,275,000, and you believe you did your job. However, all that was reduced was the station equipment—from its original inflated price! Remember, the system should have started at $1 million. How can the vendors come back and reduce the price even further? They haven't even touched the price or the profit! That's how!

The Decision-Making Process

Once you have determined that a true need exists for the purchase of any communications peripheral or system, the next step is to evaluate everything on both a financial long-term basis and a functional basis. Do not let yourself get trapped into the "love that technology" routine. Too many systems are bought on the basis of emotional appeal rather than company need. This can be the kiss of death.

Some suggestions to follow are:

■ *Buy only when necessary, but allow plenty of time to start the process.* This gives you the edge on the buying cycle. If you wait too long, then your leverage of time is gone, and you'll have to buy before you get the right deal.

■ *Remember the financial picture.* Recruit a team of others from the finance, purchasing, facilities, and legal departments to assist in the decision-making process. Bring them up to speed immediately. Other sets of eyes help to maintain focus and objectivity.

■ *Don't get too close to the salesperson.* This is business, not friendship. If things are not going their way, the vendors (salesperson and manager) will go right around you to your management and keep you out of the loop. Control your project and your vendors.

■ *Accept no verbal agreements.* Everything must be in writing! A short pencil is worth a long memory. You'd be surprised how many agreements are forgotten after the contract is signed.

■ *Keep everything above board.* Don't accept gifts or favors. This can ruin your career. Even if there is nothing going on, the perception is what will get you.

Justifying Costs

There is no one rule for the justification process. You must learn what the criteria are within your organization. Many organizations use the return on investment (ROI) method or the internal rate of return (IRR) method. Still others use a discounted cash flow (DCF) analysis to authorize expenditures. Many smaller companies might well use an emotional decision and justification process. It is really up to the management in the individual organization. So how do you know how to go about this?

■ *Get away from the technical role and become more of a business manager.* You need to speak the language of the organization. This can be in financial terms (a universal language) or in production or cost containment/cost reduction terms. Whatever the language of the management, you have to get on that platform.

■ *Ask for help.* Find a department manager or peer who has been successful in getting approval for new projects. Go with a winning style.

■ *Don't reinvent the wheel.* When you are trying to justify a system, ask others to review your information. They might well give you pointers that will help. If a similar-size project has been recently approved for any department, find out what that project manager did. Use a format that is already approved, something that management is obviously comfortable with.

■ *If you are not comfortable with the justification process, get a champion or mentor who will run with the project for you.* This could be a senior manager or financial manager. This might be a risk because the mentor or champion might try to take full control after the approval is gained. The projects are usually highly visible, so a lot of credit could be up for grabs.

Maintenance Issues

Maintenance issues should be discussed with the vendor in advance of a purchase decision. You might be replacing a system that has become obsolete or is continually breaking. The last thing you want to do is jump into a decision that might engender the same problems. You certainly do not want to buy a system that is about to be discontinued or that will not be supported before the end of its financial and useful life cycle.

Determine:

■ What the long-term maintenance costs of the system will be.

■ How much the maintenance will cost in years 2 through 5 and 6 through 10. Can you prepay maintenance at a reduced rate?

■ How long the warranty period is. It should be at least 1 year.

■ When does the warranty start: after cutover or after acceptance? There can be significant time differences between the two dates. Is this a moving window?

■ What are the normal maintenance periods (9 to 5, Monday through Friday) and what will the costs be if maintenance is required during other periods?

- What will the costs be to add one more set? Ten more? One hundred more? and so on.

- When or where are the big bumps in price?

- What will the remedies be for nonperformance?

- How will the vendor handle major and minor outages?

- Does the vendor have spares on site? If not, where are they and how available are they?

- How will the vendor keep your system current? At what cost?

- Can you train your own people? If so, for how much? If not, why not?

- What about initial user training, refresher training, new employee training? How often and how much?

- Can you trade in excess equipment? Can the trade be applied toward upgrades and new releases?

Reduced Rates Versus Improved Service

What are your objectives for the purchase of a new system: reduced rates, or improved service? Many users really never address the issue of their buying motives. If the objective is to reduce rates, then the issues might vary from those surrounding improvements in service. Of course, the best choice is both, but you might have to decide between these two issues.

If reduced rates are the primary objective, you could give up some of the features and niceties of the system. Keep the long-term benefits of a system in mind. Often, reducing rates is a short-term goal that might well prove to be more expensive in the long run. If, for example, you choose the least possible expense today by selecting a PBX with all single-line sets, you might experience dissatisfaction from the users. Ultimately, they will begin demanding multiline sets with added features and capabilities. Now you'll have a group of single-line sets in a closet, paid for but not used. Further, the add-on costs for the multiline sets might now be more expensive after the installation.

Improved service will point you in a direction of serving the end user. These are your customers. Although the initial costs might be higher for the services they need or ask for, the long-term capital outlay should be less. The acceptance of a system customized to meet their needs will also go a long way in the overall payback of the system. Increases in user satisfaction, productivity, and revenue are soft-dollar benefits for telecommunications systems. However, they are real values to the organization. Don't lose sight of the total picture.

Sample Calculations

Before you can present your case to management or approve the purchase of the system/service, you have to understand the total cost both in initial outlay and ongoing cost. To do this, you have to gather all the facts. This will start with:

- Initial costs, purchases, installation, training, preparation, and so on
- Add-on costs over the period of years, based on system and peripherals
- Maintenance costs
- Human resources needed to operate the system
- The obsolescence factor: how long the system will be adequate

Initial Costs

Assume that you are planning to buy a new PBX that will serve multiple users (1500 to 2000). You've done your homework and estimate that the system can be purchased for $1000 per line. Your initial assessment will be approximately $1.5 to $2 million to purchase the system, installed with a 1-year warranty. This system will be needed in 1 year because the existing system is too small and continually breaks down.

A further analysis should reveal additional costs to include that might not be as obvious, but must be included in the calculations to be fair:

Construction costs for PBX room (30 × 50)	$50,000
Raised flooring	$10,000
UPS system	$20,000
Fire detection/suppression system	$15,000
HVAC add-on tonnage	$20,000
Security system (card key or other)	$15,000
New conduits for new wiring	$25,000
Electrical work	$7,000
Removal of old system	$5,000
Handling of old batteries (hazardous)	$5,000
Training costs (1500 users @ 1 hr. @ $20)	$30,000
Project management	$40,000
Total	$242,000

The figures are obviously based on some industry averages. They reflect approximately 15 percent additional costs over and above the initial purchase price. These numbers can and should be included in the overall cost analysis for the system, because they represent real costs.

Some added long-term costs that should be considered over the life cycle are:

Space (rental) $4.00/sq. ft./ann.	$6,000/ann.
Insurance @ 0.50 per $1000±	$8,500/ann.
Spares (parts/sets, etc.)	$10,000/ann.
Test equipment	$20,000/one time
Heat/light/utility costs $1.00/sq. ft	$1,500/ann.
Inspection costs for HVAC/fire suppression	$500/ann.
Contingencies (5–10% total ann. exp.)	$2,650/ann.
Total other	$29,150/ann.
	$20,000/one time

Be sure to work with various departments to understand the operating costs. These might include:

- Purchasing
- Finance
- Facilities
- Insurance
- Safety/Security

Add-On Costs

Assuming that the base price for the system is as outlined above, the add-on prices might include such items as:

Automatic call distributor (100 line)	$50,000 (1)
Voice mail/automatic attendant	$80,000 (1)
Modem pooling (simple)	$10,000 (2)
LAN interface (56-Kbits/s bridge)	$15,000 (2)
T1 direct interface (5 cards @ $2800)	$14,000 (1)
Station message detail recording	$20,000 (1)
Four-wire E&M T/L cards for V-mail (4 @ $500)	$2,000 (1)
Paging access (excludes speakers)	$12,000 (2)
Power failure transfer (48 lines @ $6000)	$6,000 (2)
Total add-ons	$209,000*

*Spread over 2 years (166,000 year 1; 43,000 year 2).

TABLE 36.3

Differences of Years
Used for a Single
Decision

Person Responsible for Considering Numbers	Number of Years Expected
Telecommunications or information management	3-year payback or sooner
IRS	5-year depreciation at accelerated rate
Finance Department	7-year straight-line depreciation
Senior Management	10-year life (don't want to do this very often)

By now, you might be getting the feeling that the cost of a telecommunications system is not just something that you pick off the shelf. Nor are the prices simple to calculate, because so many other components are involved. You have to look at all of the added costs; for example, construction, maintenance, human resources, insurance, and taxes. This is the only way to truly justify the expense to management. They have to perform these calculations if you do not. Admittedly, they will use the financial department to perform these calculations; but if you have no input in that process, they might use different assumptions and disprove the benefit that you anticipated. Consequently, the onus is on the individual responsible for the technology to prove or disprove. How can management use different assumptions? Look at a simple point that is summarized in Table 36.3. This is how we compare the differences in perceptions.

CHAPTER 37

Facsimile

Facsimile Transmission

Just what is facsimile? *Facsimile* (fax, for short) is the transmission of documents, preprinted text, and graphics in paper form, converted into a graphic form and sent across a telephone line. The process is straightforward in that a prepared, printed document is scanned into the facsimile machine via a process that reads every character as a series of dots. The scanning and transmission is across the page, left to right, on a dot-by-dot (or pixel-by-pixel) basis, as shown in Fig. 37.1. Facsimile has been around

Figure 37.1
The scanner reads
and transmits dots
from left to right, top
to bottom at 192
columns and 200
lines of resolution.

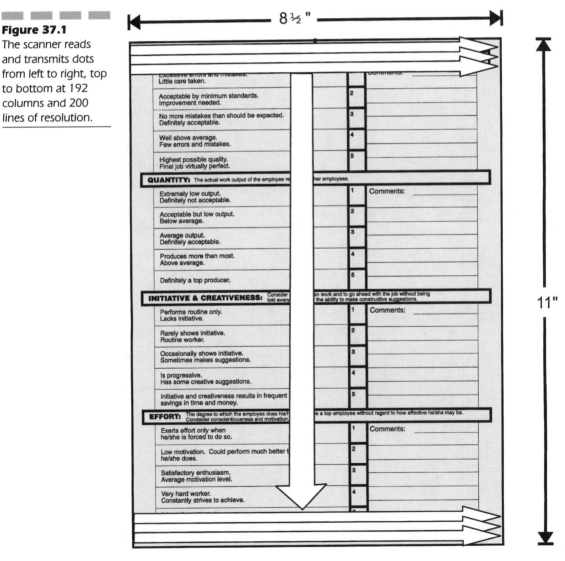

for over 120 years. Yet the technology was very slow in developing because there were some complexities in the total process.

The use of facsimile has always been regarded in the past as a tedious job. Newer machines have made the process far less painful, but it still takes a lot of patience. Part of the reason is that as facsimile evolved, the machine used the dot-by-dot scanning process to transmit a series of ever changing dots across the older analog dial-up telephone network. Knowing that the original analog network was built to carry analog voice communications, we have learned that this network was unreliable for the dot-to-dot process and left a lot to be desired. There were always problems in the transmission process. Overcoming the problems in a voice mode was easily accomplished, but for the transmission of data the problems were magnified. As the network degraded over time, the transmission process also degraded. This left a lot of information on the network that was not properly received.

Because the transmission is on a dot-by-dot basis rather than a character-by-character basis, transmission was both slow and prone to errors because of line noise. A true ASCII or EBCDIC code set, as covered in the data communications chapter in this book, was not used. Actually, the preferred alphabet in the scanning process was a modified Huffman and Read code set. This again was based on dots rather than being a true alphabet. The result was a fuzzy code set prone to line noise in the transmission process.

Types of Fax

Facsimile machines are broken down into types or groups based on the standards set by the CCITT. These groups define the scanning process, the line speed, and the modulation technique used. There have been four different groups established over the years. These have all met with different results and acceptance from end users. However, the older groups (CCITT Groups I and II) are nearly extinct. The predominant service (Group III has been widely accepted) is robust and proliferating everywhere because of price drops in these services. Newer techniques (Group IV) are slow in evolving and replacing some of the older systems.

Group I Fax

The first evolution of the fax machine was adopted as a standard under the CCITT. This was called the CCITT Group I machine. The fax machine was set up to use a rotating drum system in which a scanner reading a dot would transmit an electrical signal to a distant machine that used an electrical stylus. For those of us old enough to remember the TV series *The FBI*, the rotating drum fax machine was used extensively on that show. A picture of the 10 most wanted criminals was transmitted across the dial-up telephone network from the FBI office to any other law enforcement office. If it was necessary to transmit this picture to a multitude of locations, the process was serial and handled one site at a time. The techniques of storing and forwarding the page for multiple transmissions did not exist. This again was a very expensive and time-consuming process. Back then, the cost per minute for a long-distance call was still around $0.65.

The stylus can be described as an electric pen. As the receiver received the data, it generated an electrical pulse and sent it to the stylus. When the stylus received the electrical current from the receiver, it would move back and forth on the blank page of pretreated paper. This back-and-forth motion would scratch off an aluminum coating on the paper in the receiving machine. The aluminum oxide, when scratched off, would produce a black/gray mark on the page in the form of a circular dot. The dots were spaced closely together to form the pixels of a letter, number, or graphic image. This created the alphanumeric codes that we use in our normal business correspondence. The dots were not perfectly rounded, so the result of this transmission process could be very fuzzy (Fig. 37.2).

A serious by-product of the process was that it also produced an odor that made people want to scatter. The smell caused by the chemical reaction was so pungent that the machine might be relegated to a special room. Then only those brave souls who needed to send a page of information across the network would venture into this hidden area.

Both the receiver and the transmitter portions of these machines were slow. Using amplitude modulation, it was very common to spend 6 minutes sending or receiving a normal typed page of text. The Group I machine used an acoustic coupler to provide the input and output on the line (Fig. 37.3). The acoustic coupler is a descendant of the Hush-a-Phone

Figure 37.2
The actual output from the fax produced a fuzzy appearance due to the scribing process.

Figure 37.3

An acoustic coupler was used to send a fax in early days.

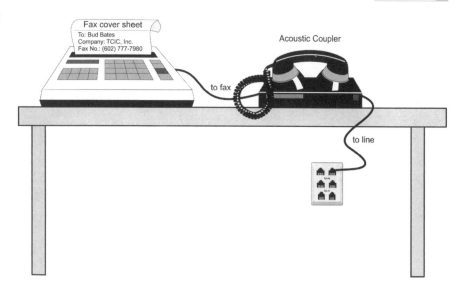

of the mid-1940s. Using the acoustic coupler, the operator would have to go through a protracted procedure to establish a connection. These systems did not answer automatically, and the process was labor intensive. The process of setting up a call using the older Group I facsimile machines was as follows:

1. First, the operator would take the first sheet of the document and wrap the sheet around the rotating drum. This would clamp the document in place.

2. Once the first sheet was on the drum, the operator would then pick up the handset of the telephone next to the facsimile machine and dial the number for the receiving end. The transmitting operator would place the machine in the transmit mode, which meant that the machine was set to scan the document (use the stylus not to scribe but to read).

3. When the receiving end got the ringing tone from the voice call, an operator would pick up the phone and answer. At this point, the transmitting operator would negotiate the connection. The transmitting operator would state that a fax needed to be sent to the receiver. The receiver would ask how many pages there would be, so that he or she could be sure that enough of the pretreated sheets were available for this reception. It would not be uncommon at this point for the receiving end to beg off from receiving the document.

A number of reasons existed for this reaction:

- The operator might not have the time to receive the document.
- Sufficient paper might not be available.
- The time of day could become a problem when sending to different time zones.

Personally, I can remember trying to send a fax to a distant location, only to be told by the receiving operator that he or she was going to lunch and I would have to call back later. When I tried to get back later, the telephone on the other end was never answered. Why? Because the operator knew that I wanted to send a multipage document that would take a long time, and he or she did not want to receive the information. Or else, the normal operator was not on hand, and some designated alternate who did not feel comfortable with the setup or did not like to operate the machine was at the other end.

4. Assuming that all was going according to plan, the receiving end agreed to accept the fax and made sure that enough paper was on hand. The operator at the receiving end had to wrap a piece of the pretreated paper around the drum of the receiver. Once the page was locked in place, the operator placed the machine in the receive mode (this meant that the receiver would use the stylus to scribe on the page). The telephone was still in the talk mode between both parties and they could converse while the process was being set up.

5. On mutual agreement, the two operators would then agree to go into the data mode. The telephone set was used to click things into place— the operators would place the telephone handset into the acoustic coupler and start the machines at both ends simultaneously. This process started the drums on both ends revolving and initiated a signaling arrangement between the two ends.

6. As the two machines came up to speed, the stylus on one end would read the data on the sheet and convert it to a series of electrical pulses. A 1 represented a change in the amplitude of the carrier tone generated between the two machines, and a 0 meant no change. The 0, after all, indicated that there was no dot (pixel), while the 1 represented the presence of a dot. This was the amplitude modulation of the carrier tone (covered earlier in this book).

7. Once the page was completely transmitted, the operators on both ends would reach over and grab the telephone handset from the cou-

pler. The drums would stop revolving and come to a stop so that the pages could be disengaged from the drum and removed from the machine. The receiver would verify that the information was received properly (a process that was not guaranteed). This paper was difficult at best to read because the aluminum oxide coating was a silver color and the dots scratched on the page were gray. However, the agreement would be handled and the next step would begin. For each page that had to be received/transmitted, steps 1 through 6 were repeated.

At any step along the way, the process could fail and the call would have to be reestablished. A page could be illegible, or the information could be trashed from glitches on the line, or any number of problems could render a transmitted page unusable. Talk about frustration! With all the manual labor and the delays inherent in page-by-page setup, a facsimile transmission of four pages could take as long as 45 minutes.

Pictures and graphic representations could take much longer because of the amount of dots needed. Because the machine was a single page wrapped around a spinning drum, the process was manually intensive—not a welcome chore for anyone using the machine. You can imagine the frustrations of having to send a multipage fax using this machine. At 6 minutes per page, plus the extra overhead for handling and reading time; telephone transmission costs of $0.65+ per minute; the need for an operator to stand by and wrap a page at a time around the drum—the cost was both expensive and undesirable. The operators were not pleased when a densely populated graphic needed to be transmitted. With all the disadvantages of this system and the reluctance of users even to try the technique, coupled with the associated costs in transmission and labor, it is a wonder that facsimile ever really survived the technology wars.

Another problem that kept creeping up in this process was incompatibilities between different manufacturers' equipment. Although the CCITT developed standards on how these machines would talk to each other, there was no guarantee that two machines manufactured by two different manufacturers would communicate with each other.

Group II Fax

To improve on the process and the acceptance of facsimile, The CCITT developed a new standard, and the manufacturers created a newer technology using a much faster process. The older analog transmissions were limited to the amplitude modulation process, but modem communications had evolved to a faster transport, using frequency modulation. Remember from the data communications section that the FM procedure doubled the rate of data transfer on the analog telephone network. In the transmission technique for Group II facsimile, frequency modulation was used, reducing the time per page from 6 minutes to 4—under special conditions, 2 minutes. You might wonder why the transmission time didn't drop to a standard 3 minutes (half that of the AM method). Obviously, with the bits-per-second rate, this would be true. But we are looking at transmitting dots rather than characters made up of bits. It takes a lot more dots to paint a character than it does bits in an ASCII code set. Furthermore, the transmission of a page using the AM technique took 6 minutes, but when you added all of the overhead, the effective rate was much longer. In facsimile transmission, a page of data is transmitted, and then the machines go through a whole new handshake to make sure that everything is still functioning properly. This handshake also makes sure that the page was received, the end of the page was recognized, and the machine is ready to receive the next page. This all takes time in the communications process; therefore the average time per page is better than half the AM rate. At 4 minutes per page, the transmission time was reduced by 33 percent. This depended on the machine-to-machine communications. Many of the manufacturers of facsimile machines had their own proprietary codes, which they used between their own machines.

Around that time period, the vendors were introducing a new feature: white space skip. If the machine scanned a line and saw no pixels, it would speed up until it recognized the next sequence of dots. This meant that all the spacing on a typical business document could be bypassed. Look at how much white space is on this page as you read it. If you have a line of spaces, then there is no sense in slowly scanning nothing.

The white space skip was one of the proprietary methods used by the vendors. Other techniques were also used, but they were dependent on compression, skips, interpage handshakes, and other features. All of these were fine if the machines on both ends were the same. If not, then the system would revert back to the standard mode issued under the CCITT guidelines.

Unfortunately, no laws required that a vendor adopt the standards, so if one end was CCITT compatible and the other was not, the communications might not go through. This still left confusion in the industry and a bitter taste in many people's mouths. They were reluctant to use faxes because of these constant complications. We recall some organizations that had various fax machines in a room; which machine was used for a given transmittal depended on the type of machine at the other end.

This ensured that at least the two ends could communicate. However, this is an expensive way to conduct business.

The printing process also changed on a Group II fax. A roll of paper using the aluminum oxide coating and the electrical scribing from a stylus was introduced. The roll of paper allowed the receiving end to get multiple pages of facsimile without requiring an operator to manually change paper on a sheet-by-sheet basis. This roll of paper also created the ability to transmit a portion of a page of information. The rolls came in two different sizes: 164 and 328 feet. This meant that approximately 179 to 358 8×11 pages could be received without any form of paper handling. This never worked out as anticipated because cover sheets are used, transmissions may still fail and trash a sheet of paper, jamming occurs ... the list of problems is endless. Each problem detracts from the actual versus expected results.

A derivative of this printing process using the aluminum-oxide-treated paper was a wet chemical technique that introduced an ammonium process. This paper reeked from the chemicals and was prone to turning black after exposure to light for extended periods of time. It was not uncommon to receive a fax and pull the page out—only to find it dripping wet from the ammonium liquid. This was a problem, because users were still uncomfortable with the smell, and now they had to contend with the use of wet chemicals. Loading the liquid chemicals and the drying agents was unpleasant. This paper was just as sensitive to light as the aluminum oxide paper was. In order to file a fax, the user had to reproduce the fax via a copier. Where a roll of paper was used, multiple receptions could be handled in sequence on a continuous sheet (Fig. 37.4). The receiving end had to cut or tear the pages off at the respective page breaks. This meant that confusion could exist when the pages were separated from each other. If a user cut the pages off at the wrong location, the result could also be lost or misplaced pages.

Another improvement in the facsimile was the connection. Many of the Group I machines were acoustically coupled to the line; the Group II machines introduced a jack to connect to the line. The machine might (or might not) have a phone built in or attached with two jacks (one for the line, one for the telset). If a call came in, the process might even use an automatic receive mode that did not require the user to place the fax into the data mode. In fact, this was one of the more significant developments on the fax machines. The ability to send or receive a fax without a human on both ends did much to improve the acceptability of this technology. Because the older machines required a user to be there, fax transmissions were often limited to business day calls only. With the automatic answer,

Figure 37.4
The roll would pro-
duce a continuous
stream of paper.
Users had to cut the
pages apart.

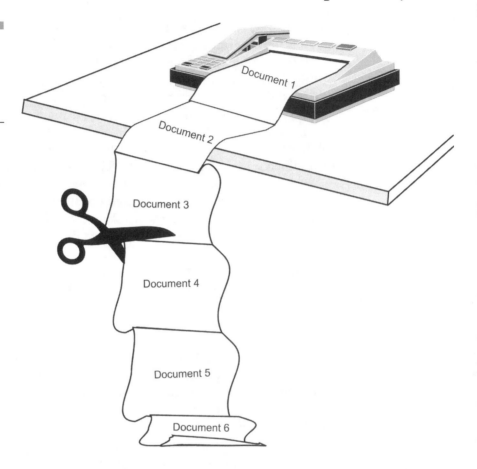

on-line capability, fax transmissions could be done during off hours to take advantage of reduced long-distance rates. Because a human did not have to be there, this could be accomplished easily. (However, many a case has been recorded where line problems occurred during the transmission and the fax machine cut off the call. This meant that the fax would require a retransmission the next day.) Furthermore, the use of the roll-fed paper eliminated the need for a human to stick around after hours just to receive a fax.

These strides helped but did not solve all the problems. Although this situation was better, it was still not universally accepted as a viable trans-mission and copy reception medium. Only those transmissions that had to be there by the next day (or in some cases instantly) were sent via fax. More users still preferred sending documents out via an alternate means such as mail, Telex, TWX, or data. However, back in the early days of fax transmission, the Telex, TWX, and data transmission services were not very comfortable to the end user, and many of the documents were already in typed form, so that the time it took to retype a document was all wasted. Graphics and pictures were relegated to the mail as the only alternative. This delayed the flow of an organization's information well beyond tolerable levels. Something still had to be done!

Group III Fax

The next development in facsimile was the CCITT Group III standard. Although this contained significant improvements over Group II, the machine was expensive in the beginning. This technique still was not readily accepted as a means of reliable transmission. Only the larger organizations could change with the standards and the newer-model equipment. Machine costs were in the range of $6000+. However, the speed of transmission was increased to 9.6 Kbps using a QAM technique, transmitting a page of information in 1 minute or less. Clearly, this was moving along with the evolution of modem communications techniques. The CCITT standards (CCITT T.30) set the stage for the ability of a fax machine to act more like a modem. The machines could be used to transmit finer and denser information. Additionally, if it had to transmit to older equipment, the equipment could sense that the receiver was a Group II machine and could reduce its speed to match the receiver's speed. This meant that the system was backward compatible.

Printing techniques also changed, although the Huffman and Read codes were still used. The graphic nature of the page was still maintained as a dot pattern. However, the introduction of a thermal printing process on the paper improved the users' acceptance of the methodology. No longer did users have to worry about the stench of the stylus-based systems with aluminum oxide or ammonium oxide. The thermal paper was easier to handle and use. To increase the throughput on the machines, a white space skip technique was used as part of the standard. No longer was this strictly a proprietary development and implementation. If no dots of information were represented in a block of space, the machine immediately skipped ahead instead of transmitting nulls (0s). This helped, especially because the transmission costs were still relatively high. Transmission costs were dropping, but the average cost per page was still $0.35 to $0.40. The paper was roll-fed in early installations of Group III fax, using a 328-foot roll of thermal paper.

Group III was slow in starting. The cost of the machines was the main impediment to acceptance. Although the past ills had been fixed, the organizations considering fax transmissions were stymied by the $6000+ price tags. Ultimately, in the early 1980s, fax began to catch on and acceptance was starting to pick up. Another factor was the drop in prices for the machines because of mass production and enhancements in scanners and chip sets. An automatic cutter was added, making things easier on the user at the receiving end, who got a pre-cut page rather than having to tear off each message. The transmitting end could use a document feeder to stack 10 to 30 pages into the machine for transmitting without someone standing by the machine all the time. The manufacturers began to incorporate a microprocessor to allow the machine to use a delay-dial feature for later transmission, retry a busy number, and confirm delivery. A newer modular jack and dial pad on the machine assisted the sender/receiver by directly attaching to the line. Built-in modems were used. Further developments were being introduced by the makers of fax, but user acceptance was still slow. Users had been through the trials and tribulations of the evolution and did not like to deal with this technique. Users had written off the fax machine, so they would not normally consider it when deciding how to send out information.

The thermal paper, although a great improvement over the oxide and wet processes, was still an area of contention with the user. Once the thermal paper was exposed to the heat treatment in the fax machine, the user still had to deal with curled paper, the result of running roll-fed and heated paper. Also, the paper would fade, so any reception had to be copied on a photocopier for long-term storage or retention. The fax copy could not be integrated into a document, so the information had to be retyped to include it into a letter, report, or other manuscript.

Printing Options

More recent options have surfaced in the ability to print with the facsimile equipment. The introduction of the microprocessor and the addition of hard disk drives in the machine allow for the received document to be stored on the hard drive for later printing. Furthermore, laser quality on plain bond paper has enhanced the printing of text and graphics in facsimile uses. The laser-quality output on bond paper allows for more acceptable quality and later storage of the paper. The user does not have to go through an added step; the paper is acceptable for inclusion in a document. For storage, the bond paper can be filed away without the added handling and cost of copying. This enhancement has clearly added to the acceptance of fax in the business environment. Not that this technique is inexpensive, but you get what you pay for. Using a plain-paper laser-quality printed output can be justified in terms of the reduced copier and personnel costs associated with the fax.

Another enhancement is the storage of outbound faxes in the hard drive for later transmission. Hard drives 20 or 40 Mbytes and larger can be used on these machines. Of course, the hard drives, laser printers, and other enhancements drove the costs of these facsimile machines up into the $14,000+ range, which makes this type of machine acceptable for larger organizations but not an attractive option for smaller companies. Who can afford it?

However, this has also helped to drive the cost of the lower-end machines down through the floor. The Group III thermal roll-fed fax machine that cost in the $2000 to $3000 range just a few years ago can now be bought for $400 to $600. Now the smaller organization can afford Group III fax with thermal printing and still communicate with the larger organizations at subminute speeds. Transmission improvements, data compression, and error-correction techniques allow us to send a page of information in as little as 20 seconds. What an improvement over the original machines, which took 6 minutes!

These improvements have led to the introduction of departmental fax machines and have created an increase in the numbers of such machines. The larger machines in the high-end price range are fine for the central-ized facsimile departments with dedicated operators. However, the casual user in a specific department might not need all the power of the high-end machines. Therefore the $400- to $600 machine provides a suitable replacement or a complementary capability in any organization. With these enhancements and price differences, fax machines began to become more common and acceptance began to increase. The smaller organiza-tions were using more and more of these services. Home-based businesses now are equipped with a telephone, a PC with modem, and a fax machine as standard equipment.

Fax Boards

Probably the single most important step in the facsimile world was the invention of the fax board. This Group III device was developed as an add-in board for a personal computer (PC). As a PC-based device, the fax card imports a true ASCII code set and transmits it to a Group II or III fax machine anywhere in the world. The same features available in the higher-end machines can be used with the PC-based fax card. Some of these features are as follows.

Store and Forward

Stores the data in the PC's file structure for later retrieval and transmission. This can be set up automatically without a human being there, as long as the PC and fax modem are powered on.

Reception Right into the PC

What used to be brought into a fax machine, then rekeyed or scanned into a PC, can now be sent directly into the PC for access later. This requires some special software because the fax received will be saved as a fax file (bit mapped graphics).

Later Printing

The document can be sent into the PC and stored. Paper has not been created initially. If a printed copy is needed later, the user can output the file to a printer.

Laser or Dot Matrix Printing

The printer attached to the PC is available for the output of a fax file. Therefore whatever printing capability is available can be used. This eliminates the need for a separate fax and PC printer.

Files Kept as Graphic or ASCII

Depending on the software used, a file can be saved as a graphic file or imported or exported as an ASCII file.

The original fax boards (1985) started out as special cards costing around $2995; but this has quickly dropped to today's price of $100 to $200, making them both economical and compatible with other fax capabilities. Newer twists in the use of the fax board provide for transmitting from a fax to a PC equipped with a fax board and importing text and graphics (such as pictures and logos) right into the PC. This compensates for the lack of a scanner at the desktop. If a user needs a logo or picture scanned and does not have this resource available, the document can be inexpensively sent via a true fax machine (this has a scanner) into the PC fax card. The scanned image can then be imported into any other file, such as word processing or desktop publishing, depending on the software available. The faxed image might be grainier; therefore, software that allows for graphics touch-up might be required. The major drawback to all this on a PC is the size of the graphics files, which can be substantial. In addition, the use of the fax board on a PC means that if the PC should fail, you also lose your facsimile capability. However, this doesn't appear to concern many who have readily endorsed this added technology.

Common Uses of Fax

Now that fax has entered the workplace as an acceptable technology, users are looking for more. Fax has become the easiest form of electronic mail (e-mail) on the market. If you can transmit images or text from a single device, why would you need a separate e-mail system?

However, another technology has also been gaining ground in the workplace at an equal or faster pace. The use of local area and wide area networks (LANs and WANs) has driven a newer approach to facsimile. Because users are getting either terminals or PCs on every desktop, the organization cannot afford to place a fax on every desk with the existing LAN-attached PC. And because the LAN is designed around the ability to share resources on the single cable system, a new demand went out to the industry: provide a fax server on the LAN! Manufacturers were ecstatic; newer needs and demands for the use of fax on the LAN introduced a new market for the existing technology.

However, the fax servers are still in their infancy, and users will either have to suffer through various growing pains or wait and see what evolves. Foremost in this development is the ability for a user to send a fax from an attached device on a LAN, but how do you get messages in from the

LAN? If you can dial into the LAN (which brings its own security risks), how do you get to the subaddress of the actual PC you want to reach? The present way entails manual or human intervention. The faxes are received on the server, and an administrator has to review and forward the fax to the intended recipient. This also makes the fax a common document, which might violate some confidentiality of the information. We'll see more in this arena over the next few years.

Future Machines

Obviously, the development and acceptance of facsimile will continue to grow. Newer capabilities will be introduced. Some of the future systems that are in various stages of development and introduction in the industry follow.

Group IV fax really isn't new. However, it has lagged in the implementation stages. Group IV operates digitally at 56/64 Kbps, transmitting a page of information in approximately 3 to 4 seconds. The cost of the machines will continue to be high, as pioneers use this technology. The backward compatibility with a Group III or II machine, which does not yet exist, has to be considered. The cost and availability of dial-up 56/64-Kbps services has to evolve further. Federal Express tried to implement this technology years ago. As a pioneer in the industry, however, this pilot, known as Zap Mail, was ahead of its time and failed.

Cellular fax also is coming. Today cellular facsimile modems operating at 9.6 Kbps are available, but not as well accepted. Until a reliable package of cellular modems becomes available at a reasonable price, this technology will basically be used by the folks on the leading (or bleeding) edge of technological implementation.

Laptop manufacturers are now developing facsimile modems to be used in conjunction with their machines. The use will depend on the costs. Because cellular communication is not inexpensive, the costs associated with the equipment will possibly hamper the widespread use.

CHAPTER 38

Cabling Systems

Introduction

Before the development of local area networks, organizations that needed within-building data communications cabling systems had a limited number of choices. Terminals used RS232C, unshielded twisted pair (UTP), or in some cases (especially block mode IBM terminals) coaxial cable. High-speed connections in an IBM environment were handled with special in-room cables (maximum distance about 200 feet) called Bus and Tag cables. (Voice always used unshielded twisted pair wires—no choice at all.)

A key characteristic of these cabling systems is that, in general, a need for a particular cable was met by pulling just enough cable to satisfy that need. The cost of installing such a cable run (to connect one device) typically ranged from about $200 to over $1000, depending on limitations of and available routes through the existing physical plant. Because moves and changes occur frequently in a business environment, the cabling cost alone for a normal data communications environment would be considerable—even after all of the initial installation was completed. This frequently occurred as soon as the initial installation was completed, because move and change activity was frozen or put off until the cables were installed. Frustration existed because users might have gone through one or two moves since the beginning of the installation, but the wires were installed at the original locations. There was no easy way to functionally change in midstream.

When local area networks were first introduced on a widespread basis, a new concept—that of a data utility—became popular. Fundamental to this approach was the idea that, given the opportunity, it would be smart to prewire every plausible work location in a building. This prewiring would allow moves and changes to consist of just movement of the user hardware, and perhaps some software changes, although some LANs allowed moves with no software changes.

The cabling of the day was either coaxial (in ARCnet's case) or heavy coaxial (in Ethernet's case); nonetheless, the idea of placing a jack in the wall for as-needed connections ("just like the electric company!" was the claim) rapidly took hold. Still, this facility was just for data communications—and only LAN data at that.

About this time, another idea began to gain in popularity: that voice and data communications should be in some way integrated. Just how they would become—and (in many cases) avoid becoming—integrated is another story. However, one area in which apparent integration could be accomplished relatively easily was in wall jacks. One could design jacks with built-in outlets for multiple types of communications. Jacks have been designed with some or all of the following:

- Voice
- Terminal
- LAN (sometimes more than one)
- Video
- Generic analog (in cases where the voice connection is for a digital PBX, analog connections are sometimes also provided to allow analog devices—for example, modems—to also gain external access)

Figure 38.1 shows a typical multipurpose outlet layout. The multiple outlets were designed to be utilitarian so that any need could be met with a single interface at the wall outlet. This was a good and noble goal, but many cases are documented proving that early implementations were not that successful. Obviously, these jacks can become quite crowded. Not so obviously, the jacks only provide the appearance of integration. After all, what is behind the jacks? Typically, each of the cabling types goes to wherever it needs to be. The office walls are neater, however.

Figure 38.1
The generic wall plate emerged as the utility function for voice, data, LAN, and video. These could have four or six outlets.

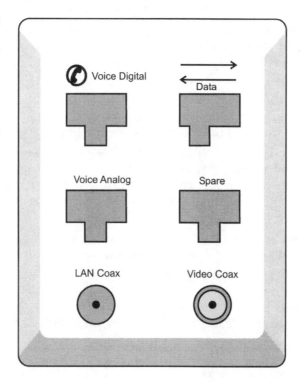

A more recent concept is that of structured cabling systems. Typically, these systems are offered by some of the major systems vendors as well as by other organizations. Some of the more successful vendors of such systems include:

- AT&T
- IBM
- DEC
- Others (such as AMP and KRONE)

The idea of a structured cabling system is to predefine the type (or types) of cable to be used for each specific purpose. Then wires of all identified types are preinstalled, coordinating the installation in order to save money and provide a consistent and neat physical plant for both current and foreseeable communications (both voice and data) needs. This approach is particularly beneficial when a new building is being constructed. But buildings with older cabling that did not measure up to modern requirements have been retrofitted on occasion—at a substantial cost.

Each vendor of a structured cabling system offers essentially all media. Company management still must decide what requirements might materialize in the future. However, once a company commits to a particular system, the requirements directly drive cable selection, configuration, and installation approaches. Fewer decisions must be made. Vendor installers can work more efficiently because they are familiar with the components of the particular system.

Of course, any given vendor ensures that at least its own products are well supported on its system. For example, when it was introduced, IBM's system had no provision for Ethernet cabling; but DEC's did offer such an option. In an environment where a company typically buys most of its data processing equipment from one vendor, selection of that vendor's cabling system might make sense. However, if a company wishes to sole-source the entire environment—and thereby have one company to point to when failures occur—then using a vendor's structured cabling system would be an excellent idea. But essentially the same system can be built by independent cabling suppliers at significantly lower cost. Also, if a multivendor environment exists in the building, the vendor-specific plans limit the opportunities to tie everything together neatly.

There are vendor-independent standards for building cabling, known as EIA/TIA,* 568 A/B, and EIA/TIA 569. The intent of these stan-

---

*EIA is the Electronics Industries Association; TIA is the Telecommunications Industry Association. These bodies set the standards for the structured wiring systems.

dards is to provide, in effect, a highest common denominator. If the standards are adhered to, they should be able to support anything—presuming that the EIA/TIA has thought of it. If you are planning to comply with the EIA/TIA guidelines, you must make your own choices as to media.

All structured cabling standards address two basic types of wiring: vertical and horizontal. Vertical wiring is cabling that connects wiring closets, each of which serves multiple users. Usually, the wiring closets to be connected are each on a different floor. With luck, they will be in the same position on each floor. Fig. 38.2 shows a sample of the wiring closets stacked above each other in a high-rise office building. Vertical wiring usually runs up and down, hence the name. Vertical wiring is always at least the same bandwidth as horizontal wiring. More often than not, higher-speed media are used for vertical wiring because connecting groups of users (with all of their traffic appearing on the vertical wiring) can demand aggregate bandwidths far exceeding those needed to satisfy individual user requirements.

Horizontal wiring runs from the wiring closet above the ceiling to the wall jack in each of the office locations. Figure 38.3 is a representation of the horizontal wiring system running to the wall jack. This includes such things as the cable, the drop inside or behind the wall, and the wall plate (jack). In many cases, the horizontal wiring will include the base cord that runs from the wall outlet and plugs into the PC or telephone set at the desk.

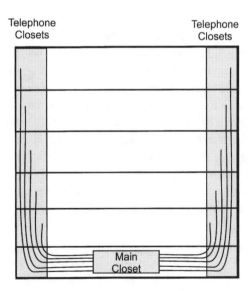

Figure 38.2
The telephone closets are typically stacked on top of each other in high-rise buildings. The vertical wiring feeds to each floor.

Telephone Closets

Telephone Closets

Main Closet

Figure 38.3
The components of
the horizontal
distribution.

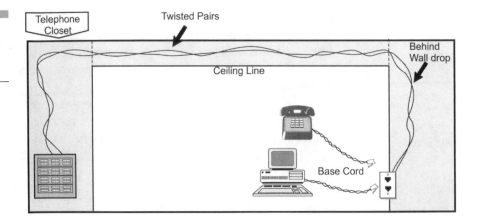

Interbuilding wiring also is often a concern to those wiring a campus or multibuilding environment. Like vertical wiring, it must carry high bandwidths (only more so). It also necessitates special requirements, such as resistance to gophers or rodents and other environmental hazards, and presents special problems, such as appropriate grounding when connecting between two different buildings' electrical systems.

The rest of this chapter presents more likely choices currently available for these types of wiring.

Twisted Pair

Twisted pair refers to telephone wiring. This type of paired wiring usually comes as two solid copper wires, each covered with a plastic insulator and twisted together. It is not the "silver satin" cable often sold in hardware stores for use with telephones. Silver satin is a flat-wire system with all the wires are laid side by side rather than twisted around each other in pairs. Twisted-pair wiring is often described as STP or UTP (shielded or unshielded twisted pair). UTP (Fig. 38.4) has been primarily used for voice communications. It is twisted to reduce induction effects in adjacent wires, otherwise known as cross talk. The more twists per foot, the better the reduction in the cross talk. The twisted-pair wiring is usually sold in various flavors, as shown in Table 38.1. In these pairings, the number of pairs is

Figure 38.4
The unshielded
twisted pair wiring.

TABLE 38.1

Typical Number of
Pairs Used for
Installation

Description	Number of Pairs	Used
Analog telephone wiring	2 pairs	1 pair
Digital telephone wiring	1–4 pairs	1–4 pairs
LAN wiring Ethernet 10BASE-T	4 pairs	2 pairs
LAN wiring token ring 4 or 16 Mbps	4 pairs	2 pairs
Horizontal telephone wires	25 pairs	1–4 for individual sets

always important because there are basic and immediate needs as well as future needs. The best approach anyone can take when installing wiring to the desk or in vertical distribution is always to add more.

The wire pairs are typically color coded so that the pairing can be kept together visually, rather than by guesswork. This pairing color code is designed so that a striped wire and a solid-color wire, or a striped wire and a striped wire of the reverse color combination, create a pair; older wiring systems, such as the old telephone wiring in households, use a solid-color wire convention. Table 38.2 shows the more common combinations. Where the convention uses a striped color code, this is represented by the two colors on the wire insulation (i.e., white/blue means a white wire with a blue stripe, blue/white means a blue wire with a white stripe).

TABLE 38.2

Color Coding for
Normal Pairings

Pair number	Color-coded Pairing
One	White wire/blue stripe and blue wire, or white wire/blue stripe and blue wire/white stripe
Two	White wire/orange stripe and orange wire, or white/orange and orange/white
Three	White/green and green or white/green and green/white
Four	White/brown and brown or white/brown and brown/white

Older analog wiring found still in some buildings uses a four-wire pairing with these color coding conventions

One	Red/green Red is the transmit and green is the receive wire
Two	Black/yellow Black is the battery and yellow is the ground for some sets; if a two-line set is used, then black is transmit 2 and yellow is receive 2

In the reverse wiring, the color code uses the two colors in the reverse; however, the normal pair contains a striped and a solid wire (i.e., *white/blue* is a white wire with blue stripes and *blue* is a solid blue wire).

If installing a telephone set (for example) that requires one pair, it would be imprudent to run the wires to the set with only one pair enclosed. Otherwise, the first thing that will happen if you choose to change the device hanging off these wires is that you will be confronted with the need to rewire with additional pairs. Telephone systems typically can be used that require from one to four pairs at the desk. Thus the minimum number of pairs to the desk is four. We have always adopted a standard that a minimum of two each of the four-pair wiring be installed. After all, the cost of adding a few hundred feet of wire at the time of initial installation is only a few cents per foot. If, however, rewiring is necessary, the labor to go back and pull more wires will cost a significant amount (Table 38.3).

UTP is a "balanced" wiring system. This means that both wires carry similar types of electrical signals and have similar characteristics. No special grounding is required. With the appropriate adapter, UTP can be used to support nonremote terminal traffic that would otherwise require RS232 cabling. Shielded twisted-pair (STP) wiring (Fig. 38.5) was endorsed by IBM at about the same time the company released the 4-Mbps Token Ring. STP was better for data communications than was early UTP—but only if installed correctly. The shield—a foil sheath covering the copper pairs—must be properly grounded. If not, the shield can create a ground loop (Fig. 38.6). This situation introduces the possibility that a wiring system will become an antenna, which is worse than if no shielding were

TABLE 38.3

Comparing the Difference Between Pulling a Single Cable Twice versus Two Cables Once Shows the Real Dollar Value of Preplanning. The Main Cost Difference Is the Labor.

Item	Cost of Pulling One Cable to Each Desk (each)	Cost of Pulling Two Cables to Each Desk (each × 2)
Materials (4-pair cable, block, RJ45, cross-connect wire, base cord)	$9.35	$18.70
Labor to pull wires, punch down, and test	$300	$320
Total cost per run	$309.35	$338.70
@ 50 pulls	$15,467.50	$16,935
Cost for additional 50 installations	$15,467.50	$0
Total cost	$30,935	$16,935
Difference		($14,000)

Figure 38.5
Shielded twisted pairs include foil wrap that is connected to a good earth ground.

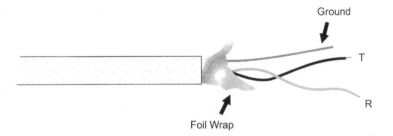

Ground

T

R

Foil Wrap

Figure 38.6
If grounded on both ends at different potentials, the shield creates a ground loop. This is worse than no ground at all.

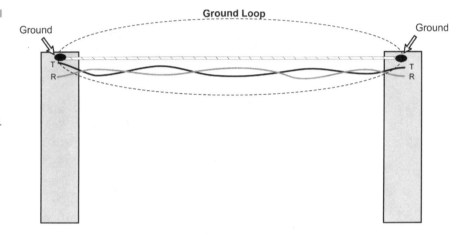

Ground Loop

Ground

Ground

T
R

T
R

used. STP is also more costly than UTP, both because of the extra foil shielding and because the jacks and plugs cost significantly more than those used with UTP. STP was generally used only for token ring installations, before appropriate UTP grades and components became available. The shield, if properly installed, will provide better cable performance, such as reduced cross talk, protection from grounds and faults, greater signal distances, and protection from EMI and RFI interference.[*]

AT&T and others experienced many queries from customers as to transmission characteristics of various UTP offerings. The company's specification sheets provided the desired information, but in a format that only an electrical engineering graduate could understand. To make buying products easier, vendors developed a set of simplified data-grade wire specifications based on the underlying detailed electrical characteristics of the wire. Initially supported by the Underwriters Laboratories (UL), this specification contained three levels numbered 1 through 3. As

---

[*]EMI refers to electromechanical interference; RFI refers to radio frequency interference. These are the conditions that are introduced into the copper on the twisted-pair wiring.

more wiring products became available and needed, the company extended the grading scheme. Now there are five defined levels or data grades for UTP wiring, each capable of handling its own requirements as well as those addressed by all lower levels:

■ *Level 1,* sometimes described as "barbed wire," will handle the requirements of voice communications, called plain old telephone service (POTS), and limited data rates that meet EIA 232 specific data rates (such as 19.2 Kbps), and that's it.

■ *Level 2* is certified to handle the requirements of a 4-Mbps token ring.

■ *Level 3* addresses Ethernet's requirements or 10 Mbps on twisted pairs. The specifications from various vendors suggest that this specification will support higher rates, such as 16 and 20 Mbps, but it is not an EIA standard.

■ *Level 4* is for 16-Mbps token ring speeds and the 20-Mbps ARCNet service. IBM is planning to introduce a 25-Mbps token-passing ring concept at duplex arrangements that will be called ATM. The specification is to use level 3 or 4 wiring to support these speeds.

■ *Level 5* is certified to function properly at up to 100 Mbps, protocol unspecified, as long as no cable run exceeds 100 meters (328-foot combined runs from the closet to the workstation). The industry is now stating that level 5 will run up to 155-Mbps speeds that are supported by ATM standards.

The term *level* applies only to the wire itself. The entire system is described as *category n,* where *n* matches the wiring level. The entire system includes jacks, plugs, and patch panels, in addition to the wiring itself. These components also need to be certified as meeting "category whatever" requirements. Because category 5 components cost somewhat more to install, people sometimes install category 5 wiring with category 3 components. Because replacing the components is much easier than rewiring a building, they expect to replace the components later if required.

Some LAN technologies require 100 Mbps. These include two varieties of Fast Ethernet, as well as a standard for FDDI that runs on category 5 UTP. No matter what LAN topology is used, UTP is almost always configured with a radial wiring plan, rarely as a physical bus. Radial (or star—radiating out from a central point) fits well into building cabling troughs, a fact that in large part accounts for the popularity of UTP data wiring.

UTP offers the lowest medium cost per foot installed. Of available technologies, it is the most limited in distance because of the nature of the medium. For every medium, there is an inverse relationship between distance and potential speed. The specific ratio varies with the media. For any given speed, UTP has the shortest distance limitation of the media discussed in this chapter.

However, there are many suppliers of UTP, and installers are familiar with it. Because of its popularity, vendors focus on getting their products to function in this type of physical plant. Unless there is a possibility that a company will need speeds greater than 100 to 155 Mbps to the desktop—or that cable runs will have to exceed 100 meters—a category 5 installation is likely to handle all requirements for at least the next 5 years. If you are planning a category 5 implementation, however, be sure to get installers experienced with category 5 and who have more than just basic telephone wiring experience. Installing category 5 is not particularly difficult—but it is different from the other wiring levels. Specific considerations must be adhered to when installing this type of wiring or all efforts will be for naught. Improperly installed, the category 5 cabling can be negated and functionally provide only speeds based on the weakest link in the system, such as category 1. You can spend a lot of money on the wiring system, only to be left with an infrastructure that doesn't meet expectations. Or, retrofits that are expensive might be required to bring the wiring up to standard. This can be a devastating financial burden to the organization.

Twisted-pair wire is typically cabled using standardized jacks and plugs. Originally standardized by AT&T, the designations of these jacks are easily understood codes (such as RJ11, RJ45, and others). We assume that someone found them easy to understand, anyway. In any case, the key differentiating factors among these jacks and plugs are:

- How many pairs of wire can terminate on the jack (two, three, or four)?
- How many pairs of wire do terminate on the jack?
- How are the pairs assigned?

Every electrical signal path on UTP requires one pair of wires. If two signal paths are required (for example, digital sending and receiving), then two pairs are required. But those pairs most likely will not be assigned in sequence (for example, left to right or right to left); that would be too easy. As an example, voice wiring on an RJ11 jack (which can hold two pairs of wire and thus handle two analog voice lines) typically uses the center pair of pins or wires on the jack for the first line, and the outer pair for the second line if present. If only one line is to be installed, sometimes the outer pair will not even be present on the jacks and plugs—even though the RJ11 jacks are still used.

Every defined use of UTP (for example, 10BASE-T for Ethernet) includes a very precise specification of which positions will be used on what jack for which transmission path. 10BASE-T is different from token ring, and both differ from the standard for FDDI* over UTP, even though all of these are defined on RJ45 jacks and plugs. Refer to Fig. 38.7 for a representation of the various ways that the wires can be used in the jacks based on differing standards.

Coax

Coaxial cable is decreasing in popularity. Before fiber was developed, it was the only practical medium to achieve high bandwidths over distances exceeding a few hundred feet. Whereas a high-speed signal (up to about 100 Mbps) can go via UTP up to 100 meters or so, some forms of coaxial

---

*FDDI stands for fiber-distributed data interface, a specification for 100 Mbps on multimode fiber optics.

Figure 38.7
The specific number of pairs and the pinouts differ on the basis of use.

RJ-45

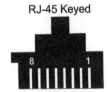

RJ-45 Keyed

RJ-11-4

RJ-11-6 Modified

RJ-11-6

cable can carry signals requiring several hundred Mbps a few thousand meters. The two basic forms of coaxial cable are baseband and broadband. Baseband is used for an Ethernet connection and broadband is the typical cable used in CATV systems. The use of baseband versus broadband systems was covered earlier in Chap. 22.

Coaxial cable is an "unbalanced" medium; the signal goes through one conductor (the inner core), while the other conductor (the outer braid) functions as the ground. Signals can be converted from a balanced to an unbalanced medium (and back) via use of a special passive device called a balun (balanced and unbalanced)—so long as the numbers of required conductors match. This technique is used to supplement coaxial cable with twisted-pair wires as an economic measure. Coaxial ends might be used on a cable run, but UTP wires are used in the middle for the vertical or horizontal run to save money in the wiring of a building. See Fig. 38.8 for a typical installation using the balun.

All coaxial cables have the same basic construction. Each has an inner conductive core, usually of copper (usually solid but sometimes stranded). Around the core is the dielectric, an insulating material. Around the dielectric is a second conductive path or sheath, usually in the form of braided copper. This basic construction is illustrated in Fig. 38.9 for baseband coaxial cable and in Fig. 38.10 for broadband coaxial cable. As with other available media, the underlying construction is varied in many ways to allow customization for specific purposes.

Variations in the thickness of the three key components, as well as in the type or types of covering used, result in a bewildering array of possible cables. Each specific type has a designation (for example, RG58A) that completely specifies its characteristics. Each type of LAN or other communications system standard typically includes a specification that designates what types of cable on which that system is supported.

Coaxial cable now is rarely used for horizontal wiring, although Ethernet can use thin-wire coax in this way. The reason is that the cost per port for thin wire is much higher than for UTP. On the other hand, coax is the second-best choice (after fiber) for vertical wiring. Its high bandwidth and its ability to handle longer distances fit the requirements of vertical wiring well. Also, it is easier to splice and repair than fiber.

Figure 38.8
The balun is used to run less-expensive twisted pairs to a coax interface.

Balun   Twisted Pair

Coaxial

Figure 38.9
The baseband coax
construction.

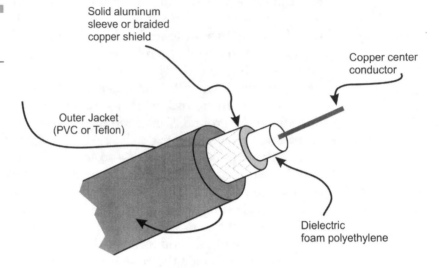

Figure 38.9
The baseband coax
construction.

Solid aluminum
sleeve or braided
copper shield

Copper center
conductor

Outer Jacket
(PVC or Teflon)

Dielectric
foam polyethylene

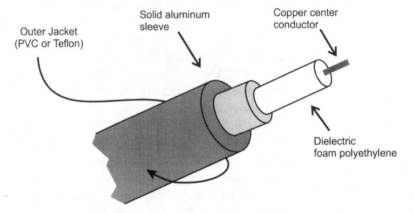

Figure 38.10
The broadband coax
construction.

Solid aluminum
sleeve

Copper center
conductor

Outer Jacket
(PVC or Teflon)

Dielectric
foam polyethylene

Coax was the first choice for interbuilding wiring in the past—that is, before fiber became generally available. But, with the exception of ease of repair and termination, fiber has now supplanted coax as the medium of choice for external wiring.

Fiber Optics

Fiber optics is the medium of the 21st century—available now. It is a technology everyone seems to be welcoming with open arms. The available bandwidth, the potential flexibility, the range of applications available, and the ring topology that all wide-area carriers appear to be installing lend credibility to the widespread use of fiber. Considering that the technology is relatively new both as regards deployment into private telecommunications networks and by the local exchange and interexchange carriers, the acceptance of the technology is positive.

Fiber can be used both as a wide area and a local area cabling medium. As a WAN medium, it has no peer. Essentially all new long-distance cable runs are now built using fiber technologies. Most of the interexchange carriers have adopted a ring topology in their networks. Because the bandwidth is so readily available to them, they can provide automatic reroute capabilities along the reverse route. This is also the way many of the local exchange carriers are going. See Fig. 38.11 for a sample layout of a carrier network with a fiber ring. This picture shows the resiliency of a ring in the event of a cable or equipment failure.

The ring topology is only one of the available possibilities. Organizations considering installing private fiber, whether LANs or WANs, must address the issue of which topology should be deployed. The three basic topologies to deliver bandwidth to the user are the star, ring, and dual counterrotating ring network.

The star network is workable as long as a diverse route exists as protection against failure. However, if the star is single threaded, the user and the carrier are at risk of failure. The use of a diverse route tends to drive up costs for the use of this topology. Therefore many users might be concerned about providing only a single feed as their primary route for local service, dedicated services, and access to the POPs.

Rings automatically provide the diverse route through a single access point. Vendor and user alike gain advantages by using a ring. In the event of the failure of some component along the way, the ring can be auto-

Figure 38.11
The carrier networks using fiber rings.

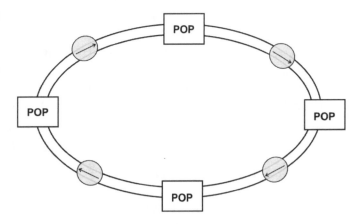

matically reversed to carry traffic in the opposite direction. Many private users have adopted this technique. The cost is higher, but the risks are minimized. The critical point of failure might well be the entrance to the building, the common cable vault where all the services come together. Thus, before choosing to deploy fiber rings, a critical look has to be cast on the total route, and choke points must be eliminated.

The dual counterrotating ring (DCRR) is the latest and greatest idea in the protection of bandwidth. If the access into the building can be diverse, and a ring is established with two diverse fiber rings running in two diverse routes to the same location, you have the DCRR. The first ring is installed normally, as any ring might be; the second is run through diverse entrances and along different conduit-, pole-, or direct-buried-routes. The most expensive solution to the network configuration, this technique is also the most versatile. Information (such as voice, data, graphics, and images) is modulated onto both fibers and simultaneously transmitted to the receiving end. The receiver, getting both transmissions, evaluates the information and selects the best signal for use. In the event that a problem occurs along the way, causing disruptions to the signal, the receiver will receive from only one source. Therefore the decision to select is null; the receiver uses the only signal it received.

Fiber comes in both glass and plastic varieties. Glass works faster, but for reasonably short distances (up to about 3 kilometers) plastic suffices. The upper bandwidth limitation on the best of the glass offerings has yet to be determined, but it is in excess of 13 Gbps (13 billion bits per second). Figure 38.12 shows a representation of the composition of the fiber. Interestingly, the gating factor limiting the speed is how quickly the electronics at the sending and receiving ends can code and decode the signals. It is less difficult to generate signals much faster than can currently be recognized and processed.

Fiber optic systems transmit digital input signals over modulated light beams (lights switched on and off work for 1s and 0s) that pass through the fibers. Except at cable ends, where electrical signals are converted into light for transmission, no electricity is involved. Thus signals on fiber cannot interfere with, transmit, or be affected by any surrounding electromagnetic or electrical signals. This implies that:

- Fiber signals are immune to interference
- Fiber signals cannot generate any electrical interference
- Fiber signals are (almost) immune to security violations, at least while on the fiber

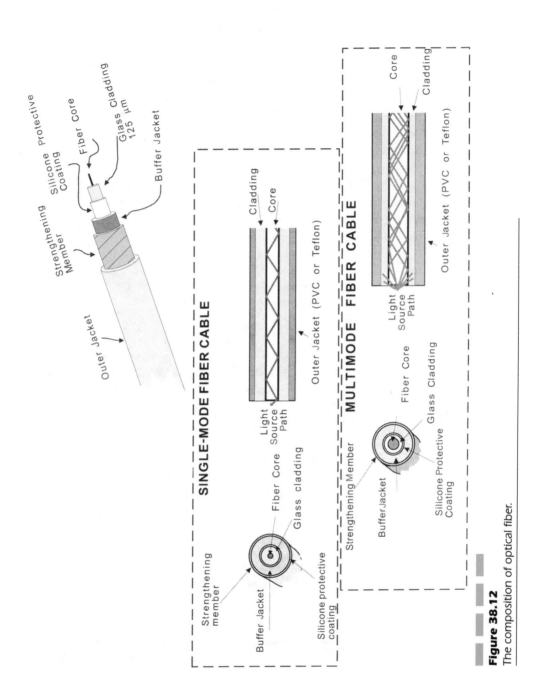

Figure 38.12

The composition of optical fiber.

Glass fiber comes in two basic types: single-mode and multimode. (All plastic fiber is multimode.)

Single-mode is better. It has no known upper bandwidth limitation.* However, it costs more and is generally overkill for non-WAN applications. Telcos and long-distance carriers use it for runs of hundreds or even thousands of miles without the need for using repeaters. The most recent single-mode Transatlantic cable was installed over three thousand miles without any repeaters. This is obviously important because the repeaters require the same (quite expensive) electronic componentry as the ends. Normally end users do not include single-mode fiber in their environments, but price decreases in the electronics and the fiber are making this far more attractive. Clearly, this is happening in major campus environments such as:

- Hospitals
- Military installations
- Corporate parks
- High-rise buildings

As the price of the glass drops and the cost of the electronics moves closer to that of multimode fiber, expect to see far more glass fiber installed. The carriers have moved to high-end electronics today that are specified to carry speed on a standard called synchronous optical networks (SONET). SONET specifies the speeds at which the equipment can multiplex signals from various sources into high-speed carrier services (Table 38.4). Today, we have commercially available products that will multiplex up to 10 Gbps onto the single-mode fiber. Now a standard for SONET is evolving that will yield a 40-Gbps throughput. Using dense wave-division multiplexing (DWDM) on the fiber, a capacity of 320 Gbps

---

*The theoretical limit of single-mode fiber is 100 terabits per second (Tbps), although no means exist to test this speed.

TABLE 38.4

Typical Speeds
Currently Specified
on Single-mode
Fiber

Single-mode Fiber									
10 MC Using SONET Standards (the range of thickness is between 8.3 and 10 microns)									
OC1	**OC3**	**OC9**	**OC12**	**OC24**	**OC36**	**OC48**	**OC96**	**OC192**	**OC255**
51.84	155.52	466.56	622.08	1.244	1.866	2.488	4.976	9.953	13.92
Mbps	Mbps	Mbps	Mbps	Gbps	Gbps	Gbps	Gbps	Gbps	Gbps

is possible immediately. However, nothing stops and waits for us to catch up. Therefore, a single mode fiber with DWDM will soon be offering speeds of 1.6 Tbps (i.e., terabits, or trillion bits, per second).

The upper speed limit on multimode cable is currently under 1 Gbps, fast enough for most applications. Table 38.5 shows a summary of the multimode speeds based on the thickness of the glass. This is something that has been around for some time now, and the thickness of the glass is continually evolving. The various thicknesses listed in the table are based on just that evolution. The current multimode fiber predominantly employed is 62.5-micron glass with a 125-micron outer cladding. Other standards exist, but the specifications for FDDI and other associated protocols call for 62.5/125. The cable costs are getting to the point of being equal these days, but the electronics for the multimode fiber are significantly less expensive. How long that will continue is a mystery. However, we do know from the previous chapters that the ATM, SMDS, and other higher-speed transport systems are all being used more and more. As this continues and other technologies roll out, you can expect that fiber will become a commodity-priced item.

Applications of Fiber

Table 38.6 is a summary of the driving applications that will cause the proliferation of more fiber in the carrier and in end user networks. Each of these applications can be satisfied by current technologies and transport systems. However, as the need for the integration of various pieces of these applications continues, the current speeds will not be enough. Thus the computers we know today will demand more speed and larger file transfers. Video to the desktop or onto a LAN will get closer and closer to reality for the masses. The wiring infrastructure must, therefore, be prepared to support raw bandwidth to the desktop rather than a slower shared medium among multiple users.

TABLE 38.5

Typical Speeds for the Various Thickness of Multimode Fiber

Multimode Fiber	
Microns/Thickness	**Typical Speeds**
100–140	40–100 Mbps
62.5–125	100–565 Mbps
50–125	466–622 Mbps

TABLE 38.6	**Uses of Fiber Systems**
Application Demands for the Speed of Fiber	*Voice.* Although voice does not inherently require the speeds of fiber, it is the multiplexing of thousands of voice calls onto a single fiber that will aid in the deployment.
	Data. Data speeds will continue to increase beyond what we currently use. At dial today, the user can get 28.8–56 Kbps depending on the network used. Future speeds in the multi-Mbps range will not be uncommon.
	Video. Real-time video requires 243 Mbps, which today is unrealistic and prohibitively priced. However, video to the desk for simultaneous conversations at dozens to hundreds of desks will require much higher aggregated bandwidth than only fiber can deliver.
	Computer conferencing. Host-to-host and client server computing systems are going to demand much more from our communications infrastructure.
	CAD/CAM. CAD files are getting larger and more complex every day. Newer CAD systems produce three-dimensional, solid graphics. Files that were once a few hundred kilobytes are now hundreds of megabytes. In the future, such files may be gigabytes or terabytes.
	LAN to LAN. Transparency across LAN boundaries at speeds approximating 155 Mbps to every desk will obviate all other technological advances and demand the fiber to the desk.
	LAN to WAN. Just as LAN boundaries need to be crossed, the wider area will be stressed to keep up with the demands of the future bandwidth.
	HOST to HOST. Already mentioned; large data transfers and real-time journaling and backup between host computers will keep up the pace on the bandwidth needs.
	Multimedia. Linking voice, data, LAN, and video traffic simultaneously will be a big hit in the future. The fiber in the backbone and to the desk will make it happen.
	Medical imaging. Teleradiology, telemedicine, and other imaging systems (CAT, MRI, etc.) will demand that the capacities of the fiber be placed to any location and at any desk for on-demand dial-up services.

Fiber Differences

There are actually two types of materials in any fiber, both glass (or both plastic) but with different indexes of refraction. The *inner core* is actually where the signal is carried; the surrounding material, called the *cladding*, has a reflective inner surface (by virtue of the different refraction index) that keeps most of the light within the inner core.

Fiber comes in various sizes, all very small. The diameters of both types of fibers are measured in microns (a micron is one millionth of a meter). The diameters are always specified with two numbers separated by a slash or dash. The first number is the diameter of the inner core; the other is the outer diameter of the cladding. The most popular and most used multimode fiber's specification is 62.5/125; the generally used monomode fiber has an 8-micron center but a similar outer diameter, so its specification is 8/125.

Fiber is the medium of choice for interbuilding wiring because of its nonelectrical nature (eliminating interbuilding grounding issues as well as being, if buried, entirely immune to the effects of lightning strikes) and the great distances it can cover without intermediate repeaters. It is also often used for vertical wiring because of both its bandwidth and its electrical immunity. Its small size also recommends it for this application—especially in buildings with risers that are almost full.

Why would one not use fiber wherever possible?

- The material still costs somewhat more than UTP.
- Fewer people know how to work with it.
- Splicing and terminating remain difficult, although this is improving, with new automated systems developed for the purpose.
- Components using it are not generally available for all types of LANs—although they are for Ethernet.

LANs implemented on fiber do not necessarily operate any more quickly than those based on UTP. For example, Ethernet runs at 10 Mbps, period. To gain the speed benefits of fiber, a different type of LAN must be used, implying the need for more expensive active components throughout the network. Use of fiber effectively eliminates the upper speed limit for future uses of the LAN cabling; only the active devices would require upgrading to increase operational speeds. Category 5 systems can operate LANs at up to 100 Mbps (but only up to about 100 meters away from a hub), faster than most communications managers think they will have to go in the near future.

Fiber Futures and Risks

A serious study is under way among many research houses, long-haul carriers, and universities to determine the impact of an all-fiber network using the various topologies outlined above. Expect more information and developments in this arena in the near future. However, a problem with some of the older fiber optic cables is starting to rear its ugly head.

The fiber installed in Florida several years ago has since started to cloud, which will impede the light source from getting through the cable. This could be a forerunner of problems to come. If all the fiber products being installed today have short life cycles, then we might have to look for a new technology for the future.

Additional work is also being done on the automatic recovery of networks. The bandwidth on fiber allows the carriers the luxury of completely protecting their networks. You might remember the disaster we wrote about earlier regarding the cable cuts in October 1990, in Illinois. Had the dual feed ring been in place, the problem would have amounted only to an operational hiccup. Instead, because the fiber ring was not completed, the outages were significant.

Research laboratories everywhere are currently looking at a process whereby they will create a solution to our "backhoe fade" problem on fiber. Free-space optics is a technology being developed that will give us the bandwidth of the fiber (1.2 to 13.9 Gbps) over an airborne media. We can expect this to take off if and when it becomes commercially available. Lucent Technologies has already introduced an airwave system which offers 2.488 Gbps on unguided laser (no glass). Lucent also promises that a 10 Gbps version of this airwave will soon be available. Plan for this to be a killer application for the use of airwaves to carry data.

Another system, Aerial Fiber, is now being used in the industry. Fiber cables wrapped in steel strands are replacing the ground wire (or guide wire) for the electrical utilities along their rights-of-way. The fiber is strong enough when wrapped in the steel-stranded outer jacket to sustain without damage a lightning hit to a pole. Extra cable is left along the route to allow enough slack for a pole to fall to the ground without breaking the cable.

INDEX

ABOUT THE AUTHOR

Regis J. (Bud) Bates Jr., President
TC International Consulting, Inc.
Phoenix, Arizona
1-800-322-2202

Regis (Bud) Bates has more than 33 years of experience in telecommunications and management information systems (MIS). He oversees the overall operation of TC International Consulting, Inc., a full-service management consulting organization. He has designed data-processing centers and has also been involved in the design of major networks including LANs and WANs. His clients span the range of Fortune 100–500 companies. His innovative ideas in implementation have been written up in many trade journals and user group magazines.

Bud also develops and conducts various public and in-house seminars, ranging from a managerial overview to very technical instruction on voice, data, and LAN/WAN and broadband communications. For the past 2 years, he has devoted much of his development and training activities to the convergence of voice and data communications. Included in these developments, Bud has been training several CLECs on the integration of voice and data. He has recommended and implemented several training programs (in-house) using all the technologies that are converging as a base model. Included in this list are several training programs that carry the organization's internal certification. His many topics include both basic and advanced courseware on Voice, Data, LAN, WAN, ATM, SONET, T1/T3, VoIP, and Voice over Data Protocols (FR, ATM, etc.).

Mr. Bates has written eight books on the technologies, many of which have been "bestsellers" for McGraw-Hill. Moreover, his *Voice and Data Communications Handbook* has led McGraw-Hill sales for three consecutive years. His recent publication, *Broadband Telecommunications Handbook* (1999), has already sold 10,000 copies. Some of his other titles include *Introduction to T1/T3 Networking; Disaster Recovery for LANs: A Planning and Action Guide; Telecommunications Disaster Recovery; Wireless Networked Communications: Concepts, Systems, and Implementation;* and *Client-Server Internetworking: A How-to Guide.*

Mr. Bates can be contacted via e-mail at bud@tcic.com or by phone at (480) 777–7992.